Lecture Notes in Computer Science 2145

Edited by G. Goos, J. Hartmanis, and J. van Leeuwen

Springer
Berlin
Heidelberg
New York
Barcelona
Hong Kong
London
Milan
Paris
Tokyo

Michael Leyton

A Generative Theory of Shape

Springer

Series Editors

Gerhard Goos, Karlsruhe University, Germany
Juris Hartmanis, Cornell University, NY, USA
Jan van Leeuwen, Utrecht University, The Netherlands

Author

Michael Leyton
Rutgers University
Center for Discrete Mathematics & Theoretical Computer Science (DIMACS)
New Brunswick, NJ 08854, USA
E-mail: mleyton@dimacs.rutgers.edu

Cataloging-in-Publication Data applied for

Die Deutsche Bibliothek - CIP-Einheitsaufnahme

Leyton, Michael:
A generative theory of shape. Michael Leyton. –
Berlin ; Heidelberg ; New York ; Barcelona ; Hong Kong ;
London ; Milan ; Paris ; Tokyo : Springer, 2001
  (Lecture notes in computer science ; 2145)
  ISBN 3-540-42717-1

CR Subject Classification (1998): I.4, I.3.5, I.2, J.6, J.2

ISSN 0302-9743
ISBN 3-540-42717-1 Springer-Verlag Berlin Heidelberg New York

Springer-Verlag Berlin Heidelberg New York
a member of BertelsmannSpringer Science+Business Media GmbH

http://www.springer.de

Typesetting: Camera-ready by author, data conversion by DA-TeX Gerd Blumenstein
Printed on acid-free paper   SPIN: 10840216   06/3142   5 4 3 2 1 0

Preface

The purpose of this book is to develop a generative theory of shape that has two properties we regard as fundamental to *intelligence* – (1) *maximization of transfer*: whenever possible, new structure should be described as the transfer of existing structure; and (2) *maximization of recoverability*: the generative operations in the theory must allow maximal inferentiability from data sets. We shall show that, if generativity satisfies these two basic criteria of intelligence, then it has a powerful mathematical structure and considerable applicability to the computational disciplines.

The requirement of intelligence is particularly important in the generation of *complex* shape. There are plenty of theories of shape that make the generation of complex shape unintelligible. However, our theory takes the opposite direction: we are concerned with the *conversion of complexity into understandability*. In this, we will develop a *mathematical theory of understandability*.

The issue of understandability comes down to the two basic principles of intelligence - maximization of transfer and maximization of recoverability. We shall show how to formulate these conditions group-theoretically. (1) Maximization of transfer will be formulated in terms of wreath products. Wreath products are groups in which there is an upper subgroup (which we will call a *control group*) that transfers a lower subgroup (which we will call a *fiber group*) onto copies of itself. (2) maximization of recoverability is insured when the control group is symmetry-breaking with respect to the fiber group.

A major part of this book is the invention of classes of wreath-product groups that describe, with considerable insight, the generation of *complex* shape; e.g., in computer vision and computer-aided design. These new groups will be called *unfolding groups*. As the name suggests, such a group works by *unfolding* the complex shape from a structural core. The core will be called an *alignment kernel*. We shall see that any complex object can be described as having an alignment kernel, and that the object can be generated from this kernel by *transferring* structure from the kernel out to become the parts of the object.

A significant aspect of all the groups to be invented in this book, is that they express the object-oriented nature of modern geometric programming. In this way, the book develops an *object-oriented theory of geometry*. For example, we will develop an algebraic theory of object-oriented *inheritance*.

Our generative theory of shape is significantly different from current generative theories (such as that of Stiny and Gips) which are based on production rules. In our theory, shape generation proceeds by *group extensions*. The algebraic theory therefore has a very different character. Briefly speaking: *Features correspond to symmetry groups, and addition of features corresponds to group extensions.*

The major application areas in the book are *visual perception, robotics,* and *computer-aided design.* In visual perception, our central principle is that an intelligent perceptual system (e.g., the human perceptual system) is structured as an n-fold wreath product $G_1 \circledW G_2 \circledW \ldots \circledW G_n$. In previous publications, we have put forward several hundred pages of empirical psychological evidence, to demonstrate the correctness of this view for the human system. We shall see that the fact that the visual system is structured as a wreath product, has powerful consequences on the way in which perception *organizes* the world into cohesive structures. Chapter 5 shows how the perceptual *groupings* can be systematically predicted from the wreath product $G_1 \circledW G_2 \circledW \ldots \circledW G_n$.

Chapter six develops a group theory of robot manipulators. We require the group theory to satisfy three fundamental constraints: (1) Perceptual and motor systems should be representationally equivalent. (2) The group linking base to effector cannot be $SE(3)$ (which is rigid) but a group that we will call *semi-rigid*; i.e., allowing a breakdown in rigidity at a specific set of points. (3) The group must encode the *object-oriented* structure.

The theory of robotic kinematics continues in two ways: (1) within the theory of mechanical CAD in Chapter 14; and (2) in the theory of relative motion (in visual perception, computer animation, and physics) given in Chapter 9.

Chapter ten begins the analysis of static CAD by developing a theory of surface primitives, showing that, in accord with the theory of recoverability, the standard primitives of CAD (and visual perception) can be systematically elaborated in terms of what we call *iso-regular groups*. Such groups are n-fold wreath products $G_1 \circledW G_2 \circledW \ldots \circledW G_n$, in which each level G_i is an isometry group and is cyclic or a one-parameter Lie group. To go from such structures to non-primitive objects, one then uses either the theory of splines given later in the book, or the theory of unfolding groups given in Chapters 11, 12 and 13.

The basic properties of an unfolding group are that it is a wreath product in which the control group acts (1) *selectively* on only part of its fiber, and (2) by *misalignment*. In many significant cases, the fiber is the direct product $G_1 \times \ldots \times G_n$ of the symmetry groups G_i of the primitives, i.e., the iso-regular groups, and any fiber copy corresponds to a configuration of objects. The fiber copy in which the object symmetry groups $G_1, \ldots, G_n$ are maximally aligned with each other is called the *alignment kernel*. The action of the control group, in transferring fibers onto each other is to successively misalign the symmetry groups. This gives an *unfolding* effect.

Chapter 14 then presents a lengthy and systematic analysis of mechanical CAD using the above theory. We work through the main stages of MCAD/CAM: part-design, assembly, and machining. For example, in part-design, we give an extensive algebraic analysis of sketching, alignment, dimensioning, resolution, editing, sweeping, feature-addition, and intent-management.

Chapter 15 then carries out an equivalent analysis of the stages of architectural CAD. Then, Chapter 16 gives an advanced algebraic theory of solid structure, Chapter 17 gives a theory of spline-deformation as automorphic actions on groups; and Chapter 18 provides an equivalent analysis for sweep structures.

Chapter 20 examines the conservation laws of physics, in terms of our generative theory, although the next volume will be devoted almost entirely to the geometric foundations of physics. Chapter 21 gives a theory of sequence generation in music.

Finally, Chapters 2, 8, and 22, examine in detail the fundamental differences between our theory of geometry and Klein's Erlanger program. Essentially, in our theory, the recoverability of generative operations from the data set means that the shape acts as a memory store for the operations. More strongly, we will argue that *geometry is equivalent to memory storage*. This is fundamentally opposite to the Erlanger approach in which geometric objects are defined as invariant under actions. If an object is invariant under actions, the actions are not recoverable from the object. We demonstrate that our approach to geometry is the appropriate one for modern computational disciplines such as computer vision and CAD, whereas the Erlanger approach is inadequate and leads to incorrect results.

June 2001 Michael Leyton

Table of Contents

1. Transfer

1.1 Introduction

The purpose of this book is to present a generative theory of shape that significantly increases the power of geometry in the computational and design disciplines, as well as the physical sciences. This is achieved by requiring the generative theory to satisfy the following two criteria which we will regard as fundamental to *intelligent* and *insightful* behavior:

(1) Maximization of Transfer. Any agent is regarded as displaying intelligence and insight when it is able to *transfer* actions used in previous situations to new situations. The ability to transfer past solutions onto new problems is at the very core of what it means to have knowledge. Thus, in our generative theory of shape, the generative sequences must maximize transfer along those sequences.

(2) Maximization of Recoverability. A basic factor of intelligence is the ability to give causal explanations. Examples include the following: (a) An agent must be able to infer the causes of its own current state, in order to identify why it failed or succeeded. (b) The fundamental goal of science is explanation, which is the inference of causes from effects. (c) Computational vision requires inferring, from the retinal image, the environmental processes that produced that image - an inference often referred to as inverse optics. (d) Computer-aided manufacturing requires inferring, from the shape of an object (on the computer screen), a means of generating the shape by simulated machining operations such as drilling and milling. (e) A similar process occurs in reverse engineering, which requires inferring, from a *physical* example of an object, the operations needed to manufacture it. All of these examples involve being presented with a data set, and inferring from the data set, a sequence of operations which will generate that data set. In other words, there must be a set of inference rules by which the generative operations can be inferred from the data set. Generativity is not enough: One must ensure the *inferentiability* of that generativity. We will usually refer to this inferentiability as *recoverability*; i.e., the generative operations must be *recoverable* from the data set.

The above two paragraphs presented our two criteria of intelligence: the maximization of transfer and the maximization of recoverability. We shall see that, if the generativity satisfies these two criteria, then it has an enormously powerful mathematical structure, to be elaborated over the course of this book. Essentially, this involves giving a new theory of geometry that incorporates intelligence into the very structure of shape.

1.2   Complex Shape Generation

The primary goal of this book is to handle *complex* shape. As an example consider the problem of human perception. The human visual system is confronted with an enormously complex environment. Yet it is able to convert this complexity into an entirely understandable form. This exemplifies the general problem that we will investigate:

(1) The conversion of complexity into understandability: *Our basic purpose is to give a generative theory of complex shape such that the complexity is entirely accounted for, and yet the structure is completely understandable.*

(2) Understandability and intelligence: *Deep consideration reveals that understandability of a structure is achieved by maximizing transfer and recoverability.*

(3) The mathematics of understandability: *A significant portion of the book will be the development of a mathematical theory of how understandability is created in a structure.*

To illustrate the theory, lengthy and detailed applications will be given in the areas listed in Table 1.1 This volume will concentrate heavily on areas 1 - 9, and a second volume will be concerned mainly with the last three areas, 10 - 12.

Table 1.1. Application areas of this book.

1. human perception	7. architectural design
2. computer vision	8. music
3. robotic kinematics	9. projective geometry
4. mechanical part design	10. general relativity
5. assembly planning	11. quantum mechanics
6. manufacturing	12. Hamiltonian mechanics

In order to describe highly complex shapes, a class of symmetry groups will be invented, which will be called *unfolding groups*. As the name suggests, an unfolding group works by *unfolding* the complex shape from a structural core. The core will be called an *alignment kernel*. We shall see that any complex object can be described as having an alignment kernel, and that the object can be generated from this kernel by *transferring* structure from the kernel out to become the parts of the object. By reducing the alignment kernel to a minimum, the object is described maximally by transfer. The very structure assigned to the object thereby embodies intelligence and insight. Furthermore, by ensuring recoverability, one ensures that the entire complexity of the object is understandable to the user. Using these groups, we will be able to give a substantial algebraic account of complex shape in computer vision, mechanical design, assembly-planning, constructive solid modeling, architectural design, and so forth.

In this book, a number of classes of symmetry groups will be invented that are particularly valuable in describing certain sub-classes of generative shape. In increasing order of complexity, they are:

$$\text{iso-regular groups} \subset \text{wreath-isometric groups} \subset \text{semi-rigid groups}$$
$$\subset \text{unfolding groups}.$$

Iso-regular groups will allow us to give a systematic classification of shape primitives in human perception and computer graphics. Wreath-isometric and semi-rigid groups will allow us to describe articulated structures such as robot manipulators and relative-motion systems in classical and quantum physics. Unfolding groups will allow us to describe arbitrarily complex objects and scenes in CAD, computer vision, assembly planning, product specification, and machining.

1.3   Object-Oriented Theory of Geometry

A crucial feature of our generative theory is that it models the *object-oriented* nature of computer-aided design and graphics. In this way, we will develop an object-oriented theory of geometry - which, to our knowledge, will be the first such theory to have been elaborated. Most importantly the object-orientedness will be formulated in the group theory to be created, and will lead to the classes of groups listed above. A particular concern will be to give a group theory of the relationship between the command structure and the internal structure of an object, as well as an **algebraic theory of inheritance** - which will underlie most of the book.

1.4   Transfer

As stated above, the generative theory is founded on two principles of intelligent behavior: the maximization of transfer and the maximization of recoverability. This chapter will give an intuitive introduction to transfer, followed, in the next chapter, by an intuitive introduction to recoverability. Then, Chapter 3 will begin the rigorous elaboration of the full theory.

A generative theory of shape characterizes the structure of a shape by a sequence of actions needed to generate it. According to our theory, these actions must maximize transfer. That is:

MAXIMIZATION OF TRANSFER. *Make one part of the generative sequence a transfer of another part of the generative sequence, whenever possible.*

We will show that the appropriate formulation of this is as follows: A situation of transfer involves two levels: a *fiber group*, which is the group of actions to be transferred; and a *control group*, which is the group of actions that will transfer the fiber group. The justification for these structures algebraically being groups will be given later, but the theory of transfer will work equally for semi-groups, which is the most general case one would need to consider for generativity.

Now, one can think of transfer as the control group moving the fiber group around some space; i.e., *transferring* it. This is illustrated in Fig. 1.1. The transferred versions of the fiber group are shown as the vertical copies, and will be called the *fiber-group copies*. The control group acts from above, and transfers the fiber-group copies onto each other, as indicated by the arrow.

A basic part of this book will be to give an algebraic theory of transfer, and to reduce complex situations down to structures of transfer. Transfer will be modeled by a group-theoretic construct called a *wreath product*. This is a group that will be notated in the following way:

Fiber Group Ⓦ Control Group.

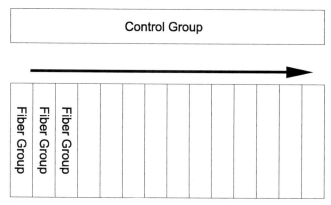

Fig. 1.1. The control group transferring the fiber group.

Intuitively, a wreath-product group contains the entire structure shown in Fig. 1.1; that is, it has an upper subgroup that will be called a control group, and a system of lower subgroups that will be called the fiber-group copies. The control group sends the fiber-group copies onto each other. The system of fiber-group copies, i.e., the entire lower block in Fig. 1.1, is the direct product of the fiber-group copies. The lower system is related to the control group above it by a semi-direct product. (Semi-direct products are explained in Appendix A.)

The entire lower system is a normal subgroup of the wreath-product group; however, any individual fiber-group copy is not. This turns out to be fundamentally important to the generative theory.

Formulating transfer in terms of wreath products has enormous advantages, as follows: (1) A wreath product is a group that contains all the transferred versions of the fiber group. Thus, rather than thinking of the transferred versions as separate algebraic entities, they are all integrated within a single algebraic structure. (2) This single algebraic structure also encompasses the control group. Thus the wreath product contains the network of algebraic connectivity that relates the control group to the fiber group. This algebraic connectivity will explain an enormous amount in human perception, computer vision, robotics, computer-aided design, navigation, manufacturing, quantum mechanics, and so on.

The full mathematical theory will begin in Chapter 3, which will elaborate in detail our claim that transfer is best modeled by a wreath product. In that chapter, the structure of a wreath product will also be described in detail. For now, the reader needs to understand only that the wreath product encodes the transfer relationship between the control group and the fiber group; i.e., that *the control group moves the fiber group around.* The purpose of the current chapter is to give the reader an intuitive description of transfer, together with several examples that will illustrate the power of transfer. Although it will not be required in the present chapter, a rigorous definition of wreath product is given in footnote[1] below.

We shall often use another term for transfer: *nested control*. That is, the group **Fiber Group⊛Control Group** will be called a *structure of nested control*; the 2-place operation ⊛ will be referred to as the *control-nesting operation*; and the fiber group will be said to be *control-nested* in the control group.

The above discussion considered a 2-level structure of transfer; i.e., the movement of a fiber group by a control group. An n-level structure of transfer will be constructed by a recursive use of the 2-place operation ⊛, and the resulting group will be written thus:

$$G_1 \text{ ⊛ } G_2 \text{ ⊛ } \ldots \text{⊛ } G_n.$$

In this group, each level G_i acts as a control group with respect to its left-subsequence $G_1 ⊛ G_2 ⊛ \ldots ⊛ G_{i-1}$ as fiber. In other words, G_i transfers its left-subsequence around some environment; but this left-subsequence is itself a hierarchy of transfer, and so on, recursively downwards. It will be argued that any intelligent shape generation is given by an n-fold hierarchy of transfer. That is, generation is structured by a group of the form $G_1 ⊛ G_2 ⊛ \ldots ⊛ G_n$.

It is now necessary for us to work through several examples so that the reader can begin to become familiar with aspects of this approach, and see also the wide range of applications.

1.5 Human Perception

In Leyton [87], [88], [89], [90], [91], [96], we put forward several hundred pages of psychological evidence that lead to this conclusion:

---

[1] Consider two group actions: the actions of groups, $G(F)$ and $G(C)$, on sets, F and C, respectively. The wreath product $G(F) ⊛ G(C)$ is the semi-direct product $\{\prod_{c \in C} G(F)_c\} \text{Ⓢ}_\tau G(C)$, where the product symbol $\prod$ means the (group) direct product, and the groups $G(F)_c$ are isomorphic copies of $G(F)$ indexed by the members c of the set C. The map $\tau : G(C) \longrightarrow Aut\{\prod_{c \in C} G(F)_c\}$ is defined such that $\tau(g)$ corresponds to the group action of $G(C)$ on C, now applied to the indexes c in $\prod_{c \in G(C)} G(F)_c$. That is, $\tau(g) : \prod_{c \in G(C)} G(F)_c \longrightarrow \prod_{c \in G(C)} G(F)_{gc}$. Finally, we have a group action of $G(F) ⊛ G(C)$ on $F \times C$ defined thus: For $\phi \in G(F)$, $\kappa \in G(C)$, and (f, c) in $F \times C$, we have $[\phi, \kappa](f, c) = (\phi_c f, \kappa c)$, where $\phi_c \in G(F)_c$.

THE STRUCTURE OF
THE HUMAN PERCEPTUAL SYSTEM

The human perceptual system is structured as an n-fold wreath product $G_1 ⓦ G_2 ⓦ \ldots ⓦ G_n$. The consequence is that perceptual organizations are structured as n-fold wreath products.

This view explains an enormous number of psychological results in the area of perceptual organization, shape representation, and motion perception. For example, all the Gestalt grouping phenomena can be explained very economically using this principle: the perceptual groupings correspond to symmetry groups G_i that are control-nested recursively in a manner described above. The principle applies not only to static shape perception but also to motion perception.

In order to illustrate this, let us begin with a very simple example. We will show how the human visual system structures a square. In a sequence of psychological experiments, Leyton [89] [90], we showed that human vision represents a square *generatively*, in the following way. It begins with the top side. Perceptually the top side is generated by starting with a corner point, and applying translations to trace out the side, as shown in Fig. 1.2.

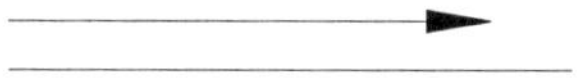

Fig. 1.2. The generation of a side, using translations.

Next, this translational structure is *transferred* from one side to the next - rotationally around the square. In other words, there is transfer of translation by rotation. This is illustrated in Fig. 1.3.

Therefore, the transfer structure is defined as:

$$\text{Translations } ⓦ \text{ Rotations.}$$

where Translations is the fiber group and Rotations is the control group. Recall that, in any transfer situation, the control group moves the fiber group around.

The translation group will be denoted by the additive group $\mathbb{R}$. The rotation group is $\mathbb{Z}_4$, the cyclic group of order 4, represented as

$$\mathbb{Z}_4 = \{\ e,\quad r_{90},\quad r_{180},\quad r_{270}\ \}$$

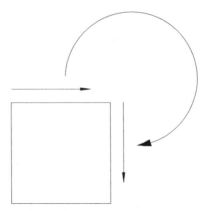

Fig. 1.3. Transfer of translation by rotation.

where r_θ means rotation by θ degrees. The successive group elements are obviously rotations by successive 90^0 increments. Thus the transfer structure illustrated in Fig. 1.3, is this:

$$\mathbb{R} \ \circledW \ \mathbb{Z}_4. \tag{1.1}$$

At first, the reader might question putting the *entire* group of translations in the fiber position, even though the group is "cut off" at the end points of a side. However, this is handled very easily in our system by placing what we will call an **occupancy group**, $\mathbb{Z}_2$ (a cyclic group of order 2), at each point along the infinite line. The group switches between two states, "occupied" and "non-occupied," for the point at which it is located. Obviously, all the points along the finite side of the square are occupied, and all the points past its end-points are unoccupied. Algebraically, we place the occupancy group as an extra level, in the structure of nested control, below the $\mathbb{R}$ group, thus:

$$\mathbb{Z}_2 \ \circledW \ \mathbb{R} \ \circledW \ \mathbb{Z}_4. \tag{1.2}$$

Occupancy structures will be investigated later, and shown to elegantly represent many phenomena, e.g., in Gestalt perception, quantum physics, etc. For the moment, however we will omit the occupancy level, to keep the discussion focussed on the geometric (spatial) structure. Observe that the occupancy group is a color group, not a geometrical group; i.e., it has no spatial action.

Thus for now, let us return to the purely geometric structure, given in expression (1.1) above. The next thing to do is show that this expression gives **generative coordinates** to the square. As before, assume, without loss of generality, that the top side is generated by translation from the left end; and that the set of sides is generated from the top side by clockwise rotations. In other words, we are using the standard scenario for drawing a square: simply trace out the sides successively in the clockwise direction. The group $\mathbb{R}\circledW\mathbb{Z}_4$

(a)   (b)

Fig. 1.4. Translation as a coordinate along side.

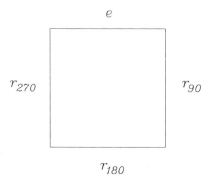

Fig. 1.5. The mapping of $\mathbb{Z}_4$ onto a square.

gives the structure of this trace, which we see is actually control-nested. The control-nested structure will give the *generative coordinates* of each point, as follows:

Because the structure is generative, one can consider the fiber group $\mathbb{R}$ as mapped onto each side. For example, consider the top side. As shown in Fig. 1.4, the zero translation, e, is mapped to the left corner on the side. Then, any other point on the side is uniquely described by the translation t that generated it from the initial point. Fig. 1.4a shows the actual translation that was applied, i.e., as an action, and Fig. 1.4b shows the action converted into the label for the point. The same structure occurs on any side.

Similarly, consider the control group $\mathbb{Z}_4 = \{\ e,\ r_{90},\ r_{180},\ r_{270}\ \}$. Again because of generativity, this group is mapped onto the set of four sides, in the manner shown in Fig. 1.5. Because, the top side is the starting side it receives the identity element e from $\mathbb{Z}_4$. Any other side receives one of the rotations in $\mathbb{Z}_4$, i.e., the generative operation that was used to create the side from the top side.

Any point on the square is therefore described by a pair of coordinates:

$$(t, r) \quad \in \quad \mathbb{R} \ \textcircled{w} \ \mathbb{Z}_4$$

where $t \in \mathbb{R}$ and $r \in \mathbb{Z}_4$. For example, as shown in Fig. 1.6, a point on the right side is labeled (t, r_{90}) where t is the translation that was used to generate it from the starting point of the side, and r_{90} was the rotation that was used to generate the right side from the top side.

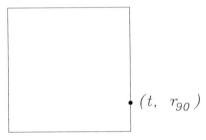

Fig. 1.6. The coordinates of a point on a square.

In order to describe the phenomenon to be investigated, it is necessary to fill in some other coordinates on the square. Consider the top left-hand corner point, as shown in Fig. 1.7. We have assumed that the entire history starts here. Therefore the amount of translation here is zero - i.e., the point is at the identity element e_1 of the fiber group $\mathbb{R}$. Furthermore, the amount of rotation that has been applied so far is also zero - i.e., the side on which the point sits is at the identity element e_2 of the control group $\mathbb{Z}_4$. Therefore, as shown in Fig. 1.7, the top left corner-point has the pair of coordinates (e_1, e_2) in $\mathbb{R}\textcircled{w}\mathbb{Z}_4$. Now consider the point at translational distance t along the top side. Its coordinates are clearly (t, e_2), as shown in Fig. 1.7. Next consider the top right-hand corner point, and consider its description as the first point on the right-hand side. As such, it has undergone no translation along that side, and is therefore at the identity element e_1 of the fiber group $\mathbb{R}$. However, the point must have coordinate r_{90} in the control group $\mathbb{Z}_4$ because the right side is achieved by a $90°$ rotation from the top side. Thus the point has the pair of coordinates (e_1, r_{90}). Finally, as seen earlier, the lower labeled point on the right-hand side has coordinates (t, r_{90}).

The crucial thing to observe is the *transfer* structure involved in this. First observe that the relationship between the two points on the top side is the translation t given by the top straight arrow in Fig. 1.8. Similarly the relationship between the two points on the right side is the translation t given by the downward straight arrow. The *transfer* effect of rotation is to send the translation on the top side to the translation on the right side. This is shown

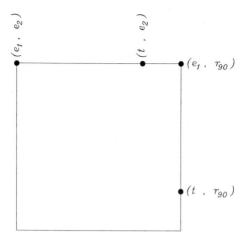

Fig. 1.7. The coordinates of four points.

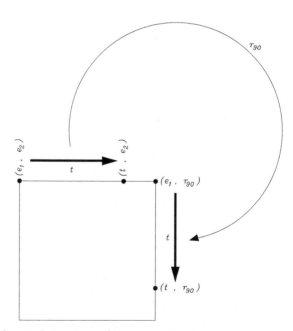

Fig. 1.8. The control-nested structure of those coordinates.

by the circular arrow in Fig. 1.8, which sends the straight arrow on the top side to the straight arrow on the right side. As was said before, the control group $\mathbb{Z}_4$, takes the fiber group $\mathbb{R}$ and transfers it from one side on to the next,

Now observe that the group being studied, $\mathbb{R}\text{ⓦ}\mathbb{Z}_4$, satisfies the following three conditions:

$\mathfrak{IR}_1$: *The group is decomposable as a control-nested structure* $G_1\text{ⓦ}G_2\text{ⓦ}\dots\text{ⓦ}G_n$.

$\mathfrak{IR}_2$: *Each level is "1-dimensional", i.e., either a cyclic group (in the discrete case) or a 1-parameter group (in the continuous case).*

$\mathfrak{IR}_3$: *Each level is represented as an isometry group.*

We will call a group that obeys the above three conditions, an **iso-regular group**. Intuitively, the three conditions will be summarized by saying that the group is a *control-nested hierarchy of repetitive isometries*. Iso-regular groups will be fundamental to our theory of shape-generation. In the theory, such groups describe non-deformed objects. The theory of recoverability will show that any shape has an underlying iso-regular group. To generate the shape, one starts by generating its iso-regular group, and then adding actions that create deformation. These actions are imposed as further levels of *transfer* on the iso-regular structure.

As a simple illustration, consider the generation of a parallelogram. Its underlying iso-regular group is the group $\mathbb{R}\text{ⓦ}\mathbb{Z}_4$ of a square. Thus to generate a parallelogram, we first generate the iso-regular group of a square, and then add the general linear group $GL(2,\mathbb{R})$, i.e., the group of invertible linear transformations, as a higher level of control, thus:

$$\mathbb{R} \text{ⓦ} \mathbb{Z}_4 \text{ⓦ} GL(2,\mathbb{R}). \tag{1.3}$$

Notice that, with this extra level, we no longer have an iso-regular group, because the iso-regularity conditions $\mathfrak{IR}_2$ and $\mathfrak{IR}_3$ have been broken; i.e., the added level $GL(2,\mathbb{R})$ is not "1-dimensional" and is not an isometry group.

Notice that the operation used to add $GL(2,\mathbb{R})$ on to the lower structure $\mathbb{R}\text{ⓦ}\mathbb{Z}_4$ is, once again, the control-nesting operation ⓦ which means that $GL(2,\mathbb{R})$ acts by *transferring* $\mathbb{R}\text{ⓦ}\mathbb{Z}_4$, as follows: Since the fiber group $\mathbb{R}\text{ⓦ}\mathbb{Z}_4$ represents the structure of the square, this means that $GL(2,\mathbb{R})$ transfers the structure of the square onto the parallelogram. In particular, it transfers the *generative coordinates* of the square onto the parallelogram. For example, recall that Fig. 1.7 showed the generative coordinates of four of the points on the square. The control action of $GL(2,\mathbb{R})$ therefore takes the coordinates of

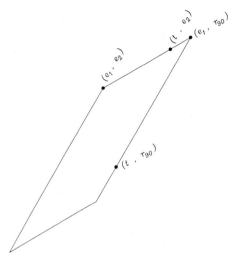

Fig. 1.9. The transferred coordinates from a square.

these four points and transfers them onto the corresponding four points on the parallelogram, as shown in Fig. 1.9.

More deeply still, the fiber group $\mathbb{R}\circledw\mathbb{Z}_4$ in expression (1.3) is itself a transfer structure, as seen in Fig. 1.8, where rotation transferred the translation process from the top side onto the right side. This transfer structure is itself transferred, by $GL(2,\mathbb{R})$, onto the parallelogram, as shown in Fig. 1.10. That is, we have **transfer of transfer**. This recursive transfer is encoded by the successive $\circledw$ operations in expression (1.3). This illustrates what was said earlier, that, given a transfer hierarchy $G_1\circledw G_2\circledw\ldots\circledw G_n$, each level G_i acts as a control group with respect to its left-subsequence $G_1\circledw G_2\circledw\ldots\circledw G_{i-1}$ as fiber. In other words, G_i transfers its left-subsequence around some environment; but this left-subsequence is itself a hierarchy of transfer, and so on recursively downwards.

The theory we give is equally applicable to 3-dimensional shape. For example, consider the structure of a cylinder. The standard group-theoretic description of a cylinder is

$$SO(2) \times \mathbb{R} \tag{1.4}$$

where $SO(2)$, the group of planar rotations around a fixed point, gives the rotational symmetry of the cross-section, and $\mathbb{R}$ gives the translational symmetry along the axis. Notice that in (1.4) the operation linking the two groups is the direct product operation $\times$.

For us, the problem with this expression is that it does not give a generative description of the cylinder. In computer vision and graphics, cylinders are described generatively as the sweeping of the circular cross-section along the axis, as shown in Fig. 1.11. To our knowledge, the group of this sweeping structure has never been given. We propose that the appropriate group is:

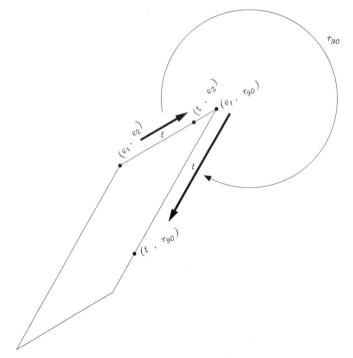

Fig. 1.10. The transfer of transfer.

$$SO(2) \; \textcircled{w} \; \mathbb{R}. \tag{1.5}$$

Notice that it uses the control-nesting operation $\textcircled{w}$ rather than the direct product $\times$, and therefore the group has a very different structure from that in expression (1.4). The operation $\textcircled{w}$ means that this new group has a *fiber-control* structure, in which $SO(2)$ is the fiber group and $\mathbb{R}$ is the control group. This is exactly what is seen in the sweeping structure shown in Fig. 1.11. The cross-section is generated first as a fiber, and then its position is controlled by translation.

NORMAL SUBGROUPS AND GENERATIVITY. *This comment is fundamental to the entire theory, and will be illustrated with the example of a cylinder. Although the direct product description $SO(2) \times \mathbb{R}$ of the cylinder is used universally, we argue that that there is a strong mathematical reason why it cannot model the cylinder as a generative structure, and is therefore inappropriate for modeling crystal growth in physics, drilling and milling in manufacturing, assembly of revolute structures in robotics, etc. The reason is as follows: In the generative representation of a cylinder, the group $\mathbb{R}$ must move the group $SO(2)$ along the cylinder. This movement must take place by the conjugation $g - g^{-1}$ of $SO(2)$ by the elements g of $\mathbb{R}$ (conjugation is the*

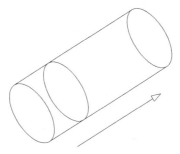

Fig. 1.11. The sweep structure of a cylinder.

group-theorists tool for movement). However, in the direct product formulation $SO(2) \times \mathbb{R}$, the rotation group $SO(2)$ is a normal subgroup[2]; which means that conjugation of $SO(2)$ by $\mathbb{R}$ will leave $SO(2)$ invariant. Therefore $\mathbb{R}$ will not be able to move $SO(2)$ along the cylinder. This means that the direct product formulation cannot model generative structure (i.e., crystal growth, drilling and milling, robot assembly, etc.). In contrast, we shall see that, in the wreath-product formulation $SO(2)\textcircled{w}\mathbb{R}$, the rotation group $SO(2)$ is not a normal subgroup. Therefore, in this latter formulation, $\mathbb{R}$ can move $SO(2)$. Indeed the fibering that occurs in a wreath product operation will ensure that $\mathbb{R}$ moves $SO(2)$ along the cylinder in the correct way.

The reader should observe that the control-nested group in (1.5) is what we call an *iso-regular group*; i.e., it satisfies the conditions $\mathfrak{IR}_1$-$\mathfrak{IR}_3$ on page 12. This fact is critical: The cylinder is an example of a standard shape primitive in graphics. In Chapter 10, it will be argued that each of the standard primitives is characterized by an iso-regular group. In fact, it will be shown that our algebraic methods lead to a *systematic classification of shape primitives*.

We will also argue that, having generated a shape primitive via an iso-regular group, one then obtains the non-primitive shapes by applying additional fiber and control levels. For example, we will show how Boolean operations and spline deformations can be algebraically formulated within this framework.

Now let us turn to the deepest problem in human perception: the problem of *perceptual organization*. This problem is standardly formulated as the following question: What are the structural principles by which the perceptual system forms *groupings*? The problem of grouping is the longest unsolved problem in perception - having been investigated for the entire 20th cen-

---

[2] Normal subgroups are explained in Appendix A

Fig. 1.12. A square distorted by projection.

tury. It underlies all aspects of perception, from image segmentation to 3D shape representation. Yet literally no progress has been made in solving this problem. However, using the theory developed in this book, the solution naturally drops out of our algebraic theory of *transfer*, as follows: We have said that the human perceptual system organizes any stimulus set generatively into a recursive hierarchy of transfer, i.e., into a control-nested hierarchy of groups: $G_1 Ⓦ G_2 Ⓦ \ldots Ⓦ G_n$. In Chapter 5, we shall show that the perceptual groupings come directly from this recursive transfer structure:

GROUPING PRINCIPLE. *Any perceptual organization is structured as an n-fold wreath product $G_1 Ⓦ G_2 Ⓦ \ldots Ⓦ G_n$. The groupings in a perceptual organization correspond to the left-subsequences $G_1 Ⓦ G_2 Ⓦ \ldots Ⓦ G_{i-1}$ of the wreath product.*

Let us conclude this initial review of human perception by considering the basic visual problem of *projection*. The visual image is the projection of some *environmental* shape onto the retina. In Chapter 22, we will argue that the appropriate approach to handle this is to describe the environmental shape generatively, and to add the projective process as an extra generative level, resulting in the shape on the image.

PROJECTION AND GENERATIVITY. *The image shape is given by an n-fold wreath product, $G_1 Ⓦ G_2 Ⓦ \ldots Ⓦ G_n$, in which the left-subsequence $G_1 Ⓦ G_2 Ⓦ \ldots Ⓦ G_{n-1}$ represents the generation of the environmental shape, and the final control group G_n represents the projective group. The image shape is therefore given a completely generative description in which the projective process is merely the last phase.*

As a simple example, consider the projection of a square onto the retina, producing the projectively distorted square shown in Fig. 1.12. We have seen that the undistorted square is represented as the group $\mathbb{R} Ⓦ \mathbb{Z}_4$. This group gives the generative structure of the square *in the environment.* Now to add the effect of projecting the square onto the retina, one merely adds the projective group $PGL(3, \mathbb{R})$ onto the generative sequence of the square thus:

$$\mathbb{R} \ \textcircled{w} \ \mathbb{Z}_4 \ \textcircled{w} \ PGL(3,\mathbb{R}). \tag{1.6}$$

Once again, notice that the operation used to add $PGL(3,\mathbb{R})$ onto the lower group $\mathbb{R}\textcircled{w}\mathbb{Z}_4$, is the control-nesting operation $\textcircled{w}$; which means that $PGL(3,\mathbb{R})$ acts by *transferring* $\mathbb{R}\textcircled{w}\mathbb{Z}_4$ from the undistorted square in the environment onto the distorted square in the image[3].

Another example is the grid of squares in Fig. 1.13. The undistorted grid of squares is generated by the wreath product:

$$\mathbb{R} \ \textcircled{w} \ \mathbb{Z}_4 \ \textcircled{w} \ \mathbb{Z}^H \ \textcircled{w} \ \mathbb{Z}^V$$

where, the first two levels $\mathbb{R}\textcircled{w}\mathbb{Z}_4$ is the square, as before; and the next level $\mathbb{Z}^H$ is the group of horizontal translations; and finally $\mathbb{Z}^V$ is the group of vertical translations. Then, to get the projective distortion, one merely adds the projective group as an extra level of control thus:

$$\mathbb{R} \ \textcircled{w} \ \mathbb{Z}_4 \ \textcircled{w} \ \mathbb{Z}^H \ \textcircled{w} \ \mathbb{Z}^V \ \textcircled{w} \ PGL(3,\mathbb{R}). \tag{1.7}$$

The important thing to notice is that the structure is exhaustively generative, from the lowest to the highest level, and that this generativity has been entirely described as a hierarchy of transfer.

Fig. 1.13. A grid of squares distorted by projection.

The approach we have defined differs substantially from the standard one in projective geometry, in the following profound way: The group sequence in (1.6) is built up from the Euclidean structure, which is given here by the iso-regular group $\mathbb{R}\textcircled{w}\mathbb{Z}_4$. This means that the undistorted square is a privileged figure in the space of quadrilaterals. The phenomenon of privileged figures in a space of distorted figures is a basic inviolable result in human perceptual psychology, as will be reviewed later. This completely violates Klein's principle that geometric objects are the *invariants* of the specified transformation group - which is the most famous principle of 20th century geometry and

---

[3] The algebraic action of $PGL(3,\mathbb{R})$ with respect to $\mathbb{R}\textcircled{w}\mathbb{Z}_4$ will be defined via the action of $PGL(3,\mathbb{R})$ on the projective plane represented intrinsically. The mathematical details will be given later.

physics. As will be seen, our generative theory of geometry is the direct opposite of Klein's approach. In our system, geometric objects are characterized by generative sequences. This means that they cannot be invariants, because invariance destroys recoverability of the applied operations, and recoverability is basic to our theory. Quite simply: You cannot characterize geometric objects generatively, if you cannot recover their generative history. But you cannot recover their generative history if they are invariants under generative actions.

1.6   Serial-Link Manipulators

A generative theory of shape encodes shape by a system of *actions*. Since we argue that the human perceptual system encodes shape generatively, this means that the perceptual system represents shapes in terms of actions. It will be argued that a major consequence of this is that the human perceptual system is structured by the same principles as the motor system, since the motor system is structured by action.

Now, we have said that actions are *intelligently* organized if they are organized by transfer; and an initial set of illustrations have been given of how the perceptual system is organized by transfer. We will now show that the motor system must also be organized by transfer. To do so, consider the most common type of motor system, the serial-link manipulator.

Review of serial-link manipulators: The most famous example of a serial-link manipulator is the human arm: Such a structure is a series of rigid links going from the base to the hand. Each link corresponds biologically to a bone. Furthermore, each successive pair of links is connected by a joint. The base end is called the proximal (near) end of the manipulator, and the hand end is called the distal (far) end of the manipulator. Standardly, a serial-link manipulator is specified by embedding a coordinate frame in each successive link. Each frame is judged relative to the next frame in the proximal direction, e.g., the frame of the hand is judged relative to the frame of the forearm, and the frame of the forearm is judged relative to the frame of the upper arm, etc. The relationship between two successive frames is given by a matrix A_i. Thus the overall relationship between the hand coordinate frame and the base coordinate frame is given by the product of matrices

$$A_1 A_2 \dots A_n \tag{1.8}$$

corresponding to the succession of links. In robotics, each matrix A_i is modeled as a rigid motion, and is therefore a member of the special Euclidean group $SE(3)$, the group generated by translations and rotations (but no reflections). Standardly, the order from left to right along the matrix sequence

(1.8) corresponds to the order from base to hand (proximal to distal). However, without loss of generality, we will choose the left-to-right order as corresponding to the hand-to-base order (distal-to-proximal). This will maintain consistency with our other notation.

An Algebraic Theory of Serial-Link Manipulators

According to the theory in this book, the basic property of serial-link manipulators is *transfer*. The hand has a space of actions that is transferred through the environment by the fore-arm, which has a space of actions that is transferred through the environment by the upper-arm, and so on down to the base (e.g., the torso). Thus, we argue that the group of a serial-link manipulator has the following *wreath-product* structure:

$$SE(3)_1 \, \circledW \, SE(3)_2 \, \circledW \, \ldots \circledW \, SE(3)_n \qquad (1.9)$$

where each level $SE(3)_i$ is isomorphic to the special Euclidean group $SE(3)$, and the succession from left to right corresponds to the succession from hand to base (distal to proximal). Thus, each matrix A_i in expression (1.8) is taken from its corresponding Euclidean group $SE(3)_i$ in (1.9). Although ordinary matrix multiplication is used between any two successive matrices in (1.8), we now see that the group product in the corresponding position in (1.9) is actually the wreath product. Thus each group $SE(3)_i$ along the sequence (1.9) acts as a control group with respect to its left-subsequence $SE(3)_1 \circledW SE(3)_2 \circledW \ldots \circledW SE(3)_{i-1}$ and this corresponds to the fact that $SE(3)_i$ transfers the action structure of its left-subsequence around the environment. As usual, the successive use of the $\circledW$ operation, is interpreted *recursively*, and it is this that defines the hierarchical nature of the motion spaces.

The entire group we have given in (1.9) for the serial-link manipulator, is very different from the group that is normally given in robotics for serial-link manipulators. Standardly, it is assumed that, because one is multiplying the matrices in (1.8) together, and therefore producing an overall Euclidean motion T between hand and base, the group of such motions T is simply $SE(3)$. However, we argue that this is not the case. The group is the much more complicated group given in expression (1.9). This group encodes the complex link-configurations that can occur between the hand and base. If the group were simply $SE(3)$, then there would be a single configuration of links between hand and base and this would remain rigidly unaltered as the hand moves. However, there are infinitely many different configurations that the links can take between hand and base, and the group in (1.9) gives the relationships between all these configurations. To put it another way: It is conventionally assumed that, because the overall relation between hand and base is a Euclidean motion, the group of motions between the hand and base is the group of *rigid* motions. However, the structure between hand and base is *not rigid*. Therefore the group is not $SE(3)$. It is the much more complicated group (1.9).

Most crucially, notice that this group was produced by considering the *transfer* relationships involved. It is this that allowed us to specify the algebraic structure. Let us finally put together what our theory says about the human perceptual and motor systems:

PERCEPTUAL-MOTOR UNIFICATION

The human perceptual and motor systems are both structured as n-fold wreath products $G_1 \circledW G_2 \circledW \ldots \circledW G_n$.

1.7   Object-Oriented Inheritance

The fact that each frame in a serial-link manipulator is judged relative to the next frame in the distal-to-proximal direction, means that serial-link manipulators are an example of what are called *parent-child structures* in object-oriented programming. Parent-child relationships express the fundamental structuring principle of object-oriented software called *inheritance*. Inheritance of properties can go from class to class (i.e., as specified statically in the software text), or can be formed at run-time by linking objects together. Let us, for the moment, consider the latter type. Such object relationships are basic, for instance, to assembly-subassembly organization in mechanical CAD. For example, most major mechanical programs such as *Pro/ENGINEER* provide menus which allow the designer to determine the parent-child relationships in an assembly hierarchy, and most part information windows in the program provide the user with the parent-child positioning of any selected part, because feasible modification of an individual part is impossible without knowing these relationships. Parent-child relationships are also a major explicit part of all animation software, such as *3D Studio Viz/Max*, where kinematic relationships between limbs are given exactly as defined in robotics. Again, all object-subobject relations in architectural CAD are parent-child relations; e.g., doors are placed relative to walls and move with them as the designer modifies the room.

The examples mentioned in the previous paragraph are all *geometric* parent-child relationships. A major part of this book will be to give an algebraic theory of such relationships in object-oriented programming. It will be claimed that the inheritance structure of parent-child hierarchies is given algebraically by wreath products $G_1 \circledW G_2 \circledW \ldots \circledW G_n$, in order to maximize transfer. This means that geometric parent-child hierarchies follow from our generative theory of shape.

1.8   Complex Shape Generation

The illustrations given in the previous sections were of relatively simple shape. However, the primary goal of the theory is to handle *complex* shape. In fact, as was said on p. 2, the goal is the representation of complex objects and scenes such that the complexity is entirely accounted for, and yet the structure is completely understandable. We argue that understandability is achieved by maximizing transfer and recoverability.

Thus, the goal will be the development of a mathematics for converting complexity into understandability. A key part of this will be the invention of a class of groups we will call **unfolding groups**. Here, the fiber group, called the **alignment kernel**, expresses the complex configuration in a maximally collapsed form. The control group is then a hierarchical process of unfolding that collapsed form outwards to become the complex configuration. The unfolding is a structure of transfer in which the complex form is seen as the unfolded image of the collapsed form.

A key factor is the size of the alignment kernel: By minimizing the size of the alignment kernel, one maximizes the transfer structure of the complex object. The very structure assigned to the object thereby embodies intelligence and insight. Furthermore, by ensuring recoverability, one ensures that the entire complexity of the object is understandable to the user.

1.9   Design

Superficially, design seems to involve the successive addition of structure. How else can one account for the fact that the design object appears to structurally grow. However, a careful analysis of how designers work, reveals there is actually very little new structure added, as design proceeds. What actually happens is that designers tend to create each additional structural component as a *transfer* of an already existing component.

Let us illustrate this by looking at the typical process by which an architect draws the floor-plan of an apartment building. The eventual plan will be extremely complex, however the architect can begin only by drawing very simple elements. Let us assume that the architect is carrying out the design on a computer using a standard drafting program.[4]

The architect begins by drawing a wall. To draw a wall, he first draws a line, representing one side of the wall. He then copies the line at some small distance to create the other side of the wall; i.e. creates an offset. An offset is a *transfer* structure in which the first line is the fiber and the movement of this fiber is the control. Next, to create the opposite wall of the room, the

---

[4] Exactly the same thing happens with a high-level solid modeling program, except that more of the stages are done by the computer rather than the user.

architect copies this pair of lines *as a single unit* to the other position. That is, he *transfers the transfer*.

This process is continued upward in the design at all levels. For example, typically in the design of an apartment house, there are studio apartments, one-bedroom apartments, two-bedroom apartments, etc. The standard procedure in an architects office is to first create the drawing for the studio apartment. This drawing would be saved in its own computer file. Next, rather than drawing the one-bedroom apartment from scratch, the architect takes the drawing for the studio apartment and modifies it until he obtains the one-bedroom apartment. This will involve copying the single room defining the studio, to make two rooms (living room and bedroom), scaling the two rooms individually, rearranging copying and scaling the closets, etc. At all stages the architect is simply *transferring structure* that exists in the studio apartment.

Chapters 11, 12, and 13 will show how to formalize this using *unfolding groups*, which were briefly mentioned in the last section. Such groups will *unfold* the studio apartment into the one bedroom apartment, and so on. Unfolding groups are a powerful way of formalizing the process of transfer.

1.10   Cognition and Transfer

What has been said here about design is true generally of cognition. Cognition seems to proceed by describing new situations as versions of old situations. For example, Rosch [132] [131] has argued that categorization proceeds by seeing objects in terms of prototypes, and Carbonell [17] has argued that learning proceeds by analogical reasoning; see also Gentner [40].

Essentially, we will say that *the mind tries to avoid creating new structure, and tries instead to adapt old structure.* This should not be viewed as a negative thing: It is the central means by which the cognitive system makes sense of the world. Thus we can restate the principle of the maximization of transfer as this:

FUNDAMENTAL PRINCIPLE OF COGNITION. *Create any new structure as the transfer of existing structure.*

Notice that we apply this not only to the receptive process of seeing and understanding the world, but also to the creative process of design. One can regard this book as offering an algebraic theory of transfer.

1.11 Transfer in Differential Equations

The next section will look at the nature of scientific laws, and show that they are structured by transfer. However, the topic of transfer in science has a more general setting within the theory of differential equations. Transfer is, in fact, fundamental to methods of solving differential equations. Most methods exploit the fact that the solutions of a differential equation can be *transferred* onto each other. This phenomenon is considerably more profound than it might at first seem. For example, it is basic to the structure of scientific laws.

Differential equations are by far the most frequently used modeling method throughout the world. It is no exaggeration to say that more than several trillion differential equations are solved per second across the world, e.g., in electrical power plants, factories, financial institutions, etc. Clearly all this depends on methods for *solving* differential equations, and a large variety of such methods have been developed. However, basic to these methods is *symmetry*. This is the modern approach that was created by Sophus Lie, and for which he formulated the machinery of Lie groups and Lie algebras. In fact, the use of symmetry to solve differential equations is very familiar to high-school students, as follows:

Consider the first-order differential equation:

$$\frac{dy}{dx} = F(x). \tag{1.10}$$

From high-school, everyone is familiar with the fact that its solution is the integral

$$y = \int F(x)dx + C \tag{1.11}$$

where C is a constant of integration. Because of this constant, one knows that there are a whole set of solution curves, each one obtained by substituting a particular number for C. This means that the solution curves are all *translations* of each other, as illustrated in Fig. 1.14.

This is the first example of Lie theory that anyone encounters. The differential equation (1.10) admits a 1-parameter Lie group of translations in the y direction. The consequence is that you can map the solutions onto each other using this translation group. The group is represented by the constant of integration C in (1.11).

Other types of differential equations can have other types of symmetry groups. For example, a first-order differential equation of this type:

$$\frac{dy}{dx} = F\left(\frac{y}{x}\right) \tag{1.12}$$

admits a 1-parameter group of *scalings*. You can then map its solution curves onto each other via this group. Thus the constant C of integration will occur

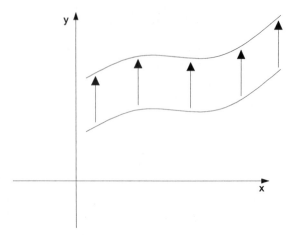

Fig. 1.14. Transfer of solutions onto solutions, in a differential equation.

not as an *addition* onto a solution, as in (1.11), but as a multiplicative factor on the solution. This constant will actually represent the scaling group involved.

Clearly therefore what is being described here is the *transfer* of solutions onto each other. This is basic to the solution of differential equations. In the next section, we will see that this is fundamental to the structure of scientific laws.

1.12   Scientific Structure

It will be argued that the very concept of a *scientific law* involves the phenomenon of *transfer*.

At the foundations of any branch of physics there is a dynamical equation, which is regarded as the fundamental dynamical *law* of that branch of physics. This law determines the evolution of a system state. For example, in Newtonian mechanics, the dynamic equation is Newton's second law, $F = ma$, which determines the trajectory of a system in classical mechanics; in quantum mechanics, the dynamical law is Schrödinger's equation which determines how a quantum-mechanical state will evolve over time; in Hamiltonian mechanics there are Hamilton's equations which determine how a point will move in phase space.

The law, being a dynamical equation, is expressed as a differential equation. Very profoundly, the *lawful* nature of the equation is given by the *symmetries* of the equation, as follows:

Consider Fig. 1.15. The bottom flow-line in the figure shows an experiment being run in a laboratory in New York. The system is set up at time 0, in

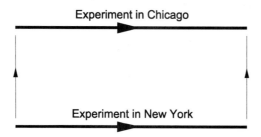

Fig. 1.15. The transfer of an scientific experiment.

initial state $s(0)$, which is the left-most point on that flow-line. The flow-line then represents the evolution of the system's state in the experiment. Suppose that the evolution is found to be governed by a particular dynamic equation.

Now, the upper flow-line in Fig. 1.15 shows the same experiment being run in a laboratory in Chicago. By this we mean that the system in Chicago is started in a translated version $T[s(0)]$ of the initial conditions $s(0)$ in New York. That is, the left-hand point of the upper flow-line is $T[s(0)]$. The upper flow-line then represents the evolution of the system's state in the Chicago experiment. Let us assume that the upper flow-line turns out to be a translated version of the lower flow-line. This translation is shown by the vertical arrows in Fig. 1.15.

The important question is this: Is the upper flow-line described by the same dynamic equation that was discovered for the New York experiment? In other words, can one say that both flow-lines are solution-curves for the same dynamic equation? If one can, then the dynamic equation begins to appear *lawful*, i.e., to apply everywhere. This lawfulness is equivalent to discovering that the equation has translational symmetry. What we mean by this is the following: A dynamical equation prescribes flow-lines; these are the solution-curves to the equation. We ask: Does the translation of one flow-line in the set of solution-curves produce another flow-line in that set? If it does, then one says that the dynamical equation has translational symmetry. This is equivalent to saying that it is a law; i.e., that it works anywhere.

The above illustrated the relation between symmetries and laws using translational symmetry as an example. However, the same argument applies to the choice of any other kind of symmetry, e.g., rotational symmetry. In physics, the basic program is to hunt for dynamical equations that have symmetries; i.e., are lawful. Conversely, one can start with a symmetry group and use it to help construct a lawful dynamical equation. For example, this was Einstein's technique in establishing the correct form of Maxwell's electromagnetic equations.

In any branch of physics, the appropriate symmetry group will be one that sends solution-curves to other solution-curves of the dynamic equation. The appropriate groups for the following branches of physics are:

$$\begin{array}{rcl}
\text{Newtonian mechanics} & \longleftrightarrow & \text{Galilean group} \\
\text{Special relativity} & \longleftrightarrow & \text{Lorentz group} \\
\text{Hamiltonian mechanics} & \longleftrightarrow & \text{Symplectic group} \\
\text{Quantum mechanics} & \longleftrightarrow & \text{Unitary group}
\end{array}$$

It is clear that the phenomenon we have been describing above is one of *transfer*. That is, a dynamical equation permits transfer if the transferred version of any solution-curve (flow-line) is also a solution-curve. It is this that makes the equation lawful. Therefore the phenomenon of transfer is equivalent to the lawful property of the equation.

$$\textbf{Lawfulness} \quad \longleftrightarrow \quad \textbf{Transfer.}$$

Now, we have said that the lawful property is due to symmetries in the equation. However, we shall see that, to describe this structure in terms of *transfer* gives a deeper description - one that captures more fully the process of scientific discovery.

Let us therefore describe the situation in terms of transfer. Observe first that the flow itself is a symmetry across the state-space. This is because a dynamical equation (differential equation) prescribes a vector field and a vector field prescribes a 1-parameter group G_1 of actions along the flow-lines of the vector field.[5]

In particular, let us now isolate any individual flow line. The group G_1 can be considered as "confined" to that flow line. For example, in Fig. 1.15, the group G_1 would be moving along any one of the horizontal lines.

Now, let us consider the symmetry discussed above: the symmetry G_2 of the differential equation. This maps flow lines to flow lines. This is illustrated by the vertical arrows in Fig. 1.15. Thus G_2 is acting *across* the flow lines, and G_1 is acting *along* any flow line. This means that we can consider the flow-lines as *fibers*, and G_2 as a control group *transferring* G_1 from one flow-line (fiber) to another. That is, we have this control-nested structure:

$$G_1 \ \textcircled{w} \ G_2. \tag{1.13}$$

Chapter 20 will show that this combined group fits the rigorous definition of wreath product. Notice that this is a richer algebraic structure than is normally used to express symmetries in physics. First, there is an independent copy of G_1 on each of the fibers. One can think of this as representing experiments that were independently done before a process of induction discovered a relation between these experiments. Then after induction had established

---

[5] For ease of discussion we are assuming that the dynamical equation is a *first-order* differential equation. A first-order equation prescribes a flow like a "fluid" directly on the space of independent variables. This is the situation for example in quantum mechanics and Hamiltonian mechanics. In Volume II, our analysis will be easily extended to *second-order* equations, which are the basis, for example, of Newtonian mechanics or Lagrangian mechanics.

the control group G_2, experiments could be coordinated and one could, for example, establish a single "wave front" of points moving along the flow. This in fact, corresponds to the "diagonal" of the wreath product. Thus all the stages of scientific discovery are contained in the wreath structure, as opposed to the conventional symmetry structure in physics. This will be fully exaborated in Chapter 3 when we deal with wreath products as hierarchies of detection.

A related reason why one searches for symmetries of the dynamic equation comes from Noether's theorem, which states that, to each continuous symmetry of the dynamic equation, there is a conservation law; i.e., a conserved quantity such as energy, linear momentum, angular momentum, etc. It is clear therefore that the possible control groups G_2 in the wreath product (1.13) above, correspond to the possible conservation laws of the system. As an illustration, let us consider quantum mechanics:

In quantum mechanics, a state of the world is given by a wave function. The space of wave functions (world states) is called physical Hilbert space; i.e., any *point* in this space is a world state. The dynamic equation tells us how the world states evolve over time. This equation is called Schrödinger's equation. Schrödinger's equation specifies a *rigid rotation* of Hilbert space. Therefore the *flow-lines* generated by Schrödinger's equation correspond to a rotation group acting on Hilbert space. This rotation group will be denoted by G_1.

Now, because one wants to identify conservation laws, one wants to find symmetry groups of the flow. These will send flow-lines onto flow-lines. Remarkably, any such symmetry group will also be a rotation group G_2 of Hilbert space. This will rigidly rotate the flow-lines of the Schrödinger equation onto each other.

Thus there are two groups: G_1, the rotation group prescribed by Schrödinger's equation, and G_2, the rotation group of symmetries. According to our generative theory of shape, one should regard these two groups, respectively, as the fiber group and control group of the wreath product in expression (1.13).

What we have just said illustrates a basic point that will be made in Chapter 20: With respect to scientific structure, there is the following correspondence.

Conservation Laws ⟷ Wreath Products.

Mathematically we will construct this by setting up a correspondence between any pair of commuting observables and the wreath product of their 1-parameter groups.

1.13   Maximization of Transfer

In the preceding sections, we have seen that several major domains are structured by transfer.

(1) Human perception: The human perceptual system is organized as an n-fold wreath product of groups, $G_1 ⓌG_2 Ⓦ \ldots ⓌG_n$.

(2) Robotics: The algebraic structure of a serial-link manipulator is of the form $G_1 ⓌG_2 Ⓦ \ldots ⓌG_n$, which is not the group that is conventionally used to connect the effector to the base.

(3) Object-Oriented Programming: In geometric object-oriented programming, a basic organizing devise called *inheritance* can be given by hierarchies of transfer, and will be algebraically formalized as wreath products.

(4) Design: In creative design, new structure is added as the transfer of existing structure. This will be modeled by classes of wreath products to be invented in Chapters 11, 12, and 13.

(5) Differential Equations: The fundamental solution methods for differential equations rely on the fact that solutions can be transferred onto other solutions.

(6) Scientific Structure: The conservation principles of physics can be formulated as structures of transfer, in which commuting observables are corresponded with wreath products of their 1-parameter groups.

In this book, these areas will be extensively investigated and mathematically formalized, so that we fully understand how they are driven by the principles of shape generation.

1.14   Primitive-Breaking

We now come to one of the fundamental rules of the generative theory: *exhaustiveness.* The theory is *generatively exhaustive* in the sense that there is no level (in a shape) that is not given a generative explanation. One consequence of this is the following:

THE NO-OBJECTS RULE. *A basic consequence of generative exhaustiveness is that there are no objects, only actions. Anything that one might want to call an object is itself generatively described, and is therefore itself a set of actions.*

This rule is crucial for navigation, manipulation, and any planning in an environment. In fact, we argue that planning generally is possible only to the extent that the no-objects rule has been put into effect.

Let us illustrate what has just been said. Recall from p. 8, that our theory gives the following generative description of a square:

$$\mathbb{Z}_2 \,\circledw\, \mathbb{R} \,\circledw\, \mathbb{Z}_4 \qquad\qquad (1.14)$$

where, from right to left, $\mathbb{Z}_4$ is the 4-fold rotation group sending sides to sides; $\mathbb{R}$ is the translation group generating a side from a point; and $\mathbb{Z}_2$ is the occupancy group which switches a point on or off (e.g., at the ends of a side). This description is generatively exhaustive, and illustrates the no-objects rule as follows:

According to this description, a square, which people usually regard as an *object*, is in fact, the 4-fold rotation of a side. But the side itself is not an object. It is the translation of a point. But the point itself is not an object. It is the action of the occupancy cycle $\mathbb{Z}_2$, causing the creation or erasure of a point.

Each one of these levels of action is useful for the agent in terms of planning, e.g., the 4-fold rotation group $\mathbb{Z}_4$ can represent movements from side-to-side around a square object; the translations $\mathbb{R}$ along a side can represent a navigator moving along the edge of a square table, or a plotter drawing the side, or a machine-cutter cutting the side; and the lowest group $\mathbb{Z}_2$ can represent the first contact of the plotter with the paper, or the cutter with the material, etc.

Failure to recognize that any level of an object is itself a space of actions, means that that level of action is lost to the agent.

One can see therefore that an essential aspect of our approach is this: Although a state-space can be viewed as an object being pushed through the alternative states, the object itself must be "opened up" and understood as a state-space of some sub-object. This sub-object must then be opened up and viewed as the state-space of some sub-sub-object, and so on. This means that objects are state-spaces that are nested downwards.

This nesting is a crucial component of our theory of geometry, as we have seen. It is equivalent to the notion of transfer. Any level is described as a level below being transferred across some space. In this sense, the no-objects rule can be regarded as equivalent to the maximization of transfer. This also means that the no-objects rule can be algebraically realized as wreath products.

The ban on objects clearly extends to a ban on such concepts as "primitives" and "features". For example, since exhaustiveness requires that any level is generatively described, then any primitive would be generatively described, and would therefore not be a primitive. The process of describing primitives generatively, will be called *primitive-breaking*. Clearly, it applies successively downward through the shape.

The following chapters will often use the term "primitive". However, this will be in order to give a generative theory of what people usually call primitives.

1.15   The Algebraic Description of Machines

For reasons that emerge in the next section, it is profoundly important to express the concepts of this book in terms of the algebraic theory of machines. This section recalls the most basic algebraic facts about machines:

A machine is a pair of functions: (1) the state-transition function σ : $I \times S \longrightarrow S$, and (2) the output function $\tau : I \times S \longrightarrow O$, where I is the set of inputs, S is the set of states, and O is the set of outputs; see Schmidt [134]. Standardly one assumes that the input and output sets are finite. The set of input *sequences* of finite length form a semigroup, and this allows one to establish the fundamental relation between the theory of machines and the theory of semigroups. If the input semigroup is a group, then the machine is called a group machine. In group machines, the input action is always a permutation of the set of states. In contrast, a *collapser* is a machine where an input can cause two distinct states to go to a single state. The fundamental theorem of the algebraic theory of machines is by Kenneth Krohn and John Rhodes, and says essentially that any machine can be decomposed into a set of group machines and four elementary types of collapsers. A basic tool of the algebraic theory of machines is *wreath products of machines*. See Krohn & Rhodes [79], and the collection of papers in Arbib [2].

1.16   Agent Self-Substitution

In Leyton [87], [88], [89], [90], [91] [96], we presented a considerable amount of psychological evidence that *human cognition structures phenomena as machines*. This view should be distinguished from the *processing machine analogy* which says that cognitive *processes* are structured as machines. In contrast, we called the view that we put forward, the *representational machine analogy*. This says that cognitive *representations* are structured as machines;

i.e., that cognition represents the environment as machines. We showed that this view explains an enormous number of psychological results from perception, categorization, linguistic semantics, and so forth. In fact, in the 120-page journal article (Leyton [88]), we reviewed six levels of the human cognitive system and showed how an algebraic machine theoretic approach explains the psychological data on each of these six levels. Indeed, the wreath products $G_1 ⓦ G_2 ⓦ \ldots ⓦ G_n$ that we have been discussing, are *wreath products of machines*. These wreath products constitute the structure of cognitive representations. This approach makes the algebraic theory of machines the foundation of cognitive science.

We now present one of the main reasons for developing a generative theory of geometry, as follows:

AGENT SELF-SUBSTITUTION. *A generative theory of shape characterizes any shape as a program running on some machine. The power of a generative theory of shape is that the agent can self-substitute for the machine, and the program thereby becomes a plan, e.g., for navigation, manipulation, design, manufacturing.*

As an example, return to the generative description of a square.

$$\mathbb{Z}_2 \; ⓦ \; \mathbb{R} \; ⓦ \; \mathbb{Z}_4 \tag{1.15}$$

This structure can be be regarded as the input group of a machine. The agent can therefore self-substitute for the machine or any of its components. For example, the agent can self-substitute for the right-most control group, $\mathbb{Z}_4$, and this corresponds to the plan of moving around the object from one side to the next. Or the agent can self-substitute for the middle control group, $\mathbb{R}$, and this corresponds to the plan of translating along an individual side. Or the agent can self-substitute for the left-most group, $\mathbb{Z}_2$, and this corresponds to switching the movement on or off. Furthermore, plans can be made from the machine structure as a whole. We will see later how to construct what will be called *canonical plans* from wreath products. A canonical plan is a systematic means of travelling cyclically through the levels, using up elements. For example, the canonical plan for the above wreath product of a square produces the standard way of drawing a square - i.e., tracing out the sides sequentially. Careful consideration reveals that this comes from a particular cyclic process through the above wreath product.

The previous paragraph is only a simple example of the theory of planning to be given in this book:

CONVERSION OF PERCEPTION INTO ACTION. *Perception is the description of phenomena as machines. The conversion of perception into action is achieved by the agent's self-substitution into the machines defined by* *its perceptual representations. This is the basis of navigation, manipulation, design, manufacturing, etc.*

In the following chapters, this will be applied to solve complex planning problems in computer-aided manufacturing, robot assembly, etc.

1.17   Rigorous Definition of Shape

Despite the fact that the term "shape" is used prolifically in mathematics, it has never been rigorously defined. We now propose a definition of shape which we will argue is the appropriate one for the physical, computational, and design sciences. Indeed, we will argue that this definition actually leads to a new understanding of what these sciences are:

> **The** *shape* **of a data set is the transfer structure of a machine whose state space is the data set.**

The first thing to observe about this claim is that we regard the concept of *machine* to be at the core of the concept of *shape*. This claim has a twofold status. First, it is a *psychological* claim: The psychological research in Leyton [96] demonstrates substantially that the concept of machine is basic to peoples' perception of shape. Second, it is a *mathematical* claim: We believe that geometry should be regarded as a branch of the algebraic theory of machines.

In fact, although we argue that the theory of shape should be regarded as a branch of the algebraic theory of machines, our claim is much more restrictive than this. We argue that shape is a specific part of the algebraic theory of machines: It is the part that deals with the *transfer* of components onto components within a machine. It is the structure of transfer that defines shape.

Finally, observe that the above definition of shape realizes our aim of ensuring that every aspect of a shape is defined by action.

1.18   Rigorous Definition of Aesthetics

Usually, the two stated goals of design are aesthetics and functionality. Functionality is basic to our system because shape is understood as the description of phenomena as machines, and functional descriptions are descriptions as machines.

However, aesthetics is also basic to our system. This book will give a rigorous mathematical theory of aesthetics. We propose the following:

Aesthetics is the maximization of transfer.

Clearly, therefore, aesthetics is connected to the very foundations of our generative theory of shape - i.e., it embodies our basic principle of the maximization of transfer. Let us now look at the role of aesthetics in design; and then at its role in science.

Any major artist, such as Beethoven or Raphael, strives for the unification of all elements in a work. A movement of a symphony by Beethoven has remarkably few basic elements. The entire movement is generated by the transfer of these elements into different pitches, major and minor forms, overlapping positions in counterpoint, etc. The equivalent is true of the paintings of Raphael - see for example our lengthy analysis of Raphael's and Picasso's paintings, in Leyton [96]. Transfer is the basic means by which an artist generates a work from a minimal set of elements.

One of the powerful consequences of understanding that aesthetics is the maximization of transfer is that, since we will give a complete mathematical theory of transfer, we will be giving a complete mathematical theory of aesthetics. It is clear that one benefit of this theory is that it will be possible to integrate aesthetics into CAD programs in a formal and explicit way.

Now let us turn to aesthetics in science. It is well known that the major innovators of scientific history declared the fundamental importance of aesthetics in determining their contributions. For example, both Einstein and Dirac strongly argued that the scientist should use *aesthetics and not truth* in determining the structure of an equation. This point is not trivial: In forming his famous relativitistic equation, Dirac violated one of the most established truths of physics, that electrons are negative. His equation predicted the existence of positive electrons, for which there was literally no evidence, and for which he was publically ridiculed. However, it was this equation that later lead to the discovery of positive electrons. In contrast, relativistic equations that had conformed to the established truth that electrons are negative, were later rejected. Dirac stated that he had consciously structured his relativistic equation by aesthetic criteria, and had deliberately ignored truth. Einstein made exactly the same statement about his own work.

It is probably the case that all uses of the term aesthetics in physics concern the symmetries of the dynamic equation (the Dirac equation is an example of this). As seen in Sect. 1.12, the symmetries of a dynamic equation correspond to the *transfer* of solutions onto solutions. This reinforces our claim that aesthetics is the maximization of transfer.

Finally, note this: All artists and scientists strive for *unity* in their work; i.e., the integration of all elements. What this book allows us to do is develop a rigorous mathematical theory of unity. Unity comes from the maximization of transfer, and transfer is expressed as wreath products.

1.19   Shape Generation by Group Extensions

The concept of *group extension* is basic to our generative theory. A group extension takes a group G_1 and adds to it a second group G_2 to produce a third, more encompassing, group G, thus:

$$G_1 \; \textcircled{E} \; G_2 \;\; = \;\; G$$

where $\textcircled{E}$ is the extension operation. See our book on group extensions, Leyton [98].

It is clear, looking back over the previous sections, that according to our theory:

Shape generation proceeds by a sequence of group extensions.

That is, shape generation starts with a base group and successively adds groups obtaining a structure of this form:

$$G_1 \; \textcircled{E} \; G_2 \; \textcircled{E} \; \ldots \textcircled{E} \; G_n.$$

This approach differs substantially from other current approaches to shape generation, such as that of Stiny & Gips [147], [44], which is based on the application of production rules. In our approach, rather than applying production rules, one is adding structure. The successive addition of structure is represented by successive group extension. Furthermore, imposing the condition of *maximization of transfer* demands that the structure $G_1 \textcircled{E} G_2 \textcircled{E} \ldots \textcircled{E} G_n$ be actually of the form $G_1 \textcircled{W} G_2 \textcircled{W} \ldots \textcircled{W} G_n$. In other words, the extension operation $\textcircled{E}$ is the control-nesting operation $\textcircled{W}$.

2. Recoverability

2.1 Geometry and Memory

A generative theory of shape represents a given data set by a *program* that generates the set. This program must be *inferrable* from the set. We shall say that the program is *recoverable* from the data set. Recoverability of the generative program places strong constraints on the inference rules by which recovery takes place, and on the programs that will be inferred. This, in turn, produces a theory of geometry that is very different from the current theories of geometry.

Essentially, the recoverability of generative operations from the data set means that the shape acts as a memory store for the operations. More strongly, we will argue that all memory storage takes place via geometry. In fact, a fundamental proposal of our theory is this:

Geometry ≡ Memory Storage.

As we shall see, this theory of geometry is fundamentally opposite to that of Klein's in which geometric objects are defined as invariant under actions. If an object is invariant under actions, the actions are not recoverable from the object. Therefore Klein's theory of geometry concerns *memorylessness*, and ours concerns *memory retention*. We argue that the latter leads to a far more powerful theory of shape.

2.2 Practical Need for Recoverability

We shall distinguish two different types of need for recoverability, the *practical* and the *theoretical* need. This section considers the first of these, by going through a number of areas where recoverability is fundamentally necessary:

(1) Computer Vision. An image is formed on the retina because light was emitted from a source, and then interacted with a set of environmental objects, and finally interacted with the retina. This is called the *image-formation process*. Computational vision is the process of recovering the image-formation process from the image. This is often referred to as inverse optics; i.e., the recovery of the optical history. It is the dominant view in computer vision, e.g., as exemplified by Berthold Horn's book [60].

(2) Machine Learning. Carbonell [17] has emphasized that an intelligent learning system must have the capacity to recover its own computations in order to understand why it failed or succeeded.

(3) Computer-Aided Design (CAD). Hoffmann [58] has emphasized the need for recovering the history of design operations in order to allow editability of that history. All major CAD programs such as *ProEnginneer* in mechanical design, and *AutoCAD* in architecture, provide full history-recovery at any stage of the design process. The same is true of solid modeling and animation software such as *3D Studio Max*, and 2D publication/image manipulation programs such as *Photoshop*. It is probably the case that the history-recovery operations in a CAD or Graphics program are amongst the most frequently used operations in design.

(4) Computer-Aided Manufacturing (CAM). When the design phase is completed, the CAD model is sent to a CAM file for machining. Computer-aided process planning (CAPP) is the means by which the CAD model can be used to generate a sequence of instructions to physically produce the part. In machining, the part is created by taking the designed shape for the part and subtracting it from the raw stock (usually a rectangular block). The remaining shape is called the *delta volume*, as shown in Fig. 2.1. The delta volume will be removed incrementally from the raw stock by a moving cutter. Therefore, the cutter actually *generates* the delta volume. This means, once again, that shape is described generatively, and the generativity must be recovered from the object data.

(5) Science. Science is ultimately the recovery of the sequence of causal interactions that lead up to the data on the measuring instruments. We shall argue that recoverability fundamentally dictates the structure of scientific laws, such as Schrödinger's equation, Hamilton's equations, etc.

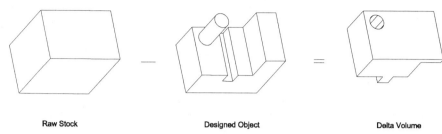

Raw Stock   Designed Object   Delta Volume

Fig. 2.1. The raw stock minus the designed object gives the delta volume.

2.3 Theoretical Need for Recoverability

The theoretical need for recoverability comes from the very nature of *generative* theories. A generative theory characterizes the *structure* of a data set as a program for generating that set. In order to produce this structural characterization one must be able to infer (recover) it from the data set. This argument is simple, but places enormous constraints on the theory, as will be seen.

2.4 Data Sets

It is now necessary to understand the constraints that recoverability places on data sets. There are four basic conditions that data sets must fulfill:

(1) The data set is structureless.

Since the generative structure is the only structure we are interested in, then, as far as we are concerned, the data set has no structure prior to the inference of a generative program. The data set will therefore be regarded as structureless.

(2) The data set is atemporal.

A particular consequence of the fact that the data set is structureless, is that it has no *temporal* structure; it is atemporal. It is only the inferred generative program that can give it temporal structure. In fact, *time* will be understood as the sequence-parameter that organizes the generative program as a succession of operations.

(3) The data set is observable within a single time-slice.

Consider a doctor who has taken a set of X-rays of a tumor over time. These X-rays are *records* of the tumor's growth. In order for the doctor to understand the development of the tumor, he needs to be able to compare all the records within a single time-slice. Therefore he does what all doctors do: He puts the entire set of X-rays up on a screen *at the same time*. From this, he can infer what change there has been over time.

This is crucial for the *inference* of generative structure: Although a data set might consist of a set of records taken at different points in time, they must all be available simultaneously within a single time-slice. In the example of the X-rays, the *screen*, onto which the X-rays are all placed simultaneously, represents the single time-slice.

In order to grasp the fundamental power of this concept, let us consider another example - the inference of *motion*. It is a logical fact that one cannot infer motion unless states at different times are viewable simultaneously within a single time-slice. As an example, consider the motion detector in the visual system of a fly. Fig. 2.2 shows the structure of the detector as given by Hassenstein & Reichardt [53], Reichardt [123]. At the top of the figure, two receptors are shown, one at position P_1 and the other at position P_2. The receptors are one unit distance apart. When a stimulus crosses a receptor, at the top of the figure, the receptor gives a response which is sent along the downward fiber from the receptor. Now consider a stimulus travelling from left to right across the two receptors, and suppose the stimulus is travelling at the speed of one unit distance per one unit time. The stimulus first passes P_1 which sends its response downward. However, this response is delayed, for one unit time, by the DELAY cell shown in the figure. Meanwhile, the moving stimulus continues along the top and passes over the second receptor P_2, which then sends its response down its own fiber. But because the response from the first receptor P_1 has been delayed by one unit time, the response from P_1 arrives at the cell labeled A at the same time as the response arrives from receptor P_2. The cell A fires in exactly this situation, i.e. when it receives a response *simultaneously* from both P_1 and P_2.

Generally, a data set consists of records. Whether the different records are points recorded at different times by a motion detector, or X-rays taken at different times by a doctor, they must all be available for comparison within the present time-slice.

(4) The box of the present is cealed.

The fact that the entire data set is available for observation within a single time-slice is crucial from the point of view of *inference*. The inference system cannot go back in time, and observe any previous data; it must observe the data as it exists *now*. The box of the present cannot be violated. Thus, one must ensure that the box of the present contains the entire data set that is

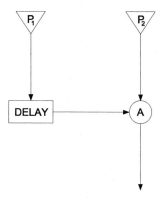

Fig. 2.2. The motion detector of a fly.

to be used. Then one must "ceal" the box, and not allow the inference to assume data outside the box. That is, the inference rules are applied to the contents of the box, and only these contents.

For example, suppose that the doctor had taken twelve X-rays of the tumor over the last twelve months. And suppose that one of these X-rays is now missing - the one taken in March. The box of the present now contains only eleven X-rays. Yet, as we shall see, a physicist will standardly make the mistake of assuming that the missing X-ray is observable. This leads to many incorrect arguments in statistical mechanics and quantum field theory.

Now consider the following situation. Suppose that, even though the March X-ray has been lost, the doctor actually remembers what the March X-ray looked like. In this case, he has a record of the tumor's state in March, in his head, right now. The box of the present therefore contains the eleven X-rays and the record he has in his head. Inference can therefore act on all twelve records. At no time, however, can he actually go back in time and see the situation in March. He must have a record in the box of the present. Thus, if he cannot remember what the state was in March, he has only the eleven records.

Therefore, in all our inference situations, the following is critical: One must always decide what data is available in the box of the present, and then, having made this decision, allow only this data to be used by the inference system. This fundamental constraint will actually determine the inference rules, as we shall see.

2.5   The Fundamental Recovery Rules

This section will state our two fundamental rules for the recovery of generative programs from data sets. The discussion will begin with simple illus-

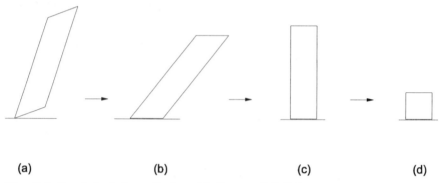

| (a) | (b) | (c) | (d) |

Fig. 2.3. Psychological results found in Leyton [89] [90].

trations, and then get successively more difficult. Later, the representational consequences of the rules will be rigorously formulated in terms of wreath products.

Let us begin by looking at how the human visual system recovers generativity from a data set. In a series of psychological experiments (Leyton [89] [90]), we found that, when subjects are presented with a parallelogram oriented in the picture plane as shown in Fig. 2.3a, they see it as a rotated version of the parallelogram in Fig. 2.3b, which they then see as a sheared version of the rectangle in Fig. 2.3c, which they then see as a stretched version of the square in Fig. 2.3d. The remarkable thing is that the only data that they are actually given is the first figure, the rotated parallelogram. The experiments found that, on being presented with this figure, their minds elaborated the sequence shown.

It is clear that what is happening is that the subjects are giving the first figure a *generative* description. In other words, given the rotated parallelogram as *data*, the subjects are saying that, at the previous generative stage, the rotated parallelogram was not rotated, and in the generative stage before that, the shape was not sheared, and in the generative stage before that, the shape was not stretched. That is, they are *recovering* the generative history by successively removing the generative operations. To emphasize: The sequence from left to right is the *reverse* of the generative program; i.e., it corresponds to *backwards time*. Thus, the forward-time direction, in the generative program, is right to left.

Now, close examination reveals that the subjects are accomplishing this recovery by applying two rules, which we will call respectively the Asymmetry Principle and Symmetry Principle:

Asymmetry Principle.

At each stage, in Fig. 2.3, the subjects are removing one asymmetry to get to the next stage (left-to-right). Let us, for the moment, define asymmetry simply to be *distinguishability* - this will be rigorized later using group theory. There are three distinguishabilities in the rotated parallelogram, Fig. 2.3a:

(D1) The orientation of the figure is distinguishable from the gravitational orientation.

(D2) The adjacent vertex angles are distinguishable in size.

(D3) The adjacent sides are distinguishable in length.

One can see that the subjects are successively removing these three distinguishabilities to create the sequence from left to right. The sequence therefore represents a process of successive symmetrization, backwards in time. Therefore one sees that, *in the forward time direction, the generative program is a succession of symmetry-breaking operations.* We shall call this the Asymmetry Principle.

Symmetry Principle.

The subjects are also using another rule in the sequence Fig. 2.3. Any symmetry in the rotated parallelogram is being preserved backwards in time. For example, the rotated parallelogram has the following two symmetries:

(S1) The opposite vertex angles are indistinguishable in size.

(S2) The opposite sides are indistinguishable in length.

Notice that these two symmetries are preserved through all the figures from left to right. In fact, the condition is even stronger than this. We saw that, by the Asymmetry Principle, each backwards transition recovers a symmetry. Remarkably, what can now be seen is that, after any symmetry has been recovered, it is also preserved backward through the remaining sequence. For example, in going from Fig. 2.3b to Fig. 2.3c, the adjacent vertex angles, which are initially different, are made the same size. This symmetry is then preserved in all the remaining figures. We shall call the backward preservation of symmetries, the Symmetry Principle.

The Asymmetry Principle and Symmetry Principle control the recovery of the generative sequence in Fig. 2.3. Note that the *data set* is the first figure (the rotated parallelogram), and the recovery takes place by applying the

Asymmetry Principle and Symmetry Princple *to the data set.* This produces the generative program in the reverse direction.

The full statements of the two fundamental rules will now be given. First, the Asymmetry Principle will be stated and discussed:

ASYMMETRY PRINCIPLE (forward-time statement). *Given a data set $\mathcal{D}$, a program for generating $\mathcal{D}$ is* **recoverable from** $\mathcal{D}$ *only if the program is symmetry-breaking on the successively generated states.*

It is important to understand that this principle concerns not simply generativity, but the actual *recoverability* of that generativity. This, in turn, depends crucially on the data set.

To sharpen the reader's mind concerning these issues, let us consider (forward-time) *symmetry-increasing* processes. For example, let us consider the basic theormodynamic situation of entropy increase. Suppose that an isolated tank of gas is in an initial state in which the gas is entirely on the left side of the tank. The gas is then allowed to settle to equilibrium in which it is uniformly distributed over the tank. This is a symmetry-increasing process. Now let us consider the *recoverability* of this process. The recoverability is dependent entirely on the data set. Suppose that the data set consists only of the gas in the final symmetric state. Then we would not be able to recover from this state, any of the preceding history. For example, this state would not tell us that the gas had been on the left rather than the right. In contrast, suppose that the data set consists of a set of photographs that we had taken over the course of the thermodynamic process. Being a data set, we can view the photographs simultaneously. For example, without loss of generality, we will assume, throughout the discussion, that the photographs are laid out on a table from left to right. Then, from this data set, we would be able to recover the process.

Now consider the *generation* of this data set. That is, the set of photographs is the set $\mathcal{D}$ in the above statement of the Asymmetry Principle, and we are considering the *program* that generated this set. The states that are successively generated by the program are the successively increasing sets of photographs over time, leading to the final full set. Each successive set is, in fact, asymmetry-increasing with respect to its preceding set: it increases the left-right directionality (assuming that the photographs are laid out from left-to-right on the table). Thus, although the thermodynamic process is symmetry-increasing, it is recoverable only because it has left a succession of data sets that are symmetry-decreasing.

SYMMETRY-INCREASING PROCESSES. *A symmetry-increasing process is recoverable only if it is symmetry-decreasing on successive data sets.*

An alternative statement of the Asymmetry Principle will now be given - one that is particularly useful.

ASYMMETRY PRINCIPLE (backward-time statement). *Given a data set $\mathcal{D}$, a program for generating $\mathcal{D}$ is* **recoverable from** *$\mathcal{D}$ only if each asymmetry in $\mathcal{D}$ goes back to a symmetry in one of the previously generated states. This symmetry will be called a* **symmetry ground-state.**

Notice, for example, in the case of the thermodynamic process considered above, the asymmetries are *between* photographs. They give the left-right asymmetry of the data set, i.e., of the set of photographs as they are laid out on a table. Notice that in accord with the above principle, these asymmetries are removed backward in time through the successively decreasing set of photographs. This situation will be fully analyzed later.

The Asymmetry Principle is the first of our two fundamental principles of recoverability. The other fundamental principle is this:

SYMMETRY PRINCIPLE. *Given a data set $\mathcal{D}$, a program for generating $\mathcal{D}$ is* **recoverable from** *$\mathcal{D}$ only if each symmetry in $\mathcal{D}$ is preserved backwards through the generated sequence.*

Let us consider an apparent counter-example to this principle. Suppose that a person is walking along a beach, and sees a perfectly rounded rock. Anyone of course knows that rocks are usually jagged (asymmetric) and therefore the person easily infers that the rock was once asymmetric, and that it became symmetric over time. This appears to violate the above Symmetry Principle. That is, a symmetry in the present data set is not being preserved backward in time.

However, the apparent violation is due to incorrectly defining the data set. One has assumed that the data set consists only of the rounded rock. It does not. It consists also of previous experiences of rocks - i.e., images in the person's head of previous rocks. The situation therefore becomes like the thermodynamic example, in which the gas is initially asymmetric and then becomes symmetric over time. Let us re-consider that example:

Recall the complete dependency of inference on the data set. If the data set consists of a set of photographs taken over time, of the tank, then the inference is made on the asymmetries *between* the photographs; i.e., these give the left-right asymmetry of photographs on the table. These asymmetries would be removed backwards in time, and this would mean the successive removal of records backwards in time.

The inference would be completely different, however, if the data set consists of only the final symmetric gas state: Here, one would not be able to infer the previous history of rightward movement - in particular the present symmetry would exclude any preference for left or right in the conjectured history. This is the Symmetry Principle. *A symmetry in the data set has to be preserved backwards in time because symmetry excludes prejudice towards any of its alternative symmetry-breakings.*

Exactly the same argument applies to the rock example, except that here we have not just left-right symmetry, but symmetry in each direction. That is, if the data set contained only the rock in its rounded state, i.e., we had no previous experience of rocks, then the symmetry of the rock would exclude prejudice towards any of its alternative symmetry-breakings. Thus no previous asymmetry is inferrable.

Throughout this discussion, we see that the correct inferences depend completely on the conditions given in Sect. 2.4, on data sets. In particular, the fourth condition (p. 38) stated that the observer must decide what is contained in the box of the present (the data set) and then *ceal* it such that the inference process acts only on the contents of the box. If the only content of the box was the rounded rock - i.e., the observer had no experience of any other rocks - then he would not be able to infer that the rock was previously asymmetric; i.e., symmetry is the only evidence.

2.6 Design as Symmetry-Breaking

The Asymmetry Principle states that the generative operations are recoverable only if they are symmetry-breaking on successively generated states.

As an example, consider the process of architectural design on a computer. The designer begins with an empty screen. Let us examine the meaning of this. The screen is a 2-dimensional flat space, and therefore has the translation group $\mathbb{R}^2$ as its symmetry group. In fact, the edge of the screen breaks this symmetry. Therefore it is best to describe this situation as an infinite flat plane, with symmetry group $\mathbb{R}^2$, that is broken by the edge of the screen. This captures the well-known description of the screen as a view *window*.

Now, at the stage at which the designer begins, the screen is empty and therefore there is still translational symmetry *within* the screen. Next, on this empty screen, the architect draws the first object, e.g., a wall. This breaks the translational symmetry within the screen. The wall however retains translational symmetry within its own border. Next, the architect draws a window within the wall, and this breaks the translational symmetry across the wall, and so on.

This illustrates the fact that the design process is one of successive symmetry-breaking. Much more complicated examples of this principle will be seen throughout the book - in the generation of extremely complex ob-

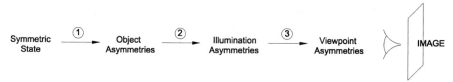

Symmetric State → (1) → Object Asymmetries → (2) → Illumination Asymmetries → (3) → Viewpoint Asymmetries → IMAGE

Fig. 2.4. The 3-stage symmetry-breaking process involved in vision.

jects. Nevertheless, the same principle will still apply - design is a process of symmetry-breaking.

Now consider the issue of *recoverability*. Notice that the preceding symmetry, at each of the stages described above, is completely recoverable. For example, after drawing the wall, the background translational symmetry of the screen is recoverable, because the designer understands that he can move the wall across screen - i.e., using the translation group of the background. In fact, we argue that the term background actually refers to the recoverable symmetry that initiates the generative process.

Throughout this book, these concepts will be seen to underly even the most sophisticated processes of design, e.g., spline manipulation, constructive solid modeling, part assembly in mechanical CAD, etc.

2.7   Computational Vision and Symmetry-Breaking

In Leyton [96], we argued that the process by which the environment is projected onto the retina is symmetry-breaking. It is this that allows perception (recoverability) to take place. Furthermore, we argued that the symmetry-breaking is decomposed into three successive stages, as shown in Fig. 2.4.

As an example, consider a cube. This would be the initial symmetry state on the far left of Fig. 2.4. The three subsequent stages are as follows: Stage 1: Certain environmental actions distort the cube's intrinsic geometry, e.g., twist it, or drill a hole in it, etc. These are a first level of asymmetries on the cube. Stage 2: The introduction of a light source causes a variation in brightness across the surface, i.e., shading. This is an additional level of asymmetries (distinguishable grey levels) that are "painted" on the surface geometry. Stage 3: The projection of the object onto the retina creates what is standardly called projective distortion, e.g., equal parallel lines project to unequal non-parallel lines, etc. These are asymmetries by virtue of viewpoint.

COMPUTATIONAL VISION AND SYMMETRY-BREAKING.
According to the theory in Leyton [96], the sequence of processes leading up to the formation of the image on the retina fall into three successive stages each of which creates a new layer of asymmetries: (1) object asymmetries,

(2) illumination asymmetries, and (3) projective asymmetries. The process of vision is therefore that of undoing the asymmetries backward in time through the generative sequence; i.e., the successive use of the Asymmetry Principle.

This explains the various components of computational vision, such as shape-from-texture, shape-from-contour, shape-from-shading, regularization, etc. For example, shape-from-texture and contour are the removal of Stage 3, backwards in time (viewpoint asymmetries); shape-from-shading is the removal of Stage 2, backwards in time (illumination asymmetries), and regularization is the removal of Stage 1, backwards in time (object asymmetries).

Let us look at this in more detail. Consider shape-from-texture. Consider a square grid in the environment, as shown in Fig. 2.5a. If it is projected onto the retina using orthographic projection, as shown in Fig. 2.5b, it becomes a rectangular grid. This breaks the 4-fold rotational symmetry of the squares. If furthermore, the projection is a perspective one, as shown in Fig. 2.5c, then the translational symmetry in the vertical direction has been additionally destroyed, as can be seen by the fact that the quadrilateral elements now decrease in size from the bottom to the top of the figure.

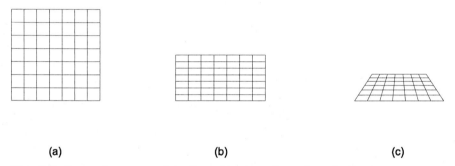

(a)   (b)   (c)

Fig. 2.5. Successive asymmetrization of texture through projection.

Now consider shape-from-shading. A standard assumption is that the light flux is uniform (symmetric) before it hits the object. This is shown on the left in Fig. 2.6. The different orientations of the object surface cause the light rays to become non-uniform (asymmetric) after leaving the surface, as shown on the right in Fig. 2.6. That is, the light-flux undergoes symmetry-breaking in the forward-time direction. Notice that the created asymmetry is what we mean by shading.

The above examples illustrate the fact that the Asymmetry Principle is fundamental to computational vision. However, our other basic recovery rule, the Symmetry Principle, is also fundamental. This principle states that the generative operations are recoverable only if each symmetry is preserved backwards through the generative sequence. We argue that this is the ba-

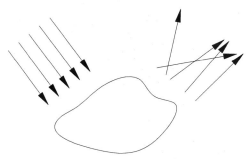

Fig. 2.6. Asymmetrization of the light-flux.

sis of all non-accidentalness rules (e.g., Kanade [72], Witkin & Tenenbaum [156]). Such rules state that properties such as parallelism, or perpendicularity, found in the image, go back to the same in the environment. Our survey of the literature concluded that all these properties are actually symmetries. Therefore, their preservation backwards from the image to the environment is a use of the Symmetry Principle.

A detailed exposition of the above theory of computational vision is given as Chapter 3 in Leyton [96].

2.8　Occupancy

As was illustrated on page 8, our theory represents any structural incompleteness in the following way: The complete structure is described by a group G. The incomplete structure is described by wreath sub-appending an occupancy group $\mathbb{Z}_2$ group below G, thus:

$$\mathbb{Z}_2 \, \textcircled{w} \, G.$$

This is a regular wreath product in which there is one copy $\mathbb{Z}_2$ for each member of G. The copy of $\mathbb{Z}_2$ then switches on or off the corresponding element of G. (Regular wreath products are described in Chapter 3.)

The crucial advantage of this method is that the group G is still present in the structure, representing the observer's sense of completion despite the incompleteness of the data set. For example, consider a straight line, with some gaps. The eye automatically fills in the gaps - which is an example of the Gestalt phenomenon of completion. Our theory models this by representing the complete line by $\mathbb{R}$, and wreath sub-appending an occupancy $\mathbb{Z}_2$ group thus:

$$\mathbb{Z}_2 \, \textcircled{w} \, \mathbb{R}.$$

In this way, the complete structure $\mathbb{R}$ is assumed even after the occupancy structure has been added.

It is important for the reader to observe that we therefore view incompleteness as a process of *asymmetrization*. One begins with the complete structure (the upper group) and then adds its incompleteness (the lower group). This accords with the Asymmetry Principle which states that, for recoverability, the generative sequence must be symmetry-breaking forward in time.

A couple of comments will be useful: (1) If the upper group is itself an n-fold wreath product, then each level can be given an incompleteness structure by wreath sub-appending an occupancy group $\mathbb{Z}_2$ to that level. (2) It is often convenient to think of occupancy as a process of occlusion. One starts with a complete structure and understands its subsequent incompleteness as due to another object that is in the way.

2.9   External vs. Internal Inference

Section 2.5 used a distinction that must now be examined carefully. We shall call this the distinction between external and internal inference. These two alternative types of inference depend on two alternative *assumptions* made by the observer:

Definition 2.1. *In* **external inference,** *also called the* **single-state assumption,** *the observer assumes that the data set contains a record of only a single state of the generative process. Any inferred preceeding state is therefore external to the data set.*

This case was illustrated, for example, in the rotated parallelogram experiment in Fig. 2.3 (p. 40). Here, the only data was the first figure, the rotated parallelogram. The observer made the assumption that this was only a *single state* of the generative process. Therefore, any inferred preceding state, was *external* to what one could see in the data set. For example, the inferred rectangle Fig. 2.3c was not contained in the data set Fig. 2.3a.

In contrast to external inference, the alternative assumption is this:

Definition 2.2. *In* **internal inference,** *also called the* **multiple-state assumption,** *the observer assumes that the data set contains records of multiple states of the generative process. A preceeding state can therefore be internal to the data set.*

This case has been illustrated a number of times in the preceding discussion: (1) The doctor who has several X-rays of a tumor taken over time, which he views simultaneously. (2) The physicist who has several photographs of a tank of gas taken over time, which he views simultaneously. (3) The individual, on the beach, who has both the visual image of the rounded rock, and the memory of past rocks. In all these three cases, the present data is assumed to contain multiple records of the generative process. Preceeding generative states can therefore be found *internal* to the data set.

Notice that we also discussed versions of these situations based on *external* inference. That is, we considered the thermodynamic situation in which one possessed only the final symmetric state. Or we considered the rock situation in which the data contained only the rounded rock and no previous experiences of rocks.

It was seen that the inference conclusions made in the external situation and the internal situation are different from each other. Nevertheless, the remarkable thing is that the inference rules are the same: the Asymmetry Principle and Symmetry Principle. What is different is the nature of the asymmetries and symmetries, as follows:

APPLICATION OF RULES. *In external inference the Asymmetry Principle and Symmetry Principle are applied to* **intra-record** *asymmetries and symmetries. In internal inference the Asymmetry Principle and Symmetry Principle are applied to* **inter-record** *asymmetries and symmetries.*

For example, consider the rotated parallelogram, which is an example of external inference. Here the Asymmetry Principle is applied, for instance, to the different angles, to make them equal. The different angles are within a single record (the rotated parallelogram). In contrast, consider the thermodynamic process, where we view a number of photographs lain out from left to right on a table. Here the asymmetries are between the records. This creates the left-right asymmetry of the data set.

2.10   Exhaustiveness and Internal Inference

As stated in Chapter 1, a primary goal of our theory is *exhaustiveness*, in the sense that the entire data set must be generatively accounted for. It will now be seen that this has powerful consequences on the issue of internal inference. To illustrate this, let us go back to the rotated-parallelogram example, Fig. 2.3 (p. 40).

In this example, the backward-time sequence terminates at the square. We saw that this sequence is produced by removing three distinguishabilities from the rotated parallelogram: (1) the difference between the orientation of the object and the orientation of the environment; (2) the difference between the sizes of the angles; (3) the difference between the lengths of the sides.

However, further distinguishabilities still remain in the square. Although the sides have been given equal length, they are nevertheless still *distinguishable by position*; i.e., there is a top side, a bottom side, a left side, and a right

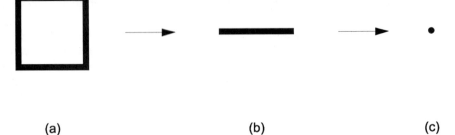

(a)   (b)   (c)

Fig. 2.7. Internal inference on a square.

side. By the Asymmetry Principle, there must have been a preceding generative state in which there was no positional distinguishability between the sides. Observe that, since the size distinguishability, between the sides, has already be removed, the removal of the positional distinguishability makes the sides completely indistinguishable; i.e., they become the same side. This means that a square, as shown in Fig. 2.7a, must have originated from a situation of only a single side, as shown in Fig. 2.7b. This is the starting side, i.e., the first side that is drawn.

Now observe that this single side (Fig. 2.7b) still contains a distinguishability: The points on the side are distinguishable by position. In fact, this is the only distinguishability between the points on a side, because they are all the same color and shape. Therefore, once the positional distinguishability is removed from the points, one obtains a single point Fig. 2.7c. This is the starting point, i.e., the first point to be drawn.

Notice that the two inference stages, shown in Fig. 2.7, are both the removal of *positional distinguishabilities.* The generative process, in the *forward time* direction (right-to-left) must therefore have been movement, since movement is the process that creates positional distinguishability. Thus, the square in Fig. 2.7a must be understood as the *trace* of the side in Fig. 2.7b; and the side in Fig. 2.7b must be understood as the *trace* of the point in Fig. 2.7c.

Now, let us consider again the backwards-time direction (left-to-right in Fig. 2.7). Notice that, in each of the two inference stages, the assumption is made that the present contains multiple states. That is, the square is assumed to be the multiple states (positions) of a side, and a side is assumed to be the multiple states (positions) of a point.

A multiple-state interpretation is, in fact, forced upon one, in any situation where one attempts to remove all distinguishabilities, achieving generative exhaustiveness. This is because, at some stage in the removal of distinguishabilities, one cannot remove any more without removing distinct parts. At this point, backward-time must therefore represent the removal of the parts and forward-time must represent the process of introducing the parts. This means that different parts must have been introduced at different times;

that is, they represent different moments in the generative process which introduces them; i.e., different states in that process. Therefore, the collection of parts is given a multiple-state interpretation.

The reader should recall from Definition 2.2 that a multiple-state interpretation is associated with *internal* inference; i.e., the preceding states are subsets of the present state. This can be seen in Fig. 2.7: Both the side and the point are subsets of the square, i.e., the inference goes to past states that are *internal* to the data set (the square).

Now let us take stock of the entire rotated parallelogram example. The inference starts with the rotated parallelogram and goes through the set of stages shown in Fig. 2.3 resulting in the square. But the inference then continues from the square, and goes through the stages shown in Fig. 2.7, resulting in the final stage, the point.

That is, there are a total of five successive transitions backwards in time, the first three being shown in Fig. 2.3, and the last two being shown in Fig. 2.7. Most crucially, one should understand that there are five transitions because there are five distinguisabilities in the rotated parallelogram. They are:

(D1) Distinguishability between the orientation of the figure and the orientation of the environment.

(D2) Distinguishability between the sizes of adjacent angles.

(D3) Distinguishability between the lengths of adjacent sides.

(D4) Distinguishability between positions of the sides.

(D5) Distinguishability between positions of the points.

These five distinguishabilities are removed successively backwards in time, in the five stages shown. The removal is required by the Asymmetry Principle (p. 43) which states that any distinguishability must be removed backwards in time.

The other thing to notice is that the first three distinguishabilities are removed by external inference (Definition 2.1), and the last two are removed by internal inference (Definition 2.2). That is, the inference sequence from a rotated-parallelogram Fig. 2.3a, starts with a single-state assumption; i.e., the rotated parallelogram is assumed to be a single state and it is seen as implying a state outside itself, a non-rotated parallelogram. The non-rotated parallelogram, Fig. 2.3b, is in turn assumed to be a single state and it is seen as implying a state outside itself, the rectangle, Fig. 2.3c. Then, the rectangle, Fig. 2.3c, is in turn assumed to be a single state and it is seen as implying a state outside itself, the square, Fig. 2.3d. Thus, a single-state interpretation is held through the successive inference stages, in Fig. 2.3, until one reaches a square. Then, in order to remove the remaining distinguishabilities, i.e., those in a square, one is forced to switch to internal inference, i.e., all subsequently inferred states are internal to the square. Correspondingly, one switches to a multiple-state interpretation; i.e., the square is seen as multiple positions of a side and the side is seen as multiple positions of a point.

The assignment of a multiple-state interpretation necessitates the square being viewed as a trace; i.e., the sides are ordered in time and so are the points. We shall later look at how traces are structured, but it is worth observing, here, that an obvious physical interpretation of the trace is that it was produced by drawing. For example, starting with the point in Fig. 2.7c, a pen traced out the points along a side, producing Fig 2.4b; and then the pen traced out the successive sides around the square, producing Fig. 2.7a. Of course, the implement need not have been a pen. It could have been a cutter in a manufacturing process; or a robot spatially navigating along the edge of a room.

2.11   Externalization Principle

One of the purposes of our generative theory of shape is to replace the different sets of laws in the different sciences by a single set of scientific laws that are universal across the sciences. This will be done by replacing the existing laws by *inferential* laws that lead to the existing laws. These inferential laws are, in fact, the laws for the recovery of generative history from data sets.

In this section, we are going to propose one of these universal scientific laws. The power of this law cannot be over-estimated, as will be seen. We first motivate it using the extremely simple example of the last section, but then show that it is entirely general. The reader will recall from the last section that the successive inferences, from the rotated parallelogram, started first with a sequence of *external* inferences - from the rotated parallelogram back to a square - and then changed to a sequence of *internal* inferences - from the square back to a point. What one should now observe is that the square is structured as an *iso-regular* group. This type of group was defined on page 12 by three conditions $\mathfrak{IR}_1$ - $\mathfrak{IR}_3$, and was intuitively described as a *control-nested hierarchy of repetitive isometries*.

Recall also that, on page 8, the generative group of a square was given as

$$\mathbb{R} \textcircled{w} \mathbb{Z}_4 \tag{2.1}$$

where $\mathbb{R}$ is the translation group along a side, and $\mathbb{Z}_4$ is the 4-fold rotation group between sides. Most crucially, this is an iso-regular group.

It can now be seen that the successive *external* inferences from a rotated parallelogram back to a square, lead to an *iso-regular* group (the group of the square). In fact, this iso-regular group describes the *internal* (trace) structure of a square. Remarkably, what is happening here turns out to be entirely general. That is, we propose this rule:

EXTERNALIZATION PRINCIPLE.   *Any external inference leads back, as much as possible, to an internal structure that is a control-nested hierarchy of repetitive isometries; i.e., an iso-regular group.*

We shall argue that each science is structured by the Externalization Principle, i.e., each science is founded on external inference back to an iso-regular group. For example, we shall show that the structure of shape primitives and shape modifications in computer-aided design (CAD) follows from the Externalization Principle. Again, we shall argue that the Schrödinger equation in quantum mechanics follows from the Externalization Principle. Again, the distinguished role of special relativity with respect to general relativity follows from the Externalization Principle. It is therefore important to understand the justification of the principle:

JUSTIFICATION FOR THE EXTERNALIZATION PRINCIPLE.
An iso-regular group has three properties: (1) It is a wreath product $G_1 \textcircled{w} G_2 \textcircled{w} \ldots \textcircled{w} G_n$; (2) each level G_i is 1-dimensional (cyclic or 1-parameter), and (3) each level G_i is an isometry group. The justification for the Externalization Principle uses all three properties. Using (1) and (2): A wreath product in which each level is 1-dimensional, maximizes transfer. Using (3): Since isometries are the most constrained of the topological actions on a space, external inference to a past state generated by isometries, maximizes recoverability. Therefore the Externalization Principle follows from the maximization of transfer and recoverability.

Let us now understand the relationship between the Externalization Principle and the other inference rules. Recall that one of the two fundamental inference rules of our theory is the Asymmetry Principle which (in the form given on p. 43) states that each asymmetry in the data set must go back to a symmetry in one of the previously generated states. Recall also that we distinguished between two realizations of the Asymmetry Principle: external inference which goes back to a past state outside the data set, and internal inference which goes back to a past state inside the data set. The Externalization Principle concerns the first of these two classes of inference.

The Externalization Principle is the Asymmetry Principle in half of all possible cases: those of external inference.

Now let us consider a number of examples of the Externalization Principle, as follows:

Visual Perception.

In Leyton [88] [89] [90] [96], we have shown that the Externalization Principle fundamentally underlies visual perception. Consider some simple examples: When an individual goes down a street and sees a bent pipe, he/she infers that the pipe was originally straight. This inference is external, i.e., the straight pipe is not actually visible within the box of the present (the data set). Notice that the straight pipe has the structure of a cylinder and is therefore the iso-regular group

$$SO(2) \circledW \mathbb{R} \tag{2.2}$$

which is the generative group that we gave on p. 14 for the cylinder. Thus the inference from the bent cylinder to the past straight cylinder is an example of the Externalization Principle.

Similarly the process by which the environment is projected onto the retina conforms to the Externalization Principle. For example, consider the retinal image of a square which has been distorted due to the projection process, as shown in Fig. 1.12 on p. 16. The inference that this is the projected image of true symmetrical square in the environment is an external inference; i.e., the symmetrical square is not visible in the data set (the image). This external inference goes from the distorted image square to the symmetrical environmental square, and the latter square is given by the iso-regular group:

$$\mathbb{R} \circledW \mathbb{Z}_4. \tag{2.3}$$

Notice, as was said on page 17, the complete generative structure of this situation is this:

$$\mathbb{R} \circledW \mathbb{Z}_4 \circledW PGL(3, \mathbb{R}) \tag{2.4}$$

where the projective group $PGL(3, \mathbb{R})$ is added to the iso-regular group $\mathbb{R} \circledW \mathbb{Z}_4$ as an extra generative level. The justification for using the wreath product to add this level was given on p. 17. The external inference is the process of going to the identity element within this level. This yields the undistorted iso-regular group given by the first two levels. Thus, one actually sees the Externalization Principle represented within the wreath hierarchy of expression (2.3). This becomes crucial in Chapter 22, when we give a theory of the relation between Euclidean geometry and projective geometry.

To fully understand the powerful role of the Externalization Principle in visual perception, it is necessary to return to our theory of computational vision summarized in Sect. 2.7. We claimed that the process by which the environment is projected onto the retina consists of three successive symmetry-breaking stages, as shown in Fig. 2.4 (p. 45). Each successive stage in Fig. 2.4 adds a layer of asymmetry, and the final image on the retina is an accumulation of these three asymmetry layers. According to this theory, the process of vision (i.e., inference) is that of undoing the asymmetries backward in time through the generative sequence; i.e., the successive use of the Asymmetry

Principle. This explains the various components of computational vision, such as shape-from-texture, shape-from-contour, shape-from-shading, regularization, etc.

Now, the use of the Asymmetry Principle on the image, to infer the previous generative stages, is the use of *external inference*. And it was said that, in all cases of external inference, the Asymmetry Principle becomes the Externalization Principle. Therefore each of the three inference stages in Fig. 2.4 is a use of the Externalization Principle. This can be seen in the examples given in Sect. 2.7. For instance, in shape-from-texture, as illustrated in Fig. 2.5 (p. 46), the succession backwards from the perspectively distorted grid Fig. 2.5c, to Fig. 2.5b, to Fig. 2.5a, recovers the original square grid which is a control-nested hierarchy of repetitive isometries, i.e., an iso-regular group. Similarly, in shape-from-shading, Fig. 2.6 (p. 47) illustrates the fact that the asymmetric light flux on the right of the figure, goes back in time to a light flux on the left which is structured by an *iso-regular group*; i.e., the fiber of this group is a light ray, and the translational symmetry across the light rays is the control group.

Notice that the two examples given earlier in this section - the bent pipe and the distorted square - fit into this analysis. The bent pipe is an example of Stage 1 in Fig. 2.5 (p. 46), and the distorted square is an example of Stage 3, i.e., shape-from-contour. As we said, both are determined by the Externalization Principle.

Computer-Aided Design.

A large part of this book will be devoted to computer-aided design (CAD). The process of design by computer usually starts with some shape primitive, then adds another primitive, and so on, with the possibility that these successive primitives can undergo distortion. There are two processes underlying this standard procedure: Boolean operations and deformation. We shall see that both of these are structured by the Externalization Principle. First, we will show that any shape primitive is characterized by a control-nested hierarchy of repetitive isometries; i.e., an iso-regular group. An example of this has already been seen with the cylinder, which is one of the standard primitives, and which was shown to be an iso-regular group on p. 14. This will allow us, in Chapter 10, to give a systematic elaboration of shape primitives - something that was not possible without the concept of an iso-regular group.

Next, given a design consisting of two primitives, e.g., a cube and a cylinder, consider the inference process that recovers the generation of this structure. Observe that the combined cube and cylinder is not an iso-regular group. However, the reverse-generation, in which one of the two primitives is removed leaving the other as the past state, does go back to an iso-regular group. Thus, shape-generation by Boolean combination conforms to the Externalization Principle.[1]

Now let us consider the other generation process: deformation. It will be a basic argument of our theory that one understands non-deformed objects to be iso-regular groups, and that the deformed object is understood as a transferred version of an iso-regular group G_{iso} under a diffeomorphism control group G_{diff} in a wreath product:

$$G_{iso} \, \textcircled{w} \, G_{diff} \qquad (2.5)$$

This deeply captures what is psychologically happening, as follows: Consider for example a bent pipe. In order for people to see it as a bent version of a straight one, they have to *transfer* onto it the structure of the straight one. However, the straight one is given by the iso-regular group $G_{iso} = SO(2)\textcircled{w}\mathbb{R}$ of the cylinder. The group that transfers this onto the bent version is some diffeomorphism group G_{diff}. Hence one obtains a wreath product of the type at (2.5). The *recovery* of that transfer (i.e., the reverse-generation) is algebraically given by moving to the identity element within the diffeomorphism control group. This movement conforms to the Externalization Principle. That is, the wreath product (2.5) *contains* the Externalization Principle within its structure.

The above considerations, with respect to Boolean operations and deformation, will allow us to give a systematic and detailed theory of part-design, assembly-planning, and machining in mechanical engineering; as well as conceptual design, design development, and construction documentation, in architectural CAD.

General Relativity.

We shall now see that the Externalization Principle is fundamental to general relativity. According to general relativity, a particle moving in a gravitational field does not accelerate but maintains an unchanging velocity, i.e., travels along a geodesic. This however means that the effect of gravity cannot be observed in the motion of a single particle. In order to observe gravitational influence, one needs two particles, like this: It follows from the structure of the curvature tensor that, for curved manifolds, pairs of geodesics that start out parallel cannot remain parallel; i.e., there is geodesic deviation. In contrast, in a flat manifold, geodesics that start out parallel remain parallel. Now

---

[1] In the case of subtraction and intersection, it is clear that backward generation is given by external inference. In the case of union, the fact that the inference is external becomes clear when one considers the bounding surfaces; i.e., the surface of the union object does not contain the surfaces of the primitive objects. Thus, to understand that the inference of Boolean union is external, we have to recognize that it is a *regularized* Boolean operation.

consider gravity. In the absence of matter, space-time is flat. When matter is introduced, it causes space-time to have curvature. This is expressed in the Einstein field equations $G = 8\pi T$, where T is the stress-energy tensor and G is the Einstein curvature tensor. In this way, one can understand why the effect of gravity is manifested in the relationship between two moving particles; i.e., gravity causes geodesic deviation.

We now describe these considerations in terms of the generative theory of shape. Consider the *inference* of gravity. The inference from curved space-time back to flat space-time is an example of *external* inference. This is because a flat manifold is not a subset of a curved manifold; i.e., not in the box of the present. Now it is easy to see that parallelism of geodesics in flat-space time can be modeled as an *iso-regular group* in which a geodesic is a fiber, and the action of translating a fiber, parallel to itself, corresponds to a control group. Flatness corresponds to the fact that such iso-regular groups can be constructed. Curvature corresponds to the fact that such iso-regular groups cannot be constructed.

Thus consider the recoverability of gravity in general relativity. Given a curved space-time manifold as a data set, the physicist's inference that gravity is responsible for the curvature, is the inference back to an origin state, flat space-time, that corresponds to an *iso-regular group*. Therefore the inference accords with the Externalization Principle.

Consider now special relativity, which is the physics of flat space-time. The above considerations enable us to understand the role of special relativity as follows: *Special relativity corresponds to the iso-regular group recovered by fully externalizing general relativity.* Most crucially we argue that special relativity arises from the need to maximize recoverability.

Quantum Mechanics.

We shall now see that the Externalization Principle is fundamental to quantum mechanics. In quantum mechanics, any state $|\psi\rangle$ is a complex function, and the space of states is a (physical) Hilbert space of such functions. Given two states $|\psi\rangle$ and $|\phi\rangle$, their inner product is defined in this way:

$$\langle\psi|\phi\rangle \;=\; \int_a^b \psi^*\phi. \tag{2.6}$$

The associated norm is obviously given by $||\phi||^2 = \langle\phi|\phi\rangle$, which is related to the probability of the state.

An observable is a differential operator on this Hilbert space, and it induces a 1-parameter group on the space. In fact, one should think of any observable as belonging to a Lie algebra of observables, and its associated 1-parameter group is created by the usual exponentiation that goes from a Lie algebra to a Lie group. In quantum mechanics, one standardly considers the Lie algebra to be a collection of Hermitian operators, and the associated Lie group to be unitary. Thus the Lie group preserves the probability metric defined at (2.6) above, and is therefore an *isometry*, in fact, a rotation. Given an observable V, its associated 1-parameter group will be denoted by G_V.

Now, measurement with respect to an observable V does not destroy the information produced by another observable W only if the two observables commute, that is, if $[V, W] = 0$ within the Lie algebra of observables. In Chapter 20, we shall argue that any commuting pair of observables should be corresponded to a wreath product of their 1-parameter groups

$$[V, W] = 0 \qquad \longleftrightarrow \qquad G_W \, ⓦ \, G_V. \qquad (2.7)$$

Most crucially, since both G_W and G_V are isometry groups, the wreath product $G_W ⓦ G_V$ is *iso-regular*; i.e., it satisfies conditions $\mathfrak{IR}_1$-$\mathfrak{IR}_3$ on p. 12. Therefore, we have this conclusion: In quantum mechanics, two observables commute only if their 1-parameter groups form an iso-regular group.

Now let us consider how physical structure is generated in quantum mechanics. One starts with a symmetric structure, e.g., an atom with a spherically symmetric Hamiltonian, and one successively adds asymmetries, in accord with our generative theory of shape. The initial symmetry corresponds to the commutation of observables, and successive addition of asymmetry corresponds to the successive breaking of the commutation. This means that the state is no longer described by the iso-regular group corresponding to the starting commutation.

For example, suppose one begins with a spherically symmetric Hamiltonian H. This means, in particular, that H and J_z commute, where J_z is the generator of rotations around the z-axis (i.e., the angular momentum observable for the z-axis). According to (2.7) above, this commutation corresponds to the wreath product $G_H ⓦ G_{J_z}$, which is *iso-regular*. The addition of an external asymmetrizing field will change H and can therefore destroy the commutation $[H, J_z]$. Thus, the asymmetrized state will not be described by the iso-regular group $G_H ⓦ G_{J_z}$.

Now consider the reverse-generation direction, i.e., the *inference* direction. When one is given an asymmetric state, the procedure in quantum mechanics is to define the state has having arisen generatively from a symmetric one, in accord with our Asymmetry Principle (p. 43). Furthermore, the above argument shows that the past symmetric state corresponds to an iso-regular group. This demonstrates that the Externalization Principle is fundamental to quantum mechanics.

2.12 Choice of Metric in the Externalization Principle

We shall argue that the Externalization Principle is the major organizing factor of any science. According to our theory, one obtains a science by selecting a particular metric for the isometries in the Externalization Principle. For example,

$$
\begin{array}{ccc}
\textbf{Visual perception} & \longleftrightarrow & \textbf{Euclidean group} \\
\textbf{General Relativity} & \longleftrightarrow & \textbf{Lorentz group} \\
\textbf{Quantum mechanics} & \longleftrightarrow & \textbf{Unitary group}
\end{array}
$$

Two comments should be made on this: (1) Note that computer-aided design (CAD) is based on visual perception. The reason is that every aspect of CAD is based on the interaction with a human agent via the visual medium. Therefore, we will understand much of CAD to be within the science of visual perception. For example, we shall show that the menu structure of a CAD program is dictated by choosing the Euclidean metric within the Externalization Principle. (2) Notice that the relativity group in General Relativity is larger than the Lorentz group: It is the general diffeomorphism group at any point on the curved space-time manifold. Nevertheless, the action of gravity is understood as creating curvature from flatness, and the physics of flatness is special relativity. Therefore, the isometry group in the Externalization Principle of General Relativity is the Lorentz group.

2.13 Externalization Principle and Environmental Dimensionality

Standardly one says that the environment is 3-dimensional. This is incorrect. For example, in relativity, it is 4-dimensional, in certain forms of string-theory it is 10-dimensional, and in quantum mechanics it is infinite-dimensional. What we argue is that the number of dimensions is determined by the Externalization Principle, in that one uses as many dimensions as will allow the data set to be accounted for by an iso-regular group.

To illustrate, Fig. 2.8 shows a 2D image on the retina. Notice that in this 2-dimensional plane, the lines are of different lengths. Now, the visual system interprets this configuration as the image of a cube in 3-space. The reason is that, in 3-space, it can be described as having edges all the same length. That is, in 3-space, it can be described by an iso-regular group.

Now let us rotate this hypothesized cube in 3-space. At some point it looks like Fig. 2.9. However, the visual system does not interpret this figure as a 3D object, but a 2D one. The reason is that two dimensions are sufficient to give an account of the data by an iso-regular group.

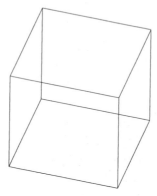

Fig. 2.8. The Externalization Principle requires three dimensions to give this data an iso-regular group.

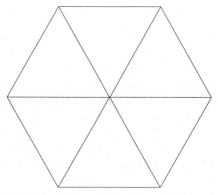

Fig. 2.9. The Externalization Principle requires only two dimensions to give this data an iso-regular group - even though it is the image of a 3D cube.

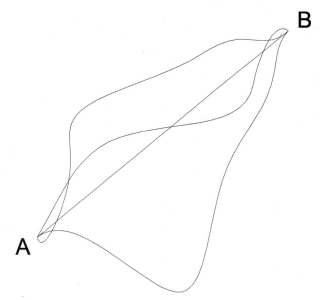

Fig. 2.10. Alternative routes from A to B.

The dimensionality of the environment is determined by the Externalization Principle: An environment is given the minimum number of dimensions that allows the data set to be accounted for by an iso-regular group.

2.14 History Symmetrization Principle

It is clear that, when people infer generative history, they *minimize* that history. Given for example, a set of alternative routes between states A and B in Fig. 2.10, they will choose the route that is most direct. In physics this is exemplified by the *principle of least action* - i.e., the route is chosen that minimizes the action integral.

Intuitively, people think of the most direct route as the line of shortest distance. In fact, the most direct route is the shortest one when the metric is Euclidean, but, for example, in the Lorentz metric of special relativity, the most direct route is the longest one. Thus to handle the general case, we need to formulate the concept of the *most direct route* without dependence on the notion of minimizing length:

HISTORY SYMMETRIZATION PRINCIPLE. *The inferred generative history should contain the minimal distinguisability.*

That is, a history is the *most direct* if it contains the *least distinguishability*. The reader should understand therefore that when we say that a history is minimal, we shall mean that it has minimal distinguishability.

Close inspection reveals that the above principle is actually the application of the Asymmetry Principle to internal history. Recall that the Asymmetry Principle (p. 43) states that each asymmetry in the data set must go back to a symmetry in one of the previously generated states. Applying this to any of the conjectured routes in Fig. 2.10, what is recovered is the most direct route.

We shall call the History Symmetrization Principle the *second* application of the Asymmetry Principle. The first is to recover a generative history. The second is to minimize that generative history. The mathematical consequences of this are profound, as will now be seen.

What one needs to understand is how groups are used in the generative structure:

USE OF GROUPS.   *Groups are used in two ways in the generative theory: (1) to characterize the structure of the starting state, and (2) to characterize the structure of the minimal history. The first use corresponds to the fiber level and the second use corresponds to the control level in a transfer structure. In an n-level structure of transfer, $G_1 ⓦ G_2 ⓦ \ldots ⓦ G_n$, these two uses become coincident in each level (except perhaps the end-levels). This means that the maximization of transfer and recoverability become essentially the same.*

The above statement is extremely important, and an illustration is needed as follows: Let us return to Fig. 2.10, which represents the use of the History Symmetrization Principle. Here the minimal history is the straight line that connects points A and B. The first thing to notice is that the history *transfers* the point at A onto the point at B. These two points are in fact two copies of the fiber group of a wreath product (i.e., the initial version and the transferred version). The structure of the fiber group (i.e., the point) has not yet been specified, so let us choose an example. Suppose that Fig. 2.10 represents the projective situation in which point A is the fully symmetric square in the environment, and point B is the projectively distorted square on the observer's retina, as in Fig. 1.12 on p. 16. Point A therefore represents the symmetry group G_{square} of the square. This is the fiber group. The straight line between points A and B represents the action of the projective group in transferring the fiber group from A onto B (as a projectively distorted representation). Therefore the straight line represents the control group. Thus one has the following structure:

$$G_{square} \; ⓦ \; PGL(3, \mathbb{R}). \tag{2.8}$$

One sees from this that the fiber group represents the starting state of the square; and the control group represents minimal history. Furthermore, consider this: The starting state G_{square} is itself a structure of transfer:

$$\mathbb{R} \ \textcircled{w} \ \mathbb{Z}_4 \tag{2.9}$$

which is the generative group given for the square. However, this itself represents the minimal history for generating the square. That is, $\mathbb{Z}_4$ is the minimal history of the side $\mathbb{R}$ around the square, and $\mathbb{R}$ is the minimal history of the point $\{e\}$ along a side. Both these levels are structures of transfer because (2.9) is really this:

$$\{e\} \ \textcircled{w} \ \mathbb{R} \ \textcircled{w} \ \mathbb{Z}_4. \tag{2.10}$$

The conclusion therefore is that the starting state G_{square} in (2.8) is really the minimal history in (2.10).

Now the starting states represent the phenomenon of recoverability, and the minimal histories represent the phenomenon of transfer. We therefore see that recoverability and transfer are forced to be coincident. This shows the enormous economy of our system:

Each algebraic level is forced to take the roles of transfer and recoverability simultaneously.

2.15 Symmetry-to-Trace Conversion

The Externalization Principle states that any external inference leads back to an internal structure that is a control-nested hierarchy of repetitive isometries; i.e., an iso-regular group. Internal structure means trace structure. Thus, for example, the external inference from a rotated parallelogram goes back to the trace structure of a square, which is given by an iso-regular group.

In fact, we now argue that the trace structure is derived from the group using the following rule:

SYMMETRY-TO-TRACE CONVERSION PRINCIPLE. *Any symmetry can be re-described as a trace. The transformations defining the symmetry generate the trace.*

The inference of traces is crucial for all spatial planning, e.g., navigation around a room, the machining of a surface, etc. We argue that the Symmetry-to-Trace Conversion Principle is fundamental to the inference of all traces. In fact, we propose:

Plans come from the symmetries of a structure. Symmetries are the "channels" along which actions take place.

This proposal will be overwhelmingly corroborated in this book, in many application areas, from manufacturing to quantum physics. It is now necessary to look at the way in which the Symmetry-to-Trace Conversion Principle is realized. The description in this section will be a very brief and intuitive summary of the theory elaborated in detail in Chapters 2 and 6 of Leyton [96].

Let us first define the basic properties of traces:

TRACE PROPERTIES.

(1) A trace consists of transformations that go between parts of the data set.

(2) One part of the data set is a distinguished starting state.

(3) The other parts of the data set follow in sequence.

These three properties are realized in the following way: Recall from sections 1.15 - 1.17 that we consider that the theory of shape should be regarded as belonging to the algebraic theory of machines. Using this concept, let us now look at how to realize the above three properties in turn:

Trace Property (1) above is realized by converting the symmetry structure of the data set into the input group of a machine. This is illustrated with a square in Fig. 2.11. Only the $\mathbb{Z}_4$ level of the square has been shown here. The figure represents the square as the state-transition diagram of a machine, in which the input group is $\mathbb{Z}_4$, and the machine states are the four states of a side. Notice therefore, that the entire input group acts on each of the sides - i.e., four arrows emerge from each side, the arrows being labeled by the four members of $\mathbb{Z}_4$.

Trace Property (2) above is realized by initializing the machine; i.e., designating one of the sides to be the starting state. This state receives the group identity element, and the other states are then characterized by the input operations that produce them from the starting state. In Fig. 2.12 the top side is chosen as the initial state.

Trace Property (3) above is realized by using the group's generative structure. For ease of exposition, let us consider discrete groups. Any discrete group can be represented by a set of generators and algebraic relations between those generators. This is called a *presentation* of the group. In the simple illustration being considered, $\mathbb{Z}_4$, a presentation is

$$r_{90} \quad : \quad (r_{90})^4 = e \qquad (2.11)$$

where, to the left of the colon, there is the list of generators (here there is only r_{90}), and to the right of the colon there is the list of relations between the generators (here there is only one relation).

Now, the information given in Fig. 2.12, does not yet have the generative structure of the square. For example, it does not tell us the order in which the elements were generated; e.g., the element r_{90} might have been produced by applying r_{270} three times to the identity element. The generative structure

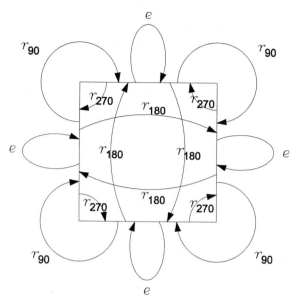

Fig. 2.11. A square as the state-transition diagram of a machine.

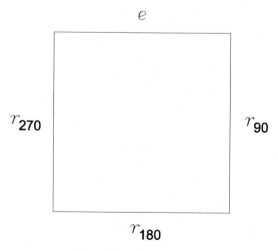

Fig. 2.12. A square as an initialized machine.

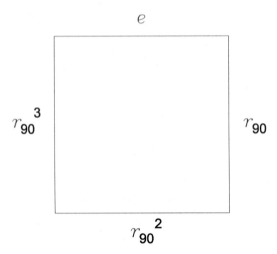

$$e$$

$$r_{90}^3 \qquad r_{90}$$

$$r_{90}^2$$

Fig. 2.13. A square as a trace.

will come from the chosen group presentation. Thus let us choose the presentation given in expression (2.11) above. This gives Fig. 2.13, which explicitly shows how the square was generated. Notice that the square is now defined as a *trace* structure.

In summary, the reader can see that the trace structure is realized in three stages:

INFERENCE OF TRACE STRUCTURE.

(1) Convert the symmetry group of the data set into an input group for a machine describing the data set.
(2) Initialize the machine.
(3) Convert the input group into one of its group presentations.

2.16   Roots

It is worth giving here a definition which will be useful later in the book. Recall that, up to isomorphism, there exist only two connected 1-parameter Lie groups: $\mathbb{R}$ and $SO(2)$. Thus, in the case where the 1-dimensional groups used in the Externalization Principle are continuous, then they will be isomorphic to $\mathbb{R}$ or $SO(2)$.

Definition 2.3. *Given an iso-regular group $G_1 \text{\textcircled{w}} G_2 \text{\textcircled{w}} \ldots \text{\textcircled{w}} G_n$, used for the Externalization Principle, any level G_i that is isomorphic to either $\mathbb{R}$ or*

$SO(2)$ *will be called a* **root** *of the iso-regular group; and* $\mathbb{R}$ *or* $SO(2)$ *will be called the roots of the externalization process.*

In other words, an iso-regular group is built up using the roots and/or cyclic groups. Roots will be particularly important in understanding computer-aided design. Notice for example that the generative group of a cylinder $SO(2)\circledW\mathbb{R}$ consists only of roots.

2.17   Inferred Order of the Generative Operations

The Asymmetry Principle (p. 42) says that the sequence of generative actions are recoverable only if they are symmetry-breaking on the successively generated states. The following concept will now be needed:

Definition 2.4. *An operation will be said to be* **structurally allowable** *to the extent to which it preserves the symmetry of a state.*

This leads to the following rule:

INTERACTION PRINCIPLE.   *The symmetry axes of an organization become the eigenvectors of the structurally most allowable operations applicable to the organization.*[2]

This rule is crucial in determining the order of the generative operations:

GENERATIVE ORDER RULE.   *The inferred order of the generative operations is the order of their structural allowability.*

In fact, this follows from the History Symmetrization Principle (p. 61) since a generative sequence having the order of structural allowability will have minimal distinguishability. Let us consider some examples.

Return first to the rotated parallelogram example Fig. 2.3 (p. 40). Consider the starting state in that diagram - i.e., the square. It is important to notice that the horizontal and vertical symmetry axes of the square are aligned with the gravitational symmetry axes of the environment. These axes are the most salient symmetry axes of the human visual system, and the

---

[2] In this principle, we shall also use the term eigenvector with respect to an invariant line of a translation group. One can think of the invariant line as corresponding to what could be called an "affine" eigenvector.

square is maximally aligned with this group. As the *square* is successively altered into the rotated parallelogram, it will successively break the *gravitational* symmetries. We shall now see that this succession accords with the Interaction Principle and Generative Order Rule:

Stretch. The stretch operation from the square to the rectangle is symmetry-breaking in that it destroys the diagonal symmetry axes. However, it preserves the horizontal and vertical symmetry axes. Most crucially, these two axes in the square become the *eigenvectors* of the stretch transformation - in accord with the Interaction Principle.

Shear. The second applied operation, shear, destroys the vertical symmetry axis, and retains the horizontal symmetry axis in a weakened form as the bisection axis. Notice that shear has a single independent eigenvector and that this is along the horizontal axis. Thus the eigenvector of shear is aligned with the axis of horizontal bisection. This once again accords with the Interaction Principle.

Rotation. The final generative operation, rotation, causes the square to break both of the gravitational symmetry axes. Notice that the rotation has 0 eigenvectors.

Thus, relative to the gravitational symmetries, the order of the three operations stretch, shear, and rotation, is the order of structural allowability. Correspondingly, the eigenspace dimensions of the three operations are successively 2, 1, 0. This accords with the Generative Order Rule.

As another example, consider the creation of a goblet, Fig. 2.14b. This is a standard example of a sweep: i.e., the goblet is generated by moving the circle cross-section upward along the vertical axis, while expanding and contracting this cross-section. Now consider the starting state, the circle. It is has two types of symmetry axes, both of which are shown in Fig. 2.14a: (1) a rotation axis perpendicular to the plane containing the circle; and (2) reflection axes along the circle's diameters. In accord with the Interaction Principle, these two types of symmetry axes become the eigenvectors of the sweeping movement: The first type of axis (the rotation axis) becomes the (affine) eigenvector of the upward sweeping motion. The second type of axis (the reflection axes along the diameters) become the eigenvectors of the expansion and contraction operations.

Thus it can be seen that the sweeping operations consist of the most structurally allowable operations on the circle (i.e., those that preserve the symmetry axes). Now if one deforms the goblet - e.g., it melts in the sun - then one would be destroying the symmetry axes. Notice therefore, if one wants

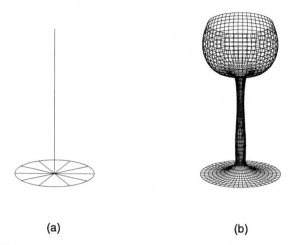

(a)   (b)

Fig. 2.14. Symmetry axes of the circle (a) become eigenvectors of the sweeping process (b).

to generate such a shape, one would first apply the structurally allowable operations of sweeping - creating the goblet - and then apply the melting deformation. One would not try to apply the operations in the reverse order. This accords with the Generative Order Rule.

2.18   Symmetry-Breaking vs. Asymmetry-Building

The Asymmetry Principle says that the generative operations must be symmetry-breaking in the forward-time direction. To maximize transfer and recoverability, we need to strengthen this concept, as follows:

Let us consider the projection of an environmental square onto its distorted image on the retina as shown in Fig. 2.15. It is clear that the environmental square has symmetry group D_4 (the dihedral group of order 8), and the image square has symmetry group $\mathbb{Z}_2$, which corresponds to its vertical symmetry. That is, the projective process creates the following transition of symmetry groups:

$$D_4 \longrightarrow \mathbb{Z}_2. \tag{2.12}$$

Notice that the second group is a subgroup of the first group. Thus the symmetry-breaking effect is given by going from the full symmetry group to one of its subgroups.

Now we come to the crucial point: Because the group operations are being lost, generative information is actually being lost. We are therefore going to consider two different methods of representing the symmetry-breaking action of the generative process. The first is the one just considered.

Fig. 2.15. An environmental square projected onto the retina.

SPECIFICATION OF GENERATIVITY: METHOD 1. *The generative process is specified as symmetry-breaking which is represented as the reduction in symmetry group.*

This method has certain advantages which will be exploited at times. However, there is a much more powerful method that we will now introduce, that does not have the fault of loosing generative information. This method uses *transfer*.

To illustrate, return to the example of the projection of a square in Fig. 2.15. Instead of characterizing the transition as one of group reduction $D_4 \longrightarrow \mathbb{Z}_2$, we will characterize it as one of group increase, using transfer thus:

$$D_4 \longrightarrow D_4 \textcircled{w} PGL(3,\mathbb{R}). \tag{2.13}$$

Notice that the group D_4 occurs both before and after the arrow. That is, the group is not lost. What happens after the arrow is that we *extend* this group by the projective group $PGL(3,\mathbb{R})$; i.e., we enlarge the original symmetry group. This is exactly what we need to express the *generativity* - i.e., the increase in generative operations over time. Most crucially, the method we choose to create the group extension is the wreath product - i.e., *transfer*. Close examination reveals that this wreath product captures all the information in the phrase "projectively distorted square." Notice that the wreath product therefore captures the *recoverable* structure. In other words, it captures both *transfer* and *recoverability*.

Observe also that the group $PGL(3,\mathbb{R})$, which is added as a control group, is responsible for the asymmetrization of the square. This is essential to the way we will set up control groups. We shall speak of any control group as the *symmetry group of the asymmetrizing process*. In fact, any wreath level can be described as such.

SPECIFICATION OF GENERATIVITY: METHOD 2. *The generative process is specified as asymmetry-building which is represented by group* *extension via wreath product; i.e., transfer. The extending group is a symmetry group of the asymmetrizing process.*

We shall see that, in the generative theory, the wreath extensions can also be built *downward*, i.e., the first group can become the control group with respect to which the added group becomes the fiber group. For example, this is basic to a very important class of generative situation we shall call *sub-local unfoldings*. Nevertheless, the statement of Method 2 also characterizes this type of situation.

Methods 1 and 2, described in this section, will be called respectively *symmetry-breaking* and *asymmetry-building*. One should think of the two approaches as occurring in parallel during the generative process. However, because the term "symmetry-breaking" is very familiar to the public, we shall, in most cases, choose this term to describe the simultaneous action of the two methods. Nevertheless, the reader should always understand that the second approach is the far more powerful one that is actually occurring.

2.19   Against the Erlanger Program

Our theory of shape affords a theory of geometry which will be called *generative geometry*. It will be seen that generative geometry is fundamentally the opposite of Klein's theory of geometry, which has been the basis of much of 20th century mathematics and physics. Klein's theory, also called the Erlanger Program, states that geometry is the study of invariants under some assumed group of transformations.

It will take us a considerable amount of discussion to understand why the two geometry theories are the opposite of each other. But, a basic sense of this can be gained from the following point:

GENERATIVE GEOMETRY vs. KLEIN'S THEORY OF GEOMETRY.   *You cannot recover a generative sequence from an object if it is invariant under the generative actions.*

The remarkable thing is that, although Klein's theory has been the basis of much of 20th century science, we shall argue that it is deeply inadequate for the scientific, computational, and design disciplines. We argue that these disciplines require the very different approach elaborated in this book. The reason is that they all fundamentally rely on the recovery of causal or generative actions from shape. Thus, with respect to the *scientific disciplines*, the entire program of a science is aimed at the recovery of the sequence of environmental events that lead up to the state appearing on the measuring instruments. Again, in the *computational disciplines*, computer vision is set up to recover the environmental structure from the image structure. Furthermore, in *machine learning*, Carbonnel [17] has argued that any advanced computational system needs to be able to recover its own computational history, in order to be able to modify that history, if the current state is not the desired one. Again, in the *design disciplines*, Hoffmann [58], Chen & Hoffmann [20], have argued that recovering the design history is essential because it allows editability of the design decisions. Again, in computer-aided manufacturing, the inference of machining operations is really the inference, from the delta shape, of operations that will generate that shape.

Thus, in complete contrast to the Klein approach, which is non-recoverable - due to invariance under operations - we argue that each of the above disciplines requires geometry to be defined in a recoverable way.

2.20   Memory

In our theory of shape, geometric objects are constructed in order to retain information of past applied operations. In Klein's theory, geometric objects - being invariant under the applied operations - do not retain that information.

We argue that this issue is at the very basis of memory. Quite simply, an object cannot be used for memory storage if it is invariant under applied operations. That is, the object is *memoryless* with respect to the applied operations. Memorylessness is equivalent to *non-recoverability*. Conversely, the capacity to store information is equivalent to *recoverability*.

One of the principle aims of this book is to give a theory of memory storage. We argue that *all memory storage takes place via geometry*. In fact, a fundamental proposal of our theory is this:

$$\text{Geometry} \quad \equiv \quad \text{Memory Storage.}$$

Certainly, one can see that our theory says that geometry implies memory storage, due to our principle of the maximization of recoverabilty. However, we also claim the reverse implication, that memory storage implies geometry. Currently we regard this reverse claim as an empirical one; i.e., any time an object is used as a memory storage it is its geometry that entirely carries that function.

This will enable our generative theory of shape to explain the structure of memory storage devises in the world. For example, consider the structure of a magnetic memory core in a computer. It consists of a grid with a $\mathbb{Z}_2$ flip element located at each node point. This structure is given by an iso-regular group, in which the translation lattice is the control group and the $\mathbb{Z}_2$ element is the fiber group. Our theory explains why an iso-regular group is chosen. Such a group best retains the effects of applied operations.

Let us also look at two scientific examples. First, consider the large-space structure of space-time. In general relativity, this is given by curvature. Curvature holds the effect of gravitation, which deforms flat space-time into curved space-time. This means that curvature is used to *recover* the deformational action of gravitation on flat space-time. Thus curved space-time is the memory store for gravitational action. We will call this the *memory structure of general relativity*.

Now consider quantum mechanics. Consider, for example, the standard method of modeling the hydrogen atom by successive perturbation. First, the principal Hamiltonian of this atom holds the effect of the Coulomb electrostatic potential; e.g., the shape of the potential energy function deviates from the translationally symmetric (constant) structure of the free particle potential. The fine structure splitting of the Colomb model then records the interaction between the electron's spin and orbital angular momentum. Then the hyperfine splitting records the interaction between the proton and electron spins. Thus, the split spectroscopic structures are memory of those perturbations. This exemplifies what we will call the *memory structure of quantum mechanics*.

Generally, we will develop a theory of science in which the following is our fundamental proposal:

THE MEMORY-RETRIEVAL THEORY OF SCIENCE. *All the laws of science have been set up for the single purpose of retrieving memory from objects in the external environment (e.g., from space-time curvature, spectroscopic data, biological tumors, etc.). In this view, science is the means of extending the memory-retrieval functions, internal to the computational system, to objects in the external environment of the computational system. This retrieval forces the external environment to become part of the computational system itself.*

2.21 Regularity

Regularity is almost never discussed as a cognitive concept. An exception is the Bayesian approach to regularity advocated by Richards, Jepson & Feldman [127]. and Feldman [35].

In contrast to that approach we develop here an algebraic approach. Conceptually, we define regularity as transfer. As a simple illustration, consider the regularity that on Monday afternoons I teach a course in a certain lecture theater. To observe (or create) this regularity in my life I must be able to *transfer* a system of actions across time, e.g., the walk to the lecture hall must be transferred from Monday to Monday, etc. Generally, some group (or semi-group) of actions must be transferred under the action of *another* group (or semi-group) of actions.

According to the mathematical theory, which begins in the next chapter, the appropriate formulation for this is wreath products. More fully, this algebraic approach will be substantiated by an intensive review of regularity in branches of physical science such as general relativity and quantum mechanics, as well as computational disciplines such as computer vision.

In this section, we point out the deep relation between the theory of regularity and the issue of *recoverability*, as follows: One of our two main rules for recoverability, the Asymmetry Principle, states that recoverability of a generative program is possible only if the program is symmetry-breaking on successive generative states. That is, the structure must become successively more asymmetric over time - i.e., successively more *irregular*.

Now a major aspect of our theory is that all irregularity is described by embedding it within a regular structure. The reader can begin to see this from the discussion in Sect. 2.18, where we distinguished between two methods of specifying the effects of the generative process. Method 1 specifies the successive generative structure by a decrease of symmetry group. Method 2, which is the method invented in this book, specifies the successive structure by group extension via wreath product; i.e., transfer. The extending group is a symmetry group of the asymmetrizing process.

Because each level is a symmetry group, it creates a regularity. Yet because it has a symmetry-breaking action on its fiber, which is also a symmetry group, one is actually creating an irregular structure. Thus the irregularity is entirely encompassed within a regularity. This is an extraordinarily powerful devise as will be seen. The purpose of this book is to give a theory of complex shape. No matter how complex, i.e., irregular, the shape becomes, it always is described as a completely regular object.

2.22   Aesthetics

In Sect. 1.18 (p. 32), we proposed that aesthetics is the maximization of transfer. In fact, this will be only half of our main proposal concerning the nature of aesthetics. The other half involves recoverability, as follows:

Consider for example physics. Current models explaining the physical constitution of the universe argue for a succession of symmetry-breakings from the underlying starting state (first to hypercharge, isospin, and color, and then to the electromagnetic guage group). There is considerable puzzlement in physics as to why backward symmetrization from the present data set should be the case, and this is linked to the general bewilderment that Wigner expressed in his famous phrase concerning the "unreasonable power of mathematics" in physics - by which he really meant the unreasonable power of symmetry in physics. However, according to our theory, symmetry generally, and the backward symmetrization, in particular, is entirely explicable. It comes from the Asymmetry Principle which states that a generative sequence is *recoverable* only if present asymmetries go back to past symmetries in the generative sequence. In other words, symmetries have an *inferential* role with respect to discovering generative structure.

Now let us return to the issue of *aesthetics*. The term aesthetics in physics is often linked to the use of symmetries to represent past generative states. Therefore, putting this notion together with the considerations of Sect. 1.18, there appear to be two uses for the term aesthetics in physics (1) the characterization of transfer, and (2) the characterization of recovered states. In fact, one can see this in all aspects of quantum mechanics; for example, in spectroscopy, e.g., in the fine/hyperfine structure of the hydrogen atom as described at the end of Sect. 2.20.

The question therefore is this: To what extent are these two situations of aesthetic judgement separate from each other? Our theory tells that they are not separate. We showed in sections 2.14 and 2.18 that each level of the wreath hierarchy necessarily takes on simultaneously the role of transfer and recoverability. To use physics as an example: The symmetry group acts both as the past state and as the operational structure that transfers flow lines of the Schrödinger equation onto each other. This is clearly evidenced for example in spectroscopy. Thus our complete definition of aesthetics can now be stated:

Aesthetics is the maximization of transfer and recoverability.

In this way, we will see that aesthetics is closely linked to functionality. Functionality comes from the transfer structure, e.g., as in the differential equations of physics. Recoverability ensures that the functionality is inferrable.

This allows us also to give a theory of art-works:

Art-works are maximal memory stores.
The rules of aesthetics are therefore the rules of memory storage.

2.23   The Definition of Shape

In Sect. 1.17, we defined the *shape* of a data set as the transfer structure of a machine whose state space is the data set. This again, is half of the definition, since it corresponds to the transfer component of our theory. The definition becomes complete when we add the recoverability component:

The *shape* of a data set is the recoverable transfer structure of a machine whose state space is the data set.

Notice, as a consequence of this, we have an equivalence between geometry, memory storage and aesthetics. For example, only this will explain the link between geometry, explanation, and aesthetics in scientific theory. This three-fold equivalence will be mathematically rigorized in this book, starting with the next chapter.

3. Mathematical Theory of Transfer, I

3.1 Shape Generation by Group Extensions

The reader will recall that a basic principle of our generative theory of shape states that one must *make one part of the generative sequence a transfer of another part of the generative sequence, whenever possible.* We are going to develop an algebraic theory of how this can be achieved.

First, our theory says this: Shape-generation proceeds by the successive addition of structural elements. Each structural element corresponds to a group. The successive addition of structural elements corresponds to successive group extensions.

$$\textbf{Structural elements} \longrightarrow \textbf{Groups.}$$

$$\textbf{Addition of structural elements} \longrightarrow \textbf{Group extensions.}$$

Next, we require that the group extensions maximize transfer. This will force the group extensions to be wreath products. We then have to invent a number of classes of groups that will allow us to represent complex shape generation only in terms of transfer. The invention of these groups will occupy us in subsequent chapters. The purpose of the present chapter however is to fully understand what is needed algebraically in any transfer situation, and to exactly correspond this to the various structural features of a wreath product (which will be fully explained).

This chapter will concern 2-level transfer. We shall argue that it takes five successive stages to build up this 2-level structure. The reason is that the 2-level hierarchy is not a simple hierarchy.

In order to illustrate the five successive stages, we shall take an extremely simple example throughout this chapter - and move onto much more complicated examples in the subsequent chapters, such as *machine assemblies*, and *entire apartment buildings.* The simple example is that of a *square.* One

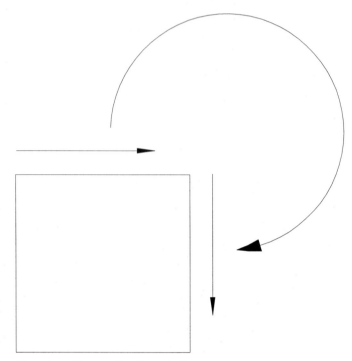

Fig. 3.1. Transfer of translation by rotation.

can consider the square to be a large obstacle, e.g., a table, around which a robot is expected to manoeuver in the environment, or the cross-section of an object, e.g., a block, which the robot is expected to manipulate. We argue that any sophisticated robot must be able to perceive and exploit the structure of *transfer* in the square, which is this: An individual side of the square allows translational movements along it; the robot should notice that any translational movement discovered on one side can be transferred to become translational movement along any other side. In fact, the structure of transfer on a square is illustrated in Fig. 3.1. Whether, the robot is moving in an environment with a square obstacle, or manipulating an object with a square structure, the robot becomes *intelligent* when it discovers that it can transfer its movements in this way.

3.2 The Importance of n-Cubes in Computational Vision and CAD

It is worth noting that squares, and more generally n-cubes, are fundamental to computational vision and computer-aided design for the following reasons.

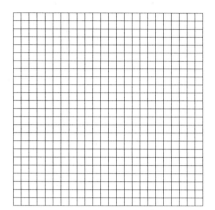

Fig. 3.2. The domain of a parametrized 2-surface.

(1) Squares and gravitational frames. We shall see that the gravitational frame controls *all* human perceptual organization. The square is the smallest-order polygon that contains the symmetry structure of the gravitational frame, and for this reason is the principal means of specifying any frame - e.g., of a picture or diagram.

(2) Bases in Projective Geometry. It takes four points, no three of which are collinear, to establish an unambiguous coordinate system for the projective plane $\mathbb{P}^2$, and therefore four such points are called a projective basis. The fundamental theorem of planar projective geometry states that, given two ordered projective bases for $\mathbb{P}^2$, then there is a unique projective transformation that maps one basis onto the other and preserves the order. In Chapter 22, we will see that projective distortion is relative to an undistorted basis corresponding to the symmetry group of the square; i.e., it is the latter group that allows recoverability.

(3) Normalized boundaries of surface patches. The most frequent example of a surface patch in geometric modeling is one where the boundary is the image of the rectangle. This is frequently *normalized* to the unit square. The domain of the parametrized surface is illustrated in Fig. 3.2. and the codomain is illustrated in Fig. 3.3. Standardly one builds up any surface as the blending of such square-based patches. Squares have a symmetry structure that can be generalized easily to n-dimensional cubes. Thus the same argument applies to higher dimensions.

(4) Control volume for free-form deformation. Free-form deformation is a standard and powerful tool in 3D solid-modeling, and was invented by Sederberg & Parry [139]. Here, one surrounds a complex object with a scaf-

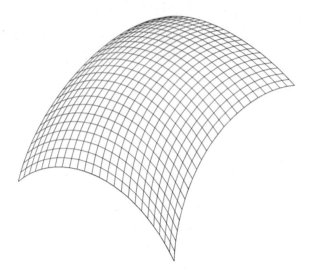

Fig. 3.3. The codomain of a parameterized 2-surface.

fold, and applies actions to selected nodes in the scaffold. The object inside the scaffold deforms accordingly. The most frequently used scaffold shape is a cube-based structure. Indeed, cubes are now being used as input tools for deformation instead of a mouse; i.e., the designer holds a cube in his/her hands and deforms it, resulting in corresponding deformations of the control volume on the screen, Murakami & Nakajima [111].

All the above examples attest to the power of squares, or generally n-cubes, as reference devises for complex structures. We need to explain why this is the case, and in order to do this we first need to give a mathematical theory of what n-cubes actually are.

3.3   Stage 1: Defining Fibers and Control

We now go through the five-stage procedure for constructing a square as a transfer structure.

Standardly, in mathematics, the square is regarded as having D_4 as its symmetry group. However, D_4 does not *characterize* a square because it is equally applicable, for example, to many types of snowflakes, etc. We want to develop a symmetry group that uniquely characterizes a square. This group will be intimately connected to the *generative* structure of a square, because our Symmetry-to-Trace Conversion Principle states that the plan for drawing a square comes from its symmetries. This basic example will occupy us for most of this chapter. When one fully grasps it, one will then be able to understand transfer in any situation.

We shall see that any transfer structure involves *five group actions*. Recall the definition of a group action from elementary group theory: If G is a group and X is a set. Then a function

$$\alpha : \begin{cases} G \times X \longrightarrow X \\ (g, x) \longmapsto gx \end{cases}$$

is called a *group action* if it satisfies:

$\mathfrak{G}_1$: $ex = x$ for $\forall x \in X$.

$\mathfrak{G}_2$: $g(hx) = (gh)x$ for $\forall g, h \in G, \forall x \in X$.

Standardly X is called a *G-set*. The function α will be called the *action equation*.

Now we saw that a square has a transfer structure in which Level 1 is given by the group of translations and Level 2 is given by $\mathbb{Z}_4$ as the group of 90^0 rotations. Both of these two levels should be expressed as group actions thus:

LEVEL 1: The Fiber Level

Expressed as a group action, Level 1 consists of a *set* - the set of points along a side - and a *group* - the group of translations along the side. Let us assume that the side is of infinite length and that finiteness can be obtained by wreath sub-appending an occupancy group $\mathbb{Z}_2$, in accord with Sect. 2.8. The occupancy structure will be ignored in this chapter.

Now the set of points along the side will be called the **fiber set**, and the translation group acting along this set will be called the **fiber group**. Generally, the following notation will be introduced:

$$\begin{aligned} F &= \text{Fiber set} \\ G(F) &= \text{Fiber group.} \end{aligned}$$

The group action of the fiber group on the fiber set will be denoted by:

$$\begin{cases} G(F) \times F \longrightarrow F \\ (T, f) \longmapsto Tf \end{cases}$$

i.e., this describes the action of translations $G(F)$ along a side F.

LEVEL 2: The Control Level

Expressed as a group action, Level 2 consists of a *set* - the set of four positions of a side - and a *group* - the rotation group $\mathbb{Z}_4$ from side to side. The set of side-positions is

$$c_1 = \text{top}, \quad c_2 = \text{right}, \quad c_3 = \text{bottom}, \quad c_4 = \text{left},$$

and the group of four rotations is

$$\mathbb{Z}_4 = \{ \ e, \quad r_{90}, \quad r_{180}, \quad r_{270} \ \}$$

where r_θ means rotation by θ degrees.

The set will be called the **control set**, and the group will be called the **control group**. Generally, the following notation will be introduced:

$$C \quad = \quad \text{Control set}$$
$$G(C) \quad = \quad \text{Control group.}$$

The group action of the control group on the control set will be denoted by:

$$\begin{cases} G(C) \quad \times \quad C \quad \longrightarrow \quad C \\ \\ (\ r \quad , \quad c \ ) \quad \longmapsto \quad rc \end{cases}$$

i.e., this describes the rotations $G(C)$ on the side-positions C.

Summarizing this section: Level 1 (the fiber level) and Level 2 (the control level) are each given by a group action.

3.4   Stage 2: Defining the Fiber-Group Product

In this section, the action of rotations will be ignored, and the action of translations will be studied further. Previously, translations were understood as acting on an individual side. However, we will also understand them as acting on the square as a whole, as follows:

There are four copies of the translations group $G(F)$, one on each side. Let us simply label the copies by their respective sides thus:

$$G(F)_{c_1} , \ G(F)_{c_2}, \ G(F)_{c_3} , \ G(F)_{c_4}.$$

Notice, most crucially, that the four copies are labeled by the four members of the *control set* C.

Now let us take the direct product of these four copies:

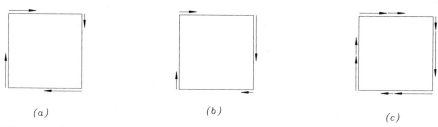

(a)   (b)   (c)

Fig. 3.4. Composition of elements from the fiber-group product.

$$G(F)_{c_1} \times G(F)_{c_2} \times G(F)_{c_3} \times G(F)_{c_4}.$$

This group will be called the **fiber-group product**. Any element from this group is a *vector* of the form

$$(T_{c_1} , T_{c_2} , T_{c_3} , T_{c_4}) \tag{3.1}$$

where $T_{c_i} \in G(F)_{c_i}$. Notice that T_{c_i} is a translation along side c_i. Thus a vector of the above form is a selection of one translation on each of the four sides, respectively. Since these four translations are selected independently, they can be translations by different amounts.

Because the fiber-group product is a direct product, multiplication of elements in the group is componentwise. That is, given two vectors $(T_{c_1} , T_{c_2} , T_{c_3} , T_{c_4})$ and $(S_{c_1} , S_{c_2} , S_{c_3} , S_{c_4})$, from the fiber-group product, multiplication is given thus:

$$\begin{aligned} &(T_{c_1} , T_{c_2} , T_{c_3} , T_{c_4}) (S_{c_1} , S_{c_2} , S_{c_3} , S_{c_4}) \\ &= (T_{c_1} S_{c_1} , T_{c_2} S_{c_2} , T_{c_3} S_{c_3} , T_{c_4} S_{c_4}). \end{aligned} \tag{3.2}$$

This is illustrated in Fig. 3.4, where the element $(T_{c_1} , T_{c_2} , T_{c_3} , T_{c_4})$ is shown as the first figure; the element $(S_{c_1} , S_{c_2} , S_{c_3} , S_{c_4})$ is shown as the second figure; and their composition is shown as the third figure. Notice the following fact about the third figure: On each side, we have added together the two translations from *that* side on the previous two figures. This means that, within each side, we are simply using the group structure on that side. Finally, observe that the multiplicative result, shown in the second line of expression (3.2) is itself a vector from the fiber-group product, which means of course that the fiber-group product is closed as a group.

Notation 3.1 *When the square is being discussed, the fiber-group product will be specifically notated as the 4-fold direct product, $G(F)_{c_1} \times G(F)_{c_2} \times G(F)_{c_3} \times G(F)_{c_4}$, defined above. When the general case of transfer is being discussed, the fiber-group product will be notated as an n-fold direct product, $G(F)_{c_1} \times G(F)_{c_2} \times \ldots \times G(F)_{c_n}$.*

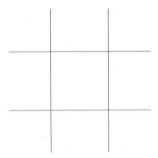

Fig. 3.5. A square before the sides have been cut down by the occupancy group.

3.5   The Fiber-Group Product as a Symmetry Group

We are now going to see that the fiber-group product $G(F)_{c_1} \times G(F)_{c_2} \times G(F)_{c_3} \times G(F)_{c_4}$ acts as a *symmetry group on the square*. Notice that we are using only the translational structure of the square, not the rotation structure.

Fig. 3.5 shows the square. In this diagram, each side is indicated as infinite. It is best to understand the square as *four infinite wires*.

Now let us take any element from the fiber-group product, that is, any vector, $(T_{c_1} , T_{c_2} , T_{c_3} , T_{c_4})$, and apply it to the square. The vector tells us to translate the top side - as an infinite wire - along itself, by the amount T_{c_1}. It also tells us to translate the right side - as an infinite wire - along itself, by the amount T_{c_2}. And so on for the other two sides. That is, each side is translated as an infinite wire along itself, by the designated translation *for that side.*

It is clear that, when these four translations are applied in the way just described, the square will simply *map to itself.* That is, the fiber-group product acts as a symmetry of the square.

3.6   Defining the Action of the Fiber-Group Product on the Data Set

The symmetry defined in the previous section is based on a *group action* of the fiber-group product on the square. We will define this group action in the present section.

First, observe that any point on the square can be specified hierarchically thus:

$$(f , c) \ = \ (\text{ position within side } , \text{ position of side around square })$$

where f is a member of the fiber set F, and c is a member of the control set C. Therefore, the set of points on the square can be defined to be this:

Fig. 3.6. A vector of translations.

$$F \times C$$

where $\times$ is the ordinary, set-theoretic Cartesian product. This set will often be referred to as the *data set*.

Now the group action that will be considered is that of the fiber-group product $G(F)_{c_1} \times G(F)_{c_2} \times G(F)_{c_3} \times G(F)_{c_4}$ on the data set $F \times C$. Thus consider, for example, an element

$$(T_{c_1} , T_{c_2} , T_{c_3} , T_{c_4})$$

from the fiber-group product, as shown in Fig. 3.6. It is necessary to understand its effect on the particular point, for example:

$$(f, c_2)$$

shown in Fig. 3.7. Notice that the coordinate c_2 of this point indicates that it is on the *second side* of the square. Its other coordinate f indicates the position *within* that side.

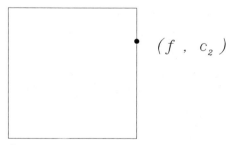

Fig. 3.7. The pair of coordinates defining a point.

Let us consider the action. Observe first that, if the four translations shown in Fig. 3.6 are applied to the *whole square*, each side would be translated along itself *using on each side only the translation assigned to that side*. The consequence of this is that, when the four translations are applied *only*

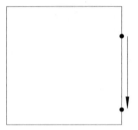

Fig. 3.8. The side selects the translation.

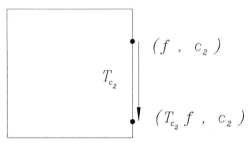

Fig. 3.9. The coordinates of a translated point.

to the point (f, c_2) on the second side, *only* the translation assigned to the second side will be used. The result of this translation in shown in Fig. 3.8. Thus, the application of $(T_{c_1}, T_{c_2}, T_{c_3}, T_{c_4})$ to (f, c_2) results in the point:

$$(T_{c_2} f, c_2)$$

where the reader can see that *only* the translation T_{c_2} has been applied. This will be called the *selective effect*. Fig. 3.9 depicts the resulting action on coordinates.

Generally, therefore, the group action of the fiber-group product on the square is given by the following action equation

$$(T_{c_1}, T_{c_2}, T_{c_3}, T_{c_4})(f, c_i) = (T_{c_i} f, c_i)$$

where one can see, on the right of this equation, that only the translation T_{c_i} has been selected to act, because this is the translation belonging to the c_i side indicated on the left of the equation. In other words,

SELECTIVE EFFECT. *In the group action of the fiber-group product on the data set, the control coordinate c_i in any data point (f, c_i), selects the component of the vector in the fiber-group product to be applied to the fiber coordinate f of the data point.*

The selective effect allows us to prove that the fiber-group product acts as a symmetry group on the square. Recall the model of the square as depicted in Fig. 3.5. Notice that each side is really a set of the form:

$$(F , c_i) \quad = \quad \text{the set of points on side } c_i.$$

Thus the entire square is of this form:

$$F \times C \;=\; (F, c_1) \cup (F, c_2) \cup (F, c_3) \cup (F, c_4).$$

The proof that the fiber-group product is a symmetry group of the square is then simply this:

$$(T_{c_1} , T_{c_2} , T_{c_3} , T_{c_4})[(F, c_1) \cup (F, c_2) \cup (F, c_3) \cup (F, c_4)]$$
$$= \; (T_{c_1} F, c_1) \cup (T_{c_2} F, c_2) \cup (T_{c_3} F, c_3) \cup (T_{c_4} F, c_4)$$
$$\text{(by the action equation)}$$
$$= \; (F, c_1) \cup (F, c_2) \cup (F, c_3) \cup (F, c_4)$$
$$\text{(by the group action of } G(F) \text{ on } F)$$

IMPORTANCE OF THE FIBER-GROUP PRODUCT.   *The fiber-group product captures that part of the full symmetry structure of the data set which is due to the fiber group alone, i.e., without adding considerations of the control group. Another way of saying this is that the fiber-group product is the full symmetry structure of the data set before the additional control symmetry has been* **detected.**

The issue of detection is crucial to our system. As we shall see, it dictates the order in which the stages are elaborated. Also note that Ishida [62] has been concerned with a similar issue of detection.

Now consider this: We have seen that the fiber-group product of the square translates each side independently. Thus, one can think of the direct-product decomposition $G(F)_{c_1} \times G(F)_{c_2} \times G(F)_{c_3} \times G(F)_{c_4}$ as allowing computation with respect to each side to occur separately and simultaneously. Generally we have this:

PARALLEL COMPUTATIONAL LEVEL.   *The fiber-group product being a direct product allows for parallel computation. The fiber-group product defines what will be called the* **parallel level** *in the computational system detecting transfer; e.g., in a perceptual system.*

3.7 Stage 3: Action of $G(C)$ on the Fiber-Group Product

The previous stage considered only the translation structure of the square. In the present stage, the rotation structure will be added. We can say therefore that this stage corresponds to the *detection* of the rotation structure.

Now, before this rotational structure is detected, the individual translational structures of the four sides are seen as independent of each other, as given by the direct-product operation of the fiber-group product. The detection of the rotational structure is equivalent to detecting a *group action* of rotations on the fiber-group product. This group action rotates the translation structure from side to side around the square. The group action is therefore equivalent to the *transfer* structure.

Section 3.3 defined the rotation structure as the control group $G(C) = \mathbb{Z}_4 = \{\, e\,,\, r_{90}\,,\, r_{180}\,,\, r_{270}\,\}$, acting on the control set $C = \{\, c_1\,,\, c_2\,,\, c_3\,,\, c_4\,\}$, the set of four positions for a side. This action is given by:

$$\beta : \begin{cases} G(C) \quad \times \quad C \quad \longrightarrow \quad C \\[2mm] (\quad r \quad,\quad c\,) \quad \longmapsto \quad rc. \end{cases} \tag{3.3}$$

However, what interests us, in this section, is the effect of rotations on the *translation structure* of the square. This will be specified by applying the group action β in (3.3), to the *indexes* of the fiber-group product

$$G(F)_{c_1} \times G(F)_{c_2} \times G(F)_{c_3} \times G(F)_{c_4}.$$

Thus, given an element g from the control group $G(C)$, its action on the sides $\{\, c_1\,,\, c_2\,,\, c_3\,,\, c_4\,\}$ is to send them to their rotated versions $\{\, gc_1\,,\, gc_2\,,\, gc_3\,,\, gc_4\,\}$. Correspondingly, its action on a vector $(\,T_{c_1}\,,\, T_{c_2}\,,\, T_{c_3}\,,\, T_{c_4}\,)$ from the fiber-group product is to send it to the vector:

$$(\,T_{gc_1}\,,\, T_{gc_2}\,,\, T_{gc_3}\,,\, T_{gc_4}\,).$$

If one thinks of $G(C)$ as permuting the members of C, then $G(C)$ will have the same permutational action on the indexes of the fiber-group product. In particular, this will mean that $G(C)$ permutes the fiber-group copies $G(C)_{c_i}$ amongst themselves. As we will see, this permutation corresponds to *transfer*.

Definition 3.1. *Consider the action of the control group on the control set. Consider correspondingly the action of the control group on the indexes of the fiber-group product. This latter action will be called the **raised** action of the control group from the control set to the fiber-group product. It is simply this: Given an element $g \in G(C)$, then its raised action on the fiber-group product is this:*

$$G(F)_{c_1} \times G(F)_{c_2} \times \ldots \times G(F)_{c_n} \quad \longrightarrow \quad G(F)_{gc_1} \times G(F)_{gc_2} \times \ldots \times G(F)_{gc_n}.$$

Simple as it is to state the particular group action of $G(C)$ on the fiber-group product in terms of indexes, it turns out that this action is mathematically very deep. The indexes are labels on algebraic structures, and the action just defined has a powerful effect on those algebraic structures, which will be explored as follows:

3.8   Transfer as an Automorphism Group

It will now be seen that any $g \in G(C)$, applied to the fiber-group product in the above way, is an *automorphism* of the fiber-group product. First of all, the effect of g is obviously one-one and onto, since it merely "permutes" the fiber-group copies. Secondly, g acts *homomorphically* on the fiber-group product; i.e., it preserves the group structure of the fiber-group product. That is, the effect of g on the multiple of two vectors from the fiber-group product is the same whether g is applied before or after the multiplication. To see this, take any two vectors (T_{c_1} , T_{c_2} , T_{c_3} , T_{c_4}) and (S_{c_1} , S_{c_2} , S_{c_3} , S_{c_4}) from the fiber-group product. If they are multiplied together, using (3.2) on p. 83, we get

$$(T_{c_1}S_{c_1} , T_{c_2}S_{c_2} , T_{c_3}S_{c_3} , T_{c_4}S_{c_4}) \quad = \quad (V_{c_1} , V_{c_2} , V_{c_3} , V_{c_4}).$$

Then applying g to this, we get

$$(V_{gc_1} , V_{gc_2} , V_{gc_3} , V_{gc_4}). \tag{3.4}$$

Conversely, applying g first to both (T_{c_1} , T_{c_2} , T_{c_3} , T_{c_4}) and (S_{c_1} , S_{c_2} , S_{c_3} , S_{c_4}), and then multiplying the result, we get

$$(T_{gc_1}S_{gc_1} , T_{gc_2}S_{gc_2} , T_{gc_3}S_{gc_3} , T_{gc_4}S_{gc_4})$$

which is the same as (3.4). Therefore, it is clear that g acts homomorphically on the fiber-group product, and is therefore an automorphism of the fiber-group product.

This means that the raised action corresponds to a particular map τ from the control group to the automorphism group of the fiber-group product. In the case of the square, we have:

$$\tau : G(C) \quad \longrightarrow \quad Aut[G(F)_{c_1} \times G(F)_{c_2} \times G(F)_{c_3} \times G(F)_{c_4}].$$

Generally, we have this:

Definition 3.2. *The representation of $G(C)$, corresponding to the raised action from the control set to the fiber-group product, will be called the **transfer representation** or τ-representation of $G(C)$. This representation is defined as follows:*

$$\tau : G(C) \quad \longrightarrow \quad Aut[G(F)_{c_1} \times G(F)_{c_2} \times \ldots \times G(F)_{c_n}]$$

where τ sends $g \in G(C)$ to the automorphism

$$\tau(g) : G(F)_{c_1} \times G(F)_{c_2} \times \ldots \times G(F)_{c_n} \quad \longrightarrow \quad G(F)_{gc_1} \times G(F)_{gc_2} \times \ldots \times G(F)_{gc_n}.$$

*The raised action of $G(C)$ will also be called the **transfer action** or τ-action and, in this capacity, $G(C)$ will be called the **transfer-automorphism group** or τ-automorphism group, and the members of the group will be referred to as the **transfer automorphisms** or τ-automorphisms.*

Obviously, the letter τ will represent the word *transfer*. Notationally, its action will be written from the left, thus:

$$\tau(g)[(T_{c_1} , T_{c_2} , \ldots , T_{c_n})] \quad = \quad (T_{gc_1} , T_{gc_2} , \ldots , T_{gc_n})$$

3.9   Stage 4: Splitting Extension of the Fiber-Group Product by the Control Group

Stage 2 detected the translational structure of the entire square, and Stage 3 detected the fact that the translational structure had a transfer structure given by rotations. We are now going to model the successive detection of these two structures as a splitting extension[1] of the first by the second.

Recall that a splitting extension is of the form

$$G \quad = \quad N \circledS_\sigma H \tag{3.5}$$

where N is a normal subgroup of G, H is a complement of N in G, and

$$\sigma : H \longrightarrow Aut[N]$$

is some chosen homomorphism which will be called a *representation* of H in $Aut[N]$. The structure of the extension depends on the particular chosen automorphic representation σ.

Now, we are interested in creating a splitting extension of the fiber-group product by the control group. In the case of the square, this means creating a splitting extension of the translation structure $G(F)_{c_1} \times G(F)_{c_2} \times G(F)_{c_3} \times G(F)_{c_4}$ by the rotation structure $G(C) = \mathbb{Z}_4$. That is, using the notation $N \circledS_\sigma H$, the components are:

$$N = G(F)_{c_1} \times G(F)_{c_2} \times G(F)_{c_3} \times G(F)_{c_4}$$
$$H = G(C)$$

---

[1] The reader who is not familiar with *splitting extensions* or *semi-direct products*, should read Appendix A before continuing.

and the splitting extension is this:

$$[G(F)_{c_1} \times G(F)_{c_2} \times G(F)_{c_3} \times G(F)_{c_4}] \; \circledS_\tau \; [G(C)]. \qquad (3.6)$$

The following facts should be noticed about this structure: (1) The fiber-group product is the normal subgroup of the extension. (2) The automorphic representation, on which the extension is based is the *transfer representation*

$$\tau : G(C) \quad \longrightarrow \quad Aut[G(F)_{c_1} \times G(F)_{c_2} \times G(F)_{c_3} \times G(F)_{c_4}].$$

given in Definition 3.2. This means that the group $G(C)$, to the right of the semi-direct product symbol $\circledS$ in expression (3.6) acts on the indexes of the group $G(F)_{c_1} \times G(F)_{c_2} \times G(F)_{c_3} \times G(F)_{c_4}$ to the left of the symbol $\circledS$. This is equivalent to rotating the translation structure from one side to the next around the square; i.e., transfer. In the general case, we will describe hierarchical detection of transfer thus:

HIERARCHICAL DETECTION OF TRANSFER. *Detection is best modeled as a splitting extension of a fiber-group product by a control group:*

$$[G(F)_{c_1} \times G(F)_{c_2} \times \ldots \times G(F)_{c_n}] \; \circledS_\tau \; G(C) \qquad (3.7)$$

where

$$\tau : G(C) \quad \longrightarrow \quad Aut[G(F)_{c_1} \times G(F)_{c_2} \times \ldots \times G(F)_{c_n}]$$

is the τ-representation given in Definition 3.2.

3.10   Wreath Products

Return to the example of the square. Although expression

$$[G(F)_{c_1} \times G(F)_{c_2} \times G(F)_{c_3} \times G(F)_{c_4}] \; \circledS_\tau \; [G(C)] \qquad (3.8)$$

is lengthy, it is built up entirely using only two components $G(F)$ and $G(C)$. Extracting these two components we write

$$G(F) \; \textcircled{w} \; G(C). \qquad (3.9)$$

The product $\textcircled{w}$ is called a *wreath product*. In other words, *there are two different notations* for the same construction, respectively (3.8) and (3.9). The first uses the semi-direct product symbol $\circledS$, and places to the left of this symbol, the entire fiber-group product. The second uses the wreath-product symbol $\textcircled{w}$, and places to the left of this symbol, only the fiber group. In the general case, the two notations are respectively:

$$[G(F)_{c_1} \times G(F)_{c_2} \times \ldots \times G(F)_{c_n}] \; \textcircled{s}_\tau \; G(C) \qquad (3.10)$$

and

$$G(F) \; \textcircled{w} \; G(C). \qquad (3.11)$$

Using the notation and terminology of the previous sections, a wreath product can now be defined in the following way: Consider two group actions: the actions of groups, $G(F)$ and $G(C)$, on sets, F and C, respectively. The wreath product $G(F)\textcircled{w}G(C)$ is the semi-direct product $[G(F)_{c_1} \times G(F)_{c_2} \times \ldots \times G(F)_{c_n}] \; \textcircled{s}_\tau \; G(C)$, where the groups $G(F)_c$ are isomorphic copies of $G(F)$ indexed by the members c of the set C. The map $\tau : G(C) \longrightarrow Aut[G(F)_{c_1} \times G(F)_{c_2} \times \ldots \times G(F)_{c_n}]$ is defined such that $\tau(g)$ corresponds to the group action of $G(C)$ on C, now applied to the indexes c_i in $G(F)_{c_1} \times G(F)_{c_2} \times \ldots \times G(F)_{c_n}$. That is, $\tau(g) : G(F)_{c_1} \times G(F)_{c_2} \times \ldots \times G(F)_{c_n} \longrightarrow G(F)_{gc_1} \times G(F)_{gc_2} \times \ldots \times G(F)_{gc_n}$. The final component of the definition of a wreath product will be given later.

WREATH PRODUCTS & INTELLIGENT SYSTEMS. *We argue that wreath products are the entire structural basis of intelligent systems. This is because systems are intelligent when they maximize transfer and recoverability, and we argue that the appropriate model for transfer and recoverability is wreath products. In particular, we have argued that the entire perceptual and motor systems of human beings are structured by wreath products.*

This section will conclude with some comments on terminology: First note that the terms *fiber group* and *control group*, with respect to wreath products, are entirely ours. These terms are not standard group-theoretic terminology. We have introduced them for the following reasons:

(1) Neurophysiological Fibers. We claim that the human perceptual and motor systems are structured in this way. Thus fiber groups in a wreath product will literally correspond to fibers or columns in the nervous system.

(2) Differential Geometry. The terminology will also allow us to relate wreath products to fiber bundles in differential geometry - which will be essential in physics and computer-aided design.[2] The same applies to our term *fiber-group product*. Conventionally, in group theory, one calls this group the *base group* of a wreath product. However, in setting up a correspondence between wreath products in group theory and fiber bundles in differential geometry, this will become misleading. It is the *control group*, in a wreath

---

[2] Note that, in the domain of CAD/CAM, Zagajac [157] and Kumar, Burns, Dutta, & Hoffmann [80] have introduced fiber-bundle models for representing composite material structures with tensor properties.

product, that corresponds to the *base manifold* in differential geometry. Thus, the fiber-group product of the wreath product does *not* correspond to the base manifold in differential geometry, but to the fiber bundle. Thus the term "base group" will be abandoned in labeling the fiber-group product.

3.11   The Universal Embdedding Theorems

According to our theory:

Cognitive systems are structured as wreath products. The process of representing the world is the process of mapping situations into the wreath products.

Therefore, critical to this must be the major theorems in mathematics concerning embeddings into wreath products. The two fundamental embedding theorems are the Krasner-Kaloujnine theorem in group theory, and the Krohn-Rhodes theorem in semi-group theory. Indeed these theorems demonstrate a remarkable universal power of wreath products as a representating structure.

The Krasner-Kaloujnine theorem states that any group extension $N\textcircled{E}H$ can be embedded in a wreath product $N\textcircled{w}H$ constructed from the same components N and H. In other words, wreath products define a universal embedding space for all group extensions. We give a lengthy explanation of this theorem in our book on group extensions, Leyton [98].

The Krohn-Rhodes theorem [79] constitutes the major decomposition theorem for machines. The applicability to machines relies on the fundamental relation between semi-groups and machines. Expressed in terms of semi-groups, the Krohn-Rhodes theorem states this: Any finite faithful semi-group action $\langle S, X, \alpha \rangle$ divides an n-fold wreath product of semi-group actions, where each is given either by a simple group or one of four elementary types of collapsers.

3.12   Nesting vs. Control-Nesting

In Chapter 1, we described transfer intuitively as *a control group pushing a fiber group around some space.* What we have done in the present chapter is taken this initial description of transfer, and given it a formulation in terms of wreath products. This has enormous advantages, as follows: (1) The wreath product is a group that contains all the transferred versions of the fiber group. Thus, rather than thinking of the transferred versions as separate algebraic structures, they are all integrated within a single algebraic structure. (2) This single algebraic structure also encompasses the control group. And thus the wreath product contains the network of algebraic connectivity that relates the control group to the fiber group.

In this section we want to look at a particularly deep aspect of the "algebraic connectivity" between the control group and the fiber group. To do this, we are going to distinguish between what we will call *nesting* and the much more powerful phenomenon we will call *control-nesting*.

We will simply say that the *nesting* of group G_1 within a group G_2 occurs when one creates a splitting extension of G_1 by G_2. Notice that this means that G_2 acts as an automorphism group of G_1. (Appendix A explains splitting extensions.)

In contrast, we will say that the *control-nesting* of a group G_1 within a group G_2 occurs when one creates a wreath product of G_1 and G_2. A critical feature emerges: G_2 no longer acts as an automorphism group of G_1; that is, it does not send G_1 *onto itself*. Instead G_2 *moves G_1 around*. In other words, it *controls* the position of the group G_1.

Nevertheless, the remarkable thing is that G_2 *does* act as an automorphism group on *something*: It acts as an automorphism group of the *fiber-group product*. It maps the fiber-group product to itself because - although it does not send any individual copy of G_1 to itself - it does send the *total* set of copies to itself. This leads to the following profound effect:

DICHOTOMOUS EFFECT. *Nested control arises from the following dichotomous effect: By acting as an automorphism group on the fiber-group product and not on the fiber group, the control group preserves the fiber-group product but not the fiber group. This forces the fiber group to be moved around within the fiber-group product. Thus the dichotomous effect is responsible for transfer.*

Notice that the reason for the dichotomous effect is that the fiber-group product is a normal subgroup of the wreath product, whereas the fiber-group is not. This relates to the comment p. 14 on normal subgroups and generativity, where, in particular, we discussed the appropriate description of the cylinder as a sweeping structure. The standard representation of a cylinder (e.g., in all of crystallography) takes the *direct product* of the cross-section group $SO(2)$ and the sweeping group $\mathbb{R}$. This is a splitting extension of $SO(2)$ by $\mathbb{R}$, and therefore does not allow for the movement of the cross-section along the cylinder - because $SO(2)$ is a normal subgroup of the extension; i.e., $SO(2)$ will be sent only to itself under the action of $\mathbb{R}$. In contrast, in the *wreath product* of the cross-section group $SO(2)$ and the sweeping group $\mathbb{R}$, the cross-section group $SO(2)$ is no-longer a normal subgroup. Therefore $SO(2)$ necessarily moves under the action of the sweeping group $\mathbb{R}$. Remarkably, this move-

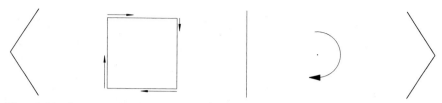

Fig. 3.10. A group element in the control-nested structure of a square.

ment takes place because the entire structure of cross-sections does form a normal subgroup, and $\mathbb{R}$ has an automorphic effect on that structure. Thus, it is the dichotomous effect that allows the *generative* process to take place. This illustrates a fundamental part of our generative theory of shape.

3.13   Stage 5: Defining the Action of $G(F) \textcircled{w} G(C)$ on $F \times C$

This is the final stage in our definition of transfer: specifying the action of the wreath-product group on the data set. It is necessary to consider the effect of the individual group elements on the data.

Again, this will be illustrated with a square. Here, the wreath product $G(F) \textcircled{w} G(C)$ has the structure:

$$[G(F)_{c_1} \times G(F)_{c_2} \times G(F)_{c_3} \times G(F)_{c_4}] \textcircled{s}_\tau [G(C)]. \tag{3.12}$$

This has the form of a semi-direct product:

$$N \textcircled{s}_\sigma H$$

where any group element is given thus:

$$\langle\, n \mid h \,\rangle \tag{3.13}$$

for $n \in N$, and $h \in H$. Thus, an element of the wreath-product group (3.12), has the form:

$$\langle\, (\, T_{c_1} ,\ T_{c_2} ,\ T_{c_3} ,\ T_{c_4} \,) \mid r_\theta \,\rangle \ . \tag{3.14}$$

where $(\, T_{c_1} ,\ T_{c_2} ,\ T_{c_3} ,\ T_{c_4} \,)$ corresponds to the element n, and r_θ corresponds to the element h in (3.13).

The wreath element (3.14) can be schematically represented by Fig. 3.10, where, on the left, the square shows the vector of translations $(\, T_{c_1} ,\ T_{c_2} ,\ T_{c_3} ,\ T_{c_4} \,)$; and, on the right, the circular arrow represents the rotation r_θ.

The application of this group element to the data set, the square, is as follows. First recall the model of the square, as shown in Fig. 3.5 (p. 84), i.e., as *four infinite wires*. The application of the element $\langle\, (\, T_{c_1} ,\ T_{c_2} ,\ T_{c_3} ,\ T_{c_4} \,) \mid r_\theta \,\rangle$ to the square has two successive phases:

(1) Translate each side c_i by the amount T_{c_i}.

(2) Rotate the square by the amount r_θ.

It is easy to see that this two-phase operation sends the square to itself; i.e., is a *symmetry* of the square.

We claim that the correct symmetry group of the square is the wreath product

$$[G(F)_{c_1} \times G(F)_{c_2} \times G(F)_{c_3} \times G(F)_{c_4}] \circledS_r [G(C)].$$

This is a much more complicated group than the usually stated symmetry group D_4, and provides the generative structure.

Let us now define the action of an individual element of the wreath-product group on an individual point of the square. To illustrate, let us take the group element to be

$$\langle\, (\, T_{c_1}\, ,\; T_{c_2}\, ,\; T_{c_3}\, ,\; T_{c_4}\,)\mid r_{180}\,\rangle$$

and the point on the square, $F \times C$, to be

$$(f, c_2)$$

which is on the second side c_2 (right side) of the square.

The two-phase operation defined above tells us to first apply the vector $(\, T_{c_1}\, ,\; T_{c_2}\, ,\; T_{c_3}\, ,\; T_{c_4}\,)$ to the point. By the selective effect, only the translation T_{c_2} on the side containing the point can be applied. The result is therefore the point:

$$(\, T_{c_2} f\, ,\; c_2\,).$$

Hence, this first phase is as shown in Fig. 3.11a. The second phase is to apply the rotation operation r_{180} to this point. This gives us the final point shown in Fig. 3.11b. Thus, the total effect is this:

$$\langle\, (\, T_{c_1}\, ,\; T_{c_2}\, ,\; T_{c_3}\, ,\; T_{c_4}\,)\mid r_{180}\,\rangle\; (f, c_2)\quad=\quad(\, T_{c_2} f\, ,\; r_{180} c_2\,)$$

Generally, therefore, for an arbitrary wreath product, we have:

ACTION EQUATION FOR WREATH PRODUCT. *For a wreath product* $G(F) \circledW G(C)$ *acting on a data set* $F \times C$, *the action equation is this:*

$$\langle\, (\, T_{c_1}\, ,\; T_{c_2}\, ,\; \ldots\, ,\; T_{c_n}\,)\mid g\,\rangle\; (f, c_i)\quad=\quad(\, T_{c_i} f\, ,\; g c_i\,)$$

This group defines the full symmetry of the data set $F \times C$, and will be called the **control-nested symmetry group**. The elements of the group will be called the **control-nested symmetries**.

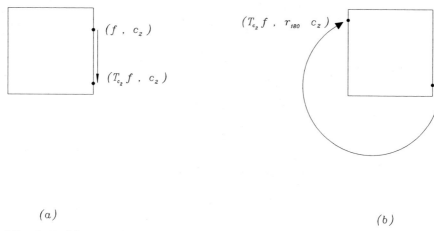

(a)

(b)

Fig. 3.11. The two-phase structure of an element in the wreath product.

3.14   Control-Group Indexes

So far, the index labels on the fiber-group copies, $G(F)_{c_i}$ have been the members c_i of the control set. However, our theory is a generative theory of shape, which means that the data points are labeled by the operations that generated them. Thus, instead of some independent control set as a labeling scheme, we should use the members of the control group itself.

To understand the consequences of this, return to the example of the square, First, rather than using the control set $C = \{\, c_1\,,\, c_2\,,\, c_3\,,\, c_4\,\}$ to label the four sides, we use the control group $G(C) = \mathbb{Z}_4 = \{e, r_{90}, r_{180}, r_{270}\}$, as shown in Fig. 3.12. Now observe the algebraic consequences of this. Whereas the fiber-group product was previously $G(F)_{c_1} \times G(F)_{c_2} \times G(F)_{c_3} \times G(F)_{c_4}$, it is now $G(F)_e \times G(F)_{r_{180}} \times G(F)_{r_{270}} \times G(F)_{r_{270}}$. Most crucially, the action of the control group on the fiber-group product now takes place by the action of the control group *on itself* as the index structure. For example, applying r_{90} to the vector

$$(T_e\,,\, T_{r_{90}}\,,\, T_{r_{180}}\,,\, T_{r_{270}})$$

produces

$$(T_{r_{90}e}\,,\, T_{r_{90}r_{90}}\,,\, T_{r_{90}r_{180}}\,,\, T_{r_{90}r_{270}}).$$

But because each index is now a multiplication of two elements from the control group, we get this:

$$(T_{r_{90}}\,,\, T_{r_{180}}\,,\, T_{r_{270}}\,,\, T_e).$$

Clearly this compositionality has enormous advantages for any of the application areas of our generative theory, e.g., robot navigation, the description of machining operations in manufacturing.

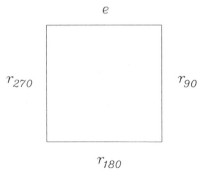

$$e$$

$$r_{270} \qquad r_{90}$$

$$r_{180}$$

Fig. 3.12. Control-group indexes.

The appropriate description of such structures is a type of wreath product called a *regular wreath product*. In contrast, the type of wreath product being described up till now uses an independent control set. This latter type is often referred to as a *permutational wreath product* because the control group tends to be understood as a permutation group of the control set. However, in a regular wreath product, the control set is the control group itself, and the action of the control group is therefore its own internal group composition.

3.15   Up-Keeping Effect of the Transfer Automorphisms

Since a basic issue of our theory of shape is recoverability of the generative sequence, we need to see how group multiplication in the wreath product supports recoverability. A wreath product is an example of a semi-direct product

$$N \text{\textcircled{s}}_\sigma H$$

and multiplication within a semi-direct product has this form:

$$\langle\, n_1 \mid h_1 \,\rangle\langle\, n_2 \mid h_2 \,\rangle = \langle\, \sigma(h_2)[n_1]\, n_2 \mid h_1 h_2 \,\rangle \qquad (3.15)$$

as explained in Appendix A.

In the particular case of a wreath product, σ is the *transfer* representation. This has a powerful role with respect to *generativity*, as we will show in this section. Let us return to the square and consider two elements from the wreath product, for example:

$$\langle\, (\, T_e \,,\, T_{r_{90}} \,,\, T_{r_{180}} \,,\, T_{r_{270}} \,) \mid r_{180} \,\rangle$$

and

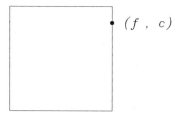

(f , c)

Fig. 3.13. The pair of coordinates defining a point.

$$\langle \, (\, S_e \, , \, S_{r_{90}} \, , \, S_{r_{180}} \, , \, S_{r_{270}} \,) \mid r_{90} \, \rangle.$$

Using expression (3.15), the multiplication of the two elements is this:

$$\langle \, \tau(r_{90}) \, [(\, T_e \, , \, T_{r_{90}} \, , \, T_{r_{180}} \, , \, T_{r_{270}} \,)] \, (\, S_e \, , \, S_{r_{90}} \, , \, S_{r_{180}} \, , \, S_{r_{270}} \,) \mid r_{180} r_{90} \, \rangle.$$

It is important now to investigate the role of $\tau(r_{90})$ within this expression.

Let the two group elements, $\langle \, (\, T_e \, , \, T_{r_{90}} \, , \, T_{r_{180}} \, , \, T_{r_{270}} \,) \mid r_{180} \, \rangle$ and $\langle \, (\, S_e \, , \, S_{r_{90}} \, , \, S_{r_{180}} \, , \, S_{r_{270}} \,) \mid r_{90} \, \rangle$, be labeled g_1 and g_2 respectivity. And let us consider a point P on the square. The question is: What is needed to ensure the simple condition:

$$g_1(g_2 P) = (g_1 g_2)P \quad . \tag{3.16}$$

Notice carefully that the left-hand side represents the *generative sequence of points*: One starts with point P, and then applies g_2 to obtain the point $g_2 P$, and then applies g_1 to obtain the point $g_1 g_2 P$. In contrast, on the right-hand side, one first multiplies the two group elements g_1 and g_2 together to obtain another group element, and then applies this to the point P. Most crucially, on the left-hand side, one does not actually use the group composition operation, only the direct effect of group elements on points. This contrasts with the right-hand side which does use the group composition operation. Thus, expression (3.16) demands that the group composition operation (right-hand side) be defined so that it accord with the generative structure on points (left-hand side).

To understand the important implications of this, we will work through a specific example. Let the point be (f, c), on the second side of the square, as shown in Fig. 3.13. The two elements g_1 and g_2 will be as before, and thus the multiplication being considered is this:

$$\langle \, (\, T_e \, , \, T_{r_{90}} \, , \, T_{r_{180}} \, , \, T_{r_{270}} \,) \mid r_{180} \, \rangle \langle \, (\, S_e \, , \, S_{r_{90}} \, , \, S_{r_{180}} \, , \, S_{r_{270}} \,) \mid r_{90} \, \rangle \, (f, c) \tag{3.17}$$

First let us evaluate this using the structure of the left-hand side in expression (3.16).

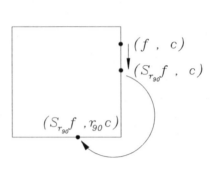

Fig. 3.14. Application of the first group element.

Left-Hand Side

The left-hand side of (3.16) tells us to work step-by-step from *right to left*, in expression (3.17). That is, first take the point (f, c). Then apply the group element, $\langle\,(\,S_e\,,\,S_{r90}\,,\,S_{r_{180}}\,,\,S_{r_{270}}\,)\mid r_{90}\,\rangle$, and then apply the element, $\langle\,(\,T_e\,,\,T_{r90}\,,\,T_{r_{180}}\,,\,T_{r_{270}}\,)\mid r_{180}\,\rangle$. Let us proceed:

The application of $\langle\,(\,S_e\,,\,S_{r90}\,,\,S_{r_{180}}\,,\,S_{r_{270}}\,)\mid r_{90}\,\rangle$ to the point (f, c) consists of a translation followed by a rotation. Since (f, c) is on the *second* side, one must select the translation S_{r90} from the *second* position of the vector $(\,S_e\,,\,S_{r90}\,,\,S_{r_{180}}\,,\,S_{r_{270}}\,)$. After applying this translation, one applies the rotation r_{90}. The result is a point on the *third* side:

$$(\,S_{r90}f\,,\,r_{90}c\,).$$

Fig. 3.14 shows the successive translation and rotation that have just been applied, as well as the resulting point.

The next stage is to apply the group element $\langle\,(\,T_e\,,\,T_{r90}\,,\,T_{r_{180}}\,,\,T_{r_{270}}\,)\mid r_{180}\,\rangle$ to this point. It is fundamentally important to understand that, whereas in the previous stage, when $\langle\,(\,S_e\,,\,S_{r90}\,,\,S_{r_{180}}\,,\,S_{r_{270}}\,)\mid r_{90}\,\rangle$ was applied, the translation used was the translation that belonged to the *second* side, we now have to change this and use the translation that belongs to the *third* side. The reason is that *the point is now on the third side!*

Algebraically, therefore the application of $\langle\,(\,T_e\,,\,T_{r90}\,,\,T_{r_{180}}\,,\,T_{r_{270}}\,)\mid r_{180}\,\rangle$ to the point $(S_{r90}f, r_{90}c)$, is

$$(\,T_{r_{180}}S_{r90}f\,,\,r_{180}r_{90}c\,)$$

where one sees that the translation $T_{r_{180}}$ from the third side has been used. The rotation is r_{180}, which has now moved the point onto the top side. This successive translation and rotation are shown in Fig. 3.15.

Now put together the successive stages that have been carried out:

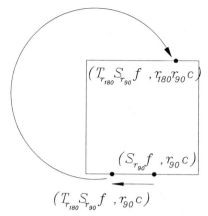

$$\left(T_{r_{180}} S_{r_{90}} f \ , r_{180} r_{90} \, c \right)$$

$$\left(S_{r_{90}} f \ , r_{90} \, c \right)$$

$$\left(T_{r_{180}} S_{r_{90}} f \ , r_{90} \, c \right)$$

Fig. 3.15. Application of the second group element.

$$\langle \, (\, T_e \, , \, T_{r90} \, , \, T_{r180} \, , \, T_{r270} \,) \mid r_{180} \, \rangle \langle \, (\, S_e \, , \, S_{r90} \, , \, S_{r180} \, , \, S_{r270} \,) \mid r_{90} \, \rangle (f, c)$$

$$= \langle \, (\, T_e \, , \, T_{r90} \, , \, T_{r180} \, , \, T_{r270} \,) \mid r_{180} \, \rangle (S_{r90} f, r_{90} c)$$

$$= (\, T_{r180} S_{r90} f \, , \, r_{180} r_{90} c \,) \tag{3.18}$$

Right-Hand Side

Let us turn to the right-hand side of (3.16), which is:

$$(g_1 g_2) P \ .$$

Here, one first multiplies the two group elements, g_1 and g_2, together, and then applies the multiple $g_1 g_2$ to the point P. The question we ask is this: *How should the multiplication of the two group elements, g_1 and g_2, be defined?*
  Using our example, the multiple $g_1 g_2$ is this:

$$\langle \, (\, T_e \, , \, T_{r90} \, , \, T_{r180} \, , \, T_{r270} \,) \mid r_{180} \, \rangle \langle \, (\, S_e \, , \, S_{r90} \, , \, S_{r180} \, , \, S_{r270} \,) \mid r_{90} \, \rangle .$$

If the multiplication is defined here via the *direct-product* operation, i.e., *componentwise*, then the result would be this group element:

$$\langle \, (\, T_e \, , \, T_{r90} \, , \, T_{r180} \, , \, T_{r270} \,) \ (\, S_e \, , \, S_{r90} \, , \, S_{r180} \, , \, S_{r270} \,) \mid r_{180} r_{90} \, \rangle .$$

Then, multiplying out the two vectors within this element, one gets:

$$\langle \, (\, T_e S_e \, , \, T_{r90} S_{r90} \, , \, T_{r180} S_{r180} \, , \, T_{r270} S_{r270} \,) \mid r_{180} r_{90} \, \rangle .$$

Most importantly, observe that the subscripts i and j on each term $T_i S_j$ are exactly the same.

The consequence of this fact is that, when the above combined element $\langle \, (\, T_e S_e \, , \; T_{r_{90}} S_{r_{90}} \, , \; T_{r_{180}} S_{r_{180}} \, , \; T_{r_{270}} S_{r_{270}} \,) \mid r_{180} r_{90} \, \rangle$ is applied to the point (f, c), the result is the point:

$$(\, T_{r_{90}} S_{r_{90}} f \, , \; r_{180} r_{90} c \,).$$

The crucial thing to notice here is that the subscripts on the TS term are both r_{90}. What can be seen therefore is that this TS term *fails to remember* the rotation that occurred when the point was moved from the second side to the third side. The equal indexes r_{90} indicate that both come from the second side, whereas only the S part should come from second side, and the T part should come from the third side. This is the deep reason why we must not use the direct-product operation: *Group multiplication in a direct product fails to remember generative succession.*

Thus, instead of using the direct-product operation, where the automorphism representation σ is the trivial map, let us see what happens if σ is actually τ, the *transfer* representation. Using this representation, the multiplication of the two elements $\langle \, (\, T_e \, , \; T_{r_{90}} \, , \; T_{r_{180}} \, , \; T_{r_{270}} \,) \mid r_{180} \, \rangle$ and $\langle \, (\, S_e \, , \; S_{r_{90}} \, , \; S_{r_{180}} \, , \; S_{r_{270}} \,) \mid r_{90} \, \rangle$ becomes this:

$$\langle \, \tau(r_{90}) \, [(\, T_e \, , \; T_{r_{90}} \, , \; T_{r_{180}} \, , \; T_{r_{270}} \,)] \, (\, S_e \, , \; S_{r_{90}} \, , \; S_{r_{180}} \, , \; S_{r_{270}} \,) \mid r_{180} r_{90} \, \rangle \tag{3.19}$$

where we see that the transfer automorphism $\tau(r_{90})$ is applied to the first vector $(\, T_e \, , \; T_{r_{90}} \, , \; T_{r_{180}} \, , \; T_{r_{270}} \,)$. When the $\tau(r_{90})$ action is carried out within the above expression, the expression becomes

$$\langle \, (\, T_{r_{90}} \, , \; T_{r_{180}} \, , \; T_{r_{270}} \, , \; T_e \,) \, (\, S_e \, , \; S_{r_{90}} \, , \; S_{r_{180}} \, , \; S_{r_{270}} \,) \mid r_{180} r_{90} \, \rangle$$

which is

$$\langle \, (\, T_{r_{90}} S_e \, , \; T_{r_{180}} S_{r_{270}} \, , \; T_{r_{270}} S_{r_{180}} \, , \; T_e S_{r_{270}} \,) \mid r_{180} r_{90} \, \rangle.$$

Notice now that the subscripts i and j on each term $T_i S_j$ are *not* the same. They have been shifted with respect to each other by 90^0.

The consequence of this fact is that, when the above group element $\langle \, (\, T_{r_{90}} S_e \, , \; T_{r_{180}} S_{r_{270}} \, , \; T_{r_{270}} S_{r_{180}} \, , \; T_e S_{r_{270}} \,) \mid r_{180} r_{90} \, \rangle$ is applied to the point (f, c), the result is the point:

$$(\, T_{r_{180}} S_{r_{90}} f \, , \; r_{180} r_{90} c \,). \tag{3.20}$$

Here one sees that the $T_{r_{180}} S_{r_{90}}$ term *remembers* the rotation that occurred when the point was moved from the second side to the third side. In fact, one can see that the point in expression (3.20) is the same as was obtained in expression (3.18) where the generative sequence of points was calculated directly. We therefore conclude: *Group multiplication in the wreath product remembers generative succession.* This is due to what we call the *up-keeping* effect of the transfer automorphisms in a wreath product.

3.16   The Direct vs. the Indirect Representation

Throughout this chapter so far, we have been using the τ-representation of the control group $G(C)$. This is the action in which $G(C)$ acts on the subscripts of the fiber-group product *in the same way* that it acts on the control set.

However, paradoxically, this has the *opposite* effect on the fiber-group product. To illustrate, consider the square, and in particular, the rotation r_{90}. This operation has the effect of rotating the members of the control set (the side positions) by 90^0 *clockwise* around the square. Now, when this action is raised to the fiber-group product, and r_{90} is applied to the subscripts, we find that it shifts the vector

$$(T_{c_1} , T_{c_2} , T_{c_3} , T_{c_4})$$

one step to the left, thus:

$$(T_{c_2} , T_{c_3} , T_{c_4} , T_{c_1}).$$

However, this is actually the *anti-clockwise* rotation of the T_{c_i} around the square. For example, T_{c_2}, which was in the second position of the original vector $(T_{c_1} , T_{c_2} , T_{c_3} , T_{c_4})$, i.e., the right side of the square, is now in the first position of the vector, i.e., the top side of the square.

Definition 3.3. *In* **direct raising***, the application of a member g of the control group to the control set $\{ c_1 , c_2 , \ldots , c_n \}$, corresponds to the application of g to the subscripts of the vectors in the fiber-group product:*

$$(T_{c_1} , T_{c_2} , \ldots , T_{c_n}) \longrightarrow (T_{gc_1} , T_{gc_2} , \ldots , T_{gc_n}).$$

This sends the vectors in the opposite direction to the control set. Direct raising corresponds to the τ-representation, $\tau : G(C) \longrightarrow Aut[G(F)_{c_1} \times G(F)_{c_2} \times \ldots \times G(F)_{c_n}]$.

In contrast to this representation, one can have the opposite representation. For example, suppose, when r_{90} is applied to the control set $\{ c_1 , c_2 , c_3 , c_4 \}$, its *inverse* r_{90}^{-1} is applied to the subscripts of $(T_{c_1} , T_{c_2} , T_{c_3} , T_{c_4})$. Then the result would be this:

$$(T_{r_{90}^{-1} c_1} , T_{r_{90}^{-1} c_2} , T_{r_{90}^{-1} c_3} , T_{r_{90}^{-1} c_4}).$$

Now, since r_{90}^{-1} is actually r_{270}, the vector just given is

$$(T_{c_4} , T_{c_1} , T_{c_2} , T_{c_3})$$

which means that the original vector has been rotated one step *clockwise* around the square. For example, T_{c_2}, which was in the second position of the original vector (T_{c_1} , T_{c_2} , T_{c_3} , T_{c_4}), i.e., the right side of the square, is now in the third position of the vector, i.e., the bottom side of the square.

Definition 3.4. *In* **indirect raising,** *the application of a member g of the control group to the control set* { c_1 , c_2 , $\ldots$, c_n }, *corresponds to the application of g^{-1} to the subscripts of the vectors in the fiber-group product:*

$$(T_{c_1} , T_{c_2} , \ldots , T_{c_n}) \longrightarrow (T_{g^{-1}c_1} , T_{g^{-1}c_2} , \ldots , T_{g^{-1}c_n}).$$

This sends the vectors in the same direction as the control set. The representation $\hat{\tau} : G(C) \longrightarrow Aut[G(F)_{c_1} \times G(F)_{c_2} \times \ldots \times G(F)_{c_n}]$ corresponding to indirect raising will be called the $\hat{\tau}$-representation.

Whether one chooses to use the τ- or the $\hat{\tau}$-representation is purely a matter of convenience. Thus, in the remainder of the book, either will be used, depending on how easy this makes the local discussion. Both representations will be referred to as the *transfer representation.*

Definition 3.5. *The terms involving the word* **transfer** *in Defintion 3.2 (p. 89), will be equally applicable if one replaces τ for $\hat{\tau}$ throughout that definition, and changes the direct raising to the indirect raising.*

3.17 Transfer as Conjugation

Generally, in a splitting extension $G = N \textcircled{s}_\sigma H$, the automorphic action of any element $h \in H$ on N is by *conjugation $h - h^{-1}$* of N (see[3] Appendix A). Now, since a wreath product is a splitting extension of the fiber-group product by the control group, we therefore conclude that the action of the control group on the fiber-group product is by conjugation. That is, *transfer is a process of conjugation.* The purpose of this section is to carefully understand how this works, since it is linked ultimately to the issue of *recoverability.*

Let us return to the illustration of a square. Consider again the rotation r_{90}. When r_{90} is represented using the $\hat{\tau}$-representation, then its effect on a translation is illustrated in Fig. 3.16. Thus a vector

---

[3] In the case of non-splitting extensions, one still uses conjugation, but since H is not contained in G, conjugation of N is not by the elements h of H, but by their transversal representatives $t(h)$ in G. See our book on group extensions, Leyton [98].

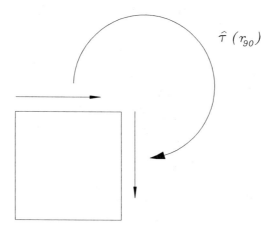

$\hat{\tau}\,(r_{90})$

Fig. 3.16. The action of $\hat{\tau}(r_{90})$.

$$(\, T_{c_1} \, , \; - \; , \; - \; , \; - \;)$$

is sent to a vector

$$(\; - \; , \, T_{c_1} \, , \; - \; , \; - \;).$$

What we are now going to see is this: Whereas T_{c_1} is a translation on the top side, its conjugate

$$r_{90} \; T_{c_1} \; r_{90}^{-1}$$

is the same translation, T_{c_1}, but now acting on the second side.

In order to show this, let us apply the conjugate $r_{90} \, T_{c_1} \, r_{90}^{-1}$ to a point on the second side, and show that the effect of $r_{90} \, T_{c_1} \, r_{90}^{-1}$ on that point is to put it through the translation T_{c_1}, *on that second side.*

The expression

$$r_{90} \; T_{c_1} \; r_{90}^{-1}$$

will be read from *right to left*, and has the three stages:

Stage 1: Apply r_{90}^{-1}
Stage 2: Apply T_{c_1}
Stage 3: Apply r_{90}.

They are illustrated in Fig. 3.17. First, in part (a) of this figure, we see the starting point on the second side. Then, in part (b), we see the effect of applying Stage 1, the rotation r_{90}^{-1}, which sends the point to its equivalent on the top side. Then, in part (c) of the figure, we see the effect of applying Stage 2, which translates the point by the amount T_{c_1}. Finally, in part (d) of the figure, we see the effect of Stage 3, which applies the rotation r_{90} sending the point back to the second side.

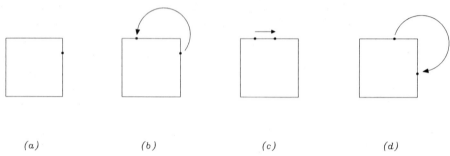

(a)   (b)   (c)   (d)

Fig. 3.17. Illustration of conjugation.

Fig. 3.18. The total effect of the conjugation in Fig. 3.17.

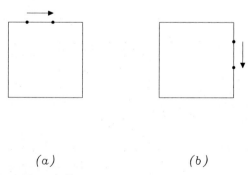

(a)   (b)

Fig. 3.19. Conjugate translations.

Thus, the total effect is simply the translation shown in Fig. 3.18.

The next figure, Fig. 3.19, illustrates two conjugate translations. If the top translation is T_{c_1}, the right translation is $r_{90}\ T_{c_1}\ r_{90}^{-1}$. The map that relates these two translations is $\hat{\tau}(r_{90})$. That is:

$$\hat{\tau}(r_{90}) \quad : \quad T_{c_1} \quad \longrightarrow \quad r_{90}\ T_{c_1}\ r_{90}^{-1} \ .$$

Now, if the above argument is applied to all the members of the fiber-group copy $G(F)_{c_1}$ on the top side, it is clear that the conjugation will send $G(F)_{c_1}$ to the fiber-group copy $G(F)_{c_2}$ on the right side:

$$G(F)_{c_1} \quad \overset{r_{90}\ -\ r_{90}^{-1}}{\longrightarrow} \quad G(F)_{c_2} \ .$$

Again, this is the same thing as the map $\hat{\tau}(r_{90})$. Generally, our conclusion is this: *Conjugation transfers the fiber-group copies from one fiber-set copy to another.* We end this section by proving this statement.

Theorem 3.2. *Transfer is conjugation.*

Proof: Strictly speaking $G(C)$ is not a subgroup of $G(F)\textcircled{w}G(C)$. It is *isomorphic* to the subgroup $G(C)^*$ of $G(F)\textcircled{w}G(C)$, where $G(C)^*$ denotes the set of all those elements of $G(F)\textcircled{w}G(C)$ that effect only the *second* coordinate in the pair (f,c), i.e., only the control coordinate c. Let g^* be a member of $G(C)^*$ and let ϕ^* be a vector (!!!) in the fiber-group product $G(F)_{c_1} \times G(F)_{c_2} \times \ldots \times G(F)_{c_n}$, where the fiber-group copies have been indexed by the members of the control set C. Let the symbol g represent the element of $G(C)$ that corresponds to g^* under the obvious isomorphism $G(C)^* \longrightarrow G(C)$. One now sees this:

$$\begin{aligned}
g^*\ \phi^*\ g^{*\ -1}(f,c) &= g^*\ \phi^*\ (f,g^{-1}c) \\
&= g^*\ (\phi_{g^{-1}c}f, g^{-1}c) \\
&= (\phi_{g^{-1}c}f, gg^{-1}c) \\
&= (\phi_{g^{-1}c}f, c)
\end{aligned}$$

Notice that, in going from the left side to the right side of the first line, the point (f,c) has been changed to $(f,g^{-1}c)$, which means that it has moved from the fiber-set copy F_c backwards to the fiber-set copy $F_{g^{-1}c}$. Then notice that, in moving from the first line to the second, we see the selection of component $\phi_{g^{-1}c}$ (in the second line) from vector ϕ^* (in the first line) , because we are now in the fiber-set copy $F_{g^{-1}c}$, that is, the control coordinate at the end of the first line is $g^{-1}c$. Finally, in the last line, we see that component $\phi_{g^{-1}c}$, originally in fiber-group copy $G(F)_{g^{-1}c}$, is now in the fiber-group copy $G(F)_c$ because c is the control-set coordinate in the last line. Clearly therefore, the fiber-group copy $G(F)_{g^{-1}c}$ has been *transferred* from the control position $g^{-1}c$ into the control position c. ∎

3.18   Conjugation and Recoverability

We shall now see that the two foundations of our generative theory - recoverability and transfer - are deeply linked. This becomes clear by considering again how transfer is represented by conjugation, as follows:

Return to the example in the last section. The translation on the second side was described as the transfer of a translation on the first side. We saw that this meant that the translation on the second side was described as the conjugate $r_{90}\, T_{c_1}\, r_{90}^{-1}$ of the translation T_{c_1} on the first side. The conjugation representation is a three-stage process: (1) apply r_{90}^{-1} to rotate backwards to the top side, (2) carry out the translation T_{c_1} on the top side, and (3) apply r_{90} to rotate the result back to the right side.

This means that the translation on the second side depends on *recovering* the translation on the first side. Therefore: *Recoverability is basic to transfer.*

Since transfer requires recoverability, it must obey the laws of recoverability established in Chapter 2. Must crucially, it must obey the Asymmetry Principle (p. 42) which states that, given a data set $\mathcal{D}$, a program for generating $\mathcal{D}$ is *recoverable* from $\mathcal{D}$ only if the program is symmetry-breaking on the successively generated states (i.e., in the forward-time direction). Equivalently, in the backward-time direction, the Asymmetry Principle, states that, given a data set $\mathcal{D}$, a program for generating $\mathcal{D}$ is recoverable from $\mathcal{D}$ only if each asymmetry in $\mathcal{D}$ goes back to a symmetry in one of the previously generated states.

The way the Asymmetry Principle is implemented depends on whether the data set supports external or internal inference (recall Sect. 2.9). In external inference, we do not require the records of past generative states within the data set. In internal inference, the data set must contain the records of past generative states.

Thus, consider a conjugacy automorphism representing transfer. It has the structure:

$$\hat{\tau}(g)\; :\; T_{c_i} \;\longrightarrow\; g\, T_{c_i}\, g^{-1}\; . \tag{3.21}$$

The right-most operation g^{-1} performs the recovery, i.e., it "back-projects" to the past state in the generative history. This operation must obey the Asymmetry Principle, i.e., it must be asymmetry-reducing, backwards in time. The data set will support either external inference or internal inference of this operation. In the case of the square, the data set supports only internal inference. This is because the square has no deformational structure, i.e., it is characterized entirely as a trace structure given by an iso-regular group. Thus, each side can have undergone only *rigid movement*. But this means that the inference of the preceding state of any side is possible only if the preceding state is present within the data set since there can be no other cues to movement (recall sections 2.10 and 2.11). Therefore, the past state of the right side must be visible within the data set (i.e., as the top side). We conclude therefore that the data set supports only an internal inference of the operation g^{-1} in expression (3.21).

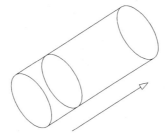

Fig. 3.20. The sweep structure of a cylinder.

In contrast, consider a parallelogram. This is not given by an iso-regular group, and therefore contains deformation. By deformation, one means that the present state implies a past state that is an iso-regular group (in accord with the Externalization Principle, Sect. 2.11). Thus, we do not need an actual record of the past state (the iso-regular group) within the data set. We conclude therefore that this data set (the parallelogram) supports external inference of the operation g^{-1} in expression (3.21).

3.19 Infinite Control Sets

Obviously, in the case of the square, the control set of side positions $\{\ c_1\ ,\ c_2\ ,\ c_3\ ,\ c_4\ \}$ is finite. However, in many cases, the control set can be infinite. An example of this is a cylinder, as shown in Fig. 3.20, where the fiber group $SO(2)$ corresponds to an individual cross-section, and the control set is the set of positions along the axis $\mathbb{R}$. Since there is one copy of the cross-section fiber group at each position along the axis, the fiber-group product must be the direct product of an uncountably infinite number of copies of the fiber group. The direct product structure in such cases is as follows:

A vector from the fiber-group product is constructed as a *formal product*, $\prod_{c_i \in C} T_{c_i}$, where any such product is a selection of one member T_{c_i} from each of the fiber-group copies $G(F)_{c_i}$. It is clear that the set of all such formal products forms a group, where the product rule $*$ is defined thus:

$$\prod_{c_i \in C} T_{c_i} * \prod_{c_i \in C} S_{c_i} = \prod_{c_i \in C} T_{c_i} S_{c_i}.$$

The group is called the *unrestricted direct product*. To see its relation to direct products generally, recall the definition of direct product: If G is a group, and $\{G_{c_i}\}_{c_i \in C}$ a set of its subgroups, indexed by the set C, then G is called the direct product of $\{G_{c_i}\}_{c_i \in C}$ if the following conditions hold:

(1) $G_{c_i} \triangleleft G, \forall c_i \in C$
(2) $G = \langle \{G_{c_i}\}_{c_i \in C} \rangle$
(3) $G_{c_i} \cap \langle \{G_{c_j} | c_j \in C \backslash c_i\}_{c_j \in C} \rangle = e, \forall c_i \in C.$

What should be observed is that, if C is infinite, then the fiber-group product, as an *unrestricted direct product*, will satisfy conditions (1) and (3) but *not* (2). In this book, all wreath products with infinite control sets will be based on a fiber-group product as an unrestricted direct product.

In any expression involving the selective effect in a wreath product, one needs to show a vector from the fiber-group product. This is impossible to notate in the situation of an uncountably infinite control set. Thus, in such a situation (and also, in any situation where the control set is countably infinite) we will often represent the vector as (T_{c_1} , T_{c_2} , ... , T_{c_n}) and the fiber-group product as $G(F)_{c_1} \times G(F)_{c_2} \times \ldots \times G(F)_{c_n}$ which is a notation that *looks* finite and certainly countable. Nevertheless, c_n will be allowed to stand for one of the infinite cardinals, and the countable appearance of the notation can be notational abuse. This will be called the **finitistic notation**, which refers only to the *appearance* of the notation, not the actual size of the control set involved.

3.20   The Full Structure

We now have all the tools to present the five-stage definition of transfer. In the particular case that the control set is infinite, the reader should understand that the finitistic notation (Sect. 3.19) is being used.

ALGEBRAIC STRUCTURE OF TRANSFER

STAGE 1: Define the Fiber Level and Control Level

The data set is decomposed into two levels:

LEVEL 1.

(1) $F = $ Fiber set
(2) $G(F) = $ Fiber group
(3) Define the **group action** of $G(F)$ on F.

LEVEL 2.

(1) $C = $ Control set $= \{ c_1$, c_2 , ... , $c_n \}$
(2) $G(C) = $ Control group
(3) Define the **group action** of $G(C)$ on C.

STAGE 2: Define the Fiber-Group Product

In the previous stage, one of the two actions defined was that of the fiber group on the fiber set. However, $G(F)$ is now used to define an action on the *entire* data set, as follows:

(1) Form the direct product $[G(F)_{c_1} \times G(F)_{c_2} \times \ldots \times G(F)_{c_n}]$ consisting of copies of the fiber group, $G(F)$, one copy for each member of the control set. This is called the *fiber-group product*.

(2) Define the **group action** of the fiber-group product on the entire data set $F \times C$, by making each copy of $G(F)_{c_i}$ act on each copy of the fiber set *independently*. The action therefore has the *selective effect*.

(3) This action is a symmetry of the data set. It is the total symmetry due to the fiber group alone.

(4) Notice that this stage represents the situation *before the detection of the control group*. It gives the results of purely *parallel computation*.

STAGE 3: Define $G(C)$ as an Automorphism Group on $G(F)_{c_1} \times G(F)_{c_2} \times \ldots G(F)_{c_n}$

The previous stage encoded the *independent* action of the fiber-group copies on the fiber sets. However, it failed to notice that these copies are equivalent to each other. That is, the copies are related to each other by the control group $G(C)$. In other words, *the control group sends the copies of the fiber group around the data set, from one fiber set copy to another*. This is achieved in the following way:

(1) There is a **group action** of $G(C)$ on the fiber-group product. The action is either raised directly (the τ-representation), or it is raised indirectly (the $\hat{\tau}$-representation).

(2) This makes $G(C)$ an *automorphism group* of the fiber-group product. The automorphisms act by *conjugation*.

(3) This action of $G(C)$ on $[G(F)_{c_1} \times G(F)_{c_2} \times \ldots \times G(F)_{c_n}]$ is called the *control-nesting* of $G(F)$ within $G(C)$.

The structure just defined is *dichotomous*: That is, in its group action, $G(C)$ preserves $[G(F)_{c_1} \times G(F)_{c_2} \times \ldots G(F)_{c_n}]$, but nevertheless sends the copies $G(F)_{c_i}$ from one fiber set, to another. This dichotomous structure is therefore the basis of *transfer*.

STAGE 4: Define a Splitting Extension of $[G(F)_{c_1} \times G(F)_{c_2} \times \ldots \times G(F)_{c_n}]$ by $G(C)$

One can regard the previous two stages as the successive detection of (i) the fiber-group product and then (ii) the control group. This successive detection can be modeled as a group extension of the fiber-group product by the control group. That is, we do this:

(1) Create the semi-direct product:

$$[G(F)_{c_1} \times G(F)_{c_2} \times \ldots \times G(F)_{c_n}] \circledS_\tau [G(C)].$$

In this structure, only *two basic groups* are used: They are $G(F)$, which is repeated a number of times to form the fiber-group product, and $G(C)$ which is then added to the fiber-group product.

(2) The creation of the entire structure from these two basic groups, $G(F)$ and $G(C)$, is called the *wreath product* of $G(F)$ and $G(C)$:

$$G(F) \, \textcircled{w} G(C).$$

STAGE 5: Define the Action of $G(F) \, \textcircled{w} G(C)$ on $F \times C$

The wreath product $G(F) \, \textcircled{w} G(C)$ is itself a group. It is the control-nested symmetry group of the data set, thus:

(1) Define the **group action** of $G(F) \, \textcircled{w} G(C)$ on the data set $F \times C$. This group action is the hierarchical composite of two previous group actions: (i) the group action of $[G(F)_{c_1} \times G(F)_{c_2} \times \ldots \times G(F)_{c_n}]$ on the data set, and (ii) the group action of the $G(C)$ on $[G(F)_{c_1} \times G(F)_{c_2} \times \ldots \times G(F)_{c_n}]$.

(2) An element of $G(F) \, \textcircled{w} G(C)$ is called a *control-nested symmetry* of the data set.

3.21   The Five Group Actions

Although, the previous section defined the algebraic structure of *two*-level transfer, this actually involves *five group actions*! These five actions contain the real power of any structure of transfer. One should think of the five actions as locked together in a single tightly-knit organization that they mutually help to construct. The actions are as follows:

THE 5 GROUP ACTIONS IN 2-LEVEL TRANSFER

Action 1. The Fiber Group on the Fiber Set
Action 2. The Control group on the Control Set
Action 3. The Fiber-Group Product on the Data Set
Action 4. The Control Group on the Fiber Group Product
Action 5. The Wreath Product on the Data Set.

The five group actions are defined in sequential order over the five stages. Each stage defines one group action, except the first stage which defines two group actions and the fourth stage which, to compensate, defines no group action. The actions are defined thus:

Action 1. The Action of the Fiber Group on the Fiber Set

$$\begin{cases} G(F) & \times & F & \longrightarrow & F \\ (\ T & , & f\ ) & \longmapsto & Tf \end{cases}$$

Action 2. The Action of the Control group on the Control set

$$\begin{cases} G(C) & \times & C & \longrightarrow & C \\ (\ g & , & c_i\ ) & \longmapsto & gc_i \end{cases}$$

Action 3. The Action of the Fiber-Group Product on the Full Data Set

$$\begin{cases} [G(F)_{c_1} \times G(F)_{c_2} \times \dots \times G(F)_{c_n}] & \times & [F \times C] & \longrightarrow & [F \times C] \\ (\ (T_{c_1}, T_{c_2}, \dots, T_{c_n}\ ) & , & (f, c_i)\ ) & \longmapsto & (T_{c_i}f, c_i) \end{cases}$$

where the selective effect is seen in the second line.

Action 4. The Action of the Control group on the Fiber-Group Product

This is the action of the control group that is raised from the control set to the fiber-group product, that is, it is the raised version of Action 2.

$$
\begin{cases}
G(C) \quad \times \quad [G(F)_{c_1} \times G(F)_{c_2} \times \ldots \times G(F)_{c_n}] \quad \longrightarrow \\
\qquad\qquad\qquad\qquad\qquad\qquad [G(F)_{c_1} \times G(F)_{c_2} \times \ldots \times G(F)_{c_n}] \\[2ex]
(\quad g \quad, \qquad\quad (\, T_{c_1},\, T_{c_2},\, \ldots,\, T_{c_n}\,)\,) \qquad \longmapsto \\
\qquad\qquad\qquad\qquad\qquad\qquad (\, T_{g c_1},\, T_{g c_2},\, \ldots,\, T_{g c_n}\,)
\end{cases}
$$

Action 5. The Action of the Wreath Product on the Full Data Set.

This is created as the hierarchical composite of Action 3 and Action 4.

$$
\begin{cases}
G(F) \textcircled{w} G(C) \qquad\qquad \times \quad [F \times C] \quad \longrightarrow \quad [F \times C] \\[2ex]
(\quad \langle\,(\, T_{c_1},\, T_{c_2},\, \ldots,\, T_{c_n}\,)\mid g\,\rangle \quad,\quad (f, c_i)\,) \quad \longmapsto \quad (T_{c_i} f,\, g c_i)
\end{cases}
$$

4. Mathematical Theory of Transfer, II

4.1 Introduction

In this chapter, we will present our *general* formalization of transfer. This extends the concept of a τ-automorphism, from the 2-level case, to the n-level case. In the latter case, the transfer automorphism is, itself, a multi-level structure in which transfer is going on within several layers simultaneously. This will become fundamental for example to understanding robotics, where limbs are being transferred simultaneously in several levels, or perceptual organization, which we argue is structured in the same way. The multi-level structure of transfer will be algebraically formalized as a *control-nested* (i.e., multi-level) τ-automorphism. An *ordinary* τ-automorphism (Definition 3.2, p. 89) has a control-nested relation with its fiber group. However, a *control-nested* τ-automorphism has a control-nested hierarchy within itself. Correspondingly the automorphism group is structured in this way. Once again, formalizing transfer in terms of an automorphism group has a number of profound advantages, the most important being the fact that all the transferred copies are contained within one group, together with the algebraic relationships involved.

4.2 The Iterated Wreath Product

Control-nested transfer automorphisms will be understood as arising in n-fold wreath products, where $n > 2$. Thus we first need a definition of an n-fold wreath product. This is obtained by iterating the wreath-product operation. This operation is a 2-level operation. Using it on two groups G_1 and G_2, produces $G_1 \textcircled{w} G_2$. Thus, to add a third group G_3, the operation is applied again, as follows:

$$(G_1 Ⓦ G_2) Ⓦ G_3.$$

The parentheses indicate that $G_1 Ⓦ G_2$ is a fiber group relative to the second wreath-product symbol Ⓦ, and G_3 is the control group of that symbol. The hierarchy of control therefore ascends from left to right.

In the n-fold case, one gets this

$$(G_1 Ⓦ G_2) Ⓦ G_3) Ⓦ G_4) Ⓦ G_5) Ⓦ \ldots) Ⓦ G_n. \tag{4.1}$$

For readability, all the parentheses have been removed from the *left* end, except one. In subsequent chapters, the parentheses across the entire sequence will usually be omitted. However, it will be useful, in the current chapter, to include those shown above.

It is important to understand that the n-fold wreath product is a group. This is because the 2-place operation Ⓦ creates a single group from two groups, and the n-fold wreath product is simply the iterated use of the 2-place operation.

Notice that any group G_i along the n-fold sequence acts to its left, in fact, on the *entire hierarchy* to its left. This hierarchy will be called the **left-subsequence** of G_i. The relation between G_i and its left-subsequence is simply that between a control group and its fiber group.

4.3   Opening Up

It is necessary to unpack the information contained in the n-fold structure. First, of course, any 2-fold wreath product

$$G(F) Ⓦ G(C) \tag{4.2}$$

is really the semi-direct product

$$[G(F) \times G(F) \times \ldots \times G(F)] \; Ⓢ_\tau \; G(C). \tag{4.3}$$

The expressions (4.2) and (4.3) are simply two different notations for the same group, and they will be referred to respectively as the **wreath product notation** and **semi-direct product notation** of the **wreath product**. The semi-direct product notation clearly reveals more of the structure of the wreath product. In the discussion, we shall often pass from one notation to the other. When passing from the first notation to the second, we shall say that we are **opening up** the wreath product symbol. Conversely, when passing from the second notation to the first, we shall say that we are **closing down** the semi-direct product symbol.

Let us now use the opening-up procedure to unpack the information in an n-fold wreath product.

$$(G_1 \textcircled{w} G_2) \textcircled{w} G_3) \textcircled{w} G_4) \textcircled{w} G_5) \textcircled{w} \ldots) \textcircled{w} G_n. \qquad (4.4)$$

When one opens up the first (left-most) wreath product symbol in this hierarchy, one gets this:

$$([G_1 \times G_1 \times \ldots \times G_1] \textcircled{s}_\tau G_2) \textcircled{w} G_3) \textcircled{w} G_4) \textcircled{w} G_5) \textcircled{w} \ldots) \textcircled{w} G_n.$$

Most crucially, there are now *three* group-product symbols in this sequence: the direct product $\times$, the semi-direct product $\textcircled{s}_\tau$ and the wreath product $\textcircled{w}$. Examining the sequence, the following rule is evident:

PRODUCT-HIERARCHY TABLE

$\times$ *links two fiber groups on the same level*
$\textcircled{s}_\tau$ *links a fiber-group product on left to a control group on right*
$\textcircled{w}$ *links a fiber group on left to a control group on right.*

Notice that the direct product operation $\times$ is non-hierarchical, whereas the other two product operations are hierarchical.

This also has consequences for the numbers i that are the *indexes* on the groups along the sequence. The relation between the indexes give a strong sense of how the three operations $\times$, $\textcircled{s}_\tau$, and $\textcircled{w}$, glue the hierarchy together:

PRODUCT-INDEX TABLE

$\times$ *index on left is same as on right*
$\textcircled{s}_\tau$ *index on left is one lower than on right*
$\textcircled{w}$ *index on left is one lower than on right.*

Now let us open up the second wreath-product symbol in the n-fold wreath product (4.4). Clearly, if all levels higher than 3 are ignored, we get:

$$[(G_1 \textcircled{w} G_2) \times (G_1 \textcircled{w} G_2) \times \ldots \times (G_1 \textcircled{w} G_2)] \textcircled{s}_\tau G_3.$$

Substituting this into the n-fold wreath product, we get:

$$[(G_1 \textcircled{w} G_2) \times (G_1 \textcircled{w} G_2) \times \ldots \times (G_1 \textcircled{w} G_2)] \textcircled{s}_\tau G_3) \textcircled{w} G_4) \textcircled{w} G_5) \textcircled{w} \ldots) \textcircled{w} G_n.$$
$$(4.5)$$

The reader is recommmended to apply the Product-Hierarchy Table and Product-Index Table above to this sequence, to fully understand it.

Now, when the first wreath product symbol in (4.5) is opened up, the result is:

$$[(([G_1 \times G_1 \times \ldots \times G_1] \textcircled{s}_\tau G_2) \times ([G_1 \times G_1 \times \ldots \times G_1] \textcircled{s}_\tau G_2) \times \ldots \times$$
$$([G_1 \times G_1 \times \ldots \times G_1] \textcircled{s}_\tau G_2)]$$
$$\textcircled{s}_\tau G_3) \textcircled{w} G_4) \textcircled{w} G_5) \textcircled{w} \ldots) \textcircled{w} G_n. \qquad (4.6)$$

Again it is worth the reader checking this expression against the two tables given earlier.

To examine further the above structure, let us assign cardinalities to the groups involved. For ease of exposition, it will be assumed that each of the group components G_i is finite, and that its order is:

$$|G_i| = m_i.$$

Also, it will be assumed that the wreath product in each case is a regular wreath product; i.e., the control set is the control group itself. (Such products were discussed in Sect. 3.14.)

Let us now return to the sequence (4.6), which we wish to study. On the far left, there is the fiber-group product $G_1 \times G_1 \times \ldots \times G_1$. This is the *first level* of fibers G_1. Notice that this product consists of m_2 copies of G_1, since there is one copy for each element in its immediate control group G_2.

Progressing one step rightward along the sequence (4.6), we encounter the semi-direct product symbol $\circledS_\tau$, which indicates that we are going up one level, because the product symbol $\circledS_\tau$ is hierarchical. That is, following $\circledS_\tau$, there is G_2. This completes the first hierarchy along the sequence; i.e.,

$$[G_1 \times G_1 \times \ldots \times G_1]\circledS_\tau G_2.$$

Now moving rightward along the sequence (4.6), it is evident that the hierarchy just given is actually copied a number of times. In fact, it is copied m_3 times because there are m_3 elements in the immediate control group G_3. We therefore conclude that the initial m_2-fold duplication of G_1 is itself duplicated m_3 times; that is, G_1 is actually duplicated $m_2 \times m_3$ times. Thus, there is a **duplication of duplication** phenomenon. This is basic to an n-fold wreath product.

Thus, progressing rightwards, each new control group G_i duplicates the entire hierarchy to its left, m_i times. In particular, the first group G_1 now has $m_2 \times m_3 \times \ldots \times m_i$ copies. In fact, generally, we have this: In an n-fold wreath product, the hierarchy up to level i, that is, $G_1\textcircled{w}\ldots\textcircled{w}G_i$, occurs a total of $m_{i+1} \times m_{i+2} \times \ldots \times m_n$ times. This is because the groups $G_{m_{i+1}}$, $G_{m_{i+2}}, \ldots, G_n$, comprising the *right-subsequence* of the fiber $G_1\textcircled{w}\ldots\textcircled{w}G_i$ have successively $m_{i+1}, m_{i+2}, \ldots, m_n$ elements. Note also: Because the semi-direct product notation exhibits the fiber-group product, we shall say that it **reveals** the duplicated copies of the fiber.

4.4   The Group Theory of Hierarchical Detection

Let us return to 2-fold nested control. In Sect. 3.9, we modeled the successive *detection* of the fiber-group product and the control group as the splitting extension of the first group by the second.

It will now be seen that there are phenomena that arise in the successive detection of n-fold nested control that did not arise in the 2-fold case, even though the n-fold structure is built entirely out of the 2-fold case. We argue that these are group-theoretic phenomena that lie at the foundations of perception organization and scientific measurement.

Consider first the 2-fold situation. The two levels will be represented by Fig. 4.1. The dots on Level 1 are the copies of the fiber group G_1 and the single dot on Level 2 is the control group G_2. It will be assumed that the wreath product is a regular one, i.e., there are as many copies of G_1, on Level 1, as there are elements within the group G_2 on Level 2. Note also that G_2 moves the copies of G_1 around onto each other under the transfer automorphims. In the remainder of this section, Fig. 4.1 will be called a *tree*.

Fig. 4.1. Relation of control level to fiber level, in a 2-fold wreath product.

With this diagram in mind, let us now consider the n-fold situation. In exact analogy with what has just been said about the 2-fold situation, let us consider two successive levels in the n-fold situation, for example, Level i and Level $i + 1$. There are a set of copies of G_i on Level i, that are connected to a single copy of G_{i+1} on the next level. In other words, the diagram is like that shown in Fig. 4.1.

However, there is an extra feature: On the upper level, there are several additional copies of G_{i+1}, one for each member of G_{i+2} on the level *above this*. So the diagram for Level i and $i + 1$ is really that shown in Fig. 4.2.

Fig. 4.2. Two successive levels within an n-fold wreath product.

In fact, there is a still further feature: We have assumed that there is only one copy of the G_{i+2} on the level above the levels shown. In fact there are several additional copies of G_{i+2} on that level, one copy for each member of G_{i+3} *two levels* above the levels shown. So the *entire* structure indicated in Fig. 4.2 is really duplicated several times *horizontally*, one time for each member of G_{i+3}.

This process keeps on going. For each level G_{i+k}, further up the hierarchy, there is an entire set of duplications of the structure horizontally, one duplication for each member of G_{i+k}. Thus, in accord with the previous section, the total number of dots on the lower level in Fig. 4.2, i.e., the total number of copies of G_i , will actually be $m_{i+1} \times m_{i+2} \times \ldots \times m_n$.

Now consider one of the individual *trees* shown in Fig. 4.2. Within each such tree, there is the single G_{i+1} on the upper level, and the copies of G_i on the lower level. Most crucially, the G_{i+1}, can move only *these* copies of G_i onto each other. The G_{i+1} cannot move the copies of G_i from one tree to another.

What can move the copies of G_i from one tree to another is a higher control group G_{i+k}. Let us look carefully at this. There are $m_{i+1} \times m_{i+2} \times \ldots \times m_n$ copies of G_i along the bottom of Fig. 4.2. These are partitioned into trees as shown in Fig. 4.2, by the individual copies of G_{i+1} on the next level. On the level above this (not shown), the trees themselves are partitioned into blocks of trees by the individual G_{i+2}, to which they are connected. Using G_{i+2}, one can now move the G_i, on the bottom level, out of one tree into another within the same block.

If one wants to move the G_i horizontally out of the block of trees into another block of trees, one has to go to a still higher level, and so on. To move a G_i horizontally from *any* position x to *any other* position y, enough levels have to be added above to get a tree high enough to encompass both positions x and y.

Also observe that each G_i shown along the bottom of Fig. 4.2 is itself the head of its own tree. This is the tree $G_1 ⓦ \ldots ⓦ G_i$ which is the *fiber group* of the control group G_{i+1}, a node on the upper level of Fig. 4.2. Thus the reader should think of the tree $G_1 ⓦ \ldots ⓦ G_i$ as hanging down from each lower node in Fig. 4.2; and the upper G_{i+1} as moving these trees around onto each other within the set dominated by G_{i+1}.

Note therefore that, because there are $m_{i+1} \times m_{i+2} \times \ldots \times m_n$ copies of G_i along the bottom of Fig. 4.2, there are $m_{i+1} \times m_{i+2} \times \ldots \times m_n$ copies of the fiber group $G_1 ⓦ \ldots ⓦ G_i$ along the bottom of Fig. 4.2.

Now, it is part of the definition of nested control - in fact in Stage 1 of that definition - that a *fiber group acts on a fiber set*. Furthermore, in Stage 2, we said that each copy of the fiber group acts on its own copy of the fiber set. That is, *the data set is partitioned into copies of the fiber set* and, on each copy of the fiber set, there acts only the copy of the fiber group that has been assigned to it.

The implications for Fig. 4.2 are as follows: It was seen that there are $m_{i+1} \times m_{i+2} \times \ldots \times m_n$ copies of the fiber group $G_1 ⓦ \ldots ⓦ G_i$ along the bottom of Fig. 4.2. Each copy has its own copy of the fiber set on which it acts. The conclusion therefore is that the data set is partitioned into $m_{i+1} \times m_{i+2} \times \ldots \times m_n$ copies of this fiber set, each with a copy of $G_1 ⓦ \ldots ⓦ G_i$ acting on it. In terms of Fig. 4.2, each node along the bottom level shown corresponds to its own particular subset of the data set. The node acts on that subset and only that subset.

Observe that, once the upper level shown in Fig. 4.2 has been detected, this level can move the lower-level subsets onto each other, if these lower-level subsets occur within the same tree. However, before this detection, the only action is that of a lower-level node on its own independent subset.

Let us therefore consider the detection problem. Suppose that we have only detected up to (and including) Level i, which is the lower level shown in Fig. 4.2. This means that we do not yet have Level $i+1$, that binds these nodes themselves into larger subsets. That is, by removing the upper level of Fig. 4.2, we are simply left with the lower level, and this is depicted in Fig. 4.3. Each node in Fig. 4.3 can be regarded as representing the fiber group $G_1 ⓦ \ldots ⓦ G_i$, and each acts *independently* on its own particular subset of the data set.

Fig. 4.3. Detection only up to, and including Level i.

How shall we *group-theoretically* describe the structure shown in Fig. 4.3? Because there are $m_{i+1} \times m_{i+2} \times \ldots \times m_n$ independently acting copies of $G_1 ⓦ \ldots ⓦ G_i$, along Fig. 4.3, there is a *direct product* of these copies. This direct product will be notated thus:

$$[G_1 ⓦ \ldots ⓦ G_i]^{m_{i+1} \times m_{i+2} \times \ldots \times m_n}.$$

Most crucially, before any higher level groups G_{i+1}, G_{i+2}, $\ldots$, G_n are detected, the direct-product group just defined must be the *entire* detected structure of the full data set. We therefore have:

Definition 4.1. *The direct product*
$Det_i = [G_1 ⓦ \ldots ⓦ G_i]^{m_{i+1} \times m_{i+2} \times \ldots \times m_n}$ *will be called the* **detected symmetry group** *up to Level i.*

Notice some simple features about the detected symmetry groups. When we detect the next level above Det_i, the detected symmetry group becomes

$$[G_1 ⓦ \ldots ⓦ G_i ⓦ G_{i+1}]^{m_{i+2} \times \ldots \times m_n}$$

where the wreath product sequence has *added* a new factor G_{i+1}, but the power has *lost* the factor m_{i+1}.

The difference between the two *successive* detected symmetry groups Det_i and Det_{i+1} is the single horizontal row of G_{i+1} along Level $i+1$. This row is the direct product of $m_{i+2} \times \ldots \times m_n$ copies of G_{i+1}, that is

$$G_{i+1}^{m_{i+2} \times \ldots \times m_n}.$$

Thus, it is easy to see, group-theoretically, that this row can be obtained as the factor group:

$$Det_{i+1}/Det_i = G_{i+1}^{m_{i+2} \times \ldots \times m_n}.$$

Clearly, this is what was detected in order to raise the detected symmetry group from Det_i to Det_{i+1}.

GROUP-THEORETIC CHARACTERIZATION OF DETECTION. *The detection of Level $i+1$ is a splitting extension of* Det_i *by* $G_{i+1}^{m_{i+2} \times \ldots \times m_n}$.

We will show later that this characterizes detection in scientific experiments and perceptual organization.

It is easy to show that each detected symmetry group Det_i is a *normal subgroup* of the next higher detected symmetry group Det_{i+1}. To see this, observe that each node G_{i+1} in the upper row in Fig. 4.2, acts via conjugation as an automorphism group on a copy of the fiber-group product $[G_1 \textcircled{w} \ldots \textcircled{w} G_i]^{m_{i+1}}$, which is formed from the set of nodes it dominates. This automorphism action is easily extended to an automorphic action of the entire upper row $G_{i+1}^{m_{i+2} \times \ldots \times m_n}$ on the entire (partitioned) lower row. Therefore Det_i is a normal subgroup of Det_{i+1}.

Thus the hierarchy of detected symmetry groups

$$Det_1 \lhd Det_2 \lhd Det_3 \lhd \ldots \lhd Det_n$$

forms a *subnormal series*.[1] In fact, further consideration reveals that each Det_i is normal in the full group Det_n. That is, simply by extending the conjugation argument, each successively higher sequence $[G_{i+1} \textcircled{w} \ldots \textcircled{w} G_k]^{m_{k+1} \times \ldots \times m_n}$, for k increasing up to n, must be acting via conjugation as an automorphism group of Det_i where the automorphic action must be control-nested. Therefore we conclude:

HIERARCHY OF DETECTED SYMMETRY GROUPS. *In an n-fold control-nested hierarchy $G_1 \textcircled{w} \ldots \textcircled{w} G_n$, the detected symmetry groups form a group-theoretic* **normal series,** *thus:*

$$Det_1 \lhd Det_2 \lhd Det_3 \lhd \ldots \lhd Det_n.$$

Now observe the following: Before Level $i+1$ has been detected, each of the $m_{i+1} \times m_{i+2} \times \ldots \times m_n$ copies of the $G_1 \textcircled{w} \ldots \textcircled{w} G_i$ in Det_i are observed

---

[1] In group theory, the definitions of subnormal series and normal series are as follows: Let G be a group. Then the sequence

$$H_1 \lhd H_2 \lhd H_3 \lhd \ldots \lhd H_n$$

is called a *subnormal series* of G, if each H_i is a subgroup of G, and each H_i is a normal subgroup of H_{i+1}. The subgroups H_1 and H_n, at the ends of the sequence, are the trivial subgroups $\{e\}$ and G respectively. The series is called a *normal series* if, in addition to these conditions, each H_i is normal in the full group G.

as unrelated to each other, i.e., it is only the higher level groups G_{i+1}, G_{i+2}, ..., G_n that will detect the relationships between the copies. Furthermore, any copy of a lower fiber group $G_1 ⓦ \ldots ⓦ G_j$, where j is below the highest detected Level i within Det_i, is related to its associated copies within $G_1 ⓦ \ldots ⓦ G_{j+1}$ (which is within Det_i). Therefore, the only unrelated objects within Det_i are the copies of G_i (these are the top level of Det_i). The copies of G_i are thus the only objects within Det_i that correspond to computations that are purely *parallel*. We therefore have this definition:

Definition 4.2. *In the detected symmetry group Det_i, the direct product $G_i^{m_{i+1} \times \ldots \times m_n}$ constitutes the* **parallel computation level.**

4.5   Control-Nested τ-Automorphisms

In this section, we will give our general formalization of transfer. This extends the concept of a τ-automorphism, from the 2-level case, to the n-level case. In the latter case, the transfer automorphism is, itself, a multi-level structure in which transfer is going on within several layers simultaneously. This will become fundamental for example to understanding robotics, where limbs are being transferred simultaneously in several levels, or perceptual organization, which we argue is structured in the same way. The multi-level structure of transfer will be algebraically formalized as a *control-nested* (i.e., multi-level) τ-automorphism. An *ordinary* τ-automorphism (Definition 3.2, p. 89) has a control-nested relation with its fiber group. However, a *control-nested* τ-automorphism has a control-nested hierarchy within itself. Correspondingly the automorphism group is structured in this way. Once again, formalizing transfer in terms of an automorphism group has a number of profound advantages, the most important being the fact that all the transferred copies are contained within one group, together with the algebraic relationships involved.

A control-nested τ-automorphism is illustrated in Fig. 4.4. Each node at the bottom level of this figure represents a copy of the fiber group $G_1 ⓦ \ldots ⓦ G_i$. On the level above this, each node represents a copy of the group G_{i+1}; on the level above this, each node represents a copy of the group G_{i+2}; and so on ... till the single node at the top level which represents the final control group G_n. In this section, for notational convenience, the permutational wreath product will be used, in which each group G_j acts on a control set C_j. Thus, pick any node in Fig. 4.4. It is a copy of a group G_j. It dominates a set of nodes below it. These are the copies of G_{j-1} corresponding to the members of the control set C_j. The copy of G_j can permute this set of copies of G_{j-1}. This permutational action is indicated by the circular arrow relating these copies. Of course, on level j there are several copies of G_j, each with its own copy of the control set C_j. Therefore each dominates a set of copies of G_{j-1}, where the set corresponds to a *copy* of the control set C_j.

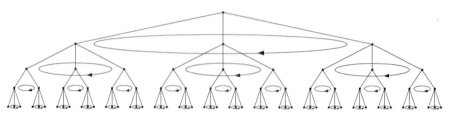

Fig. 4.4. The action of a control-nested τ-automorphism.

In fact, let each circular arrow in the hierarchy be some chosen *element* from the control group above it. Then the entire collection of circular arrows, shown in Fig. 4.4 (one element selected from each node), represents what we will call a *control-nested* τ-automorphism. Clearly, it is acting on several levels. This will become important for robot manipulation, perceptual organization, etc.

On what does this collection of arrows act automorphically? On the bottom level, which represents the detected symmetry group Det_i. It is this automorphic action that will allow us to move things around an environment.

Let us now rigorously formalize this. Choose any level j (above the bottom level). The set of nodes in this level is simply the collection of copies of G_j. In fact, the algebraic structure of the level is the direct product of these copies, that is:

$$G_j^{m_{j+1} \times \dots \times m_n}.$$

Now, from each of these copies of G_j, we have selected one element, indicated by the circular arrow below it, in Fig. 4.4. Thus an entire vector of elements has been selected from the direct product $G_j^{m_{j+1} \times \dots \times m_n}$. The vector is of length $m_{j+1} \times \dots \times m_n$. It will be denoted by $\bar{g}_j$,

$$\bar{g}_j \ \in \ G_j^{m_{j+1} \times \dots \times m_n}.$$

Fig. 4.4 shows one such vector for each level above the bottom level. Putting all these vectors together, we obtain a single overall vector, which will be written thus $\circledS\bar{g}_{i+1}\circledS\bar{g}_{i+2}\circledS \dots \circledS\bar{g}_n$. This is a member of the group $G_{i+1}\circledW \dots \circledW G_n$. That is, we have

$$\circledS\bar{g}_{i+1}\circledS\bar{g}_{i+2}\circledS \dots \circledS\bar{g}_n \ \in \ G_{i+1}\circledW \dots \circledW G_n.$$

On the one hand, the semi-direct product symbols $\circledS$ on the left are an abuse of notation, because a semi-direct product is a product between groups, not elements. Furthermore, the symbol $\circledS$ here is actually simply a comma in a vector. On the other hand, there is some value to this notation. It represents the following: Let us open $G_{i+1}\circledW \dots \circledW G_n$, the group on the right, in its semi-direct product notation. This will reveal all the copies of all the control groups, each with the appropriate duplication of duplications on its level.

Each vector $\bar{g}_j$ on the left is a vector from a level in the semi-direct product. Thus the number of components in the vector $\bar{g}_j$ is the number of group copies in the semi-direct product, and the indexes also correspond. Therefore, the semi-direct product symbol will be used in the vector to indicate this fact.

Now let us define the action of the control-nested τ-automorphism on an individual element in the bottom row of the figure. As an example, let us choose the element at the far right end of the bottom row. When the automorphism $\text{⑤}\bar{g}_{i+1}\text{⑤}\bar{g}_{i+2}\text{⑤}\ldots\text{⑤}\bar{g}_n$ shown in Fig. 4.4 is applied to this element, only certain group elements (circular arrows) are relevant, as shown in Fig. 4.5. They are those that "dominate" the chosen bottom node. This is due to the *selective* effect that occurs generally in wreath products, as described on p. 86.

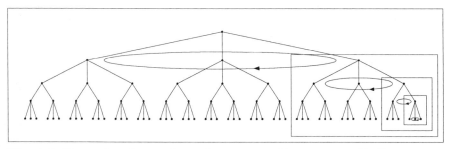

Fig. 4.5. The action of control-nested τ-automorphism on a specific fiber.

Let us notate the selective effect in the multi-level structure now being considered: A node on the bottom row is a copy of the fiber $G_1\text{ⓦ}\ldots\text{ⓦ}G_i$. Its position along the bottom is determined by its position in the hierarchy of control sets upwards from the bottom. Thus, let us label the node like this:

$$[G_1\text{ⓦ}\ldots\text{ⓦ}G_i]_{c_{i+1},c_{i+2},\ldots,c_n}$$

where c_{i+1} is its position within the control set C_{i+1} (level $i+1$), c_{i+2} is its position within the control set C_{i+2} (level $i+2$) above that, ..., and so on upwards.

Now apply the control-nested τ-automorphism, $\text{⑤}\bar{g}_{i+1}\text{⑤}\bar{g}_{i+2}\text{⑤}\ldots\text{⑤}\bar{g}_n$, shown in Fig. 4.4. The automorphism is a vector of all the group elements (circular arrows) shown in this diagram. The first level of this vector is the vector $\bar{g}_{i+1}$, which is the vector of circular arrows along the bottom of the figure. The selective effect ensures that only one element is chosen from this vector - the circular arrow on the far right, which will be labeled g_{i+1}. Now, go up to the next level, and make the appropriate selection (far right); and so on upward. On each level, which is the vector $\bar{g}_j$ the selected element will be labelled g_j. Notice that this is just an abbreviation. The element g_j comes from a *copy* of the group G_j. The copy is indexed by its position in the successive control sets upwards from it. That is, the copy is $[G_j]_{c_{j+1},c_{j+2},\ldots,c_n}$.

Therefore, the element g_j is really $[g_j]_{c_{j+1},c_{j+2},...,c_n}$. However, in this context, the abbreviation g_j will be used.

$$g_j = [g_j]_{c_{j+1},c_{j+2},...,c_n} \in [G_j]_{c_{j+1},c_{j+2},...,c_n}.$$

We are now ready to define how a control-nested τ-automorphism works:

Definition 4.3. *Define a* **control-nested** τ-**automorphism** $\circledS\bar{g}_{i+1}\circledS\bar{g}_{i+2}\circledS\ldots\circledS\bar{g}_n$ *of the group* Det_i *thus:*

$$[G_1\circledW\ldots\circledW G_i]_{c_{i+1},c_{i+2},...,c_n} \circledS\bar{g}_{i+1}\circledS\bar{g}_{i+2}\circledS\ldots\circledS\bar{g}_n$$

$$= \tau(g_n)\ldots\tau(g_{i+2})\tau(g_{i+1})[G_1\circledW\ldots\circledW G_i]_{c_{i+1},c_{i+2},...,c_n}.$$

$$= [G_1\circledW\ldots\circledW G_i]_{g_{i+1}c_{i+1},g_{i+2}c_{i+2},...,g_n c_n}. \tag{4.7}$$

where each $\bar{g}_j$ *is a vector in* Det_j/Det_{j-1} *and each* g_j *is the component in* $\bar{g}_j$ *selected from the group copy* $[G_j]_{c_{j+1},c_{j+2},...,c_n}.$

Observe that the selective effect has occurred in going from the first line to the second of Equation (4.7); that is, each g_j has been selected from its vector $\bar{g}_j$. The selected element g_j has then been converted into a transfer automorphism $\tau(g_j)$ using the transfer representation τ (one might wish to label τ for each level and control set copy). Now, since $\tau(g_j)$ is the raised action from the control set C_j, to the subscripts on the fiber-group copies, this means that, to apply $\tau(g_j)$, one merely applies g_j to the corresponding subscript c_j.

Most crucially, the third line in Equation (4.7) gives the location to which the node has been moved along the bottom of Fig. 4.5 under the control-nested τ-automorphism. Notice that the automorphism has moved the node successively: first within the smallest box shown in Fig. 4.5, then out of the smallest box, then out of the next larger box, then out of the next larger box, ..., and so on.

This, of course, gives us the general method of *transferring* a node along the bottom row: To move it from its position X to any other position Y (assuming transitivity), one has to use an operation g_{i+1} from the node G_{i+1} immediately above X in order to move X *within* the tree dominated by G_{i+1}, and then one has to use an operation g_{i+2} from the node G_{i+2} immediately *above that* in order to move X out of the tree but *within* the block of trees dominated by G_{i+2}, and then one has to use an operation g_{i+3} from the node G_{i+3} immediately *above that* in order to move X out of the block but *within* the block of blocks of trees dominated by G_{i+3}, and so on upwards.

This illustrates the *control-nested* effect defined in the term *control-nested* τ-*automorphism*; i.e., as opposed to an *ordinary* τ-automorphism (Definition 3.2, p. 89).

Associated with the above definition, we have this:

Definition 4.4. *The* **control-nested** τ**-conjugates** *of* $[G_1 \textcircled{w} \ldots \textcircled{w} G_i]_{c_{i+1}, c_{i+2}, \ldots, c_n}$ *are all the expressions of the form*

$$[G_1 \textcircled{w} \ldots \textcircled{w} G_i]_{c_{i+1}, c_{i+2}, \ldots, c_n} \; \textcircled{s} \bar{g}_{i+1} \textcircled{s} \bar{g}_{i+2} \textcircled{s} \ldots \textcircled{s} \bar{g}_n$$

for all elements $\textcircled{s} \bar{g}_{i+1} \textcircled{s} \bar{g}_{i+2} \textcircled{s} \ldots \textcircled{s} \bar{g}_n \;\in\; G_{i+1} \textcircled{w} \ldots \textcircled{w} G_n.$

This becomes basic to understanding work-spaces in robotics and grouping in perceptual organization.

Given a collection of elements g_j, as in Definition 4.3, it will be often useful to construct a canonical vector $\textcircled{s} \bar{g}_{i+1} \textcircled{s} \bar{g}_{i+2} \textcircled{s} \ldots \textcircled{s} \bar{g}_n$ from these elements. Two such canonical candidates are as follows:

Definition 4.5. *For a specific choice of* $c_{i+1} \in C_{i+1}, \ldots, c_n \in C_n$, *consider a set of elements* $g_j \in [G_j]_{c_{j+1}, c_{j+2}, \ldots, c_n}$, *for* $i + 1 \leq j \leq n$. *A* **canonical control-nested** τ**-automorphism** *constructed from the* g_j, *will be defined as follows: (1) All entries in the vector, except* g_j *are the identity element in the group-copies corresponding to those entries. (2) All entries in the vector are the "same" element* g_j *in the corresponding group copies (i.e., this is a multiple version of what is called the "diagonal" in wreath products). A canonical control-nested* τ-*automorphism constructed from the* g_j *will be notated as* $\textcircled{s} g_{i+1} \textcircled{s} g_{i+2} \textcircled{s} \ldots \textcircled{s} g_n$ *(no bars).*

We have seen that the definition of control-nested τ-conjugation divides the n-fold wreath product thus:

$$[G_1 \textcircled{w} \ldots \textcircled{w} G_i] \textcircled{w} [G_{i+1} \textcircled{w} \ldots \textcircled{w} G_n]$$

where the control-nested τ-conjugation operation $\textcircled{s} \bar{g}_{i+1} \textcircled{s} \bar{g}_{i+2} \textcircled{s} \ldots \textcircled{s} \bar{g}_n$ comes from the right half $G_{i+1} \textcircled{w} \ldots \textcircled{w} G_n$ of the partition. This half should be regarded as the full control group of its left-subsequence $G_1 \textcircled{w} \ldots \textcircled{w} G_i$. Thus, a distinction will be made between the full control group $G_{i+1} \textcircled{w} \ldots \textcircled{w} G_n$, and the immediate control group G_{i+1} of the same left-subsequence. The immediate control group will be referred to usually as just the control group of the sequence. And the full control group will often be called the *control-nested* τ-*automorphism group*, to emphasize that its actions are control-nested.

Group-theoretically, one should observe that the division just given with the wreath-product symbol at the dividing point:

$$[G_1 \textcircled{w} \ldots \textcircled{w} G_i] \textcircled{w} [G_{i+1} \textcircled{w} \ldots \textcircled{w} G_n]$$

is the same as

$$Det_i \textcircled{s} [G_{i+1} \textcircled{w} \ldots \textcircled{w} G_n]$$

with the semi-direct product symbol at the dividing point. The Det_i structure is the direct product of *all* copies of the fiber group $G_1 \textcircled{w} \ldots \textcircled{w} G_i$, that is, *all* the nodes along the bottom of Fig. 4.5, not just the nodes at the bottom of one sub-tree. Its control-nested τ-automorphism group is $G_{i+1} \textcircled{w} \ldots \textcircled{w} G_n$, the entire system of groups above it. This full group acts by conjugation on the bottom nodes, and therefore the appropriate group product between the bottom level, Det_i , and everything above it, $G_{i+1}Ⓦ...ⓌG_n$, is the semi-direct product.

Definition 4.6. *The* **control-nested** τ-**automorphism group** *of Det_i is* $G_{i+1}Ⓦ...ⓌG_n$.

We are now able to state this:

GENERAL STRUCTURE OF TRANSFER

The transfer of action spaces around an environment is carried out by control-nested τ-automorphism groups. The set of transferred action spaces are the control-nested τ-conjugates of their left-subsequences within the associated wreath product.

This proposal becomes fundamental, for example, to our theory of perceptual organization. For instance, we argue that what the Gestalt psychologists meant by symmetry is really a control-nested τ-automorphism group. Again, the proposal is basic to our theory of robot manipulation and navigation. We shall argue that work-spaces are related by control-nested τ-automorphism groups. The same applies to the parts of any design. The same applies to quark multiplets in quantum mechanics.

4.6   The Wreath Modifier

A number of times in this book, structures will be characterized in the following way: If Ω is a property of certain groups, then the term *wreath-Ω* will be used as a label for some "corresponding" class of wreath-product groups. Usually the correspondence will be one of two alternative forms:

(1) If Ω is a property of certain groups, then a class of wreath-product groups $G_1ⓌG_2Ⓦ...ⓌG_n$ will be called *wreath-Ω* if its *components* G_i have the property Ω.

(2) If Ω is a property of a certain class of *group extensions* $G_1ⒺG_2Ⓔ...ⒺG_n$, then the corresponding class of wreath-product groups $G_1ⓌG_2Ⓦ...ⓌG_n$ will be called *wreath-Ω*.

The placement of the word *wreath* in front of some group property Ω will be called the *wreath modifier*. It will often be the case that the class, wreath-Ω, will not have the property Ω. This will be enormously useful to us. To illustrate, consider the most basic example possible: that in which wreath-Ω is the property of being a wreath product. In this case, Ω is the property that a group is a extension $G_1 \mathrm{\textcircled{E}} G_2$. The corresponding property, wreath-Ω, can be called the *wreath-extension* $G_1 \mathrm{\textcircled{W}} G_2$, which is the wreath product. Then let us assume the usual convention that, in an extension $G_1 \mathrm{\textcircled{E}} G_2$, the component G_1 is a normal subgroup of the extension. Observe that, in the corresponding wreath-extension $G_1 \mathrm{\textcircled{W}} G_2$, the component G_1 is *not* a normal subgroup. Therefore one must conclude that a wreath-extension of G_1 by G_2 is not an extension of G_1 by G_2. Notice that this corresponds to the dichotomous effect (p. 94); i.e., in a wreath product, the subgroup G_2 acts automorphically on the fiber-group product but not on any fiber-group copy, because the latter is not a normal subgroup. It is this that allows transfer and generativity to take place. Thus, it is valuable to us that a wreath-extension of G_1 by G_2 is not an ordinary extension of G_1 by G_2; i.e., in the present example, the properties, Ω and wreath-Ω, do not coincide. In relation to this example, the reader should re-read the comment on (p. 14) called Normal Subgroups and Generativity, which contrasts the ordinary extension $SO(2) \times \mathbb{R}$ with the wreath extension $SO(2) \mathrm{\textcircled{W}} \mathbb{R}$ and shows that only the latter allows generativity.

The next section will illustrate some further issues concerning the relation between properties Ω and properties wreath-Ω.

4.7 Iso-Regular Groups

In Chapter 1, a class of groups was defined which will be fundamental to our generative theory: iso-regular groups (p. 12). We shall see that these groups characterize the standard shape primitives of CAD and computer vision. The justification for the link between iso-regular groups and shape primitives is our Externalization Principle, which states that any external inference applied to an arbitrary shape leads back to a shape that is generated by an iso-regular group.

The concept of the wreath modifier (last section) will now be used to give a new statement of the iso-regularity conditions, initially presented on p. 12. We start with the following definition:

Definition 4.7. *A group is* **c-cyclic** *if it is either a cyclic group or a connected 1-parameter Lie group.*

Recall from standard group theory that a group is *polycyclic* if it has the structure $G_1 \mathrm{\textcircled{E}} G_2 \mathrm{\textcircled{E}} \ldots \mathrm{\textcircled{E}} G_n$, where each G_i is cyclic (Segal [140], Sims [146]). Analogously, we now define:

Definition 4.8. *A group is* **c-polycyclic** *if it has the structure* $G_1 \textcircled{E} G_2 \textcircled{E} \ldots \textcircled{E} G_n$, *where each* G_i *is c-cyclic.*

Now we use the wreath modifier (2) in Sect. 4.6, and obtain this:

Definition 4.9. *A group is* **wreath c-polycyclic** *if it has the structure* $G_1 \textcircled{w} G_2 \textcircled{w} \ldots \textcircled{w} G_n$, *where each* G_i *is c-cyclic.*

Comment 4.1 *In Sect. 4.6, we discussed the possibility of a group having a property wreath-Ω but not the corresponding property Ω. The definition just developed, gives such an example: There are groups that are wreath c-polycyclic but not c-polycyclic. An example is $\mathbb{Z} \textcircled{w} \mathbb{Z}$ which is not even poly-cyclic.*[2]

The reader can now see that the iso-regularity condition $\mathfrak{IR}_2$ (p. 12) is the condition that a group is wreath c-polycyclic.

Now let us consider isometry conditions. If we say that Ω is the property that G is an isometry group, then using the wreath modifier (1) in Sect. 4.6, we have:

Definition 4.10. *A group is* **wreath-isometric** *if it has the structure* $G_1 \textcircled{w} G_2 \textcircled{w} \ldots \textcircled{w} G_n$, *where each* G_i *is an isometry group.*

Comment 4.2 *Again we have a situation in which a group having a property wreath-Ω does not necessarily have the corresponding property Ω. There are many groups that are wreath-isometric but are not isometric. For example, our symmetry group $\mathbb{R} \textcircled{w} \mathbb{Z}_4$ for the square, is wreath-isometric but not isometric in its action on the square.*

Finally, iso-regularity condition $\mathfrak{IR}_1$ (p. 12) is simply the application of wreath modifier (2) in Sect. 4.6 to the property Ω that G is a group extension. This was discussed as an example in that section.

Using the above, we can now state the three iso-regularity conditions, each as the application of the wreath modifier. This will be an important formulation in the book:

$\mathfrak{IR}_1$: *The group is a wreath-extension* $G_1 \textcircled{w} G_2 \textcircled{w} \ldots \textcircled{w} G_n$.

---

[2] The proof is as follows. A group G is said to satisfy the maximal condition if every non-empty set S of subgroups of G contains a subgroup that is maximal in S. A theorem states that a group is polycyclic iff it is soluble and satisfies the maximal condition. Now, $\mathbb{Z} \textcircled{w} \mathbb{Z}$ is soluble, because it is an extension of an abelian normal subgroup (the fiber-group product) by an abelian group (the control group). However, the set of subgroups

$$\{e\} \quad < \quad \langle x_0, x_1 \rangle \quad < \quad \langle x_0, x_1, x_2 \rangle \quad < \quad \ldots$$

does not have a maximum. Therefore $\mathbb{Z} \textcircled{w} \mathbb{Z}$ not polycyclic.

$\mathfrak{IR}_2$: *The group is wreath c-polycyclic.*

$\mathfrak{IR}_3$: *The group is wreath-isometric.*

4.8   Canonical Plans

Wreath c-polycyclic groups are fundamental for our generative theory not only because they are required for iso-regularity, but because they yield, very directly, a generative structure, as will now be seen.

First observe that, since each level in a wreath c-polycyclic group is "1-dimensional," it can be regarded as corresponding to the time parameter, i.e., the parameter along the generative sequence. Second, the wreath organization of such a group allows for a canonical means of unpacking that time parameter from the group, as follows:

The procedure can be illustrated with a square. We shall see that the standard way of drawing a square - i.e., simply drawing the sides successively around the figure - comes from the wreath c-polycyclic group characterizing the square. Recall that we gave the *symmetry group* of the square to be

$$\mathbb{R} \textcircled{w} \mathbb{Z}_4.$$

This is a wreath c-polycyclic group. It yields the standard plan for drawing a square as follows: Start with the identity element of the control group $\mathbb{Z}_4$; move down one level to the fiber group, $\mathbb{R}$, and apply this lower group as a whole to generate the first side. Then go back up to the control group and apply the next element r_{90}. Move back down to the fiber level, $\mathbb{R}$, and apply this group as a whole again to generate the second side. And so on, oscillating in cycles between the control level and the fiber level. In each cycle, one applies only a single element from the control level, but the entire fiber level. This procedure successively rotates around the square and draws a side at each individual rotation position.

For a wreath c-polycyclic group with n levels, one goes through an analogous n-level procedure, which is this:

GENERATION OF CANONICAL PLAN FROM A WREATH C-POLYCYCLIC GROUP. *Given a wreath c-polycyclic group $G_1 \textcircled{w} G_2 \textcircled{w} \dots \textcircled{w} G_n$, one repeats a cycle consisting of an entire downward movement through the levels. In each cycle, one uses only a single element from each successive downward level G_i, except on the bottom level G_1, where one uses the entire* *group. The order of elements within any level is the order prescribed by the generator of that level. One fills levels from the bottom up.*[3]

Observe that this procedure exploits the Symmetry-to-Trace Conversion Principle (p. 63). That is, the generativity comes from the symmetries involved, as can be seen from the example of the square. In fact, one is using the inference rules given in Chapter 2, principally, the Asymmetry Principle and Symmetry Principle, as explained in that chapter.

From the above procedure, one can see that wreath c-polycyclic groups correspond to nested do-while loops in computer programming. Control moves down from an element in an upper loop, and the loop below is performed "while" that upper element has been selected (and so on downward). According to our theory, do-while loops come from symmetries in the data, and are inferred using the recoverability rules of Chapter 2. As we have said before, there should be no difference between the theory of programming and the theory of memory objects. Both are structured by symmetries, in fact, by geometry, as it is defined in this book.

4.9   Wreath Poly-X Groups

Conceptually it is valuable to have both wreath modifiers. However, sometimes this can lead to conflicting terminology. An example is stated in the present section. We want to define a class of groups that will turn out to be extremely important for perception and robotics, due to the recursive nature of the latter two areas:

Definition 4.11. *Given a group X, a wreath product of the form*

$$X \circledw X \circledw \ldots \circledw X$$

i.e., where each level is isomorphic to X, will be called a **wreath poly-X group.**

For example, the group we propose for a serial-link manipulator will be the wreath poly-$SE(3)$ group:

$$SE(3) \circledw SE(3) \circledw \ldots \circledw SE(3).$$

---

[3] Since each level is c-cyclic, it is given by a single generator - either a discrete element in the case of a cyclic group, or a vector in a Lie algebra in the case of a 1-parameter group. One can therefore think of the loop on any level as the repetition of the generator. In the continuous case, the repetition is at infinitessimally close intervals, and therefore one makes the usual discrete approximation to program the structure.

Similarly, we shall argue that a group that is fundamentally important for the human perceptual system is the wreath poly-G group

$$G \textcircled{w} G \textcircled{w} \ldots \textcircled{w} G$$

where G is the symmetry group of a reference frame; i.e., the reference frame is recursively substituted into itself to many levels. Again, a group that we shall argue is fundamental to musical structure is the wreath poly-$\mathbb{S}$ group:

$$\ldots \textcircled{w} \mathbb{S} \textcircled{w} \mathbb{S} \textcircled{w} \mathbb{S} \textcircled{w} \mathbb{S}.$$

where $\mathbb{S}$ is the group of a scale.

The term wreath poly-X is really a use of the wreath modifier (2). However, the term will be used judiciously because it can lead to conflicts with the use of wreath modifier (1). For example, the type of group we have called a wreath-isometric group, using modifier (1), might be called a wreath poly-isometric group, using modifier (2). One could of course simply eliminate the use of modifier (1). However, both modifiers capture conceptually important phenomena, and, for this reason, both will be used.

4.10   Wreath Covering

The definition of a wreath product requires the fiber-set copies to be independent sets. However, consider for example a typical situation where the work-spaces of a robot are overlapping within an environment. Here, we would want the work-spaces to be fiber-set copies.

The solution to this problem is to define an abstract wreath product, together with a set of functions, one for each fiber-set copy, mapping each onto the environment. Note that there is no way that one would have a group action of the wreath product on the environment itself. The group action would be defined, as usual, with respect to fiber structure of the wreath product, and the effects of the action would then be mapped down onto the environment. However, with respect to the environment these effects might be multi-valued. This actually accords with the way an intelligent system works. The multi-valuedness corresponds to the different intentions of the intelligent system with respect to the environment, and a single point in the environment can serve several intentions. Thus we have this:

Definition 4.12. *Let S be a set. A* **wreath-covering** *of S is a wreath product together with a set of maps $\phi_{c_i} : F_{c_i} \longrightarrow S$, one for each fiber-set copy F_{c_i}, for all c_i in the control set. The wreath product without the maps will be called the* **abstract wreath product,** *and the maps will be called the* **covering maps.**

In nearly all cases considered in this book, the wreath products will be wreath-coverings of some set S. This will be so obvious, in most cases, that we will assume it without explicitly pointing it out.

5. Theory of Grouping

5.1 Introduction

Our fundamental claim about human perception is that the perceptual system is structured as a wreath product, $G_1 \textcircled{w} G_2 \textcircled{w} \ldots \textcircled{w} G_n$. This has powerful consequences on the way in which perception *groups* the world into cohesive structures. In the current chapter, we will show how the groupings can be systematically predicted from the wreath product $G_1 \textcircled{w} G_2 \textcircled{w} \ldots \textcircled{w} G_n$. This theory applies equally to organization in physics, for example, to the structuring of quark multplets in quantum mechanics, or to software cohesion in object-oriented design.

5.2 Grouping from Wreath Products

By far the largest body of research on perceptual grouping was provided by the Gestalt school in the first part of the twentieth century. However, despite the discovery of many empirical phenomena, no actual theory of grouping has ever been put forward. A complete mathematical theory of grouping will be presented over a number of chapters in this book. Nevertheless, the basic principles will be given in the present chapter.

In the conventional literature, the term *grouping* is closely related to the term *perceptual organization*. The stimulus array is a set of independent data points, and therefore highly ambiguous with respect to grouping. *Perceptual organization* concerns how the perceptual system decides to choose one organization of the stimulus set as opposed to another. It was discovered by the Gestalt psychologists that the choice of organization was determined by criteria for *grouping*. Examples of such criteria were these: The perceptual

(a)　　　　　　　　　　　　　　　　　　(b)

Fig. 5.1. A classic Gestalt example.

system chooses to put stimuli together in the same grouping if they are symmetric to each other, or if they are near to each other, etc. Thus, stimuli that are not symmetric to each other, and not near to each other, etc., are not put in the same grouping.[1]

In order to develop our theory of grouping let us begin by analyzing one of the most famous examples from the Gestalt literature. It is shown in Fig. 5.1. In *both* Fig. 5.1a and 5.1b, there are the same number of squares, arranged six vertically and six horizontally. However, in Fig. 5.1a, the perceptual system groups the squares into rows, whereas, in Fig. 5.1b, the perceptual system groups the squares into columns. The reason given for this, by Gestalt psychologists, is that the perceptual system uses the criterion of *proximity*: Nearer stimuli are grouped together - hence the grouping into rows in Fig. 5.1a and the grouping into columns in Fig. 5.1b.

Our main proposal will be this: *All Gestalt phenomena arise from the maximization of transfer and recoverability.*

Let us begin by examining Fig. 5.1a. This configuration is in fact a *5-level hierarchy of transfer*. That is, it is given by a *5-fold wreath product*

$$G_1 \text{ⓦ} G_2 \text{ⓦ} G_3 \text{ⓦ} G_4 \text{ⓦ} G_5.$$

The levels are as follows: First, the lowest level G_1 is the individual point, which will be represented by the trivial group $\{e\}$. The point is transferred

---

[1] In spite of the fact that the Gestalt school showed that symmetry is enormously important to human vision, there has been almost no psychological research on symmetry in the last 60 years. The exceptions have been Leyton [87]-[97], Carlton & Shepard [18], [19], Shepard [142], Ishida & Kotovsky [63], Lenz [83], See also Foote, Mirchandani, Rockmore, Healy & Olson [36] which uses wreath products in the domain of signal-processing. Finally, an outstanding paper on the relationship between Gestalt theory and computational vision is Vishwanath, [151].

along an individual side by the translation group and therefore G_2 is $\mathbb{R}$. Next, a side is transferred around an individual square by the 4-fold rotation group, and therefore G_3 is $\mathbb{Z}_4$. Next, a square is transferred along a particular row by integral translations $\mathbb{Z}$, and therefore G_4 will be given as $\mathbb{Z}^H$, where the superscript H means horizontal. Finally, a row is transferred in the vertical direction by the group of integral translations, and therefore G_5 will be given as $\mathbb{Z}^V$, where the superscript V means vertical.

The 5-fold wreath product describing Fig. 5.1a is therefore,

$$\{e\} \ \textcircled{w} \ \mathbb{R} \ \textcircled{w} \ \mathbb{Z}_4 \ \textcircled{w} \ \mathbb{Z}^H \ \textcircled{w} \ \mathbb{Z}^V.$$

Now our theory is that grouping - the cohesiveness across perceptual elements - is created by transfer across the elements. This means that one can enumerate the perceptual groupings by enumerating the left-subsequences of the wreath product that models the stimulus set. The following table elaborates the left-subsequences of the above 5-fold hierarchy. They are

$\{e\}$	a point
$\{e\} \ \textcircled{w} \ \mathbb{R}$	a side
$(\{e\} \ \textcircled{w} \ \mathbb{R}) \ \textcircled{w} \ \mathbb{Z}_4$	a square
$(\{e\} \ \textcircled{w} \ \mathbb{R}) \ \textcircled{w} \ \mathbb{Z}_4) \ \textcircled{w} \ \mathbb{Z}^H$	a row
$(\{e\} \ \textcircled{w} \ \mathbb{R}) \ \textcircled{w} \ \mathbb{Z}_4) \ \textcircled{w} \ \mathbb{Z}^H) \ \textcircled{w} \ \mathbb{Z}^V$	the whole.

It is clear that this list corresponds exactly to what people would regard as the groupings, or "objects," in Fig. 5.1a. Later in this book, when we investigate much more complex shape, we shall see that exactly the same rules apply.

Page 138 gives our basic law of grouping, and the reader should work through each of the statements on that page, before continuing.

Note the way in which perceptual levels are linked according to this law, as follows: Express the grouping $G_1 \ \textcircled{w} \ ... \ \textcircled{w} \ G_i$ in its semi-direct product notation:

$$[(G_1 \ \textcircled{w} \ ... \ \textcircled{w} \ G_{i-1}) \times (G_1 \ \textcircled{w} \ ... \ \textcircled{w} \ G_{i-1}) \times ... \times (G_1 \ \textcircled{w} \ ... \ \textcircled{w} \ G_{i-1})] \circledS_\tau [G_i].$$

Observe that the perceptual elements, which are the copies of $(G_1 \ \textcircled{w} \ ... \ \textcircled{w} \ G_{i-1})$ along this sequence, are linked by the *direct-product* operation, $\times$. As listed in the Product-Hierarchy Table p. 117, groups that are linked by $\times$ are on the *same level of the hierarchy*. Therefore the sequence shown above acknowledges that the perceptual elements, that are bound together, are on the same level of the hierarchy.

Now move one step to the right along the above sequence. To the right of the fiber-group product, there is the semi-direct product symbol $\circledS_\tau$. As listed on the Product-Hierarchy Table, this symbol *is* hierarchical. It links one level (left of $\circledS_\tau$), to the *next higher level* (right of $\circledS_\tau$). Indeed, on the right of $\circledS_\tau$, in the above sequence, there is the control group G_i, which, as the Law of Grouping states, is the *grouping factor*. Thus the semi-direct

LAW OF GROUPING

(1) Grouping is created by transfer.

(2) Therefore the groupings in an organization are enumerated as the left-subsequences $G_1 \text{ⓦ} \ldots \text{ⓦ} G_i$ in an n-fold wreath product (for all $i \leq n$).

(3) Consider a particular grouping $G_1 \text{ⓦ} \ldots \text{ⓦ} G_i$. Because grouping is created by transfer, the *cohesive structure* of this grouping is the expression of the grouping as its wreath product $(G_1 \text{ⓦ} \ldots \text{ⓦ} G_{i-1}) \text{ⓦ} G_i$, where the relevant wreath product symbol is that after the parentheses. The various factors in the cohesive structure can then be read directly from this wreath product as follows:

(4) A perceptual *element* of the grouping is the fiber $G_1 \text{ⓦ} \ldots \text{ⓦ} G_{i-1}$.

(5) The *set of elements grouped* is the set of fiber copies $(G_1 \text{ⓦ} \ldots \text{ⓦ} G_{i-1})_{g_1}, \ldots, (G_1 \text{ⓦ} \ldots \text{ⓦ} G_{i-1})_{g_n}$.

(6) The *grouping factor* is transfer, i.e., the control group G_i.

(7) The *grouping action* is the τ-automorphic action of G_i on its fiber-group product.

Grouping $= G_1 \text{ⓦ} \ldots \text{ⓦ} G_i =$ left-subsequence
Cohesive structure $= (G_1 \text{ⓦ} \ldots \text{ⓦ} G_{i-1}) \text{ⓦ} G_i =$ wreath product
Individual Element $= (G_1 \text{ⓦ} \ldots \text{ⓦ} G_{i-1}) =$ fiber
Set of elements grouped $= (G_1 \text{ⓦ} \ldots \text{ⓦ} G_{i-1}), \ldots, (G_1 \text{ⓦ} \ldots \text{ⓦ} G_{i-1})$
$=$ fiber copies
Grouping factor $= G_i =$ control group
Grouping action $= \tau$-automorphic action of G_i on its fiber-group product

---

product symbol ⓢ_τ links the *set of elements* on the left to the *grouping factor* on the right. This exhibits the algebraic structure of transfer.

5.3   Grouping as Algebraic Action

According to Sect. 3.17, transfer is achieved by *conjugation*; i.e., given a member g of the control group, its action, in sending the fiber-group copy $G(F)_{c_i}$ onto the fiber-group copy $G(F)_{gc_i}$, is given thus:

$$G(F)_{c_i} \xrightarrow{g - g^{-1}} G(F)_{gc_i}.$$

This means that the second fiber-group copy $G(F)_{gc_i}$ is really the conjugate $gG(F)_{c_i}g^{-1}$ of the first fiber-group copy $G(F)_{c_i}$.

Now, our Law of Grouping states that grouping is transfer and that the elements that are grouped are the fiber-group copies. Thus we are lead to the following conclusion:

GROUPING IS CONJUGATION.   *The cohesive action in grouping is carried out by conjugation. The elements that are grouped are conjugate copies of each other.*

There is a deep consequence of the above statement: As argued in Sect. 3.18, conjugation is really a *memory* structure. The definition $gG(F)_{c_i}g^{-1}$ of the second fiber-group copy $G(F)_{gc_i}$ is really a three-stage definition, which says that, if you want to carry out any action within the second fiber-group copy, you should go through the following three steps: (1) apply g^{-1} to get you from the second fiber-group copy back to the first, (2) then carry out the required action within the first fiber-group copy $G(F)_{c_i}$, and (3) then apply g to transfer the result forwards to second fiber-group copy. Therefore, the conjugation process *recalls* the action within the first fiber-group copy $G(F)_{c_i}$.

We therefore conclude that, since grouping works by conjugation, it is actually a memory structure. Therefore it relies on *recoverability*, which is the other main component of our theory, besides transfer. The relation to recoverability, particularly, to external and internal inference was given in Sect. 3.18. The example of Fig. 5.1a uses only internal inference, since all levels are traces. This is because the group defining this particular figure is an iso-regular group (Sect. 4.7). However, the theory applies equally well to the case where external inference is required. Such a case would occur if, for example, the figure were deformed, e.g., by the projective process or by some spline-based action.

5.4   Generative Crystallography

It is worth characterizing the particular Gestalt example in Fig. 5.1a by defining a *class* of groups, as follows:

Fig. 5.1a has a type of regularity that is evidenced in *crystallographic* structures. Such structures are based on lattices, and are important in several areas. Examples are *image screens* such as the human retina, and most *memory-storage devises*. Furthermore, crystallographic methods have been used to solve shape-from-texture problems in computer vision, Liu & Collins [100]; to characterize ornamental structures, Jablan [64], Washburn & Crowe [155]; and for solving robot assembly problems, Liu [99], Liu & Popplestone [101], Popplestone, Liu & Weiss [120].

Standardly, a crystallographic group is defined as

$$G \ = \ [\mathbb{Z} \times \mathbb{Z} \times \ldots \times \mathbb{Z}] \; ⓔ \; H \tag{5.1}$$

where the normal subgroup $\mathbb{Z} \times \mathbb{Z} \times \ldots \times \mathbb{Z}$ is the translation group of the lattice and H is the point group, which is a subgroup of the orthogonal group $O(n)$. The extension is not necessarily splitting. However, because the normal subgroup is maximal abelian in G, the map

$$\sigma \ : \ H \longrightarrow Aut[\mathbb{Z} \times \mathbb{Z} \times \ldots \times \mathbb{Z}]$$

on which the extension is based, is a momomorphism. See Ascher & Janner [3], [4], Shubnikov & Koptsik [143], Schwarzenberger [138].

In contrast to this approach, we want to characterize crystallographic structure *generatively*. To do this, we argue that one must use a different kind of extension from that given in expression (5.1). We do this by proposing the following definition:

Definition 5.1. *A* **generative crystallographic group** *is an iso-regular group (wreath c-polycyclic, wreath-isometric) of the form*

$$G \ = \ H \, ⓦ \, [\mathbb{Z} \, ⓦ \, \mathbb{Z} \, ⓦ \ \ldots \ ⓦ \, \mathbb{Z}]. \tag{5.2}$$

Note that the component H, being iso-regular, has itself a wreath c-polycyclic wreath-isometric decomposition $G_1 ⓦ G_2 ⓦ \ldots ⓦ G_n$.

The *upper* group in expression (5.2) is now the translation group. Observe that it is a wreath product $\mathbb{Z} ⓦ \mathbb{Z} ⓦ \ldots ⓦ \mathbb{Z}$ in contrast to the direct product $\mathbb{Z} \times \mathbb{Z} \times \ldots \times \mathbb{Z}$ in standard crystallography (5.1). This means that it is now a generative structure, and can therefore model physical processes such as crystal growth. For example, consider the 2-dimensional case $\mathbb{Z} ⓦ \mathbb{Z}$, which can model, for instance, an image screen created by a moving scan-line. Notice the important role of the dichotomous effect in the wreath product: Because the fiber $\mathbb{Z}$ is not a normal subgroup, is moved by the control group $\mathbb{Z}$. This means that $\mathbb{Z} ⓦ \mathbb{Z}$ correctly models the successive movement of the scan-line down the screen. If instead the group were a direct product $\mathbb{Z} \times \mathbb{Z}$, the scan-line could not move because it would correspond to a normal subgroup. The same issue would arise in crystal physics where one would wish to model the successive accumulation of layers in a crystal structure, i.e., in crystal growth. The wreath product would model this, and the direct product would not.

The group in extension (5.2) should be understood as a lattice translation group moving around an iso-regular group H from lattice position to lattice position. In an intuitive sense, one can understand the extension in (5.2) as reversing the positions of the translation group and point group H in (5.1). This however, should not be emphasized too strongly, for the following reasons: First, the point group in (5.1) is literally an automorphic property of the lattice, and this is not the case in (5.2). There is no restriction in (5.2) that H should pick up any symmetries in the translation group. Second, the H group in (5.2) corresponds more to what crystal physicists would call the asymmetric element - a molecule or configuration of molecules, which is repeated throughout the array. What we have done therefore in (5.2) is to express this repetition as a process of *transfer*, i.e., the group of the "molecular element" is transferred from lattice site to lattice site. Furthermore, the entire structure is given as an iso-regular group. This means that one can generate the entire structure via a canonical plan, as described in Sect. 4.8.

Finally, returning to the Gestalt example in Fig. 5.1a, one can see that it is an instance of what we have called here a *generative crystallographic group* (the above definition). Such groups will appear in this book a number of times, e.g., in mechanical and architectural CAD.

5.5   Using the Law of Grouping

Our basic proposal is that grouping is transfer, i.e., that the *environment is unified by transfer*. The Law of Grouping tells us exactly how this is achieved. In this section, we learn more about this law by applying it carefully to the 5-fold wreath product defining the grid of squares. Throughout, we will use the regular wreath product, i.e., there is one copy of the fiber group for each member of the control group. Also, we will use the finitistic notation (p. 110) in which infinite direct products are shown in a finite notation.

Let us go left-to-right through the 5-fold wreath product for the grid of squares

$$\{e\} \text{Ⓦ} \mathbb{R} \text{Ⓦ} \mathbb{Z}_4 \text{Ⓦ} \mathbb{Z}^H \text{Ⓦ} \mathbb{Z}^V. \tag{5.3}$$

The first left-subsequence is $\{e\}$ which is the trivial grouping, a point.

Moving to the first non-trivial left-subsequence

$$\{e\} \text{Ⓦ} \mathbb{R}$$

one gets the grouping that people call a "side". Opening up this wreath product, one gets:

$$[\{e\} \times \{e\} \times \ldots \times \{e\}] \circledS_\tau [\mathbb{R}]$$

where there is one copy of the fiber group $\{e\}$ for each element in the control group $\mathbb{R}$. The set of copies of $\{e\}$ in the fiber-group product $\{e\} \times \{e\} \times \ldots \times \{e\}$ is the set of points within a side. It is this set of elements that are perceptually *grouped* together. In accord with the Law of Grouping (p. 138), cohesiveness is created by the control group $\mathbb{R}$ (in the above sequence), which we call the *grouping factor*. Furthermore the law states that the way in which $\mathbb{R}$ achieves the cohesiveness is via its *transfer* of points $\{e\}$ onto each other. This is carried out by the τ-automorphic action of $\mathbb{R}$ on the fiber-group product. The action takes place by conjugation of the copies of $\{e\}$, thus:

$$\{e\}_{c_i} \xrightarrow{g - g^{-1}} \{e\}_{gc_i}.$$

In other words, the second element $\{e\}_{gc_i}$ is really the conjugate $g\{e\}_{c_i}g^{-1}$ of the first element $\{e\}_{c_i}$. Thus the points along a line are conjugate copies of each other.

RECURSIVENESS OF THE VISUAL SYSTEM. *We shall show that the visual system is recursive, and therefore a large non-trivial wreath group $H_1 \circledW H_2 \circledW \ldots \circledW H_n$ can be substituted for the point $\{e\}$ at the left-end of any wreath sequence $\{e\} \circledW G_1 \circledW G_2 \circledW \ldots \circledW G_n$. For example, an entire Cartesian frame can be substituted for each fiber copy of $\{e\}$. The ability to perceptually map between these frames is then carried out by the τ-automorphic conjugacy structure that we just described.*

Now move one step to the right in the full sequence and obtain the next left-sub-sequence:

$$(\{e\} \circledW \mathbb{R}) \circledW \mathbb{Z}_4.$$

This correponds to the grouping people call a *square*. Much time was spent in Chapter 3 understanding the wreath-product structure of a square, but some extra comments should now be made:

According to the Law of Grouping, the cohesive structure is given by the final wreath product symbol $\circledW$ in the above left-subsequence. The various factors in this grouping can be read directly from this wreath product: An individual element in this grouping is given by the *fiber* $(\{e\} \circledW \mathbb{R})$, a side. The set of elements that are grouped are revealed when the wreath-product symbol is opened up in its semi-direct product notation, thus:

$$[(\{e\} \circledW \mathbb{R}) \times (\{e\} \circledW \mathbb{R}) \times (\{e\} \circledW \mathbb{R}) \times (\{e\} \circledW \mathbb{R})] \circledS_\tau [\mathbb{Z}_4]. \qquad (5.4)$$

To the left of $\circledS_\tau$, there is the fiber-group product $(\{e\} \circledW \mathbb{R}) \times (\{e\} \circledW \mathbb{R}) \times (\{e\} \circledW \mathbb{R}) \times (\{e\} \circledW \mathbb{R})$, which consists of the four sides. The Law of Grouping states that the elements are bound together by the *grouping factor*, which is the (right-most) control group $\mathbb{Z}_4$. The grouping is achieved by the τ-automorphic action of $\mathbb{Z}_4$ on the fiber-group product. Since this action corresponds to *transfer*, we see that grouping is achieved by transfer. Again: Environmental unification is achieved by transfer!

Note also that the τ-automorphic action is conjugation:

$$(\{e\} \,\circledW\, \mathbb{R})_{c_i} \xrightarrow{g - g^{-1}} (\{e\} \,\circledW\, \mathbb{R})_{gc_i}.$$

In other words, the second fiber-group copy $(\{e\} \,\circledW\, \mathbb{R})_{gc_i}$ is really the conjugate $g(\{e\} \,\circledW\, \mathbb{R})_{c_i} g^{-1}$ of the first fiber-group copy $(\{e\} \,\circledW\, \mathbb{R})_{c_i}$. Thus *the elements that are grouped in a cohesive structure are conjugate copies of each other.*

Note that the sequence (5.4) involves all three group-product symbols, $\times$, $\circledS$, and $\circledW$. It is worth the reader checking the *hierarchical* information that these products give in (5.4), in accord with the Product-Hierarchy Table (p. 117).

In the sequence (5.4), notice that the first product that occurs is a wreath product $\circledW$. From this we see that the grouping actions are control-nested. That is, within the fiber-group product of the sequence, there are copies of the previous left-subsequence $(\{e\} \,\circledW\, \mathbb{R})$, in which the grouping factor is $\mathbb{R}$, which groups the points within a side.

Thus opening up each of the wreath products $(\{e\} \,\circledW\, \mathbb{R})$, in the sequence (5.4), one gets this:

$$
\begin{aligned}
&[(([\{e\} \times \{e\} \times \ldots \times \{e\}] \,\circledS_\tau\, [\mathbb{R}]) \\
&\times ([\{e\} \times \{e\} \times \ldots \times \{e\}] \,\circledS_\tau\, [\mathbb{R}]) \\
&\times ([\{e\} \times \{e\} \times \ldots \times \{e\}] \,\circledS_\tau\, [\mathbb{R}]) \\
&\times ([\{e\} \times \{e\} \times \ldots \times \{e\}] \,\circledS_\tau\, [\mathbb{R}])] \\
&\quad \circledS_\tau\, [\mathbb{Z}_4].
\end{aligned}
\tag{5.5}
$$

This sequence explicitly shows the effect of the first two levels of the binding structure: Each of the first four lines gives the grouping of the points by the translation control group. Then the fifth line gives the grouping of the previous four lines by adding a transfer level $\mathbb{Z}_4$ across them.

Let us now move one step further to the right in the full 5-fold sequence (5.3), and thus obtain the following left-subsequence.

$$(\{e\} \,\circledW\, \mathbb{R} \,\circledW\, \mathbb{Z}_4 \,\circledW\,) \, \mathbb{Z}^H.$$

Because this is a left-subsequence, the Law of Grouping implies that it is a *perceptual grouping*. And indeed it is. It is the grouping people call a *row*.

According to the law, the *cohesive structure* of this subsequence is given by the final wreath-product symbol $\circledW$, and the various grouping factors can be read directly from this wreath product, as follows: An *individual element* in this grouping is given by the *fiber* $(\{e\} \circledW \mathbb{R} \circledW \mathbb{Z}_4)$, which represents a *square*. The *set of perceived elements* in the grouping are *revealed* when the wreath product is opened up in its *semi-direct product notation*, thus:

$$[(\{e\} \circledW \mathbb{R} \circledW \mathbb{Z}_4) \times (\{e\} \circledW \mathbb{R} \circledW \mathbb{Z}_4) \times \ldots \times (\{e\} \circledW \mathbb{R} \circledW \mathbb{Z}_4)] \circledS_\tau [\mathbb{Z}^H]. \quad (5.6)$$

Observe that, to the left of the semi-direct product symbol $\circledS_\tau$, there is the fiber-group product which is the direct product of the perceived elements (squares), one element for each member of the control group $\mathbb{Z}^H$, shown to the right of $\circledS_\tau$. The group $\mathbb{Z}^H$ is the group of horizontal movements along a row. This is the *grouping factor*. It works by its τ-automorphic action - i.e., *transferring* one square onto another horizontally. Again, the action is conjugation: Any member g of the control group $\mathbb{Z}^H$ sends the square $(\{e\} \circledW \mathbb{R} \circledW \mathbb{Z}_4)_{c_i}$ to the square $(\{e\} \circledW \mathbb{R} \circledW \mathbb{Z}_4)_{gc_i}$, via the action $g - g^{-1}$:

$$(\{e\} \circledW \mathbb{R} \circledW \mathbb{Z}_4)_{c_i} \xrightarrow{g-g^{-1}} (\{e\} \circledW \mathbb{R} \circledW \mathbb{Z}_4)_{gc_i}.$$

Thus, the squares within a row are really conjugates of each other.

Notice, in the grouping structure (5.6), that even though the direct-product symbols $\times$ link the individual fibers, the fibers themselves are internally structured by wreath-product symbols $\circledW$. This indicates that the fibers are themselves groupings, in fact, the groupings we have studied previously. Thus, within any such fiber $(\{e\} \circledW \mathbb{R} \circledW \mathbb{Z}_4)$, let us open its *second* wreath-product symbol, in its semi-direct product notation. Expression (5.6) then becomes this:

$$[([(\{e\} \circledW \mathbb{R}) \times (\{e\} \circledW \mathbb{R}) \times (\{e\} \circledW \mathbb{R}) \times (\{e\} \circledW \mathbb{R})] \circledS_\tau [\mathbb{Z}_4])$$
$$\times ([(\{e\} \circledW \mathbb{R}) \times (\{e\} \circledW \mathbb{R}) \times (\{e\} \circledW \mathbb{R}) \times (\{e\} \circledW \mathbb{R})] \circledS_\tau [\mathbb{Z}_4])$$
$$\times \ldots \times ([(\{e\} \circledW \mathbb{R}) \times (\{e\} \circledW \mathbb{R}) \times (\{e\} \circledW \mathbb{R}) \times (\{e\} \circledW \mathbb{R})] \circledS_\tau [\mathbb{Z}_4])]$$
$$\circledS_\tau [\mathbb{Z}^H]. \quad (5.7)$$

In the top line, the fibers, $(\{e\} \circledW \mathbb{R})$ represent the sides of the square, and the control group $\mathbb{Z}_4$, at the end of the line, is the grouping factor that binds the sides together. This structure is repeated on the next two lines. The final line is the grouping factor that binds all the previous lines together.

Notice, in the above structure, that one wreath-product symbol is still not opened, that in the fiber $(\{e\} \circledW \mathbb{R})$. Opening this, one gets:

$$[\{e\} \times \{e\} \times \ldots \times \{e\}] \circledS_\tau [\mathbb{R}]$$

which is the grouping of points in a side. This opened structure can then be substituted in the sequence (5.7), thus revealing all elements.

Finally, one comes to the highest level of grouping in Fig. 5.1a: the binding of the set of rows into the full percept. This is given by the *final* left-subsequence, which is the full 5-fold wreath product itself:

$$(\{e\} \, ⓦ \, \mathbb{R} \, ⓦ \, \mathbb{Z}_4 \, ⓦ \, \mathbb{Z}^H \, ⓦ) \, \mathbb{Z}^V.$$

According to the Law of Grouping, the cohesive structure of this left-subsequence is given by the final wreath-product symbol, and thus an individual element in this grouping is the *fiber* $(\{e\} \, ⓦ \, \mathbb{R} \, ⓦ \, \mathbb{Z}_4 \, ⓦ \, \mathbb{Z}^H)$, which represents a *row*. The set of perceived elements in the grouping are the *copies* of the fiber *revealed* when the wreath product is opened up in its *semi-direct product notation*:

$$[(\{e\} \, ⓦ \, \mathbb{R} \, ⓦ \, \mathbb{Z}_4 \, ⓦ \, \mathbb{Z}^H) \times (\{e\} \, ⓦ \, \mathbb{R} \, ⓦ \, \mathbb{Z}_4 \, ⓦ \, \mathbb{Z}^H)$$
$$\times \ldots \times (\{e\} \, ⓦ \, \mathbb{R} \, ⓦ \, \mathbb{Z}_4 \, ⓦ \, \mathbb{Z}^H)]$$
$$ⓢ_\tau \, [\mathbb{Z}^V].$$

The *grouping factor* that binds the elements together is the (right-most) control group in this sequence. This is $\mathbb{Z}^V$, the group of vertical movements. The grouping action is its τ-automorphic action on the fiber-group product to its left. Observe that this is the means by which each row is transferred vertically through the visual configuration. Since transfer is algebraic conjugation, $g - g^{-1}$, the rows are conjugate copies of each other.

Finally, by opening up the wreath-product symbols ⓦ, in the above sequence, one *reveals* the grouping hierarchies *below* the highest level just described. This shows that the cohesive structures are control-nested in each other.

5.6 Hierarchical Detection in Grouping

The reader will recall from Sect. 4.4 that, given an n-fold wreath product $G_1 ⓦ \ldots ⓦ G_n$, we defined the detected symmetry group up to Level i as

$$Det_i = [G_1 ⓦ \ldots ⓦ G_i]^{m_{i+1} \times m_{i+2} \times \ldots \times m_n}$$

(where $|G_j| = m_j$). At this level, all fibers $G_1 ⓦ \ldots ⓦ G_i$ are independent, not just the fibers within a particular fiber-group product of the next level. Then the group

$$Det_{i+1}/Det_i = G_{i+1}^{m_{i+2} \times \ldots \times m_n}.$$

is what is detected in order to raise the detected symmetry group from Det_i to Det_{i+1}. Thus the detection of Level $i + 1$ is a splitting extension of Det_i by $G_{i+1}^{m_{i+2} \times \ldots \times m_n}$. Successive hierarchical detection therefore corresponds to a group-theoretic *normal series*, thus:

$$Det_1 \lhd Det_2 \lhd Det_3 \lhd \ldots \lhd Det_n.$$

We shall now see that the issue of hierarchical detection relates fundamentally to grouping. To illustrate, consider Fig. 5.1a, *after* the points have been grouped into sides, but *before* the sides have been grouped into squares. At this stage, the perceptual organization is a set of independent sides. However, this set consists not only of the sides within a particular square, but the sides in the *entire* configuration. This means that the symmetry group of the entire set is not simply the fiber-group product of the next level but

$$Det_2 = [\{e\} \ \textcircled{w} \ \mathbb{R}]^{m_3 \times m_4 \times m_5} \tag{5.8}$$

where the sequence $\{e\} \ \textcircled{w} \ \mathbb{R}$ is the group of an individual side, and the power $m_3 \times m_4 \times m_5$ is the cardinality of the right-subsequence $\mathbb{Z}_4 \ \textcircled{w} \ \mathbb{Z}^H \ \textcircled{w} \ \mathbb{Z}^V$ of $\{e\} \ \textcircled{w} \ \mathbb{R}$ in the full 5-fold sequence $\{e\} \ \textcircled{w} \ \mathbb{R} \ \textcircled{w} \ \mathbb{Z}_4 \ \textcircled{w} \ \mathbb{Z}^H \ \textcircled{w} \ \mathbb{Z}^V$ (using finitistic notation on the cardinals). What (5.8) expresses is the entire grouping that has taken place up to and including the level of the sides, but not beyond. This views the entire grid as a purely *parallel* structure of independent sides.

Next, when one detects the squares, what one has detected is this:

$$Det_3/Det_2 \;=\; (\mathbb{Z}_4)^{m_4 \times m_5}.$$

In accord with our group-theoretic characterization of detection (p. 122), the detection of the squares is the splitting extension of Det_2 by $(\mathbb{Z}_4)^{m_4 \times m_5}$. This extension gives the symmetry group of the entire set of independent squares, which is

$$Det_3 = [\{e\} \ \textcircled{w} \ \mathbb{R} \ \textcircled{w} \ \mathbb{Z}_4]^{m_4 \times m_5}. \tag{5.9}$$

Notice that the symmetry group of the set of sides is a normal subgroup of the symmetry group of the set of squares.

Thus, if we consider the full Gestalt created by Fig. 5.1a, the successive detection of the groupings is given by the following *normal series*:

$$Det_1 \;\triangleleft\; Det_2 \;\triangleleft\; Det_3 \;\triangleleft\; Det_4 \;\triangleleft\; Det_5.$$

5.7 Perceptual Relationship between Similar Groupings

It is necessary to understand the relationship between similar groupings in an organization. As an example, consider the two sides that have been emphasized in Fig. 5.2. Assume now that all perceptual levels have been detected. The two sides are groupings. However, because all higher levels have been detected, the two sides now have a perceptual relationship between each other. They are not simply independent. What is the perceptual relationship?

Fig. 5.2. Calculating the Gestalt relationship between two elements.

The reader will recall the difference between a τ-automorphism and a control-nested τ-automorphism (Sect. 4.5). In the first case, one considers the control group of a left-subsequence $G_1 \textcircled{w} G_2 \textcircled{w} \ldots \textcircled{w} G_i$ to be the next level, G_{i+1}. In the second case, one considers the control group to be the entire right-subsequence $G_{i+1} \textcircled{w} G_{i+2} \textcircled{w} \ldots \textcircled{w} G_n$. The τ-automorphisms in the former case are single-level, coming from G_{i+1}, and the τ-automorphisms in the second case are multi-level, coming from $G_{i+1} \textcircled{w} G_{i+2} \textcircled{w} \ldots \textcircled{w} G_n$.

Thus we argue that the perceptual relationship between the two sides in Fig. 5.2 is a *control-nested* τ-automorphism. Notice that such an automorphism moves one side onto the other via a hierarchical action that uses the successive control groups above the side. Since the side is

$$\{e\} \textcircled{w} \mathbb{R}$$

and the full 5-fold sequence is

$$\{e\} \textcircled{w} \mathbb{R} \textcircled{w} \mathbb{Z}_4 \textcircled{w} \mathbb{Z}^H \textcircled{w} \mathbb{Z}^V$$

the successive control groups above the side are $\mathbb{Z}_4$, $\mathbb{Z}^H$, and $\mathbb{Z}^V$.

Thus, let us see how this control-nested movement works. Note that each side exists in a particular *orientation*, on a particular *square*, on a particular *row*. To move one of the sides onto the other, one must first use an element from $\mathbb{Z}_4$ to rotate it until it has the same position within its own square as the other side. Then one must use an element from $\mathbb{Z}^H$ to move its square horizontally, till this is in the same position on its row as the other square. Finally, one must use an element from $\mathbb{Z}^V$ to move its row onto the other row.

This three-fold hierarchical operation is a control-nested τ-automorphism

$$\bar{g}_3 \text{\textcircled{S}} \bar{g}_4 \text{\textcircled{S}} \bar{g}_5$$

which comes from the group $\mathbb{Z}_4 \text{\textcircled{W}} \mathbb{Z}^H \text{\textcircled{W}} \mathbb{Z}^V$.

Now the first of the two sides is a *copy* of the fiber group $\{e\} \text{\textcircled{W}} \mathbb{R}$. One knows which copy it is by the control indexes that appear on it, thus

$$[\{e\} \text{\textcircled{W}} \mathbb{R}]_{c_3,c_4,c_5}$$

where c_3 is the position of the side within the square, c_4 is the position of the square within the row, and c_5 is the position of the row within the whole.

When one applies the control-nested automorphism $\bar{g}_3 \text{\textcircled{S}} \bar{g}_4 \text{\textcircled{S}} \bar{g}_5$ to this, one obtains

$$[\{e\} \text{\textcircled{W}} \mathbb{R}]_{c_3,c_4,c_5} \bar{g}_3 \text{\textcircled{S}} \bar{g}_4 \text{\textcircled{S}} \bar{g}_5$$

which is the other copy of the side. Evaluating this expression, one gets

$$[\{e\} \text{\textcircled{W}} \mathbb{R}]_{g_3 c_3, g_4 c_4, g_5 c_5}$$

in accord with Definition 4.3 (p. 126). Since the successive elements g_3 , g_4 , and g_5 , each act by conjugation, this second side is the *control-nested τ-conjugate* of the first side.

Now, each of the two sides is a grouping in its own right. Furthermore, these two groupings are on the same level as each other. With this in mind, we are now conceptually ready to make the following statement:

EXTENSION OF LAW OF GROUPING.   *All occurrences of a grouping $G_1 \text{\textcircled{W}} \dots \text{\textcircled{W}} G_i$ in a perceptual organization, are of the form*

$$[G_1 \text{\textcircled{W}} \dots \text{\textcircled{W}} G_i]_{c_{i+1},c_{i+2},\dots,c_n} \text{\textcircled{S}} \bar{g}_{i+1} \text{\textcircled{S}} \bar{g}_{i+2} \text{\textcircled{S}} \dots \text{\textcircled{S}} \bar{g}_n$$

where $\text{\textcircled{S}} \bar{g}_{i+1} \text{\textcircled{S}} \bar{g}_{i+2} \text{\textcircled{S}} \dots \text{\textcircled{S}} \bar{g}_n$ is a member of the control-nested τ-automorphism group $G_{i+1} \text{\textcircled{W}} G_{i+2} \text{\textcircled{W}} \dots \text{\textcircled{W}} G_n$.

The Law of Grouping, as stated on p. 138, concerns the structure downwards in a grouping. The extension just given concerns the structure above a grouping.

This proposal becomes crucial, for example, to our theory of robot workspaces. A grouping can be considered to be a workspace, because it is defined by actions. The control-nested τ-automorphisms transfer workspaces around the environment. The system of transfer actions is itself a workspace. Thus, one gets what we call a *transfer diagram*:

$$\textit{Workspace-1} \xrightarrow{\textit{Workspace-2}} \textit{Workspace-1}.$$

where to the left and right of the arrow there are two copies of *Workspace-1*, and over the arrow, creating the transfer, there is *Workspace-2*, which is the system of control-nested τ-automorphisms.

As a simple example, a side is really a workspace - e.g., it could be the edge of a block along which a robot finger can move. Furthermore, the system of transfer actions from side to side is also a workspace. Thus, as an example of the above transfer diagram, one gets the following mapping between sides:

$$[\{e\} \,ⓦ\, \mathbb{R}]_{c_3,c_4,c_5} \xrightarrow{\mathbb{Z}_4 \,ⓦ\, \mathbb{Z}^H \,ⓦ\, \mathbb{Z}^V \overset{\bar{g}_3\text{Ⓢ}\bar{g}_4\text{Ⓢ}\bar{g}_5}{\in}} [\{e\} \,ⓦ\, \mathbb{R}]_{g_3c_3,g_4c_4,g_5c_5}.$$

Of course, the view that groupings are workspaces (and vice versa) is an example of our view that cognition organizes the environment as machines, and that intelligence consists of maximizing transfer of machines. A workspace, or grouping, is a machine. The transfer structure is given by wreath products of machines.

5.8 Product Ordering

We have so far examined Fig. 5.1a. Let us now establish what underlies the difference between the perceptual organization of Fig. 5.1a and Fig. 5.1b.

Observe first that the generative structure of Fig. 5.1a and 5.1b use the *same groups*. This is because the configurations have the same symmetries, and the generative structure comes from the symmetries, by the Symmetry-to-Trace Conversion Principle. However, what is different about the two configurations is this: In Fig. 5.1a, the group of vertical movements, as control, acts on a horizontal row as fiber; and in Fig. 5.1b, the group of horizontal movements, as control, acts on a vertical column as fiber. This means their respective wreath products have a *different order*, as follows:

$$\{e\} \,ⓦ\, \mathbb{R} \,ⓦ\, \mathbb{Z}_4 \,ⓦ\, \mathbb{Z}^H \,ⓦ\, \mathbb{Z}^V \tag{5.10}$$

$$\{e\} \,ⓦ\, \mathbb{R} \,ⓦ\, \mathbb{Z}_4 \,ⓦ\, \mathbb{Z}^V \,ⓦ\, \mathbb{Z}^H \tag{5.11}$$

i.e., the last two factors are reversed in the two sequences.

We shall now study the issue of product ordering. A basic fact will be important to us: A wreath product is an asymmetric construction, in the sense that, the control group makes copies of the fiber group, but the fiber group does not make copies of the control group.

Let us now ask why it is that the perceptual system chooses the specific wreath-product orderings that it does. The answer consists of applying the History Symmetrization Principle, which says that the history must contain minimal distinguisability (Sect. 2.14). When the metric being considered is Euclidean, as it is in the visual domain, the principle can be regarded as the History Minimization Principle.

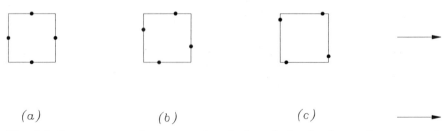

(a)   (b)   (c)

Fig. 5.3. Incorrect groupings are produced when the levels of control are reversed.

We now claim that this principle determines the ordering of the wreath components. First let us illustrate this claim by considering a square, and then give a general proof of the claim.

The square has two non-trivial levels: $\mathbb{R}$ describing a side, and $\mathbb{Z}_4$ describing the rotations between the sides. The wreath product of these two groups has two *alternative orders*:

$$\mathbb{R} \textcircled{w} \mathbb{Z}_4$$
$$\mathbb{Z}_4 \textcircled{w} \mathbb{R}$$

Each of these orders is of course preceded by the group $\{e\}$, representing a point.

The preceding sections have studied the *first* of these two orders. Let us however now consider the *second* order, $\mathbb{Z}_4 \textcircled{w} \mathbb{R}$. In this case, the fiber group becomes $\mathbb{Z}_4$ and the control group becomes $\mathbb{R}$. This means that $\mathbb{R}$ now makes copies of the fiber $\mathbb{Z}_4$. What do these fiber copies look like?

To answer this, note that the wreath product, now being examined, is really

$$\{e\} \textcircled{w} \mathbb{Z}_4 \textcircled{w} \mathbb{R}.$$

Thus, working from left to right, along this sequence, there is first a point, $\{e\}$, which (without loss of generality) can be assumed to be the central point on the top side. To this point, one applies the four rotations $\mathbb{Z}_4$. This results in the four points shown in Fig. 5.3a. Then, at the end of this wreath product one applies $\mathbb{R}$, the translation group, to produce the copies of $\mathbb{Z}_4$. There is some ambiguity about how to apply translations to Fig. 5.3a, but for the sake of argument, one can consider that the copies of $\mathbb{Z}_4$ produced by $\mathbb{R}$ are the sets shown in Fig. 5.3b, c, ..., etc. That is, these are the translations of the four points along their respective sides, by equal amounts on each side.

The crucial fact however occurs at the level of the fiber $\mathbb{Z}_4$ below $\mathbb{R}$. In Fig. 5.3a, one sees that the trajectory taken by $\mathbb{Z}_4$ around the square is four points that are separated by large distances. Most importantly, although $\mathbb{Z}_4$ has gone all around the square in Fig. 5.3a, it has accounted for very little in the square, i.e., only four points. The same is true of each of the other fiber copies of $\mathbb{Z}_4$, shown in Fig. 5.3b, c, ..., etc.

The consequence is that this wreath product produces a history that is very long: The history consists of the trajectories shown in Fig. 5.3a, b, c, ..., etc. That is, one starts by drawing the successive four points around Fig. 5.3a, then one draws the successive four points around Fig. 5.3b, then one draws the successive four points around Fig. 5.3c, and one continues like this till one has produced all the points on the square. This means that, in completing the history, one will have gone *many times around the square*, because each time, one could put down only four points.

Let us now contrast this with the alternative ordering for the wreath product of a square - the ordering which was studied in Chapter 3:

$$\{e\} \ \textcircled{w} \ \mathbb{R} \ \textcircled{w} \ \mathbb{Z}_4.$$

We will call this, the *standard ordering*. Working from left to right, along this sequence, there is first a point, $\{e\}$. To this one applies the translations, $\mathbb{R}$. This results in a single side. Then, at the end of the wreath product one applies, $\mathbb{Z}_4$, the four rotations. This merely rotates the side around the square.

Concerning this history, observe the following: (1) The fiber group $\mathbb{R}$, of translations, takes extremely *small* steps (infinitessimal) from one point to the next along an individual side. These steps are then duplicated by the control group, and therefore the history consists of copies of small steps. This contrasts with the history in Fig. 5.3, where the fiber (e.g., Fig. 5.3a) involved large steps, and these large steps were duplicated in the other fiber copies (Fig. 5.3b, c, a, etc.), thus resulting in a history consisting of copies of large steps. (2) Next, in the *standard ordering*, $\mathbb{Z}_4$ is the control group, and is therefore used *only once*; i.e., in any wreath product, it is the *fiber* that is copied, *not* the *control* group. Thus the entire standard history involves going around the square *only once*. This contrasts with the history in Fig. 5.3, which involved going around the square many times; i.e., because $\mathbb{Z}_4$ is the *fiber* in Fig. 5.3, and, in a wreath product, it is always the fiber that is copied.

The above argument shows that the standard ordering gives a much shorter history than the alternative ordering. We now generalize this argument to all cases, and thus establish the relationship between wreath-product orderings and the History Minimization Principle. To do so, consider two arbitrary groups that must be brought together in a wreath product. The goal is to understand which ordering satisfies the History Minimization Principle, as follows:

The first important point to understand is that the stimulus set (data set) is the same for both of the alternative wreath-product orderings. The individual groups involved represent *trajectories* through the same set. The group *elements* are the steps in these trajectories.

Now, let us suppose that the elements in one group take short steps in the stimulus set, and the elements in the other group take long steps in the stimulus set. Call the first group, the *short-step* group, G_S, and call the second group, the *long-step* group, G_L. Our question is this: In which order should these two groups be put together, in the wreath product? That is, should the order be $G_L \textcircled{w} G_S$, or should it be $G_S \textcircled{w} G_L$?

Let us consider the first alternative, $G_L \textcircled{w} G_S$. That is, in this alternative, the assumption is that the *fiber* group is the *long-step* group. It will generate a trajectory in which the steps are long, and this will be called a *long-step trajectory*. The *control* group then makes copies of this long-step trajectory. In other words, it creates further long-step trajectories. So, the *history* that generates the stimulus set is a concatenation of long-step trajectories!

Now consider the other ordering for the wreath product, that is $G_S \textcircled{w} G_L$. Most crucially, the *fiber* group is now the *short-step* group G_S. It will generate a trajectory in which the steps are short, and this will be called a *short-step trajectory*. The *control* group then makes copies of this short-step trajectory. In other words, it creates further short-step trajectories. So the history that generates the stimulus set is a concatenation of short-step trajectories! Furthermore, in this history, the long-step group is the *control* group, and is therefore used *only once*; i.e., in any wreath product, it is the *fiber* group that is copied, *not* the *control* group. This contrasts with the other wreath-product ordering, which involves a number of copies of the long-step trajectory, because, in this case, the long-step trajectory was the *fiber* group. Thus we conclude:

CONTROL-NESTING: ORDERING RULE.   *The order in which components are control-nested is determined by the following rule: Smaller steps are assigned to lower levels of control, and larger steps to higher levels of control. Only this will satisfy the History Minimization Principle.*

5.9   Local-to-Global in a Wreath Product

The argument given in Sect. 5.8 on product-ordering can be regarded as related to the argument that we will now give:

Ultimately, there is only one issue that determines the ordering in a wreath product: Higher levels must transfer lower levels. The following statement can be considered to be a corollary of this:

LOCAL-TO-GLOBAL ORDERING.   *Because the fiber group acts only within a fiber, and the control group acts across the fibers, the fiber-to-control ordering, along a wreath product $G_1 \textcircled{w} G_2 \textcircled{w} \ldots \textcircled{w} G_n$, tends to correspond to a local-to-global ordering.*

Examples so far in this book include:

$$a \quad b \quad c \quad d \quad e \quad f \quad g \quad h$$

Fig. 5.4. Another classic Gestalt example.

(1) Structure of square: This is given by $\mathbb{R}\,\widehat{w}\,\mathbb{Z}_4$. The lower level $\mathbb{R}$ is the local action within side. The higher level $\mathbb{Z}_4$ is the global action across sides.

(2) Gestalt grouping: Rows of squares in a vertical array. This is given by $\mathbb{Z}_H\,\widehat{w}\,\mathbb{Z}_V$. The lower level $\mathbb{Z}_H$ is the local action within fiber (a row). The higher level $\mathbb{Z}_V$ is the global action across rows.

(3) Robotics: On p. 19, we gave the structure of a serial-link manipulator to be $SE(3)_1\,\widehat{w}\,SE(3)_2\,\widehat{w}\ldots\widehat{w}SE(3)_n$. The highest control level is with respect to world frame. The successive levels downwards are successively more local because they belong to the successively more dependent smaller limbs; e.g., they are successively down the arm from torso to finger.

5.10 Perceptual Effect of Inclusion and Omission of Levels

Another way of showing that the perception of a grouping depends crucially on the arrangement of the transfer structure, is by causing the omission of one of the levels in the wreath hierarchy. To illustrate this, let us consider a classic Gestalt effect due to Wertheimer (cited in Rock [128] p273). In the sequence of dots in Fig. 5.4, the separation between pairs ab, cd, ef and gh is only slightly less than between bc, de, fg and hi. Depending on whether the observer sees this difference in spacing, a different percept arises: The dots are perceived either as a line partitioned into pairs, or as a homogeneous line (without pairing). According to our theory, the visual difference is determined by the difference in the respective structures of *transfer*.

Let us look at the *non-homogeneous* case first (i.e., where the pairing occurs). There are three levels of symmetry in this configuration. These are given by the following three groups:

$\{e\}$, which maps a single point to itself.

$\mathbb{Z}_2$, which describes the fact that an individual *pair* of points is reflectionally symmetric about its bisecting vertical axis.

$\mathbb{Z}$, which describes the fact that the *pairs* of points are indistinguishable from each other under integer translations.

Observe that these symmetries are arranged in a hierarchical transfer structure: $\mathbb{Z}_2$ *transfers* the symmetry structure $\{e\}$ from one point to another point within a pair, and $\mathbb{Z}$ *transfers* the symmetry structure $\mathbb{Z}_2$ from one pair to the next. Thus, the full structure is the following hierarchy of nested control:

$$\{e\} \ⓦ\ \mathbb{Z}_2 \ⓦ\ \mathbb{Z}.$$

Now, according to the Law of Grouping, the perceptual groupings must be given by the left-subsequences, which, in this case, are:

$\{e\}$	a point
$\{e\} \ⓦ\ \mathbb{Z}_2$	a pair
$\{e\} \ⓦ\ \mathbb{Z}_2 \ⓦ\ \mathbb{Z}$	the whole.

We now use the Extension to the Law of Grouping (p. 148) to enumerate the entire set of occurrences of these groupings in the organization. According to the Extension, all the occurrences of a grouping $G_1 \ⓦ\ \dots \ⓦ\ G_i$, in a perceptual organization, are of the form

$$[G_1 \ⓦ\ \dots \ⓦ\ G_i]_{c_{i+1},c_{i+2},\dots,c_n} \ⓢ\bar{g}_{i+1}\ⓢ\bar{g}_{i+2}\ⓢ\dots\ⓢ\bar{g}_n$$

where $\ⓢ\bar{g}_{i+1}\ⓢ\bar{g}_{i+2}\ⓢ\dots\ⓢ\bar{g}_n$ is any member of the control-nested τ-automorphism group $G_{i+1} \ⓦ\ G_{i+2} \ⓦ\dots\ⓦ\ G_n$. That is, the occurrences of the grouping are its *control-nested τ-conjugates* with respect to the latter group. Thus, let us now take each of the above left-subsequences, in turn, and elaborate the conjugates:

(1) The control-nested τ-conjugates of $\{e\}$

Choose a particular point. It is of the form

$$\{e\}_{c_2,c_3}$$

where the two indexes c_2 and c_3 are from the two levels *above* the level of points in the hierarchy. The control-nested τ-conjugates of this point must be of the form

$$\{e\}_{c_2,c_3}\ⓢ\bar{g}_2\ⓢ\bar{g}_3$$

where $\bar{g}_2\ⓢ\bar{g}_3$ comes from the control-nested τ-automorphism group $\mathbb{Z}_2 \ⓦ\ \mathbb{Z}$.

There is something important that must be understand about this structure. We are actually considering how a point is moved along the sequence of points, i.e., transferred onto any other point in the organization. The above expression tells us that the point cannot be moved *directly* onto any other point. This is because there is an intervening level $\mathbb{Z}_2$ between $\{e\}$ and $\mathbb{Z}$ in the wreath structure $\{e\}ⓌZ_2Ⓦ\mathbb{Z}$. Thus for example, $\mathbb{Z}$ cannot move a point to *its adjacent point within a pair*. It is only $\mathbb{Z}_2$ that can do this. Thus if one wants to move the point to an arbitrary point along the row, one has to use $\mathbb{Z}_2$ to move the point within its pair, and then use $\mathbb{Z}$ to move the point out of the pair to the target point in some other pair. That is, movement along the line of points is *control-nested*, exactly as illustrated in Fig. 4.5 (p. 125). Therefore the above expression gives each point a *control-nested position* along the line. This is different from simply its linear position along the line.

(2) The control-nested τ-conjugates of $\{e\} \textcircled{w} \mathbb{Z}_2$

Choose a particular *pair*. It is of the form

$$[\{e\} \textcircled{w} \mathbb{Z}_2]_{c_3}$$

where the index c_3 comes from the level *above* the level of pairs in the hierarchy. The conjugates of this point must be of the form

$$[\{e\} \textcircled{w} \mathbb{Z}_2]_{c_3} \textcircled{s} g_3$$

where g_3 comes from $\mathbb{Z}$. The set of such conjugates must be the set of pairs in the organization. Each is transferred onto each other by whole-number steps in $\mathbb{Z}$.

(3) The control-nested τ-conjugates of $\{e\} \textcircled{w} \mathbb{Z}_2 \textcircled{w} \mathbb{Z}$

There is only one control-nested conjugate of $\{e\} \textcircled{w} \mathbb{Z}_2 \textcircled{w} \mathbb{Z}$ in the sequence

$$\{e\} \textcircled{w} \mathbb{Z}_2 \textcircled{w} \mathbb{Z}.$$

It is the sequence $\{e\} \textcircled{w} \mathbb{Z}_2 \textcircled{w} \mathbb{Z}$ itself. This is of course the whole line. Thus a consequence of the Law of Grouping is that the whole is a grouping itself - which accords with perceptual experience.

The reader can easily check that any subsets of the data set, other than the control-nested τ-conjugates of left-subsequences, do not produce perceptual parts. This is because any such subsets break out of the transfer structure.

Let us now turn to the homogeneous case. As in the non-homogeneous case, we first consider the levels of symmetry. There are two such levels:

$\{e\}$, which maps a single point to itself.

$\mathbb{Z}$, which describes the fact that the points are indistinguishable under translations by integer amounts.

These symmetries are arranged in a hierarchical structure, thus:

$$\{e\} \, \widehat{\mathbb{W}} \, \mathbb{Z}.$$

Now let us look at the control-nested τ-conjugates of left-subsequences. The first left-subsequence is $\{e\}$, a point. Any particular point is a copy

$$\{e\}_{c_2}$$

where c_2 comes from the control level above $\{e\}$. The conjugates of this copy are all of the form

$$\{e\}_{c_2} \widehat{\mathbb{S}} g_2$$

where g_2 comes from $\mathbb{Z}$. This means that $\mathbb{Z}$ now acts *directly* on a point, and can therefore move the point onto any other, without an intermediate group. This accounts for one's sense that the percept is homogeneous.

In this section, we have seen the critical role of intermediate transfer levels in determining the percept. Such levels cause non-accessibility of control groups above the intermediate levels to fibers below the intermediate levels. In Chapter 9, we shall see that the same mathematics expresses what the Gestalt psychologists called the separation of systems in *motion perception*.

5.11 Non-iso-regular Groups

The illustrations used so far in this chapter have been given by iso-regular groups. However, the Law of Grouping applies equally to structures that are not iso-regular, as will be seen many times in the book. Nevertheless, the iso-regularity condition has an important general status with respect to grouping as follows: The Externalization Principle (p. 53) implies that any shape is perceived as having an underlying iso-regular group, e.g., a deformed shape is perceived as having an underlying un-deformed shape, and the latter is characterized by an iso-regular group.

The relation between this and grouping is that the groupings of the underlying iso-regular group are preserved upwards into the non iso-regular structure. The reason is that the iso-regular structure is a left-subsequence of the non iso-regular structure.

As an example, consider a cylinder, as shown in Fig. 5.5a. We have said (p. 14) that it is structured as the iso-regular group

$$\{e\} \, \widehat{\mathbb{W}} \, SO(2) \, \widehat{\mathbb{W}} \, \mathbb{R}. \tag{5.12}$$

Notice that the Law of Grouping therefore predicts that the cross-sections are seen as visual groupings because they correspond to the left-subsequence

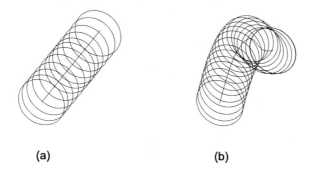

(a)   (b)

Fig. 5.5. A cylinder and a deformed version.

$\{e\}$ Ⓦ $SO(2)$. In fact, the set of cross-sections are the τ-conjugates of this fiber in the overall group.

Now let us apply a deformation to the cylinder, e.g., as shown in Fig. 5.5b. One method of producing a deformation is to apply a spline-shaping process (e.g., to the axis). This is formalized in Chapter 17 by adding spline-control as a group $\mathcal{H}$ of tensors above the group of the cylinder, thus:

$$\{e\} \text{ Ⓦ } SO(2) \text{ Ⓦ } \mathbb{R} \text{ Ⓦ } \mathcal{H}. \tag{5.13}$$

The group $\mathcal{H}$ then deforms the cylinder below it in the wreath hierarchy.

Now notice, in expression (5.13), that the left-subsequence $\{e\}$ Ⓦ $SO(2)$, corresponding to the cross-sections, still exists. Therefore, one still understands the deformed cylinder as having cross-sections as groupings. This is a general result: The groupings deduced from the iso-regular structure are preserved upwards through the deformation. Notice the relation between this and the issue of recoverability. The constraint of recoverability forces the generative sequence to be what we called *asymmetry-building* rather that *symmetry-breaking* (Sect. 2.18). Symmetry-breaking proceeds by the destruction of symmetry groups, whereas asymmetry-building proceeds by increasing the symmetry group by group extension via wreath product; i.e., transfer. The extending group is a symmetry group of the asymmetrizing process. In this way, one ensures recoverability.

Notice that this process of extension increases the number of groupings, exactly as one would want. For example, the cross-sections of the non-deformed cylinder now have additional conjugates within the deformed cylinders. This is because the group $\mathcal{H}$ has been added as an extra level of conjugation; thus, in expression (5.13), the cross-section $\{e\}$ Ⓦ $SO(2)$ now has a control-nested τ-automorphism group given by the right sequence $\mathbb{R}$ Ⓦ $\mathcal{H}$. This exactly captures the perceptual situation.

Notice also that the addition of the new level $\mathcal{H}$ means that we have a unification of the class of deformed cylinders within one grouping. This is

Fig. 5.6. A square distorted by projection.

because the cylinder is a left-subsequence of expression (5.13), and any of its deformed versions

$$[\{e\} \ⓌＷ \ SO(2) \ⓌＷ \ \mathbb{R}]_{c_4} \ⓈＳ \ g_4 \tag{5.14}$$

for all $g_4 \in \mathcal{H}$ are conjugates via $\mathcal{H}$. This captures the psychological sense that the cylinders are perceptually related to each other.[2]

Thus grouping applies not just to the structure of a particular shape but to systems of shapes. As another example, consider a projectively distorted square shown in Fig. 5.6. We saw on p. 17 that this percept is structured by the following wreath product:

$$\{e\} \ⓌＷ \ \mathbb{R} \ⓌＷ \ \mathbb{Z}_4 \ⓌＷ \ PGL(3, \mathbb{R}) \tag{5.15}$$

where the first three factors give the structure of the square (the iso-regular group), and the last gives the projective group creating the distortion. Notice that the left-subsequence $\{e\} \ⓌＷ \mathbb{R} \ⓌＷ \mathbb{Z}_4$ has conjugates

$$[\{e\} \ⓌＷ \ \mathbb{R} \ⓌＷ \ \mathbb{Z}_4]_{c_4} \ⓈＳ \ g_4 \tag{5.16}$$

where g_4 is any element in the projective group $PGL(3, \mathbb{R})$. An example of expression (5.16) is Fig. 5.6 itself. In other words, the expressions of the form (5.16) for all $g_4 \in PGL(3, \mathbb{R})$ give the set of projectively distorted squares. According to our theory of grouping, any such distorted square must be a grouping, because it is a left-subsequence. Furthermore, the perceptual system unifies the class of distorted squares within one grouping via $PGL(3, \mathbb{R})$. This again illustrates our theory that transfer is the process that creates cognitive unification.

---

[2] For consistency, we shall often use the semi-direct product symbol for situations like expression (5.14) where the control-nested τ-automorphism has only one level. This makes the notation easy to extend when higher groups are added to this structure. Notice that, in any situation like (5.14), a bar is not needed on the group element above the semi-direct product symbol; i.e., because the vector, in this case, has only one component (g_4, in the present case). This is generally the case for the highest control level. However, when levels higher than this are added, the bar will be needed.

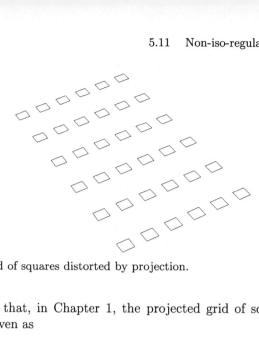

Fig. 5.7. A grid of squares distorted by projection.

Recall also that, in Chapter 1, the projected grid of squares shown in Fig. 5.7 was given as

$$\mathbb{R} \, \textcircled{w} \, \mathbb{Z}_4 \, \textcircled{w} \, \mathbb{Z}^H \, \textcircled{w} \, \mathbb{Z}^V \, \textcircled{w} \, PGL(3, \mathbb{R}). \qquad (5.17)$$

We can characterize groups of this type in the following way:

Definition 5.2. *A* **projective generative crystallographic group** *is a group of this form*

$$G \; = \; H \, \textcircled{w} \, [\mathbb{Z} \, \textcircled{w} \, \mathbb{Z} \, \textcircled{w} \, \dots \, \textcircled{w} \, \mathbb{Z}] \, \textcircled{w} \, PGL(n+1, \mathbb{R}). \qquad (5.18)$$

where H is iso-regular, and the wreath-lattice $[\mathbb{Z} \, \textcircled{w} \, \mathbb{Z} \, \textcircled{w} \, \dots \, \textcircled{w} \, \mathbb{Z}]$ is n-fold.

The groupings can then be read off the full wreath sequence in accord with the Law of Grouping. This can easily be checked with the example given by expression (5.17) above.[3]

---

[3] A final level of complexity will concern unfolding groups, to be defined later. In such groups, a control level can act selectively on some lower level which we will call an *unfolding fiber*. In these cases, we replace an ordinary fiber by an unfolding fiber, in the theory of grouping. A grouping is then defined by its transfer, as is basic to our theory.

6. Robot Manipulators

6.1 Three Algebraic Conditions

A generative theory of shape encodes shape by a system of *actions*. An important part of our approach has been to develop a theory of how actions are organized intelligently. This theory is equally applicable to perceptual systems as it is to motor systems. In the present chapter, we apply this theory to motor systems. We will return to this topic a number of times in the book to develop it further, e.g., in Chapter 14 on mechanical design.

The present chapter concentrates on the structure of manipulators. Significant work on manipulators has been done in robotics. Our theory differs from current approaches in that it attempts to fulfill three conditions:

(1) We argue that perceptual systems and motor systems are structured in the same way. As indicated above, this follows from our generative theory.

(2) The second condition concerns rigidity. It is standardly assumed that, because the successive matrices that relate successive links from the hand to the manipulator-base are members of the special Euclidean group $SE(3)$, then the group linking the hand to base must be $SE(3)$. However, we argue that this is not the case. If the group were $SE(3)$ then the manipulator (e.g., arm) would be rigid, i.e., frozen. But it is not. The very fact that there are joints means that the manipulator is not rigid. We will call structures that are rigid, except at a discrete set of points, *semi-rigid*. The goal is to develop groups to describe such structures. We shall call such groups, *semi-rigid groups*.

(3) The third condition is to express the object-oriented structure of manipulators. Manipulators have an inheritance hierarchy. We need to develop a group theory that describes such hierarchies. This will be done in Chapter 7.

Thus our three conditions are as follows:

**THREE CONDITIONS ON A GROUP THEORY OF MANIPU-
LATORS.** *The group theory of manipulators should express:*

   (1) Perceptual-Motor Equivalence.
   (2) Semi-Rigidity.
   (3) Object-Oriented Structure.

Satisfying these three conditions will allow us to establish what we will call
the *full group of a manipulator*. For example, in a later section, we will systematically elaborate a single symmetry group that captures the entire joint
structure of the human body.

6.2   Object-Centered Frames as Transfer

All robot motion and manipulation exploits *object-centered frames*: i.e., coordinate frames fixed to the object being moved (including to the robot itself).
In this section, we show that the power of such frames is that they maximize
transfer.

Notation 6.1 Any point in 3-space, described relative to a coordinate frame
R, will be specified by coordinates (x^R, y^R, z^R).

Suppose first that one has only a single Cartesian reference frame, W, the
world-frame fixed to the environment. In this environment, there is an object
to which one applies a *translation T*, for example, moving it from left to right
as shown in Fig. 6.1. Suppose that one knows the coordinates (x_1^W, y_1^W, z_1^W)
of each point in the object, prior to the translation. In order to specify the new
position of the object, there is no alternative but to specify the coordinates
(x_2^W, y_2^W, z_2^W) of each point of the object in the second position. The only
way to arrive at these new coordinates is to perform the calculation for each
point in the object *individually*, e.g., by adding the translation vector to each
of the points in the previous position.

   However, let us now consider an alternative strategy. If, in addition to the
world-frame W, one uses a *second* frame which is defined *relative* to the first
frame, then the calculation can be simplified enormously, as follows: Let the
second frame be a Cartesian frame F that is embedded in the object such
that it is *fixed* relative to the object, as shown in Fig. 6.2. Then all points
in the object are now described relative to frame F thus: (x^F, y^F, z^F).
Furthermore, this description is the *same* before and after the translation. It
remains only to establish the *relationship* between the object-frame F and

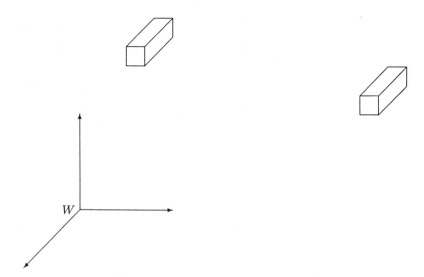

Fig. 6.1. Use of only global frame.

the world-frame W. This relationship is given simply by a vector translating the origin of the world-frame W to the origin of the object-frame F. To anticipate the later discussion, we shall call any such vector a *control* vector. The control vectors for the object in the first and second position are v_1 and v_2, respectively, as shown in Fig. 6.2.

Obviously the translational effect on the object is given by the vector

$$v = v_2 - v_1.$$

However, a considerable simplification has occurred. Rather than having to apply this vector to *all* the points in the object, one needs to apply it to only the origin of frame F. This is because the points in the object have a fixed relation to the frame F.

Generally, the calculation of the position of any object-point, relative to the world-frame W, is obtained by simply adding the point's fixed description within the object-frame F to the control vector going from W to F. Thus, there is a two-step decomposition from the object-point back to the origin of W. This means that, in any movement, one can *re-use* the previous object-centered description. The reader can see that this **re-usability** is actually an example of *transfer*; i.e., the object-centered description is *transferred* across locations. The only additional thing required is the calculation of the two vectors, v_1 and v_2, which relate the positions of F back to W. This itself embodies the notion of transfer because it relies on the fact that all the

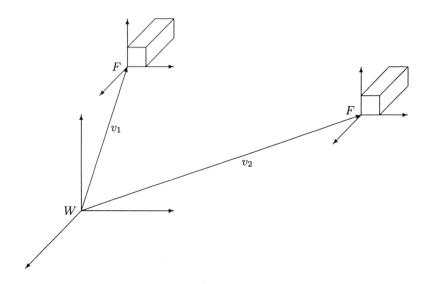

Fig. 6.2. Use of object-centered frames.

changes go on in the control vectors, not in the object-frame. We argue that this is best formalized using a wreath product, thus:

OBJECT-CENTERED FRAMES & WREATH PRODUCTS. *The use of an object-centered frame, to specify object movement, maximizes transfer. The transfer structure is given by the wreath product*

$$G_1 \textcircled{w} G_2 \tag{6.1}$$

where the fiber group G_1 specifies the relationship of object-points to the object-frame, and the control group G_2 specifies the relationship between the object-frame and the world-frame.

The main form of the two groups G_1 and G_2 will be defined as follows:
So far only translations have been considered. However, these considerations can easily be extended by adding rotations. The group generated by translations and rotations is the special Euclidean group, $SE(3)$. Define a *configuration* of an object-frame F with respect to a world-frame W to be a member of $SE(3)$. Note that $SE(3)$ gives the space of alternative configurations of F with respect to W. Observe that two ammendments need to be made to the purely translational situation: First, the two vectors v_1 and v_2 in Fig. 6.2 need to be replaced by two members g_1 and g_2 of the group $SE(3)$.

Second, the relationship *between* the two placements of F is no longer simply the difference between the two vectors

$$v = v_2 - v_1$$

but the "difference" between the two group members g_1 and g_2, thus:

$$g = (g_2)(g_1)^{-1}$$

where, by group closure, g must also be a member of $SE(3)$, and must therefore also be a Euclidean motion.

Everything else we said about the translation case still applies. Most crucially, what was said about the object-centered frames as embodying transfer, still applies. Thus, we now have the wreath product

$$SE(3) \, ⓦ \, SE(3).$$

Clearly, the second $SE(3)$ is the group of *configurations* of the object frame. One could leave the first $SE(3)$ as simply the group of vectors characterizing the *positions* of points within F. However, as we will argue many times in this book, *the perceptual/motor system is recursive*, and one can substitute whole frames for individual points, and therefore it is useful to consider the first group also to be $SE(3)$. In this case, the individual points each carry a copy of the Cartesian frame. For example, this feature is useful for generative crystallographic groups which were defined in Sect. 5.4, and which we argue are fundamental to perception.

6.3   The Serial-Link Manipulator

The main purpose of our discussion of robot motion is to show that the representation of this motion maximizes transfer. As a mathematical consequence, we will be able to establish the full group of the serial-link manipulator. This will be accomplished in the next section.

The present section will briefly review the standard attachment of frames to a serial-link manipulator. Any reader who is familiar with this topic can go directly to the next section.

A serial-link manipulator is exemplified by its most important example - the human arm. It is a sequence of links (limbs) each connected to the next by an actuated joint. The sequence starts with what is called the *base* (e.g., the torso), and progresses outwards from the base to the end-effector (e.g., the hand or finger-tip). Each link corresponds to a bone in human biology, and it has a *proximal* end, that nearer the base, and a *distal* end, that nearer the effector.

Fig. 6.3. Frame-assignment in serial-link manipulator (2D).

The most common means of assigning Cartesian frames to a serial-link manipulator is based on the work of Denavit & Hartenberg [26]. In this description, there is one Cartesian coordinate frame attached to each link. Standardly, it is attached to the *distal* end of the link. This is illustrated on the left in Fig. 6.3, where the ith link is shown as the thick line labelled i. Its own Cartesian frame labelled F_i is attached to its distal end (the proximal to distal direction is left to right). For readability, this diagram shows the Cartesian frame to consist of only two orthogonal axes, but we will always assume a 3D orthogonal frame.

The right of Fig. 6.3 shows the addition of the next link in the distal direction, which is labelled $i - 1$. Its own frame F_{i-1} is also shown attached to its own distal end. For purely notational reasons that will emerge later, the links (and frames) are numbered in decreasing order from the proximal to the distal end - which is the reverse order from that usually used in the literature.

Most crucially, the relationship between frame F_i and frame F_{i-1} is given by a special Euclidean motion (rotation and translation), i.e., a member of $SE(3)$. This motion is standardly represented by a matrix called the *A-matrix* linking those two frames. Usually, this is given as a *homogeneous* 4×4 matrix:

$$\begin{pmatrix} R_{11} & R_{12} & R_{13} & t_1 \\ R_{21} & R_{22} & R_{23} & t_2 \\ R_{31} & R_{32} & R_{33} & t_3 \\ 0 & 0 & 0 & 1 \end{pmatrix} \tag{6.2}$$

where the upper-left 3×3 block represents the rotation component, and the upper-right block (t_1, t_2, t_3) represents the translation component.[1]

---

[1] The particular representation depends on the particular way a frame is aligned with the end of the limb. The most famous such alignment method is given by an algorithm of Paul [117], where the z-axis of the frame is made coincident

6.4   The Full Group of a Serial-Link Manipulator

A basic proposal of this book is that both the perceptual system and motor system are structured by the attempt to maximize *transfer of action spaces*. In this section, we are going to demonstrate this proposal with the serial-link manipulator. This will then allow us to develop the full group of the manipulator, which, to our knowledge has not previously been developed. Using this, we will be able to establish a rigorous algebraic definition of transfer in a manipulator.

Consider an n-chain serial-link manipulator. Let the transformation matrix from the base frame B to the effector frame E be labelled $_E T^B$. Standardly, in robotics, one decomposes this transformation in accord with the serial-link structure thus:

$$_E T^B \;=\; _E T^1 \;*\; _1 T^2 \;*\; \dots \;*\; _{n-2} T^{n-1} \;*\; _{n-1} T^B \qquad (6.3)$$

where each symbol $_{i-1} T^i$ denotes the Euclidean transformation from frame F_i (fixed to link i) to frame F_{i-1} (fixed to link $i-1$). Notice that we are writing the direction from base to effector as right-to-left in the above sequence, which is the opposite of the usual notation in the literature - this is done to provide consistency with the notation used throughout this book (as will soon become clear). Thus the frames are labelled n to 1, from proximal to distal. Each of the component transformations is given by an 4×4 homogeneous A-matrix, and forward matrix multiplication will have to be right-to-left.

Now, it is a matter of considerable elegance in robotics that the same mathematical language is used to describe the movement of (1) an object being manipulated and (2) any link within the manipulator itself. This is the language of object-centered coordinate frames. To illustrate, return to Fig. 6.2 (p.164). The diagram shows the object frame F being expressed in

---

with the joint axis (e.g., the rotation axis of a revolute joint), the x-axis is made parallel to the common normal between the two successive joint-axes, and the y-axis is then determined by the right-hand rule. Only four parameters are needed to determine the relationship between two successive frames: *two types of translation* - associated respectively with (1) the length of the link and (2) the distance between links - and *two types of rotation* - associated respectively with (3) the angle between links and (4) the twist angle between joint-axes. These are the four transformations that connect one frame to the next. When one multiplies these transformations together, one gets a single Euclidean transformation that relates one frame to the next. This is what is called the A-matrix. For example, the particular Euclidean transformation that goes from frame F_i to frame F_{i-1} in Fig. 6.3 is given by an A-matrix. Here we give some basic group-theoretic work on robotics and locomotion: Lie-group theoretic methods can be found in Karger & Novak [73], Brockett [16], Chevallier [21], Murray, Li and Sastry [112], Park, Bobrow & Ploen [115]. Also, interesting group-theoretic work has been done on locomotion by Collins & Stewart [24], Golubitsky, Stewart, Buono & Collins [46], and related work on coupled systems with wreath products by Dionne, Golubitsky, & Stewart [28], and Dias [27]. See also Gallistel [37] [38] for extensive analysis of coordinate systems in animal motion.

relation to the world-frame W, the relationship being given by a Euclidean transformation from W out to F. Now, exactly the same method is used to express the relationship between two successive links in a serial-link manipulator; that is, one makes the link frame F_i equivalent to the world-frame W - so frame F_i will be called the *referent* frame - and one makes the link frame F_{i-1} equivalent to the object-frame F.

For us, the power of this approach resides in the conclusion, in Sect. 6.2, that object-centered frames express *transfer*. Thus the sequence of transformations (6.3) expresses the serial-link manipulator as a *hierarchy of transfer*.

Based on this, we are now ready to give the full group of the serial-link manipulator. Each transformation $_{i-1}T^i$ in the above sequence comes from the group $SE(3)$. Furthermore, each expresses a structure of transfer, which we gave in Sect. 6.2 as a wreath product $SE(3) \wr SE(3)$, where the *control* group $SE(3)$ represents the outward Euclidean transformations from the referent frame to the object frame, and the *fiber* group $SE(3)$ represents the outward relation from the object frame to the object points. Using this relation recursively, we conclude:

FULL GROUP OF A SERIAL-LINK MANIPULATOR.   *The full group of the n-chain serial-link manipulator is the n-fold wreath-product:*

$$SE(3) \wr \ldots \wr SE(3) \wr SE(3).$$

i.e., a wreath poly-$SE(3)$ as in Definition 4.11.

This means, of course, that the Euclidean groups along this sequence are *control-nested*. That is, any group along this sequence acts as a control group on its entire left-subsequence. This embodies the physical fact that any limb can transfer the action space of its lower limbs through the environment. We will now go on to define transfer rigorously in the manipulator.

6.5   Transfer in the Serial-Link Manipulator

In order to algebraically define the structure of transfer in a serial-link manipulator, let us index each of the groups along the sequence $SE(3) \wr \ldots \wr SE(3) \wr SE(3)$ by the level on which it occurs; i.e., by the limb to which it corresponds. In particular, consider the group $SE(3)_{i+1}$ along this sequence, i.e., the group corresponding to limb $i+1$. First note that this group acts as a control group on its entire left-subsequence $SE(3)_1 \wr \ldots \wr SE(3)_i$. However, observe also that the entire right-subsequence $SE(3)_{i+1} \wr \ldots \wr SE(3)_n$ acts on the left-subsequence, moving it from one part of the environment to the other.

This gives us a rigorous expression of transfer as follows: Consider an element $Ⓢ\bar{g}_{i+1}Ⓢ\bar{g}_{i+2}Ⓢ\ldots Ⓢ\bar{g}_n$ from the right-subsequence. Transfer is the application of this element to a specific *copy* of the fiber below it. That is, transfer is the conversion of the copy at one position into the copy at another position.

Thus, let the copy at the first position be

$$[SE(3)_1 \ⓌĖ \ \ldots \ \ⓌĖ \ SE(3)_i]_{c_{i+1},c_{i+2},\ldots,c_n}.$$

The copy corresponds to a particular node on the bottom of the tree in Fig. 4.5 (p. 125). The sequence of indexes $c_{i+1}, c_{i+2}, \ldots, c_n$ give the chosen sequence of nodes that dominate the bottom node. Transfer is the action of moving the bottom node across the bottom level of the diagram. The transferring action of $Ⓢ\bar{g}_{i+1}Ⓢ\bar{g}_{i+2}Ⓢ\ldots Ⓢ\bar{g}_n$ is represented thus:

$$[SE(3)_1 \ Ⓦ \ \ldots \ Ⓦ \ SE(3)_i]_{c_{i+1},c_{i+2},\ldots,c_n}Ⓢ\bar{g}_{i+1}Ⓢ\bar{g}_{i+2}Ⓢ\ldots Ⓢ\bar{g}_n.$$

To find out where this node is moved to, we algebraically evaluate this expression, as follows: From each level $\bar{g}_j$ on the right, the relevant component g_j is really a control τ-automorphism $\tau(g_j)$. Therefore, the above expression actually means this:

$$\tau(g_n)\ldots\tau(g_{i+2})\tau(g_{i+1})[SE(3)_1 \ Ⓦ \ \ldots \ Ⓦ \ SE(3)_i]_{c_{i+1},c_{i+2},\ldots,c_n}.$$

However, recall that each $\tau(g_j)$ is the raised action from the control set C_j, to the subscripts on the fiber-group copies - note, the members of the control set C_j are the configurations of the jth limb. This means that, to apply $\tau(g_j)$, one merely applies g_j to the corresponding subscript c_j. Therefore the above expression becomes

$$[SE(3)_1 \ Ⓦ \ \ldots \ Ⓦ \ SE(3)_i]_{g_{i+1}c_{i+1},g_{i+2}c_{i+2},\ldots,g_nc_n}.$$

To summarize:

TRANSFER IN A SERIAL-LINK MANIPULATOR. *Transfer in a serial-link manipulator is defined by the control-nested τ-automorphisms; that is, as follows:*

$$[SE(3)_1 \ Ⓦ \ \ldots \ Ⓦ \ SE(3)_i]_{c_{i+1},c_{i+2},\ldots,c_n}Ⓢ\bar{g}_{i+1}Ⓢ\bar{g}_{i+2}Ⓢ\ldots Ⓢ\bar{g}_n.$$

$$= [SE(3)_1 \ Ⓦ \ \ldots \ Ⓦ \ SE(3)_i]_{g_{i+1}c_{i+1},g_{i+2}c_{i+2},\ldots,g_nc_n}.$$

The group we have given for the serial-link manipulator is considerably more complicated than the standard $SE(3)$, but embodies the information needed to express the hierarchy of limb control in the manipulator. In particular, the limb structure $SE(3)_1 \ Ⓦ \ \ldots \ Ⓦ \ SE(3)_i$ is pushed around the environment by its control-nested τ-automorphism group, $SE(3)_{i+1}$ Ⓦ ... Ⓦ $SE(3)_n$. Note that this means that transferred action spaces in the serial-link manipulator are control-nested τ-conjugates of each other.

The reader will recall, from our theory of perceptual grouping in Chapter 5, that groupings consist of control-nested τ-conjugates. The crucial connection between our theory of grouping and our theory of robot motion is the fact that, according to the theory of grouping, environmental cohesion is provided by the transfer of action spaces. Thus our approach achieves the first of the three group-theoretic conditions given on p. 162, that of making perception and action structurally equivalent.

6.6   The Full Group of a General-Linked Manipulator

So far, in this chapter, only serial-linked manipulators have been examined. We now generalize the discussion to producing the full group of general-linked manipulators. An example of such a manipulator is the human body where, for instance, the two arms are not in serial relation to each other.

The way we handle the general case is to introduce this rule:

PRODUCT RULE.   *The groups of independent units are joined by direct products, the groups of dependent units are joined by wreath products.*

We will now give the full group of the human body. This will be done by choosing one of the body parts as the base frame, and generating the full group by using the above product rule. The structure of the group is therefore entirely dependent on the choice of base frame. In this illustration, the torso will be chosen as base frame. Therefore we get this:

$$[[G_{Arm} \text{ Ⓦ } \mathbb{Z}_2] \times [G_{Leg} \text{ Ⓦ } \mathbb{Z}_2] \times G_{Head}] \text{ Ⓦ } G_{Torso}. \qquad (6.4)$$

Let us examine this sequence from right to left. On the far right, there is the group of motions of the torso. These are motions *relative* to the world frame (the torso is the base frame *within* the body). Immediately to the left of G_{Torso}, in the above sequence, there is a wreath-product symbol which indicates that everything to the left of this is judged relative to the torso. There are three groups to the left, connected by direct products, which means that they are independent. Let us go through these three groups from left to right. First, G_{Head} is the group of motions of the head *relative* to the torso. One knows that this is relative to the torso, because it is connected rightward to the group of the torso by a *wreath* product.

The second group is $[G_{Leg} \circledW \mathbb{Z}_2]$, which is the group for the pair of legs. The legs are reflectionally related to each other as spaces of action, and therefore we have the group $\mathbb{Z}_2$ mapping the group G_{Leg} of one leg onto the group G_{Leg} of the other leg.

The same structure holds for the two arms, thus giving the group $[G_{Arm} \circledW \mathbb{Z}_2]$ for the pair of arms.

Now, each arm and leg is a serial-link manipulator, and we have already worked out the group of such structures. Thus, one can make the following substitutions into the group sequence shown at (6.4):

$$G_{Arm} = G_{Hand} \circledW G_{ForeArm} \circledW G_{UpperArm} \tag{6.5}$$

$$G_{Leg} = G_{Foot} \circledW G_{Shin} \circledW G_{Thigh}. \tag{6.6}$$

Furthermore, one can continue to expand this as follows: First, the hand consists of a palm and five fingers. Thus into (6.5), make the substitution:

$$G_{Hand} = [G_{Finger_1} \times G_{Finger_2} \times \ldots \times G_{Finger_5}] \circledW G_{Palm} \tag{6.7}$$

where the *wreath* product to the left of G_{Palm} indicates that the frames of the fingers are linked referentially upwards to the palm; and the *direct* products between the groups G_{Finger_i} indicate that the fingers are independent of each other,

There remains only one final substitution stage: Each finger is a serial-link manipulator consisting of three segments. Therefore, into the sequence (6.7) we make the substitution:

$$G_{Finger} = G_{Segment_1} \circledW G_{Segment_2} \circledW G_{Segment_3} \tag{6.8}$$

where the wreath-product symbols indicate that the finger-segments are serially linked. Notice that this three-fold wreath product (6.8) is substituted for each of the five finger groups in (6.7), and therefore the direct products in (6.7) will be between three-fold wreath products.

Now, what we have done here, in successively expanding G_{Hand} in (6.5), should also done for G_{Foot} in (6.5). When all the successive substitutions are made back into (6.4), we obtain the full group of the human body.

Note the following about reading this group sequence: Each of the G-labeled groups is an instance of $SE(3)$. Furthermore, each relates a higher Cartesian frame to a lower one. The lower one is indicated by the suffix on the G-group. The higher frame is indicated by the suffix on the next higher G-group in the *wreath* hierarchy. It is worth considering an example: Return to the sequence:

$$[[G_{Arm} \circledW \mathbb{Z}_2] \times [G_{Leg} \circledW \mathbb{Z}_2] \times G_{Head}] \circledW G_{Torso}.$$

There are two kinds of group here: the G-labeled groups, and the $\mathbb{Z}_2$-labeled groups. To understand the reference-frame structure, consider as an example, the group G_{Arm} on the far left. It relates a higher frame to a lower frame. The lower frame is that indicated by its suffix "Arm". To find the higher frame, move right-ward to the next G-group in the *wreath* sequence; i.e., ignore the $\mathbb{Z}_2$ groups, and any other groups linked to G_{Arm} by direct products, and find that the next higher group is G_{Torso}. The lower reference frame of G_{Torso} is indicated by its suffix "Torso". Thus group the G_{Arm} relates the frame "Arm" indicated by its own suffix, to the frame "Torso" indicated by the suffix on the next higher group G_{Torso}.

Two final comments: First, we have developed a single group that captures the full motion of the human body - which is a highly complex object. This is the general power of the theory being elaborated in this book: giving *a single symmetry group for any arbitarily complex object*. The group theory will be developed much further in the following chapters.

Second, in Leyton [88], we gave an initial formulation of this theory, and based on this formulation, Kirupaharan & Dayawansa [77] developed an approach to posture control. Their approach is to elaborate successive stabilizations from the base frame through the hierarchy. Thus, an example which they consider in detail is posture control in a human infant, where the base is in fact, the head. Kirupaharan and Dayawansa give a dynamical equation for the head, and argue that the infant must first master a control law to stabilize that equation. Then, they formulate the larger problem of controlling the head and torso in such a way that the previous solution to head stabilization is used in solving this expanded problem. This process continues as the successive frames are added. Note that, in terms of the theory in the present section, this corresponds to successive expansions of the group sequences downward through the wreath hierarchy.

6.7   Semi-Rigid Groups

The group given in the previous section, for the human body, illustrates the class of groups we now wish to define:

Definition 6.1. *A group will be called a* **semi-rigid group** *if it is a wreath product, each of whose levels is a direct product of isometry groups or wreath products ... of isometry groups. The dots mean repeat, a number of times, the phrase "each of whose levels is a direct product of isometry groups or wreath products".*

Notice the way in which the group of the human body, given in expression (6.4), is a semi-rigid group: The fiber $[[G_{Arm} \, \textcircled{w} \, \mathbb{Z}_2] \times [G_{Leg} \, \textcircled{w} \, \mathbb{Z}_2] \times G_{Head}]$ is a direct product where the three components are G_{Head}, which is an isometry group $SE(3)$, and $[G_{Leg} \, \textcircled{w} \, \mathbb{Z}_2]$ and $[G_{Arm} \, \textcircled{w} \, \mathbb{Z}_2]$ which are each wreath products of isometry groups.

In fact, in this chapter, we have developed a group theory in which the semi-rigid structure of the manipulator comes from its *object-oriented struc-ture*. This will become clear in the next chapter, where we give an algebraic theory of object-oriented inheritance. We now have the following inclusion hierarchy of the classes of groups so far proposed in this book:

$$\text{iso-regular} \quad \subset \quad \text{wreath-isometric} \quad \subset \quad \text{semi-rigid}.$$

For example, notice that the group we gave for the square is semi-rigid.

6.8   Including Manipulator Shape

In this chapter, what was meant by the "full" group of a robot manipulator was the full group of its reference frames. In Chapter 14, this group will be extended to include manipulator shape. To obtain the full group in this extended sense, we will need to define a class of still more complex groups, which we will call *unfolding groups*. This will be added to the right of the above group inclusion hierarchy.

7. Algebraic Theory of Inheritance

7.1 Inheritance

The term inheritance in object-oriented programming refers to the passing of properties from a parent to a child. The child incorporates these parent properties, but also adds its own. The former properties will be called the inherited ones, and the latter will be called the personal ones.

This kind of structure covers two types of situation. The first is **class inheritance**. Here, a class (e.g., the class of squares) can inherit a property (e.g., having four sides) from a more extensive class (e.g., the class of rectangles). The former class is called the child class, and the latter class is called the parent class. Thus one says that child is related to the parent by the *is-a* relationship (e.g., a square *is-a* rectangle). This form of inheritance is specified in the actual software which defines the classes; i.e., it is independent of the particular run-time sessions in which the software is used. For example, a rectangle and square can be two classes of objects specified in the software text of a design program (see e.g., Meyer [108]).

The second type of inheritance will be called **(run-time) linking inheritance**. Here, parent-child relationships are set up within the run-time session. The sequence of events is as follows: At run time, the user creates specific examples from classes (e.g., specific examples of rectangles from the class of rectangles). These run-time examples are called *objects*; i.e., objects are run-time instances of classes. Once created, objects can be linked in parent-child relationships. For example, a sphere representing the moon can be linked, as child, to a sphere representing the earth, as parent. The moon will inherit properties of the earth. This linking is created in *3D Studio Max* by clicking the "link" button, and dragging the mouse from an object to another object, thereby defining them as linked in a child-parent relationship. This type of inheritance is basic to design. For example, in architectural design, a door can be inserted in a wall and inherit the position of the wall - so that if the wall is moved (in the design process), then the door will move with it. All major design programs allow one to form parent-child linking between objects, and to view these created hierarchies in detailed information windows as the design process continues. Design would be impossible without this facility. This type of inheritance is also the basis of serial-link manipulators, where the successive manipulator links are linked upward (distal-to-proximal) in successive child-parent relationships; i.e., each inheriting the transform of the parent above it, and adding its own.

Terminology 7.1 *Throughout this book, the term* **inheritance** *will mean (run-time) linking inheritance; unless explicitly stated as class inheritance.*

7.2   Geometric Inheritance

Any use of the term inheritance in this book will refer to what one can call *geometric* inheritance - inheritance with respect to shape or kinematic structures. Inheritance of this type rests ultimately on relationships between local coordinate frames because each object is created with its local coordinate frame. The object hierarchy is described as a hierarchy of nodes. At each node, there is a local coordinate system and a transformation matrix. Each node *inherits* the transformation of its parent and adds its own transformation.

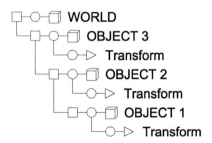

Fig. 7.1. The representation of parent-child relations in *3D Studio Max.*

A good diagrammatic representation of this is used by *3D Studio Max*, as illustrated in Fig. 7.1. Here, inheritance is represented by indentation - i.e., an indented object is a child of the next object above with respect to which it is indented. A line connects the child object to the parent object. In this illustration there are four objects, the World object at the top, and three successive objects below. The indentation sequence, in this example, indicates that each successive object is a child of the object above it. Now observe that each object, except the World object, has a transform shown just below it. The transform relates the coordinate frame of the object to the coordinate frame of its parent. This transform is the "personal" transform of the object. In addition, the object inherits the transform of its parent. The object therefore adds its personal transform to its inherited transform. This means, of course, that via its parent, it inherits the transform of its parent's parent, and so on.

7.3 Theory of Inheritance

We shall now give an algebraic theory of inheritance in geometric structures. Notice first that the structure of inheritance is unidirectional; i.e., inheritance is only downward in the hierarchy. To quote the manual of *3D Studio Viz*, "The linkage is unidirectional in that superior objects control subordinates but subordinates have no effect on their superiors."

We argue that this unidirectionality makes inheritance extremely appropriate for description in terms of wreath products. Thus we propose the following theory:

ALGEBRAIC THEORY OF INHERITANCE. *Inheritance in geometric structure corresponds to wreath products, as follows:*

$\mathfrak{I}_1$ **(Parent):** *A parent corresponds to a control group $G(C)$. A property of a parent is a member g of that group.*

$\mathfrak{I}_2$ **(Child):** *A child corresponds to a fiber-group $G(F)$. A property of a child is a member γ_g of a fiber-group copy $G(F)_g$, where $\gamma \in G(F)$ and $g \in G(C)$. Its personal property is γ. Its inherited property is g.*

$\mathfrak{I}_3$ **(Inheritance):** *Inheritance therefore arises from the wreath structure: It is the labeling of members of the fiber-group copies by members of the control group.*

As an example, consider any successive pair of links in a serial-link manipulator. According to our theory, this is algebraically characterized as follows: The parent property, i.e., the position/orientation of the parent frame, is a member g of a control group $G(C)$. A property of the child link is a member γ_g of a fiber-group copy $G(F)_g$. The copy index g is the inherited member from the control group. This sets the referent with respect to which γ is the personal transformation of the child's frame.

Now consider the example of a square $\mathbb{R} \textcircled{w} \mathbb{Z}_4$. A point is a child of the parent, a side. It is judged relative to the side on which it occurs. The orientation of the side sets the referent with respect to which the point is judged.

The *point* has two properties: (Inherited property) the orientation it inherits from the control group $\mathbb{Z}_4$; and (Personal property) the point's position within its particular side, i.e., the translation within the fiber group $\mathbb{R}$.

What is powerful about the theory given in $\mathfrak{I}_1$-$\mathfrak{I}_3$ above, is that it gives an algebraic theory of inheritance structure, as will now be seen. First observe this:

INHERITANCE AS TRANSFER. *Inherited properties $h \in G(C)$ send child properties γ_g onto child properties γ_{hg}.*

This is conceptually very important according to our theory. In our approach to shape and kinematics, transfer is what inheritance is about.

Notice that the group generated by the child properties corresponds to the fiber-group product. Thus we conclude that inherited properties $h \in G(C)$ act as automorphisms on that group. Generally, therefore, we conclude this:

INHERITANCE AS CONTROL-NESTED
τ-AUTOMOROPHISMS. *Since our theory implies that multi-level inheritance is given by an n-fold wreath product $G_1 \textcircled{w} G_2 \textcircled{w} \ldots \textcircled{w} G_n$, one can see that the inherited properties of any level i correspond to the members of the control-nested τ-automorphism group $G_{i+1} \textcircled{w} \ldots \textcircled{w} G_n$ of that level. That is:* **Inherited properties correspond to control-nested τ-automorphisms.**

Every shape and kinematic example in this book will illustrate this proposal.

Finally, notice that a parent can have several children. To handle the group theory of this situation, we offer this rule:

PRODUCT RULE FOR INHERITANCE. *The transforms G_i of children $1 \leq i \leq n$ are connected to the transform of a parent by a wreath product; whereas the transforms of the children are connected to each other by a direct product. That is, one has:*

$$[G_1 \times \cdots \times G_n] \textcircled{w} G.$$

7.4 Relating Inheritance Diagrams to Algebra

It will be useful to understand the relation between an inheritance diagram of the type shown in Fig. 7.1 (p. 176), and the algebraic structure we have given for inheritance. First consider the particular example shown in that figure. Here, there are *four* objects starting with the world frame, and each successive object downward is a child of the object above it. Notice there are therefore *three* transforms, because a transform relates the frame of an object to the frame of its parent object, and the World object cannot have a transform, because there is no object above it. The transform is the *personal* transform of the object to which it is connected. Let us assume that the transform for Object i, is the group G_i.

Now, to relate this diagram to the algebra. Our algebraic method is to take the successive transforms downward and form their wreath product, thus:

$$G_1 \text{ⓦ} G_2 \text{ⓦ} G_3.$$

To code, on this expression, the successive frames used, let us use the symbol W for the world frame, and the symbol F_i for the frame of Object i. Then the frames are indicated in the wreath structure thus:

$$_{F_1}G_1^{F_2} \text{ⓦ} \; _{F_2}G_2^{F_3} \text{ⓦ} \; _{F_3}G_3^{W}.$$

Here we have adapted the notation used earlier for frames in robot manipulators: The group G_i relates the frame of its upper index to the frame of its lower index. The relation between the indexes is given by the general case, as follows:

GROUP OF ENTIRE TRANSFORM STRUCTURE. *Consider a set of $n+1$ objects: Object 1 to n, and the World. Suppose that they are linked such that Object i is the child of Object $i+1$, and Object n is the child of the World. Then the group of the entire transform structure is the wreath product:*

$$_{F_1}G_1^{F_2} \text{ⓦ} \; _{F_2}G_2^{F_3} \text{ⓦ} \; \cdots \; \text{ⓦ} \; _{F_n}G_n^{W}$$

where

(1) Object i has personal transform G_i and frame F_i.

(2) Personal transform G_i relates frame F_{i+1} of the parent (upper index) to the personal frame F_i (lower index).

Notice that the subscript i of the group G_i is the same as the subscript i of its lower index F_i; that is, G_i and F_i are both personal to the Object i defining that level.

7.5   Class Inheritance

Let us now turn to class inheritance. Consider a typical class-inheritance hierarchy for closed figures, based on Meyer [108] p528. It is shown as Fig. 7.2.

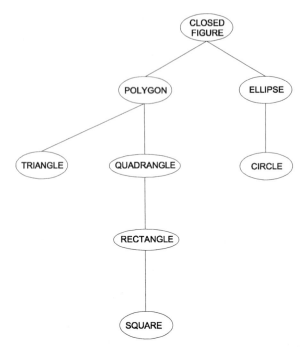

Fig. 7.2. A typical class-inheritance hierarchy based on Meyer [108] p528.

In object-oriented programming there is currently no systematic way of explaining such a hierarchy. However, our generative theory of shape very clearly explains it. The two basic principles of the theory are the maximization of transfer and recoverability. Consider first recoverability. This is ensured by the Asymmetry Principle which recovers symmetries from asymmetries. In particular, the Externalization Principle says that any use of the Asymmetry Principle for external inference must eventually lead back to an iso-regular group.

Our theory in fact predicts the class-inheritance hierarchy shown in Fig. 7.2, as follows: We observe that, as one descends through the hierarchy, one is reaching successively more symmetrical states - in accord with the Asymmetry Principle. Furthermore, the end-point of any downward branch is an iso-regular group - in accord with the Externalization Principle. Thus, we view class inheritance as a *recovery* procedure.

Let us now formulate an algebraic theory of the *is-a* relation that is basic to class-inheritance. Notice that *is-a* means *sub-class of* rather than *member of*; that is, as is often stated, it really means *is-a-kind-of*. To illustrate our algebraic theory of this, consider the inheritance descent from the node *QUADRANGLE*. This is shown in Fig. 7.3. In fact, an extra node, *PARALLELOGRAM*, has been added here, since our psychological studies, Leyton [88] [89], show that this should be included.

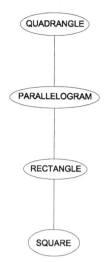

Fig. 7.3. A downward branch in a class-inheritance hierarchy.

At the top, there is the class $QUADRANGLE$. One can consider its class text, in the software, as including an invariant stating that there are four sides, and a feature stating that the four side-lengths are real numbers.

Let us now ask what the symmetry group of this structure is. We claim it is

$$\mathbb{R} \textcircled{w} \{e\}$$

as a permutational wreath product in this way: Let the control set be c_1, c_2, c_3, c_4, corresponding to the four sides. The fiber-group product is therefore $\mathbb{R}_{c_1} \times \mathbb{R}_{c_2} \times \mathbb{R}_{c_3} \times \mathbb{R}_{c_4}$. The permutational action of the control group $\{e\}$ is to leave each side where it is. This corresponds to the fact that, on an arbitrary quadrangle, there is no symmetry group that carries the sides onto each other, because, typically, the sides have different lengths and the vertices have different angles.

Now move one step down in the class-inheritance hierarchy (Fig. 7.3) to the next node $PARALLELOGRAM$. This class inherits the invariant (four sides) and feature (side-lengths are real numbers) from the class above. However, the symmetry group now increases. It is

$$\mathbb{R} \textcircled{w} \mathbb{Z}_2$$

where the control group $\mathbb{Z}_2$ represents 180^0 rotation of the parallelogram (about its center). This sends the figure onto itself.

Now let us move one step further down the class-inheritance hierarchy to the next node $RECTANGLE$. The symmetry group increases still further, thus:

$$\mathbb{R} \textcircled{w} [\mathbb{Z}_2 \times \mathbb{Z}_2]$$

where the control group $[\mathbb{Z}_2 \times \mathbb{Z}_2]$ is the Klein-four group; i.e., this sends a rectangle onto itself.

Finally move one step further down to the bottom node which is the class $SQUARE$. Here, the symmetry group increases still further, thus:

$$\mathbb{R} \, \textcircled{w} \, [[\mathbb{Z}_2 \times \mathbb{Z}_2] \, \textcircled{s}_\tau \, \mathbb{Z}_2].$$

where the control group $[\mathbb{Z}_2 \times \mathbb{Z}_2] \, \textcircled{s}_\tau \, \mathbb{Z}_2$ is actually D_4 the dihedral group of order 8.

Therefore, at each successive class downwards, the symmetry group increases. In other words, the downward hierarchy is a sequence of group extensions:

$$\mathbb{R} \, \textcircled{w} \, \{e\}$$
$$\mathbb{R} \, \textcircled{w} \, \mathbb{Z}_2$$
$$\mathbb{R} \, \textcircled{w} \, [\mathbb{Z}_2 \times \mathbb{Z}_2]$$
$$\mathbb{R} \, \textcircled{w} \, [[\mathbb{Z}_2 \times \mathbb{Z}_2] \, \textcircled{s}_\tau \, \mathbb{Z}_2]. \tag{7.1}$$

In relation to this, the reader should see our book on group extensions, Leyton [98].

We will call the group on each level (i.e., each group listed above), the **internal symmetry group** of that level. Furthermore, we will regard each of these groups as a fiber group in relation to the *command* operations (modifications) on that level; e.g., the stretches, shears, rotations, translations, etc., that can be applied to an object on that level. Typically, in a modeling program, the command operations would be the same for each level. For example, in *3D Studio Max*, they appear as operations on the tool bar and can be applied to the coordinate frame of any object. Technically, one sets this up by making the command operations part of the specification text of a root class such as $GRAPHIC\_OBJECT$. Thus the various shape classes (being subclasses of the root) inherit these operations.[1] Let us therefore call the group of command operations, the **command group**, and denote it by $G(C)$. By our transfer-based theory, the command group has a wreath relation to the internal symmetry group of the object. That is, the command operations transfer the internal symmetry group onto each of the shapes resulting from those command actions, e.g., the deformations.[2] Therefore, we wreath super-append the command group to each of the internal symmetry groups in the above list (7.1), thus:

---

[1] In fact, at the level of $GRAPHIC\_OBJECT$ the operations would be abstract (partial implementations), and their implementations would be completed in the individual subclasses in a way that is appropriate for each of those respective subclasses.

[2] This theory of command and internal groups was first presented by us in Leyton [87]-[90], where they were referred to respectively as the external and internal groups.

$$\mathbb{R} \,\textcircled{w}\, \{e\} \,\textcircled{w}\, G(C)$$
$$\mathbb{R} \,\textcircled{w}\, \mathbb{Z}_2 \,\textcircled{w}\, G(C)$$
$$\mathbb{R} \,\textcircled{w}\, [\mathbb{Z}_2 \times \mathbb{Z}_2] \,\textcircled{w}\, G(C)$$
$$\mathbb{R} \,\textcircled{w}\, [[\mathbb{Z}_2 \times \mathbb{Z}_2] \,\textcircled{s}_\tau\, \mathbb{Z}_2] \,\textcircled{w}\, G(C). \tag{7.2}$$

Observe that the total group of each level is a subgroup of the next level down in the inheritance hierarchy. Notice that this conforms to the *is-a* relationship necessary in class inheritance. That is, the class of objects satisfying the group on one level is a subclass of the objects satisfying the group on the level above. Thus we conclude:

THEORY OF CLASS INHERITANCE.   *Each geometric class is given by a wreath product:*

$$G_{sym} \,\textcircled{w}\, G(C)$$

where G_{sym} is the internal symmetry group of a figure in the class, and $G(C)$ is the group of command operations. Class-inheritance is given by a sequence of group extensions of the internal symmetry group:

$$G_{sym} \longrightarrow G_{sym} \,\textcircled{E}\, G.$$

8. Reference Frames

8.1 Reference Objects

In all aspects of perception and action, the generation of shape involves reference objects - points, axes, and Cartesian frames - with respect to which the other objects are judged and constructed. For example, in *human perception*, reference frames strongly determine the visual organization of a stimulus set: The same stimulus set can have different perceptual representations depending on how the frames are oriented with respect to the set; see Mach [102], Goldmeier [45], Leyton [96]. In *categorization*, Rosch [130], has shown that cognitive categories have reference points with respect to which the remaining stimulus members are judged, e.g., diagonal orientations are seen in terms of vertical or horizontal orientations, the number 99 in terms of 100. In *animal motor control and navigation*, Gallistel [37] [38] has put forward substantial neurophysiological and behavioral evidence demonstrating the significant role of reference frames in determining the animal's computation of directed action. In *computer-aided design*, the design process moves fowards by the successive use of reference objects, starting with the construction plane. In mechanical design, several types of reference objects are explicitly available and are usually called datum objects. These are employed for every phase of the design - from initial sketching, to full part design, to assembly planning. They are crucial in establishing constraints, which are required in all aspects of design. In *3D solid modeling*, systems of reference objects are continually used; e.g., in free-form deformation, a shape is framed by a reference scaffold which is distorted, thereby distorting the embedded shape. A corresponding process occurs in warping. In *robotics*, we saw that reference frames determine the representation of kinematic structure. The same applies to *3D animation*, where kinematics works on the same principles as in robotics.

This chapter will give a theory of what reference objects actually are, and how they function in the generative process. We shall argue that they arise from the *group theory* of shape generation. This will be used, in the following chapters, to significantly increase the understanding of the cognitive and design disciplines.

Throughout the discussion, the term reference *frames* (2D or 3D) rather than reference *objects* will usually be used. However, the theory will apply equally to lower dimensional reference objects such as points and lines.

8.2   Non-coordinate-free Geometry

Section 2.19 began to show that our generative theory of shape is fundamentally opposite to Klein's Erlanger program, which says that the objects of geometry are invariants under operations. The oppositeness of the generative theory is due to the fact that it is based on *recoverability* of the generative operations, and one cannot recover generative operations from objects if the objects are invariant under those operations.

This issue links to that of *reference frames*, which can be regarded as coordinate systems. Klein's program is coordinate-free. That is, invariance under groups of transformations is invariance under changes of coordinate system - which means that the geometric properties must be those that are *coordinate-free.*

The program of geometry and physics for the entire 20th century is the program of making objects coordinate-free. It is the basis for example of Einstein's approach to physics: Einstein's fundamental proposal was that *a physical object is one that is coordinate-free.* In special relativity, coordinate-free means invariant with respect to Lorenz transformations. In general relativity, coordinate-free means invariant with respect to more general transformations: local diffeomorphisms. The general-relativistic program was so powerful that it lead substantially to the development of tensor geometry, which is the foundation of differential geometry. The author remembers the great differential geometer S.S. Chern once saying: "Differential geometry is the study of objects with transient coordinates."

We are going to argue in this book that the coordinate-free approach is fundamentally wrong for the real needs of geometry. We will argue what one actually wants is *complete non-freedom from coordinates.* This is required not only for the computational and design sciences, but ironically for the very physical sciences which promoted the coordinate-free approach to geometry.

We call the geometry developed in this book *generative geometry,* and argue that generative geometry is necessarily non coordinate-free. One of the basic reasons is this:

NON COORDINATE-FREEDOM OF GENERATIVE GEOMETRY. *A reference frame cannot be removed from an object because it is* *part of the generative structure of the object. A change of reference frame changes the structure of the object.*

8.3   Processes and Phases

Before giving a theory of reference frames, it is first necessary to define *processes* and *phases* in the generative history. It will be seen that the most economical generative history proceeds by repeating an operation as many times as required, then repeating another operation as many times as required, and so on. This obviously helps to minimize the history because one does not have to go through the proceedure of setting up a particular operation several times if one groups together all similar occurrences of that operation. Thus, for example, in drawing a line in CAD, one does not draw a few points on the line, and then return to it several times to add more points, in between working on parts of another line elsewhere. Grouping the operations into repetitions, whenever possible, is a fundamental means of minimizing history. In fact, this signficantly determines the psychological structuring of a shape, whether in the domain of human perception, or CAD, or theoretical physics, as follows:

In Leyton [96], we argued that, psychologically, any *process* is understood as the repetition of a generator. Recall that a c-cyclic group was defined earlier as a group on one generator, i.e., a cyclic group, in the discrete case, or a 1-parameter Lie group, in the continuous case. Thus we have this:

DEFINITION OF PROCESS.   *Psychologically, a process is a c-cyclic group; i.e., the discrete or continuous repetition of a generator.*

We said that the need for grouping into repetitions comes from history minimization, but more strictly in comes from the History Symmetrization Principle (p. 61), i.e., maximal *symmetry* across the history. It is this that implies repetition of a generator. We claim that this is why, in physics, the most advanced formulations, e.g., Hamiltonian mechanics and general relativity, represent trajectories as geodesics; i.e., continuous repetitions of a generator along itself. Similarly, this is why Schrödinger's equation in quantum mechanics is the continuous repetition of the Hamiltonian operator as generator.

In fact, the data situation might allow one only to *decompose* the history into a sequence of processes. One tries, as much as possible, to force the asymmetry to be concentrated at a point, and let the remainders of the history be simply repetitions of generators.

Definition 8.1. *A **phase** is a sequence of processes (repetitions of generators).*

An example is the design process in CAD. It proceeds by repeating a generator till some stop-point, at which one then repeats a new generator till some stop-point, at which one then repeats a new generator ... etc. Consider some examples: In the sweep creation of a cylinder, one repeats the rotation generator to obtain the circular cross-section. One then repeats the translation generator to sweep the cross-section through space. Thus, by the above definition, the cylinder is a *phase* in the design process. Notice that this applies not just to simple objects such as cylinders. For example, in mechanical CAD (e.g., in *ProEngineer* and also *AutoCAD Mechanical Desktop*), one creates a mechanical *part* by first drawing a 2D polyline, which is a sequence of straight-line segments, hinged at turning points. One then extrudes the polyline in the perpendicular direction, thus obtaining a 3D part. By the above definition, the initial 2D polyline is a *phase*. Furthermore, the definition implies that the *entire part* is also a phase, since it is a sequence of processes (repetitions of generators).

8.4   Theory of Reference Objects

Let us return to using the term reference *objects*, and use the term reference *frames* later when considering specific examples of 2D and 3D coordinate frames. This section will give our basic theory of reference objects, and the remainder of the chapter will show how this theory works.

It is obviously the case that a reference object *begins* a phase of the shape-generation procedure. This is because it is used as a *reference* for the phase.

Now the previous section started to give a theory of *phases*. A phase was understood as a sequence of processes, and a process as a repetition of a generator. The crucial factor to now bring into consideration is the Asymmetry Principle (p. 42). This says that the generation of a shape is recoverable from a data set only if it is symmetry-breaking on successively generated states. Thus recoverable shape-generation must proceed from a symmetry ground-state to an asymmetrization of that ground-state. All apparent counter-examples are due to a failure to correctly define the boundary of the data set, as shown in Chapter 2.

Thus by using the Asymmetry Principle, we conclude that any *phase* of shape-generation must be a progression from symmetry to asymmetry. This leads us to propose the following:

THEORY OF REFERENCE OBJECTS.   *Any reference object corresponds to the symmetry ground-state of a phase. When there is a sequence of phases, a reference object occurs at a phase-transition. It corresponds to the point of maximal symmetry (start point) of the new phase and the point of maximal asymmetry (end point) of the previous phase.*

8.5   The Necessity of Reference Frames

Let us now return to the issue of coordinate freedom. The Erlanger program is coordinate free. In contrast, generative geometry is necessarily non coordinate-free. We can now see why:

NON COORDINATE-FREEDOM OF GENERATIVE GEOMETRY.   *A reference frame cannot be removed from an object because it is the symmetry ground-state which defines the object as an asymmetrization of the ground-state. This means that it is part of the generative structure of the object, and that a change of reference frame will change the generative structure of the object.*

8.6   Structure of the 2D Reference Frame

The conventional theory of Cartesian frames does not *explain* the enormous importance of these frames in all shape computation, from human perception, to CAD, to navigation. We are now going to give a theory of the 2D Cartesian frame which will become fundamental to understanding these areas as they are elaborated in rest of the book.

It is first necessary to analyze more fully the structure of the square because it is closely related to the 2D frame. Up to now we have taken the control group of the square to be the rotation group $\mathbb{Z}_4$. This is a subgroup of D_4, the dihedral group of order 8. One can construct D_4 from $\mathbb{Z}_4$ as a splitting extension thus:

$$D_4 \;\; = \;\; \mathbb{Z}_4 \; \circledS \; \mathbb{Z}_2$$

where $\mathbb{Z}_2$ is a reflection group. Then one can add this as a control group to the group of a side $\mathbb{R}$, thus:

$$\mathbb{R} \; \circledW \; \mathbb{Z}_4 \; \circledS \; \mathbb{Z}_2$$

to give the full symmetry group of the square. However, the problem with this expression is that it is not wreath c-polycyclic, which means that one cannot extract from it a *canonical plan* for drawing a square (Sect. 4.8).

The obstruction is the semi-direct product decomposition occurring in the D_4 component. Thus, the problem would be solved if D_4 itself had a wreath c-polycyclic decomposition. Remarkably it does. In order to understand it, one must take a look at the subgroup diagram of D_4, shown in Fig. 8.1.

Notice that there are four levels in this diagram. The single subgroup on the top level has eight elements (the total group); the subgroups on the

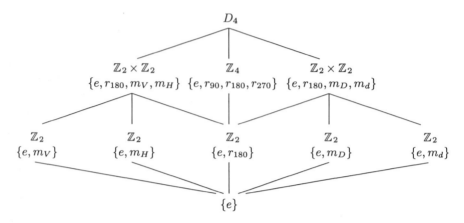

Fig. 8.1. Subgroup diagram of D_4.

second level each have four elements; the subgroups on the third level each have two elements; and the subgroup on the bottom level has one element. The elements themselves are represented as actions on a square. As usual, each element r_θ represents rotation by θ degrees; obviously there are four of them in 90^0 increments. The remaining four elements m_i are reflections, where m means "mirror", and the subscript means the axis about which the reflection is taken. The subscripts are V = vertical; H = horizontal; D = one diagonal; and d = other diagonal.

Now constructing D_4 as a splitting extension $N \, \circledS \, H$ of two of its subgroups means taking a normal subgroup N from the group, and adding a complement subgroup H. The complement condition is that $N \cap H = \{e\}$, and $NH = D_4$. To satisfy this, notice that the previous decomposition $D_4 = \mathbb{Z}_4 \circledS \mathbb{Z}_2$ took the normal group $N = \mathbb{Z}_4$ from the second level down in Fig. 8.1, and took its complement $H = \mathbb{Z}_2$ to be one of the subgroups on the third level - a subgroup that must intersect $\mathbb{Z}_4$ only at the identity element.

We now want a new decomposition of D_4, one that is a wreath product. Of course a wreath product is also a semi-direct product. To construct it, take as normal subgroup N, one of the $\mathbb{Z}_2 \times \mathbb{Z}_2$ groups on the second level down, in Fig. 8.1. In fact, take the one on the far left, which can be written as

$$\mathbb{Z}_2 \times \mathbb{Z}_2 \;\; = \;\; \{ \; e, \; m_V \; \} \;\; \times \;\; \{ \; e, \; m_H \; \} \tag{8.1}$$

i.e., it is generated by the vertical and horizontal reflections.

The complement H of this normal subgroup N will be taken to be

$$\mathbb{Z}_2 \;\; = \;\; \{ \; e, \; m_D \; \} \tag{8.2}$$

from the third level down, in Fig. 8.1.

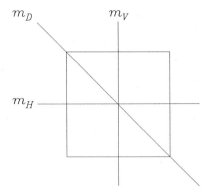

Fig. 8.2. Three of the four reflection axes of a square.

Then form the semi-direct product of these two subgroups thus:

$$D_4 \;=\; [\mathbb{Z}_2 \times \mathbb{Z}_2] \, \circledS \, \mathbb{Z}_2.$$

Of course this is not yet fully defined because the way in which the complement $\mathbb{Z}_2$ acts as an automorphism group on the normal subgroup $[\mathbb{Z}_2 \times \mathbb{Z}_2]$, has not yet been specified.

The automorphic action one needs to choose can best be understood by considering a diagram of the square, Fig. 8.2. In this figure, the vertical and horizontal reflection axes define the normal subgroup $[\mathbb{Z}_2 \times \mathbb{Z}_2]$, by equation (8.1). Obviously this is the Klein-four group. In contrast, the diagonal axis gives the complement subgroup, by equation (8.2). Notice that the diagonal reflection exchanges the vertical reflection axis and the horizontal reflection axis. This is exactly what will be used as the automorphic action. The diagonal group $\mathbb{Z}_2 = \{\ e,\quad m_D\ \}$ will simply exchange the vertical group $\mathbb{Z}_2 = \{\ e,\quad m_V\ \}$ and the horizontal group $\mathbb{Z}_2 = \{\ e,\quad m_H\ \}$. Notice that this means that it will simply reverse the two reflection components of expression (8.1).

This automorphic action actually *transfers* the vertical and horizontal reflection structures onto each other. Thus the automorphic action is the τ-automorphic action. The normal subgroup and the complement are therefore glued together as follows:

$$D_4 \;=\; [\mathbb{Z}_2 \times \mathbb{Z}_2] \, \circledS_\tau \, \mathbb{Z}_2 \tag{8.3}$$

which explicitly shows that the transfer structure τ is being used; i.e. it reverses the two reflection components within the $[\mathbb{Z}_2 \times \mathbb{Z}_2]$ structure. Expression (8.3) is now a wreath product, written in its semi-direct product notation. In its wreath product notation, it is:

$$D_4 \;=\; \mathbb{Z}_2 \, \circledW \, \mathbb{Z}_2. \tag{8.4}$$

Table 8.1. D_4 in two different decompositions.

$[\mathbb{Z}_2 \times \mathbb{Z}_2] \textcircled{S} \mathbb{Z}_2$				$\mathbb{Z}_4 \textcircled{S} \mathbb{Z}_2$	
$\langle\ e$, e $\mid e\ \rangle$	e	e	identity
$\langle\ m_H$, e $\mid e\ \rangle$	m_H	$m_V r_{180}$	H-reflection
$\langle\ e$, m_V $\mid e\ \rangle$	m_V	m_V	V-reflection
$\langle\ m_H$, m_V $\mid e\ \rangle$	$m_H m_V$	r_{180}	rotation by 180^0
$\langle\ e$, e $\mid m_D\ \rangle$	m_D	$m_V r_{90}$	D-diagonal reflection
$\langle\ m_H$, e $\mid m_D\ \rangle$	$m_D m_H$	r_{270}	rotation by 270^0
$\langle\ e$, m_V $\mid m_D\ \rangle$	$m_D m_V$	r_{90}	rotation by 90^0
$\langle\ m_H$, m_V $\mid m_D\ \rangle$	$m_D m_H m_V$	$m_V r_{270}$	d-diagonal reflection

This can be somewhat confusing. This is a *wreath* product of $\mathbb{Z}_2$ and $\mathbb{Z}_2$. Therefore, it should not be confused with the Klein-four group, which is a *direct* product of $\mathbb{Z}_2$ and $\mathbb{Z}_2$. However, the wreath product contains the Klein-four group as its *fiber-group product*!

We shall see that, to analyze the applications of reference frames to human perception, CAD, robotics, and navigation, it is necessary to understand what the elements of D_4 look like in the wreath-product decomposition. For this, one needs to use the semi-direct product notation of the wreath product, that is:

$$[\mathbb{Z}_2 \times \mathbb{Z}_2] \textcircled{S} \mathbb{Z}_2.$$

Here, each element must have the triple-form:

$$\langle\ z_1\ ,\ z_2\ \mid z_3\ \rangle$$

where each z_i comes from the corresponding $\mathbb{Z}_2$ in the expression directly above this. There are eight elements in D_4. Using the triple-form, the elements are presented down the left column in Table 8.1.

Let us concentrate on this left-hand column. The first thing to understand is that, in the triple-form, each of the elements is expressed only in terms of reflections - no rotations! It is of course the case that, when the reflections are multplied out, some of the multiple reflections will actually be equivalent to rotations. However, all entries in this column use only the mirror operations m_V, m_H, and m_D.

Next, observe the middle horizontal line in the table. It divides the top set of four elements from the bottom set. The top set is the Klein-four group; i.e., the fiber-group product. Notice that, in the left-hand column, these four elements all have e as the control component, thus:

$$\langle\ -\ ,\ -\ |\ e\ \rangle.$$

In contrast, the lower four elements all have m_D as the control component:

$$\langle\ -\ ,\ -\ |\ m_D\ \rangle.$$

Therefore, the lower set of elements is a coset of the Klein-four group, with coset leader m_D.

Let us now move on to the second column. Here, the reflections in the first column have been multiplied together. Also, the control-group reflection has been brought up to the front of each triple. This has certain benefits.

Now for a crucial point. The entire table shows two decompositions of D_4. The wreath-product decomposition $[\mathbb{Z}_2 \times \mathbb{Z}_2]\ \circledS\ \mathbb{Z}_2$ is shown in the first column; and the alternative decomposition $\mathbb{Z}_4\ \circledS\ \mathbb{Z}_2$ is shown in the third column. In this third column, the following notation has been used:

$$D_4 = \{\ e, \qquad r_{90}, \qquad r_{180}, \qquad r_{270}, \qquad (8.5)$$
$$m_V, \qquad m_V r_{90}, \qquad m_V r_{180}, \qquad m_V r_{270}\ \}.$$

Notice, in this expression, that the upper line is the rotation subgroup $\mathbb{Z}_4$, which is the normal subgroup of this decomposition. The lower line is the coset $m_V \mathbb{Z}_4$, which is the set of four reflections, each written as a multiplication of m_V with a rotation.

Returning to Table 8.1, one can see that this latter coset decomposition does not correspond to the coset decomposition of the Klein-four group, i.e., the division given by the middle horizontal line. That is, the rotations $\mathbb{Z}_4$ distribute across the two cosets of the Klein-four group.

The final column gives the overall effect of each element. For example, it gives the overall effect of each of the triple-reflections in the first column.

8.7 Canonical Plan from the 2D Reference Frame

The wreath decomposition $D_4 = \mathbb{Z}_2 \circledW \mathbb{Z}_2$ becomes basic to drawing a square, i.e., its *generative* structure. We must first add the side-level $\mathbb{R}$, thus:

$$\mathbb{R}\ \circledW\ \mathbb{Z}_2\ \circledW\ \mathbb{Z}_2. \qquad (8.6)$$

Notice that this is a *wreath c-polycyclic group*. Therefore it can be used for a *canonical plan* to generate a square. Recall from p. 131, that one derives a canonical plan from a wreath c-polycyclic group $G_1 \circledW G_2 \circledW \ldots \circledW G_n$, by repeating a cycle consisting of an entire downward movement through the levels. In each cycle, one uses only a single element from each successive downward level G_i, except on the bottom level G_1, where one uses the entire group. The order of elements within any level is the order prescribed by the generator of that level. One fills levels from the bottom up.

When this is applied to the particular wreath c-polycyclic group given above for a square, we obtain one of the most common scenarios for drawing a square: draw the top side, then the bottom side, then the left side, and finally the right side. This is a control-nested structure, as follows: The first two sides (top and bottom) are drawn as reflectional copies of each other about the horizontal axis. The drawing structure of the first two sides is then reflected, as a *whole*, about the diagonal to obtain the drawing structure of the second two sides (left and right). In other words, the left and right sides are a *transferred* version of the top and bottom sides, under the control action of diagonal reflection. In terms of group elements, we have this: The first two sides are e and m_H respectively, both with control parameter e. The second two sides are exactly the same elements, but with control parameter m_D. Thus, the control parameter m_D has *tranferred* e and m_H from one fiber-group copy $\mathbb{Z}_2$ to the other fiber-group copy $\mathbb{Z}_2$, within the fiber-group product $\mathbb{Z}_2 \times \mathbb{Z}_2$.

Finally observe this: The two most frequent scenarios for drawing a square come from the two alternative *wreath c-polycyclic groups* we have given for the square:

$$\mathbb{R} \textcircled{w} \mathbb{Z}_4 \quad \text{and} \quad \mathbb{R} \textcircled{w} \mathbb{Z}_2 \textcircled{w} \mathbb{Z}_2.$$

The first one, discussed in Chapter 3, uses a rotation control structure, i.e., it traces the pen around the square. The second one uses a reflection control structure, i.e., drawing the sides in opposite pairs. These two scenarios are respectively the most common, and the second most common, ways in which a person draws a square. One can therefore clearly see the power of our method of assigning to the data set a symmetry group that has a wreath c-polycyclic structure, and deriving from the wreath c-polycyclic structure its canonical plan.

8.8   Organizing Role of the Cartesian Reference Frame

The symmetry structure of the 2D Cartesian frame is closely related to the symmetry structure of the square. Thus, we will often take the symmetry structure of the Cartesian frame to be

$$\mathbb{R} \textcircled{w} \mathbb{Z}_2 \textcircled{w} \mathbb{Z}_2,$$

which will be called the **hyperoctahedral wreath hyperplane group**, to be discussed fully in Chapter 16. In fact, it will become clear that the most significant component of this structure is the wreath reflection component $\mathbb{Z}_2 \text{ⓦ} \mathbb{Z}_2$. We shall argue that this reflection group is the single most important structuring factor in all of human perception, CAD, robotics, navigation, etc. It explains numerous phenomena in perception, the choice of datum and sketch planes in mechanical CAD, basic aspects of assembly planning, feature-extraction in NC-machining, the structure of buildings, the representation of path structures in navigation, and so on. The reason is that this group derives from the gravitational structure of the environment and the way the gravitational structure is manifested in the organization of the moving agent and its moving parts. We argue that virtually any real-world use of geometry begins with the imposition of the reflection wreath product $\mathbb{Z}_2 \text{ⓦ} \mathbb{Z}_2$. This will become very evident, as this book proceeds.

Most crucially, this will emerge in the theory of *complex* shape generation to be developed over the following chapters. However, the next section will give some simple examples relating to human perception. Simple as these examples are, they profoundly undermine the belief that Klein's Erlanger program is correct, and will become the basis of our attack on the Erlanger program in Chapter 22.

8.9   Orientation-and-Form

The *orientation-and-form effect* is perhaps the single most important phenomenon in human perception. It is completely ignored by the computer vision community, and we believe that this is ultimately the reason why computer vision systems do not work, despite the enormous research in that area.

The orientation-and-form phenomenon is this: The same figure in different orientations can be perceived as two entirely different figures. Fig. 8.3 shows an example due to Goldmeier: Subjects fail to notice that the two shapes shown in this figure are actually the same - but in different orientations. It should be pointed out that this effect comes from Goldmeier's PhD thesis [45], which was written in 1936, but published in English almost 40 years later because of the extraordinary power of its ideas. One can quite categorically say that anyone who has not read Goldmeier's thesis will never get a computer vision system to work. Consider for example the standard optics/camera model used in computer vision: No optics model will explain why the two shapes in Fig. 8.3 are seen as completely different figures. Consider also the invariants program in computer vision. If invariant descriptions were the actual representations used by the visual system, then the two figures would be seen as the same, since they are exactly congruent under the 2D rotation group. However, the figures are given entirely different representations. We shall argue that this is because the wreath reflection structure $\mathbb{Z}_2 \circledw \mathbb{Z}_2$ is assigned completely differently to the two shapes.

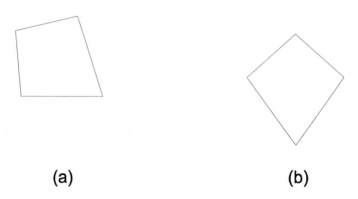

(a)   (b)

Fig. 8.3. One of Goldmeier's orientation-and-form examples.

The orientation-and-form phenomenon was in fact discovered by Ernst Mach [102]. His famous example is called the square/diamond effect. It is considerably weaker than the Goldmeier example, but is nevertheless extremely important to understand. Mach pointed out that the two shapes shown in Fig. 8.4 appear to be somewhat different, despite the fact that they are merely the same shape in two different orientations. He called the left figure a *square*, and the right figure a *diamond*. Whereas most people are convinced about the Goldmeier effect Fig. 8.3, they are less convinced about the Mach effect Fig. 8.4. However, the following will show that the Goldmeier effect actually exploits the Mach effect, and therefore it is necessary to understand the Mach effect first.

The theory of the Mach effect was given by us in Leyton [96]. We argued, using a large amount of empirical data, that the human visual system contains a double reflection group $\mathbb{Z}_2 \times \mathbb{Z}_2$, with its two reflection components aligned with the two axes of the *gravitational* Cartesian frame. In fact, we argued that this $\mathbb{Z}_2 \times \mathbb{Z}_2$ group is the most significant structuring component in the human visual system - fundamentally determining the hierarchical organization of every visual representation. Notice that it is the fiber-group product of $\mathbb{Z}_2 \circledw \mathbb{Z}_2$.

Thus, returning to the Mach effect Fig. 8.4, we argued that the difference between the left and right figures, is that, in the left figure, the $\mathbb{Z}_2 \times \mathbb{Z}_2$ reflection axes are aligned with the side-bisectors; whereas, in the right figure, the $\mathbb{Z}_2 \times \mathbb{Z}_2$ reflection axes are aligned with the angle-bisectors. This means that the two figures have different generative structures. That is, in the left figure, the *opposite* sides are generatively related by reflection, whereas, in the right figure, the *adjacent* sides are generatively related by reflection.

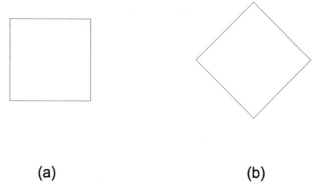

(a)   (b)

Fig. 8.4. Ernst Mach's Square/Diamond Effect.

Table 8.2. The theory of the square/diamond effect given by Leyton [96].

INTERNAL STRUCTURE

Square: $\mathbb{Z}_2 \times \mathbb{Z}_2$ axes aligned along side-bisectors
Diamond: $\mathbb{Z}_2 \times \mathbb{Z}_2$ axes aligned along angle-bisectors

INTERACTION PRINCIPLE: *The symmetry axes of an organization become the eigenvectors of the most structurally allowable actions on that organization.*

PURE DEFORMATION STRUCTURE

Square: eigenvectors aligned along side-bisectors
Diamond: eigenvectors aligned along angle-bisectors

Our theory continued in this way: Not only are the two figures distinguished by this major difference in "internal" structure, but also by a major difference in "external" structure. To see this, one first requires the Interaction Principle (p. 67) which states that the symmetry axes of an organization become the lines of flexibility (eigenvectors) of the structurally most allowable operations applicable to the organization. Thus the first figure (the square) is seen as changeable easily into a rectangle - since its symmetry axes are along side-bisectors, and these axes become the directions (eigenvectors) of most allowable stretch. In contrast, the second figure (the diamond) is seen as changeable easily to elongated diamonds - since its symmetry axes are along angle-bisectors, and these axes become the directions (eigenvectors) of most allowable stretch. This puts the two figures in two different categories - the square belongs to the category of rectangles, as shown in the top

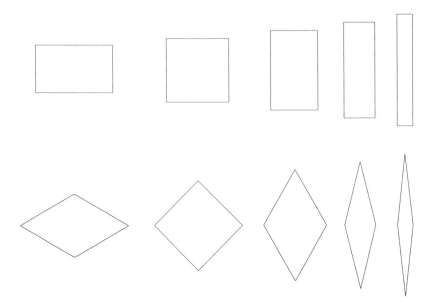

Fig. 8.5. A square and a diamond belong to different deformation spaces.

row of Fig. 8.5; whereas the second figure belongs to the category of general diamonds, as shown in the bottom row of Fig. 8.5. In particular, a square necessarily has 90^0 angles, whereas in a diamond, the 90^0 angles are hardly even noticed and are regarded as inessential. The theory is summarized in Table 8.2.

Now let us turn to the Goldmeier shapes in Fig. 8.3. Clearly, the transformations responsible for obtaining these shapes from the square and diamond are less structurally allowable than those just described, since neither shape in Fig. 8.3 can be obtained by linear deformation along the symmetry axes of the square or diamond; i.e., greater deformation is involved. We shall regard the minimal group containing these new transformations as the projective group $PGL(3, \mathbb{R})$.

According to our theory, the projective group is then added as a control group *transferring* $\mathbb{R}\wr\mathbb{Z}_2\wr\mathbb{Z}_2$ from the undistorted square or diamond onto the distorted square or diamond. This means, that one has the 4-fold wreath product:

$$\mathbb{R} \wr \mathbb{Z}_2 \wr \mathbb{Z}_2 \wr PGL(3, \mathbb{R}). \tag{8.7}$$

It is worth the reader checking how each level is transferred by the next level above it in the hierarchy.

Most crucially, expression (8.7) tells us that the wreath reflection group $\mathbb{Z}_2\wr\mathbb{Z}_2$ is transferred by the projective group onto each figure. According to p. 189, this is the *meaning* of assigning a Cartesian reference frame; i.e., reference-frame assignment defines the symmetry ground-state as the reflec-

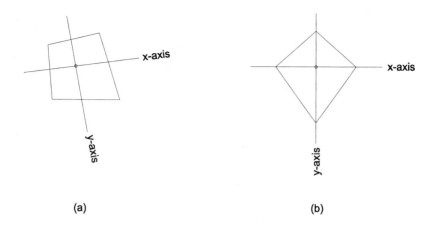

(a)   (b)

Fig. 8.6. The assignment of axes in the Goldmeier figures.

tion wreath product $\mathbb{Z}_2 \textcircled{w} \mathbb{Z}_2$. Since the generative theory says that the presented state must be an asymmetrized version of the ground-state, and that asymmetrization must take place by transfer, the reflection wreath product must be transferred onto the presented state.

How does this transfer take place? The answer comes from the History Minimization Principle, i.e., one must choose the shortest generative sequence that goes from the ground-state to the presented state. Now, the ground-state $\mathbb{Z}_2 \textcircled{w} \mathbb{Z}_2$ has its fiber-group product $[\mathbb{Z}_2 \times \mathbb{Z}_2]$ positioned such that one of the $\mathbb{Z}_2$ fibers is aligned with the gravitational vertical axis and the other is aligned with the gravitational horizontal axis. The minimal projective transformation going from this assignment to Fig. 8.3a, must be one in which these axes were aligned with the side bisectors, i.e., the ground-state was a *square*. The result of the transfer is shown in Fig. 8.6a. Secondly, the minimal projective transformation going from the gravitational assignment to Fig. 8.3b, must be one in which the reflection axes were aligned with the angle bisectors, i.e., the ground-state was a *diamond*. The result of the transfer is shown in Fig. 8.6b.

Finally notice this: In this section, our inference rules were used to *recover* the generative history of the two shapes. The inferential proceedure concluded that the Cartesian axes were the reflection axes of the symmetry ground-state of the generative process. We shall see that this theory of reference frames is entirely general, and explains the use of reference frames throughout all of human perception, CAD, assembly planning, manufacturing, robotics, navigation, etc.

8.10   Cartesian Frame Bundle

We have seen that the reflection structure, defining the Cartesian reference frame, is extremely salient. In fact, we now propose that the salience indicates that this structure is actually hard-wired into the visual system. Our claim is that the nervous system uses these specific hardwired units to impose the reflection structure on anything it possibly can, and at every level of scale. For example, a square receives the structure not only globally, as described above, but along its edges, as shown in Fig. 8.7, where each disc indicates the placement of the Klein-four group $\mathbb{Z}_2 \times \mathbb{Z}_2$, which is that part of the wreath reflection structure that the visual system can instantiate at each of these points. That is, locally, the edge possesses two axes of reflection, one along the edge and one perpendicular to the edge. The same applies to any curved edge.

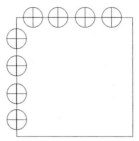

Fig. 8.7. Part of a Cartesian frame bundle.

Recall also that we have proposed (p. 142) that the human visual system is *recursive*, with the recursion taking place via wreath products - to maximize transfer. We shall call any recursive use of the reflection wreath product, a **Cartesian frame bundle**. The visual salience of such bundles is enormous as will now be demonstrated.

Fig. 8.8. Attneave's pointing triangle.

To build up these examples, it is first necessary to consider the phenomenon of multistability discovered by Attneave [5], [6]. In Fig. 8.8 there is an equilateral triangle oriented so that none of its three bisectors are aligned with the gravitational axes. Attneave discovered that, despite the fact that the triangle is equilateral, it appears to point in only one direction, and that this direction changes periodically from one bisector to another. Attneave also made the discovery that this extends to fields of equilateral triangles as shown in Fig. 8.9. All the triangles in the field are perceived as pointing in the same direction simultaneously. They then flip and point simultaneously in one of the other two directions.

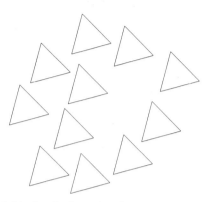

Fig. 8.9. Attneave's field of pointing triangles.

The importance of these multi-stability phenomena is that they show fundamental aspects of the structure of the human visual system, which have been ignored in computer vision, and indicate why computer vision systems actually don't work. Consider for example the standard optics/camera model used in computer vision: No optics model will explain why an equilateral triangle is seen as pointing, i.e., as having a prejudice not contained in the optic array. Consider also the invariants program in computer vision. If invariant descriptions were the actual representations used by the visual system, then a triangle would not point. Notice that the pointing phenomenon violates Klein's Erlanger program.

We are now going to explain the Attneave phenomena. Consider first the single pointing triangle. If the only group that were operating here were the symmetry group of the triangle, then there would be no prejudice. However, we argue that the Cartesian frame is being imposed on the triangle - specifically, the Klein-four group $\mathbb{Z}_2 \times \mathbb{Z}_2$ is being imposed. The frame has two *perpendicular* reflection axes which the triangle does not possess. Therefore the imposition of the frame cannot actually use the center of the triangle. The two reflection axes are imposed like this: (1) One reflection axis is instantiated as one of the bisectors of the triangle, i.e., through one of the vertices. (2) The perpendicular reflection axis is instantiated as the triangle-side opposite to that bisected vertex. Notice that the two reflection axes pick out two different levels of the triangle structure. The first axis is a global axis of the triangle, and the second axis is a reflection axis of one of the sides (along that side). The salience of the $\mathbb{Z}_2 \times \mathbb{Z}_2$ structure is so great for the visual system, that the system must find it where it can.

Now let us consider the *field* of pointing triangles. To describe this, we simply control-nest the Klein-four group within the planar translation group, thus:

$$[\mathbb{Z}_2 \times \mathbb{Z}_2] \ \text{ⓦ} \ [\mathbb{R} \times \mathbb{R}]. \tag{8.8}$$

That is, the translation component acts as a control group transferring the double-reflection structure from one point to another. This means that all the triangles can point in only the same direction at the same time.

In the example just given, the reflection group was control-nested within the translation group. However, in Leyton [96], we have shown that the reflection group can be control-nested within itself. To demonstrate this, we devised the stimuli shown in Fig. 8.10. Here, we combined the pointing triangle with the square-diamond effect. Notice that Fig. 8.10a and b are exactly the same stimulus in two different orientations. However, the figures are perceived differently. In Fig. 8.10a, the visual system groups together the triangles on *opposite* sides and therefore the triangles are seen as parallel: two pointing vertically and two pointing horizontally. In contrast, in Fig. 8.10b, the triangles are not seen as parallel. The two triangles on the top half are grouped together and are therefore seen as pointing away from each other; correspondingly the two triangles on the bottom half are grouped together and are therefore seen as pointing towards each other.

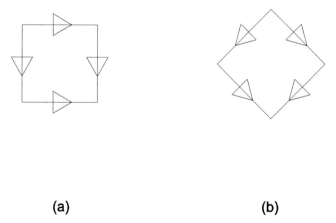

(a)   (b)

Fig. 8.10. The orientation-and-form phenomenon of Leyton [96] p362.

The two figures give strong evidence for the recursive nature of the human visual system. Each figure consists of a 2-level wreath product in which *both* the lower level and upper level are given by the double reflection group. Therefore the structure is this:

$$[\mathbb{Z}_2 \times \mathbb{Z}_2] \ \textcircled{w} \ [\mathbb{Z}_2 \times \mathbb{Z}_2]. \tag{8.9}$$

This is an example of what we call a a wreath poly-$\mathbb{Z}_2 \times \mathbb{Z}_2$ group as in Definition 4.11. We claim that such groups are fundamental to the human perceptual system. Considerably more has been written by us on this; e.g., see particularly p324-363 in Leyton [96].

8.11　External Actions on Frames: Decomposition

The rotated parallelogram example in Fig. 2.3 (p. 40) is particularly significant with respect to reference frames in the human perceptual system. The rotated parallelelogram goes back to the square which can essentially be identified with the Cartesian frame as the hyperoctahedral wreath group $\mathbb{Z}_2 \ \textcircled{w} \ \mathbb{Z}_2$.

One needs to encode the fact that, although the overall transformation from the square to the rotated parallelogram is a single linear transformation, i.e., a member of the general linear group $GL(2, \mathbb{R})$, the subjects decompose it into a stretch, shear, and rotation. This decomposition corresponds to what is called, in Lie groups, the *Iwasawa decomposition*. We have written several hundred pages about the Iwasawa decomposition in human perception Leyton [87], [88], [89], [90], [91], [96], because we have demonstrated the enormous significance of this decomposition in revealing the underlying structuring principles of the human perceptual system. Perhaps the most comprehensive account is the 200-page Chapter 6 in Leyton [96].

It is clear that the subjects are using the Iwasawa decomposition in the rotated parallelogram example. Additional experiments showed that subjects take either of two strategies: They either try to preserve area, or they try to preserve width. In either case, one is dealing not with all of $GL(n, \mathbb{R})$, but with a 3-parameter subgroup; i.e., one parameter for each of stretch, shear, and rotation. Without loss of generality, let us assume they are using the first strategy - that of preserving area. This means that they are using the 3-parameter subgroup $SL(2, \mathbb{R})$, the group of area-preserving linear transformations. Its Iwasawa decomposition is this:

$$SL(2, \mathbb{R}) \quad = \quad A.N.SO(2) \tag{8.10}$$

where A is the group of stretches, N is the group of shears, and $SO(2)$ is the group of rotations. Notice that the order - stretch, shear, rotation - in this sequence, corresponds to the generative order (right-to-left) in Fig. 2.3 (p. 40).

The expression of the group in this form does not encode the parent-child hierarchy on Cartesian frames. Also, a parent-child hierarchy on frames corresponds to a global-to-local hierarchy on frames (Sect. 5.9). Consider the global-to-local structure in the current situation. At one end of the sequence (8.10), the stretch group A is relative to the intrinsic structure of frame of the square, i.e., the symmetry axes of the square become the eigenvectors (directions) of the stretch transformation in accord with the Interaction Principle (p. 67). Therefore, the frame is local at this end of the sequence. At the other end of the sequence, the rotation group $SO(2)$ relates the object to the global frame of the environment; i.e., the world frame. Therefore, the order of frames in the sequence (8.10), is local-to-global in the left-to-right direction.

Using our correspondence between parent-child hierarchy, global-to-local hierarchy, and wreath hierarchy, replace the sequence in (8.10) with this sequence:

$$SL(2,\mathbb{R})_{\circled{w}} \quad = \quad A \circled{w} N \circled{w} SO(2). \tag{8.11}$$

The group $SL(2,\mathbb{R})_{\circled{w}}$ is formed out of the same components as $SL(2,\mathbb{R})$, except that the components are now put together using the wreath product operation. Generally, it is valuable for us to define the following:

Definition 8.2. *Let a group G have a decomposition into components G_1, G_2, ..., G_n. Then a wreath product $G_{\circled{w}} = G_1 \circled{w} G_2 \circled{w} \ldots \circled{w} G_n$, will be called a* **wreath-reconstituted** *version of G.*

Thus expression (8.11) gives a wreath-reconstituted version of $SL(2,\mathbb{R})$. The advantage of this version is that it encodes the parent-child structure and global-local structure of the reference frames. (Note that the group *Special-Linear* used in Chapter 6 of Leyton [96] is actually the wreath-reconstituted version of $SL(2,\mathbb{R})$).

Now expression (8.11), as it stands, can be considered to involve only the ordinary Cartesian frames of conventional mathematics. However, let us now use our formulation of the Cartesian frame as the hyperoctahedral wreath group $\mathbb{Z}_2 \circled{w} \mathbb{Z}_2$, which also represents the square. We wreath sub-append this group to the sequence (8.11) as follows:

$$[\mathbb{Z}_2 \circled{w} \mathbb{Z}_2] \circled{w} A \circled{w} N \circled{w} SO(2). \tag{8.12}$$

Thus, the higher groups A, N, $SO(2)$, act successively on this structure, which is exactly the generative effect; i.e., left-to-right in this sequence we have: square $\longrightarrow$ rectangle $\longrightarrow$ parallelogram $\longrightarrow$ rotated parallelogram.

The reader should observe that the symmetry axes of the hyperoctahedral fiber become the eigenvectors of the structurally most allowable group, A, that is, the group that is next to it in the sequence. In Leyton [96], we encode this fact as follows:

$$[\mathbb{Z}_2 \circled{w} \mathbb{Z}_2] \wedge A \circled{w} N \circled{w} SO(2) \tag{8.13}$$

where the symbol, $\wedge$, stands for the wreath product in which the most salient symmetry axes of the fiber group are aligned with eigenvectors of the immediately higher control group. The reader should recall Sect. 2.17 on the ordering of the successive control groups.

One can add higher levels of action above the linear actions, A, N, $SO(2)$. For example, consider *warping*, which is a facility offered in 3D solid modeling. In warping, the world frame is itself deformed. Within our formulation, this is easily encoded by wreath super-appending a diffeomorphism group G_{warp} to sequence (8.11) thus:

$$[\mathbb{Z}_2 \textcircled{w} \mathbb{Z}_2] \textcircled{w} A \textcircled{w} N \textcircled{w} SO(2) \textcircled{w} G_{warp}. \tag{8.14}$$

Again, this sequence correctly encodes the parent-child and global-local ordering on frames. The reader should note that this ordering is exactly that found in the geometry pipelines of 3D solid modeling programs such as *3D Studio Max*. Thus we have succeeded in giving a systematic means of algebraically formulating the generative hierarchies in both human perception and solid modeling.

However, the above has presented only a fraction of the complete theory elaborated in Chapter 6 of Leyton [96]. That theory gives much deeper insight into the sequence (8.14), exploiting the recursive nature of the Cartesian frame and the Externalization Principle.

8.12 The 3D Reference Frame

It was argued above that the most important factor in the 2D gravitational frame is the reflection structure, and that this is given by a wreath product. Correspondingly it will now be argued that the most important factor in the 3D gravitational frame is the reflection structure, and that this is also given by a wreath product. Later, the enormous significance of this wreath product will be demonstrated in all areas of human perception, CAD, assembly planning, manufacturing, robotics, and navigation.

First one needs to understand how the 3D wreath product works. Just as the 2D frame is related to the square, the 3D frame is related to the cube. Also: just as the control group in the 2D case is a 2-fold cyclic group, the control group in the 3D case is a 3-fold cyclic group, which implies that the cube has a 3-fold axis which acts as a control group in a wreath product. To understand this axis, we need to go through the following two stages: Draw a hexagon, as shown in Fig. 8.11a, with a pair of opposite vertexes aligned *vertically*. Then draw a line from every alternate vertex to the center, starting with the *bottom* vertex. The result is shown in Fig. 8.11b, which gives a compelling impression of a cube.

Now observe the following: The *outline* of Fig. 8.11b is a *hexagon*. However, the figure has only 3-fold symmetry because of the central 3 lines. This means that the figure has the symmetry group D_3 of a *triangle*, not D_6 of a *hexagon*.

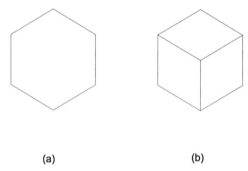

(a)　　　　　　　　　　(b)

Fig. 8.11. Understanding the 3-fold axis of a cube.

Note that D_3 is isomorphic to the permutation group Σ_3 on three elements. In the geometric situation illustrated in Fig. 8.11b, it will be more convenient to write the symmetry group as Σ_3 rather than D_3, because this will generalize to n dimensions.

The structure Σ_3 acts as a control group in a wreath product constructed in the following way: The cube has six faces. These form three reflectionally opposite pairs. The two members of each pair are related by the reflection group $\mathbb{Z}_2$. Thus, there are three copies of $\mathbb{Z}_2$, one for each pair of opposite sides. Therefore, form the wreath product:

$$[\mathbb{Z}_2 \times \mathbb{Z}_2 \times \mathbb{Z}_2] \;\circledS\; \Sigma_3. \tag{8.15}$$

The fiber-group product $[\mathbb{Z}_2 \times \mathbb{Z}_2 \times \mathbb{Z}_2]$ combines the three reflection structures as independent entities. The control group Σ_3 then *transfers* the three reflection structures onto each other by its τ-automorphic action on the fiber-group product.

Notice that the wreath product (8.15) is

$$\mathbb{Z}_2 \;\circledW\; \Sigma_3. \tag{8.16}$$

Observe also that this is a *permutational* wreath product rather than a *regular* wreath product. That is, there are three copies of the fiber (one for each member of the set on which Σ_3 acts), rather than six copies of the fiber (one for each member of Σ_3 itself).

The group $\mathbb{Z}_2 \;\circledW\; \Sigma_3$ can be generalized to n-dimensions, obtaining

$$\mathbb{Z}_2 \;\circledW\; \Sigma_n \tag{8.17}$$

which is called the *hyperoctahedral group* of degree n. This is a symmetry group of the n-dimensional cube. The main two examples in this book are the hyperoctahedral group of degree 2 and degree 3. The case of degree 2 is the dihedral group D_4 (order 8), since this is the permutational wreath product $\mathbb{Z}_2 \;\circledW\; \mathbb{Z}_2$ which is $\mathbb{Z}_2 \;\circledW\; \Sigma_2$. Therefore this is the reflection group we gave for the 2D Cartesian frame. The case of degree 3 was defined in this section.

Whereas the 2D case $\mathbb{Z}_2 \textcircled{w} \Sigma_2$ is wreath c-polycyclic, the 3D case $\mathbb{Z}_2 \textcircled{w} \Sigma_3$ is not, because Σ_3 is not cyclic. Wreath c-polycyclic groups are important to us because canonical plans can be derived from them. In the 2D case, the wreath c-polycyclic nature of the frame allowed us to derive a standard means of drawing a square. We want a wreath c-polycyclic group for the 3D case in order to obtain a standard means of generating a cube. Although $\mathbb{Z}_2 \textcircled{w} \Sigma_3$ is not wreath c-polycyclic, it has a maximal normal subgroup that is. This is simply:

$$\mathbb{Z}_2 \textcircled{w} \mathbb{Z}_3 \;=\; [\mathbb{Z}_2 \times \mathbb{Z}_2 \times \mathbb{Z}_2] \textcircled{s} \mathbb{Z}_3. \tag{8.18}$$

The control level only *rotates* each reflection pair onto each other. Given this wreath structure, one can now derive the following canonical plan for generating a cube: First place one side, then its opposite side. Then rotate and do the same. Finally rotate and do the same. Notice that the cube is thereby generated purely by *transferring* previous actions.

Of course each side is given by the double-translation group. Therefore to complete the frame we add this group below the hyperoctahedral group via a wreath product (i.e., wreath sub-append the translation group). Often we shall take the translation component for granted and omit it from the notation, to emphasize the reflection structure. As a point of terminology, the hyperoctahdral group will often be referred to as the hyperoctahedral *wreath* group, to emphasize the wreath-product decomposition of the group. There are of course other decompositions.

8.13   Assigning Triple-Reflection Structures to Surfaces

In human perception, CAD, assembly planning, manufacturing, robotics, and navigation, it is often very important to attach a 3D Cartesian frame to a surface. We shall show that a basic aspect of this is the assignment of the hyperoctahedral group to the surface. The assignment is as follows:

The hyperoctahedral group (degree 3) contains a triple-reflection structure $\mathbb{Z}_2 \times \mathbb{Z}_2 \times \mathbb{Z}_2$ as its fiber-group product. The three component reflection groups are assigned as illustrated in Fig. 8.12. One mirror plane is made coincident with the object plane - *because the object plane is reflectionally symmetric about itself.* The other two mirror planes are assigned perpendicularly to the object plane because the plane also has these symmetries. Observe that, if the object plane were the *boundary* of a solid object, then the reflection plane that lies in that plane would also have to reverse occupancy; i.e., the reflection operation m in this $\mathbb{Z}_2$ would also be a *color* operation.

Now, a frequently-used technique in shape generation (e.g., human perception) is to assign a Cartesian frame to each of the surfaces of an object, as

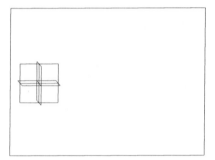

Fig. 8.12. Triple-reflection structure assigned to a plane.

shown in Fig. 8.13. For example, in CAD, one might need to add a sky-light window to the slanting roof shown in this figure. This is easy to draw in the coordinate system defined by the slanting roof, and very difficult to draw in any of the other coordinate systems. Most crucially, we argue that this means that the *triple-reflection* structure is being assigned to the surface; i.e., the visual system is picking out the local reflection structure of the surface.

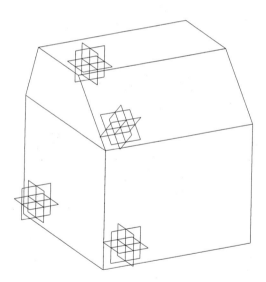

Fig. 8.13. Triple-reflection structure assigned to several planes of a house.

Recall our argument that the visual system is recursive, i.e., that a large non-trivial wreath group $H_1 \textcircled{w} H_2 \textcircled{w} \ldots \textcircled{w} H_n$ can be substituted for the point $\{e\}$ at the left-end of any wreath sequence $\{e\} \textcircled{w} G_1 \textcircled{w} G_2 \textcircled{w} \ldots \textcircled{w} G_n$. For example, an entire Cartesian frame can be substituted for each fiber copy of $\{e\}$. The ability to perceptually map between these frames is then carried out by the τ-automorphic conjugacy structure within the higher sequence $G_1 \text{ⓦ} G_2 \text{ⓦ} \ldots \text{ⓦ} G_n$.

In particular, let us consider any one of the planes in Fig. 8.13. We can regard it as structured like this:

$$\{e\} \text{ⓦ} [\mathbb{R} \times \mathbb{R}]. \tag{8.19}$$

The left end of this sequence, $\{e\}$, represents the individual point on the surface, as fiber, duplicated by the translation control group. Using the above concept of recursion, one can substitute the entire 3D Cartesian frame for $\{e\}$. Thus one obtains a Cartesian frame at each point on the surface. This is particularly significant in human perception, where we argue that a triple-reflection structure is detected at each point of the surface, as illustrated in Fig. 8.14. The structure is an example of what we call a frame bundle, and is analogous to the assignment of a 2D Cartesian frame at each point along the side of the square, as shown in Fig. 8.7 (p. 200).

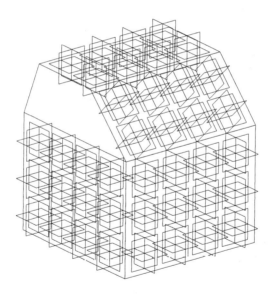

Fig. 8.14. Triple-reflection structures are assigned massively in human perception.

Now the frames in Fig. 8.14 are related to each other by the τ-automorphic conjugacy structure provided by the translation control group. This exploits the wreath-product symbol in the sequence (8.19) into which the frames have been substituted. Notice that there must be a particular frame which is the fiber-group copy associated with the identity element in the translation control group. This is a reference starting frame, with respect to which the other frames on the surface are generated. Our theory of reference objects (p. 188) says that a reference object is a phase-transition and therefore is either a point of maximal symmetry or maximal asymmetry. Therefore, on each of the faces of the house shown in Fig. 8.14, there is a distinguished reference frame located either at the center of the face - i.e., point of maximal symmetry - or at one of its corners - i.e., point of maximal asymmetry.

8.14   Construction Plane

In CAD and solid modeling, the creation of an object involves the crucial use of a *construction plane*, which is a base reference plane with respect to which the object is generated. Standardly the drawing of the object is decomposed into two phases, in which the first phase is a drawing on the plane and the second is a sweeping outward of the drawing, e.g., by translation, revolution, etc. The two phases can be given by the movement of the cursor or numerical specification.

For example, Fig. 8.15 illustrates the standard procedure for drawing a block in solid modeling, using a cursor. The two-phase procedure is this: The first phase, labeled by 1 in the diagram, is the movement of the cursor that creates the first face of the object. The second phase, labeled by 2, is the movement of the cursor that translates the face in the perpendicular direction. In more complex object design, e.g., part design in mechanical CAD, phase 1 is the drawing of a polyline or spline on the construction plane, and phase 2 is again a sweeping of the drawing in the perpendicular direction. Using the terminology of sweeps, the shape in the construction plane is the *profile*, and the perpendicular movement is the *path*.

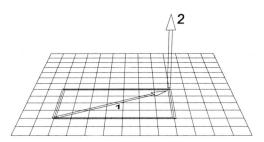

Fig. 8.15. Drawing a block using two phases.

Sweeping structures will be algebraically formalized in Chapter 18 of this book. Such structures exemplify our claim that shape-generation is intelligent when it maximizes *transfer*. Here, the 3D shape is generated by the *transfer* of a profile along a path. Algebraically we will formulate this via a wreath product, in which the profile is the *fiber* and the path is the *control*:

$$\textbf{Sweep} \;=\; \textbf{Profile} \;\textcircled{w}\; \textbf{Path}.$$

Notice that the wreath-product symbol $\textcircled{w}$ predicts correctly that there is one copy of the profile for each member of the path.

The purpose of the present section is to show that the construction plane conforms with our theory of reference frames. The theory says that a reference frame defines a *phase-transition*. Clearly, this is seen in the above discussion: The construction plane defines the first phase of the drawing process, and, by the transfer structure, becomes the basis for the second phase.

Most crucially, the theory says that the phase-transition must define the symmetry ground-state of a phase. This is also evident here: The construction plane is reflectionally symmetric about itself. Therefore, a *profile* on a construction plane possesses the same reflectional symmetry.

Now observe that, from this initial symmetry, sweeping will create an asymmetry: it will create *positional distinguishability* between the successive cross-sections, i.e., internal history (the term internal is defined in Sect. 2.9). This accords with our theory of shape-generation, in the following fundamental ways: The sweeping asymmetrization actually goes along a symmetry of the final object, i.e., in conformance with our view that plans, and therefore asymmetrizations, go along symmetries as channels. Furthermore, the symmetry group of the asymmetrization process is used as the control group moving the fiber - again in accord with the theory of asymmetry-building (Sect. 2.18).

Notice also that the construction plane containing the profile is matched with the reflectional symmetry of a face (or cross-section) within the final object. Thus the symmetry of the construction plane is not actually lost in the asymmetrization process. This is because it is a fiber of the final object. This, as we said, is the power of describing generativity using asymmetry-building by transfer: The previous structure is not lost - it is part of the representation of the final structure, and is recoverable from it because the control group is an asymmetrization.

Finally observe this: The construction plane is a reflectional fiber taken from the hyperoctahedral group defining the Cartesian frame. In fact, in major programs such as *ProEngineer*, the user is asked to select a construction plane from the three perpendicular Cartesian planes given on the screen. The crucial thing is that this fiber from the Cartesian frame is then matched to a profile fiber of the goal object. Our full theory is given by the panel on p. 212.

THEORY OF CONSTRUCTION PLANES

(1) A construction plane is a reflectional $\mathbb{Z}_2$ fiber taken from the hyperoctahedral wreath group defining the Cartesian frame.

(2) Any drawing on a construction plane is reflectionally symmetric about the plane.

(3) In standard CAD and 3D modeling, the drawing is swept into the perpendicular direction. This creates a structure of transfer, and therefore a wreath product, in which the profile drawing is the fiber, and the sweeping is the control.

(4) The drawing fiber contains not only a reflectional symmetry of the construction plane, but a reflectional symmetry of the fiber within the final 3D object - i.e., usually the reflectional symmetry of a face (or cross-section) of the final goal object.

(5) Therefore, in setting up the drawing, the construction plane's reflectional symmetry is chosen to coincide with the reflectional symmetry of one of the faces (or cross-sections) of the goal object, and the sweeping is chosen to coincide with a perpendicular continuous symmetry of the goal object (e.g., translation or rotation).

(6) The construction plane defines the first phase of the drawing process, and, by the transfer structure, becomes the basis for the second phase. It therefore accords with our theory that reference frames (a) constitute phase transitions in the generative history, and (b) arise from symmetries of the object - since a phase transition corresponds to a symmetry ground-state for generation.

(7) In particular, the use of the construction plane is a way of matching a reflectional $\mathbb{Z}_2$ fiber from Cartesian hyperoctahedral wreath group to a reflectional fiber of the goal object, so that the fiber of Cartesian frame can become the symmetry ground-state in the generation of the goal object.

9. Relative Motion

9.1 Introduction

As was said in Chapter 1, the major purpose of our theory is the **conversion of complexity into understandability**. Any intelligent system is faced with an environment, e.g., a 3D scene, of enormous complexity, and must convert this into an understandable structure. We argue that the conversion of complexity into understandability is achieved by maximizing transfer and recoverability. Furthermore, we show how to give transfer and recoverability a well defined mathematics. Therefore this book gives a mathematical theory of understandability.

Relative motion is an important phenomenon in a number of areas such as human perception, robot manipulation, computer animation, and physics. It is in fact used to simplify the representation of motion that would otherwise be complex. This chapter will show that relative motion accords exactly with our theory; i.e., it works by maximizing transfer and recoverability. This allows us to give a group-theoretic representation of relative motion that has not so far been given in the literature, and deepens the understanding of what relative motion is.

In a sense, Chapter 6 on robot manipulators can be regarded as concerning relative motion, and the present chapter can be regarded as continuing that chapter. Whereas that former chapter concerned relative motion in robot kinematics, the present chapter will concern relative motion in human perception, computer animation and physics.

9.2   Theory of Relative Motion

In this section, we will propose a theory of relative motion. In order to understand it, an important example of relative motion will first be studied in human perception. It is important because it shows the relevance of our hyperoctahedral theory of reference frames, as well as our algebraic theory of grouping, to the phenomenon of relative motion.

The example is a motion experiment carried out by Johansson [65]. The stimulus used in this experiment is illustrated in Fig. 9.1a which shows two dots moving perpendicularly, along the arrows M_1, M_2, backwards and forwards in phase. Johansson found that subjects did not see the configuration like this; i.e., as perpendicular motion. Instead, they saw the dots as moving to and from each other along the *diagonal* line marked $R_1 R_2$ in Fig. 9.1b. Furthermore, they saw this diagonal line as moving, as a whole, along the opposite diagonal - that given by the arrow marked C in Fig. 9.1b.

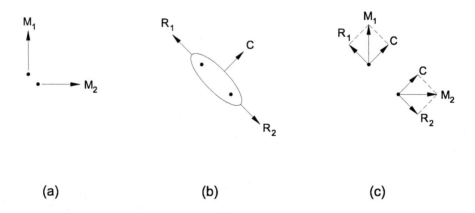

(a)   (b)   (c)

Fig. 9.1. The Johansson 1950 motion phenomenon.

Johansson's explanation was that the motion, as shown in Fig. 9.1a, is decomposed into two factors: (1) the motion that the two dots have *relative* to each other, i.e. to and from each other, given by the arrows R_1 and R_2 in Fig. 9.1b; and (2) the motion that the two dots have in *common*, given by the arrow C in 9.1b. The two components were called, respectively, the *relative motion*, and the *common motion*. This vector decomposition is shown in Fig. 9.1c.

It will now be argued that the Johansson motion phenomenon is explained by our generative theory of shape. Let us begin by examining the two dots themselves, without the motion, as shown in Fig. 9.2a. Observe that the pair of dots is reflectionally symmetric about two axes: (1) the line that joins the dots, shown as line m_1 in Fig. 9.2b; and (2) the line that perpendicularly bisects them, labeled m_2 in Fig. 9.2b. That is, reflecting the dots about either of these lines sends the dot-pair to itself. This means that the internal structure of the dots is the Klein-four group. In fact, using occupancy, one can say that the group structuring this situation is the hyperoctahedral wreath group $\mathbb{Z}_2 \textcircled{w} \mathbb{Z}_2$, which we have claimed can be identified with the Cartesian frame.

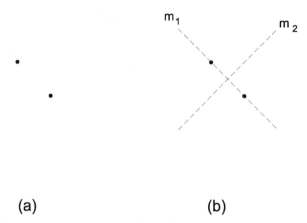

(a)   (b)

Fig. 9.2. The symmetry structure of two dots.

Now examine the motion of the dot-pair. Consider first the relative motion shown as R_1, R_2 in Fig. 9.1b. The motion R_1 can be described by the translation group $\mathbb{R}$ (in the direction R_1). If this motion is put below the hyperoctahedral wreath group $\mathbb{Z}_2 \textcircled{w} \mathbb{Z}_2$, it will be reflected to produce the motion of the other dot R_2. Thus the relative motion structure is given by

$$\mathbb{R} \textcircled{w} \mathbb{Z}_2 \textcircled{w} \mathbb{Z}_2$$

which is the *hyperoctahedral wreath hyperplane group*, introduced in Sect. 8.8 as a fuller version of the Cartesian frame.

Now turn to the common motion, shown as C in Fig. 9.1b. This translation moves the entire previous structure along the axis C. Therefore, it is given by wreath super-appending the translation group $\mathbb{R}$ above the previous structure thus:

$$[\mathbb{R} \textcircled{w} \mathbb{Z}_2 \textcircled{w} \mathbb{Z}_2] \textcircled{w} \mathbb{R}.$$

Notice that the fact that this added motion is *common* motion is captured exactly by the wreath product. It codes the fact that the common motion is a structure of *transfer*; i.e., the higher $\mathbb{R}$ transfers copies of the hyperoctahedral wreath reflection structure from one position to the next along its path.

Finally, the entire structure is rotated relative to the World frame, and thus we wreath super-append the rotation group $SO(2)$ to the entire previous structure thus:

$$[\mathbb{R} \ \textcircled{w} \ \mathbb{Z}_2 \ \textcircled{w} \ \mathbb{Z}_2] \ \textcircled{w} \ \mathbb{R} \ \textcircled{w} \ SO(2). \tag{9.1}$$

Examining this sequence, one can see that it is constructed by taking the group for the Cartesian frame $[\mathbb{R} \ \textcircled{w} \ \mathbb{Z}_2 \ \textcircled{w} \ \mathbb{Z}_2]$ and wreath super-appending successively the translation group $\mathbb{R}$ and then the rotation group $SO(2)$. Therefore, the Johansson motion phenomenon corroborates our theory that perceptual organization is fundamentally based on the hyperoctahedral wreath group.

Notice also that the full group (9.1) is wreath c-polycyclic. Furthermore, the symmetry axes correspond with the lines of flexibility (affine eigenvectors) of the motion, in accord with our Interaction Principle (p. 67).

Now let us turn to the inheritance structure: According to our theory of inheritance, inherited properties are the control-nested τ-automorphisms (Sect. 7.3). In order to see the inheritance structure, let us first use a diagram similar to the type used in *3D Studio Max* (described on p. 176), and then use the method described in Sect. 7.4 (p. 178) for converting such inheritance diagrams into algebra.

For the Johansson situation, the diagram would be as shown in Fig. 9.3. It shows three frames: the World, Frame 2, and Frame 1. By the indentation structure of this diagram, Frame 2 is a child of the World, and Frame 1 is a child of Frame 2. The transform that relates Frame 2 to the World is 45^0 rotation (which is assumed to be clockwise): It re-orients the horizontal (x) and veritcal (y) axes of the World into the diagonal orientation shown in Fig. 9.2. The transform that relates Frame 1 to Frame 2 is the common motion translation C. It moves Frame 1 along the y-axis of Frame 2.

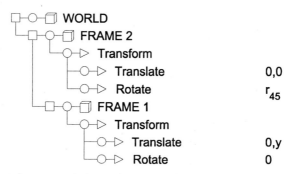

Fig. 9.3. The inheritance diagram for the Johansson phenomenon.

Using the symbols F_1, F_2, and W, for Frame 1, Frame 2, and World, respectively, the group structure in expression (9.1) is now given as:

$$[\mathbb{R} \ \textcircled{w} \ \mathbb{Z}_2 \ \textcircled{w} \ \mathbb{Z}_2] \ \textcircled{w} \ _{F_1}\mathbb{R}^{F_2} \ \textcircled{w} \ _{F_2}SO(2)^W. \tag{9.2}$$

Observe that, in going from right-to-left in this group sequence, the frame indexes go from W to F_2 to F_1. The question we therefore ask is: What is the frame F_1, which appears as the final (left-ward) index along this group sequence? The answer is this: F_1 is the *fiber group* to the left of it in the sequence, i.e., it is the hyperoctahedral reference frame group $[\mathbb{R} \, ⓦ \, \mathbb{Z}_2 \, ⓦ \, \mathbb{Z}_2]$.

With this in mind, let us now look at the inheritance structure. Consider the reference frame group $[\mathbb{R} \, ⓦ \, \mathbb{Z}_2 \, ⓦ \, \mathbb{Z}_2]$. It is a fiber group within the full group sequence, and its fiber-group copies are given thus:

$$[\mathbb{R} \, ⓦ \, \mathbb{Z}_2 \, ⓦ \, \mathbb{Z}_2]_{y,r}$$

where y comes from its immediate control group $\mathbb{R}$; and r comes from the final control group $SO(2)$. According to our algebraic theory of inheritance (p. 177), the inherited properties are therefore y and r. These are, respectively, (1) the distance along the common motion direction, and (2) the amount of rotation relative to the World frame. Notice that our theory predicts that this inheritance is hierarchical: The common motion is translation along the y direction *within* the frame that has been rotated from the world frame.

Similarly, consider the next larger left-subsequence: $[\mathbb{R} \, ⓦ \, \mathbb{Z}_2 \, ⓦ \, \mathbb{Z}_2]$ $ⓦ \, {}_{F_1}\mathbb{R}^{F_2}$. It is a fiber group within the full group sequence, and its fiber-group copies are given thus:

$$[[\mathbb{R} \, ⓦ \, \mathbb{Z}_2 \, ⓦ \, \mathbb{Z}_2] \, ⓦ \, {}_{F_1}\mathbb{R}^{F_2}]_r$$

where r comes from the final control group $SO(2)$. The fiber-group copy therefore has one inherited property r. This is the rotation from the World frame.

Our theory therefore predicts the correct inherited properties at each level of the hierarchy. Notice that, since the inherited properties come from the right-subsequences of the respective fibers, they correspond to control-nested τ-automorphisms, as our theory states.

So far we have not mentioned what Johansson called the relative motion R_i in Fig. 9.1. This is given by the group $\mathbb{R}$ at the left-most end of the group sequence. Its inheritance is similarly correct. This motion occurs *within* the y-translated, rotated, world frame, where the y-translation and rotation are those that have just been discussed. Therefore, within the y-translated rotated world frame, this motion is actually the x-translation.

We are now ready to give our general theory of relative motion:

THEORY OF RELATIVE MOTION. *A motion is decomposed into a hierarchy of relative motion systems in order to maximize transfer and recoverability. Therefore it is represented by an n-fold wreath product $G_1 ⓦ G_2 ⓦ \ldots ⓦ G_n$ in which any left-subsequence is a relative motion system, which itself undergoes relative motion within its control-nested τ-automorphism group.*

Notice the similarity between this and our theory of grouping (Chapter 5), which says that groupings are left-subsequences and their occurrences are given by their control-nested τ-automorphism groups.

The remainder of this chapter will continue to investigate this theory.

9.3   Induced Motion

Let us now consider the following phenomenon discovered by Duncker [29]. Suppose the visual field is completely empty, except for a rectangular frame and a dot within the frame, as shown in Fig. 9.4. Then, if the frame is moved and the dot is fixed, the viewer perceives the motion to be the *opposite*: That is, the frame is seen as fixed and the dot is seen as moving. The explanation offered was that the larger surrounding system is chosen to be a referent for a smaller system within, rather than the reverse.

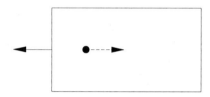

Fig. 9.4. Duncker effect: Rectangle moving and dot fixed, is seen as opposite by observer.

Duncker observed that the dot-frame illusion is a phenomenon with which people are quite familiar in their daily lives: When one looks up at the sky on a stormy night, one sees the moon moving quickly through a large cloud. Of course, relative to the viewer the moon is actually still and the cloud is actually moving. Nevertheless, one perceives the movement to be the opposite way round.

Duncker also observed that several such systems can be embedded within each other. For example, if one sees the moon and cloud between tall buildings, as shown in Fig. 9.5, then the moon is seen as moving relative to the cloud, and the cloud is seen as moving relative to the buildings. This of course is simply an embedding of the dot-frame pair within itself. That is, on one level, the moon corresponds to the dot and the cloud corresponds to the frame; and on the next level, the cloud corresponds to the dot and the buildings correspond to the frame.

Notice that, within the time-scale of observation, the moon and buildings are *both* at rest; i.e., should be assigned the *same* velocity as each other.

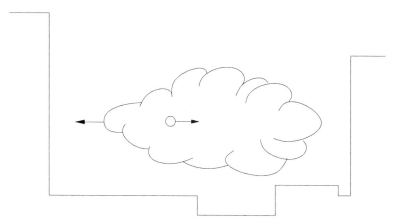

Fig. 9.5. Successive embeddings of the Duncker effect.

Nevertheless, the moon is seen as moving because it is judged relative to the clouds, but the buildings are seen at rest. Thus the moon and buildings are assigned different velocities. The Gestaltists put forward the following principle to explain this phenomenon: *The motion of an object is judged relative to the next larger surrounding system but not the system beyond that.* This is called the **Principle of the Separation of Systems**.

Of course, if one is viewing the buildings from the window of a passing ship, then the buildings appear to move. But their movement is seen as relative to the surrounding window frame, not relative to anything within the scene. This 4-level structure (moon-cloud-buildings-window) is shown schematically in Fig. 9.6, where the largest rectangle is the window frame, and the successive inward objects are the buildings, the cloud, and the moon.

Now examine this structure in terms of our algebraic theory of relative motion: Start with the original dot-frame structure in Fig. 9.4. Recall that, for this experiment, the visual field was entirely empty except for the rectangle and dot. In particular, there were no visual features outside the rectangle. This ensures that the rectangle is not seen as moving. Most crucially, it ensures that the rectangle is identified with the World frame. Generally, therefore, we conclude that, even in the multi-level embedded examples, the largest surrounding frame is identified with the World frame.

Now let us consider the inheritance diagrams. Return to the basic dot-rectangle example. The inheritance diagram is shown in Fig. 9.7. Here the rectangle is identified with the World frame, and the dot is a child of the rectangle, as indicated by the indentation in this diagram. Notice that there is no transform under the rectangle itself. This is because it is identified with the World frame, and the World frame is not transformationally related to anything above it, because the World has no parent. In contrast, the dot is the child of the World, and therefore has a transform (translation) giving the relation of its frame to the frame of the World.

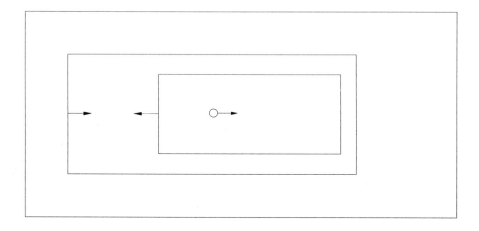

Fig. 9.6. The ship example.

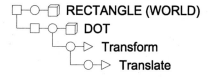

Fig. 9.7. The inhertance diagram of the dot-rectangle illusion.

Now let us go to the 4-level structure of the ship example (moon-cloud-buildings-window). The inheritance diagram for this situation is shown in Fig. 9.8. Here, the window is identified with the World frame (since it is the largest rectangle). Its child is the buildings, and the child of the buildings is the cloud, and the child of the cloud is the moon. Again, the top level, the window, does not have an associated transform, but each of the children underneath do have a transform, and this relates the frame of each child upwards to the frame of its parent. The transform in each case is translation, which can be taken to be $\mathbb{R}$ as translation in the x direction.

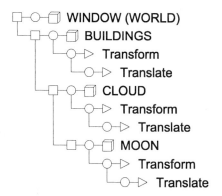

Fig. 9.8. The inhertance diagram for the ship example.

Using the theory developed in Sect. 7.4 (p. 178) for converting inheritance diagrams into algebra, we form the group of the situation by taking the *wreath product* of the transforms (in the diagram), and assigning to each component group G_i of the wreath product, an lower index giving the frame on level i, and an upper index giving the frame of the parent. The wreath product is therefore:

$$_{F_1}\mathbb{R}^{F_2} \ \textcircled{w} \ _{F_2}\mathbb{R}^{F_3} \ \textcircled{w} \ _{F_3}\mathbb{R}^{W} \tag{9.3}$$

where F_3 is the frame of the buildings, F_2 is the frame of the cloud, and F_1 is the frame of the moon.

This structure supports our view that a relative motion system is given by an n-fold wreath product in which any level undergoes a relative motion hierarchy given by its control-nested τ-automorphism group.

9.4 Inheritance via Extra Frames

In relative motion structures, it is often valuable to create extra frames, in order to simplify the motion. This should be understood in terms of our generative theory of shape, as follows:

As an example, consider the representation of the moving solar system by computer animation. Let us confine ourselves to the sun-earth-moon structure: i.e., the earth orbiting around the sun and the moon orbiting around the earth. This example is often used to illustrate the importance of creating dummy objects in animation (e.g., Elliot et al., [31]). The argument is this: There are two contrasting methods for representing the structure:

Method 1. In this method, one makes the moon a child of the earth, and the earth a child of the sun. However, one moves the center of the moon's coordinate system to the center of the earth. Then, when one applies the rotation to the moon-object, the latter will rotate around the center of the earth because a transform is always applied to the coordinate frame of an object, and because the coordinate frame of the moon is centered in the earth.

In order to establish the movement of the earth around the sun, one does the analogous thing: One moves the center of the earth's coordinate system to the center of the sun. Then, at animation time, one applies a rotation to the earth-object, which will result in the earth rotating around the sun, because the coordinate frame of the earth is centered in the sun.

Method 2. The problem with the previous method is that it does not allow additional rotation that can typically be required: For example, one might want a planet to rotate around its own axis, which would be difficult to set up if its coordinate frame is centered in another planet. Thus, much more convenient than Method 1, is to use extra coordinate frames. Standardly, one obtains them by creating "dummy objects". Since any object carries with it a coordinate frame, one has therefore created extra coordinate frames. These frames will be called *dummy frames*.

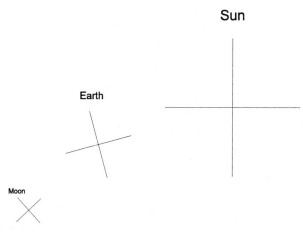

Fig. 9.9. Planet frames in a solar system.

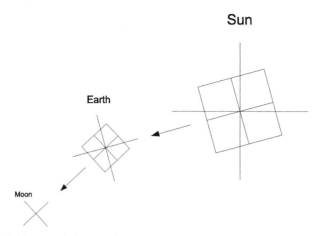

Fig. 9.10. Addition of dummy frames in a solar system.

Fig. 9.9 shows the frames that belong to the actual physical planets. Unlike Method 1, each remains at the center of its own planet. Next, in Fig. 9.10, two dummy frames are added: Each is represented by a square with cross-hairs. The dummy objects will be called the sun-dummy and the earth-dummy. The former is located at the center of the sun and the latter is located at the center of the earth.

Now let us specify the inheritance structure. This is shown in Fig. 9.11. There are six objects including the World. Each successive object downward is a child of the object directly above it, as indicated by the indentation structure. The successive transforms are as follows: The transform of the sun is translation by a single amount t from the World's coordinate frame. The transform of the sun-dummy is continuous rotation r relative to the sun. The transform of the earth is translation by a single amount t' from the coordinate frame of the sun-dummy. The transform of the earth-dummy is continuous rotation r' relative to the earth. And finally, the transform of the moon is a translation by a single amount t'' from the coordinate frame of the earth-dummy.

Now let us use our method for converting inheritance diagrams into algebra. That is, form the group of the situation by taking the *wreath product* of the transforms (in the inheritance diagram), and assign to each component group G_i of the wreath product, a lower index giving the frame on level i, and an upper index giving the frame of the parent. The wreath product is therefore:

$$_m T^{ed} \ \textcircled{w} \ _{ed}SO(2)^e \ \textcircled{w} \ _e T^{sd} \ \textcircled{w} \ _{sd}SO(2)^s \ \textcircled{w} \ _s T^w. \tag{9.4}$$

The symbols are as follows: There are two groups involved: T, the 2D translation group, and $SO(2)$ the rotation group. These two groups alternate down the sequence. Also, going down the sequence, the respective frame-symbols are w = world; s = sun; sd = sun-dummy; e = earth; ed = earth-dummy; m = moon. (Here, lower case letters have been used for frames, to help readability.)

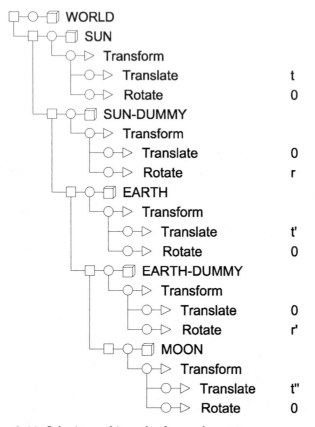

Fig. 9.11. Inheritance hierarchy for a solar system.

In order to complete the structure, now add the group of the frame itself, i.e., the hyperoctahedral structure $[\mathbb{R} \, \textcircled{w} \, \mathbb{Z}_2 \, \textcircled{w} \, \mathbb{Z}_2]$. This is wreath subappended to the sequence (9.4), to obtain:

$$[\mathbb{R} \, \textcircled{w} \, \mathbb{Z}_2 \, \textcircled{w} \, \mathbb{Z}_2] \, \textcircled{w} \, _mT^{ed} \, \textcircled{w} \, _{ed}SO(2)^e \, \textcircled{w} \, _eT^{sd} \, \textcircled{w} \, _{sd}SO(2)^s \, \textcircled{w} \, _sT^w. \quad (9.5)$$

Notice that all the copies of the frame are of the form:

$$[\mathbb{R} \, \textcircled{w} \, \mathbb{Z}_2 \, \textcircled{w} \, \mathbb{Z}_2]_{t'',r',t',r,t} \quad (9.6)$$

where the five subscripts come from the five respective control groups.

Notice that the fiber-group copy

$$[\mathbb{R} \ \textcircled{w} \ \mathbb{Z}_2 \ \textcircled{w} \ \mathbb{Z}_2]_{e,e,e,e,e}$$

corresponds to the World frame, and

$$[\mathbb{R} \ \textcircled{w} \ \mathbb{Z}_2 \ \textcircled{w} \ \mathbb{Z}_2]_{e,e,e,e,t}$$

corresponds to the sun frame, and

$$[\mathbb{R} \ \textcircled{w} \ \mathbb{Z}_2 \ \textcircled{w} \ \mathbb{Z}_2]_{e,e,e,r,t}$$

corresponds to the sun-dummy frame, and so on.

The crucial thing to observe is that the action of each of the control group components is to cause a symmetry-breaking of the frame. This can easily be seen by looking at the successive actions in Fig. 9.10. For example, each dummy frame has been drawn as a square. This follows the convention in 3D computer animation (e.g., in *3D Studio Max*), of drawing a dummy object as a cube. In other words, dummy objects - the default objects that carry coordinate frames - are given a hyperoctahedral structure. What can be seen therefore is that the successive actions in animation break the symmetries of this structure.

Generally then, what the group in expression (9.5) illustrates for us is this:

THEORY OF ANIMATION.   *Animation follows precisely the rules of our generative theory of shape: That is, animation is given by a wreath product of symmetry-breaking phase-transitions, where coordinate frames are the symmetry ground-states of the successive phase-transitions.*

9.5   Physics

In physics there is the following well-known result concerning angular momentum: The total angular momentum of a system of particles about an origin O is decomposable into two components: The first is the angular momentum of the particles with respect to the center of mass. The second is the angular momentum of the center of mass with respect to the origin O, all masses being considered to be concentrated at the center of mass. The decomposition is this:

$$\overbrace{\sum \mathbf{r}'_i \times m_i \mathbf{v}'_i} \ + \ \overbrace{\bar{\mathbf{r}} \times M\bar{\mathbf{v}}}. \tag{9.7}$$

where the $\mathbf{r}'_i$ and $\mathbf{v}'_i$ are the position vectors and velocity vectors of the *particles in the center of mass frame*; and $\bar{\mathbf{r}}$ and $\bar{\mathbf{v}}$ are the position vector and velocity vector of the *center of mass in the origin frame*. An example is of course the decomposition of the earth's angular momentum into its "personal" rotation about its own center of mass, and its orbit around the sun.

Clearly the angular momentum decomposition fits exactly a wreath structure; i.e., the angular momentum about the center of mass corresponding to a fiber group and the angular momentum of the center of mass about the origin O corresponding to a control group. The control group transfers the fiber group around the environment. The groups involved can be those as given in Sect. 9.4.

Now consider a system of two particles, where the position vectors of the particles are $\mathbf{r}_1$ and $\mathbf{r}_2$ with respect to the origin O, and the potential energy $V(|\mathbf{r}_1 - \mathbf{r}_1|)$ depends only on the radial distance between the two particles. Let the conjugate momenta of the two particles be $\mathbf{p}_1$ and $\mathbf{p}_2$. Then the (classical) two-particle Hamiltonian is

$$\mathcal{H}(\mathbf{r}_1, \mathbf{p}_1; \mathbf{r}_2, \mathbf{p}_2) \;=\; \frac{\mathbf{p}_1^2}{2m_1} \;+\; \frac{\mathbf{p}_2^2}{2m_2} \;+\; V(|\mathbf{r}_1 - \mathbf{r}_1|). \tag{9.8}$$

Most crucially, notice that, on phase space, the coordinate system ($\mathbf{r}_1$, $\mathbf{p}_1$; $\mathbf{r}_2$, $\mathbf{p}_2$), used here, corresponds to the two particles.

Now apply a coordinate transformation:

$$(\overbrace{\mathbf{r}_1,\ \mathbf{p}_1}^{Mass1};\ \overbrace{\mathbf{r}_2,\ \mathbf{p}_2}^{Mass2}) \;\longrightarrow\; (\overbrace{\mathbf{r}_{CM},\ \mathbf{p}_{CM}}^{MassCM};\ \overbrace{\mathbf{r}_{Rel},\ \mathbf{p}_{Rel}}^{MassRel}) \tag{9.9}$$

where ($\mathbf{r}_{CM}$, $\mathbf{p}_{CM}$) are the center-of-mass motion coordinates, and ($\mathbf{r}_{Rel}$, $\mathbf{p}_{Rel}$) are the relative motion coordinates. These are defined as follows:

Center-of-Mass Motion:

$$\mathbf{r}_{CM} \;=\; \frac{m_1\mathbf{r}_1 + m_2\mathbf{r}_2}{m_1 + m_2}$$
$$\mathbf{p}_{CM} \;=\; \mathbf{p}_1 + \mathbf{p}_2. \tag{9.10}$$

Relative motion:

$$\mathbf{r}_{Rel} \;=\; \mathbf{r}_1 - \mathbf{r}_2$$
$$\mathbf{p}_{Rel} \;=\; \frac{m_2\mathbf{p}_1 - m_1\mathbf{p}_2}{m_1 + m_2}. \tag{9.11}$$

These two pairs of coordinate vectors can be thought of as corresponding to two fictitious masses. The first is called the *total particle*, and the second is called the *relative particle*. The two associated masses are, respectively:

$$m_{CM} = m_1 + m_2$$

$$m_{Rel} = \frac{m_1 m_2}{m_1 + m_2}. \tag{9.12}$$

The first is called the *total mass*. The second is called the *reduced mass*; i.e., it is the geometrical mean of the two masses m_1 and m_2.

The Hamiltonian can now be written in terms of this new coordinate system, thus:

$$\mathcal{H}(\overbrace{\mathbf{r}_{CM}, \mathbf{p}_{CM}}^{MassCM}; \overbrace{\mathbf{r}_{Rel}, \mathbf{p}_{Rel}}^{MassRel})) = \overbrace{\frac{\mathbf{p}_{CM}^2}{2m_{CM}}}^{\mathcal{H}_{CM}} + \overbrace{\frac{\mathbf{p}_{Rel}^2}{2m_{Rel}} + V(\mathbf{r}_{Rel})}^{\mathcal{H}_{Rel}}. \tag{9.13}$$

Compare this coordinization of the Hamiltonian to the previous one given in (9.8), i.e., where the two particles were the actual ones. Inspection of the previous Hamiltonian (9.8) reveals that it cannot be separated into two components corresponding to the two actual particles, because the potential energy has terms dependent on both particles. In contrast, in the new coordinization, (9.13), the Hamiltonian can be separated into two components corresponding to the two fictitious particles.

Notice that there is no $\mathbf{r}_{CM}$ term in the new equation, which means that $\mathbf{r}_{CM}$ is a cyclic coordinate and that the momentum of the center of mass must be constant. This means of course that the center of mass moves in uniform rectilinear motion, i.e., that of a free particle.

Physics has long recognized the importance of the above fictitious particle decomposition. According to our theory, this importance is as follows:

IMPORTANCE OF DECOMPOSITION INTO FICTITIOUS PARTICLES. *The above decomposition into fictitious particles corresponds to a structure of* **transfer**, *i.e., a wreath product. The motion of the total particle corresponds to the control group; and the motion of the relative particle corresponds to the fiber group.*

Notice that one can regard the above decomposition as analogous to the decomposition in the Johansson motion phenomenon (Sect. 9.2).

10. Surface Primitives

10.1 Defining and Classifying Primitives

It is generally accepted that complex objects are created by deforming and combining primitives. However, what are primitives? No one has precisely defined them, or given them a systematic classification. In this chapter, we give a rigorous theory of primitives. Primitives, we argue, arise from the maximization of transfer and recoverability. When we fully understand how they fulfill these conditions, then we will know what they are and how to classify them.

To begin with basics: It is generally accepted that primitives are simpler than the objects that are created from them. For example, standardly in CAD and solid modeling, one uses Boolean operations to combine solid primitives into some required shape, as illustrated in Fig. 10.1. The bottom nodes of this figure show the primitives, and their successive combination upwards is achieved through the Boolean operations (union, subtraction, intersection) to achieve the top shape. The progression upwards is clearly one of greater complexification.

Thus, one has the following progression in the generative direction:

$$\text{simplicity} \longrightarrow \text{complexity}.$$

In fact, although it is never stated like this, the above arrow actually means:

$$\text{symmetry} \longrightarrow \text{asymmetry}.$$

There are two advantages to the latter formulation: (1) We can use group-theory to rigorously understand what building-with-primitives actually is. (2) We discover that building-with-primitives is a form of *symmetry-breaking*, or more strongly, *asymmetry-building*.

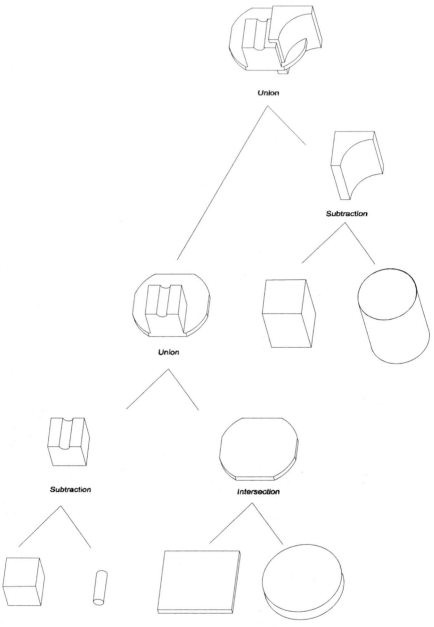

Fig. 10.1. Boolean operations used to combine solid primitives into a complex shape.

This chapter develops a theory of primitive *surfaces*. These are basic also to primitive *solids*, because they constitute the bounding surfaces of the latter. Primitive solids will be classified later in the book.

The most crucial fact is that when one uses the principles of maximization of transfer and recoverability, one obtains a rigorous theory of primitives and primitive-building.

10.2 Level-Continuous Primitives

A standard classification of the surface primitives (e.g., Heo, Kim, & Elber [57] p33) is the following:

plane
natural quadrics (sphere, circular cylinder, circular cone)
torus

In order to achieve our aim of ensuring that primitives maximize transfer and recoverability, we should first express the above list as wreath c-polycyclic groups. We do so as follows:

STANDARD PRIMITIVES: LEVEL-CONTINUOUS

Plane:	$\mathbb{R} \textcircled{w} \mathbb{R}$
Sphere:	$SO(2) \textcircled{w} SO(2)$
Circular cylinder:	$SO(2) \textcircled{w} \mathbb{R}$
Circular cone:	$\mathbb{R} \textcircled{w} SO(2)$
Torus:	$SO(2) \textcircled{w} SO(2)$

This table will now be discussed in detail. But first, the following definition is required:

Definition 10.1. *An n-fold wreath product will be called* **level-continuous** *if each of its levels is a continuous group. If at least one of its levels is discrete it will be called* **level-discrete.**

Clearly, each of the primitives in the above table is represented as a level-continuous wreath product. Since each level in any of these wreath products is c-cyclic, each level is either $SO(2)$ or $\mathbb{R}$. Observe also that each primitive is described as a 2-level hierarchy. Thus according to the above description, each of the primitives is a *2-level wreath c-polycyclic, level-continuous, group.*

In order to achieve maximization of transfer and recoverability, we now need to examine relationships between the members of the above list.

10.3   Sphere and Torus

Observe that the wreath structure given above for the sphere and the torus are the same, that is, both are $SO(2) \wr SO(2)$. This implies that there is a higher-order symmetry involved, and this higher level can be expressed as an extra level of *transfer* that changes the sphere into the torus, as follows:

To understand the symmetry, let us describe a sphere as an example of a torus. For ease of exposition, let us assume that the torus lies horizontally. In a torus, one has two planar rotation groups $SO(2)$, one around the central vertical axis of the torus, and the other around the horizonal rotation axis of the any of the tube cross-sections. The former group acts as the control group rotating the latter group as a fiber around the torus. Hence we obtain the wreath structure, given above: $SO(2) \wr SO(2)$.

Now consider one of the cross-sections of the tube, i.e., a fiber $SO(2)$. Its horizontal axis is perpendicular to the vertical axis of the control group, and some distance away from the latter axis. This distance will be called the *control radius*.

The control radius can be varied giving smaller or larger tori. *When this distance is zero, i.e., the two axes intersect, the torus becomes a sphere.* Notice that the sphere, described like this, is structured exactly as the standard description of the globe: The fiber $SO(2)$-groups on the torus tube become the longitudes on the sphere, and the control $SO(2)$-group of the torus is represented by the latitudes on the sphere. It is in this way that both the sphere and the torus have the structure $SO(2) \wr SO(2)$. In fact, the sphere can be considered to be a singular case of a torus - the case where the control radius is zero.

Now let us represent the control radius by the additive group $\mathbb{R}$ (negative values produce an American football structure). This group $\mathbb{R}$ will be added as a third level of control to the structure $SO(2) \wr SO(2)$ of a torus, thus:

$$SO(2) \wr SO(2) \wr \mathbb{R}. \tag{10.1}$$

Within this structure, the sphere is the copy of the fiber $SO(2) \wr SO(2)$ that corresponds to the zero point in the control group $\mathbb{R}$. As the control radius is increased within $\mathbb{R}$ from zero, one gets "proper" tori as the other fiber copies of $SO(2) \wr SO(2)$.

Now consider the issue of maximization of transfer and recoverability. With respect to *transfer*, observe that a great benefit of the structure just given is that the organization of the sphere is transferred onto the torus. This is extremely valuable for applications: For example, notice that the standard wire-frame description of a sphere can now be seen to be related to the standard wire-frame description of a torus. Again, the *action* structure associated with the sphere - that is, the two orthogonal rotational actions - can be transferred onto the torus, e.g., in relating the structure of spherical and toroidal cutting tools.

With respect to *recoverability*, notice that the operation of moving the control radius away from zero within $\mathbb{R}$ can be regarded as symmetry-breaking on the sphere; i.e., the sphere allows its latitude-longitude description in any direction, whereas the torus does not. This exemplifies our use of control groups as symmetry-breaking operations. In fact, this third level $\mathbb{R}$, in the sequence (10.1), is *asymmetry-building*, in accord with the theory in Sect. 2.18.

Finally, observe this: level $\mathbb{R}$ is used here as an *external* group. That is, any shape (e.g., a sphere or a torus) elaborated in that structure, corresponds to the occupancy of only one point in the third level $\mathbb{R}$. Thus, we should regard only the fiber $SO(2) \textcircled{w} SO(2)$ as the primitive and add the action of $\mathbb{R}$ when deformations of that primitive are required. Notice that since the $\mathbb{R}$ is symmetry-breaking, then to allow recoverability, the primitive must be chosen to be the sphere; i.e., in accord with the Asymmetry Principle. We therefore conclude this:

To maximize transfer and recoverability, two of the standard primitives, the sphere and torus, must be replaced by just the sphere. The torus must be classified as a derived shape under the external symmetry-breaking action of increasing control radius.

10.4 Cylinder and Cone

Notice that the wreath structure given for the circular cylinder and the circular cone in Sect. 10.2 are reverse structures of each other, $SO(2) \textcircled{w} \mathbb{R}$ and $\mathbb{R} \textcircled{w} SO(2)$ respectively. The description given there of the cylinder is that of a circular cross-section (as fiber) undergoing translation (as control). This will be called the *cross-section cylinder*. In contrast the description given of the cone is that of a line (as fiber) undergoing rotation, such that the line and rotation axis intersect; i.e., forming the apex. In this way, the cone is described as a *ruled-surface*.

Now it is often useful to describe a cylinder also as a ruled surface; i.e., as a line that is rotated around an axis. Observe, however, that the cylinder given this alternative representation can be understood as a *cone* where the intersection point of the surface-line and axis (i.e., the apex) is at infinity. This will be called the *ruled cylinder*.

Therefore, what is emerging from this discussion is the fact that an extra group is needed to control the angle of the surface-line to the axis in the cone. Let us describe this in the following way. Select a surface-line on the cone. Insert in it a copy of $SO(2)$ with its center-point on the line, and its rotation plane containing the line and the cone axis. This will control the angle of the line to the axis. The angle will be called the *control angle*, and the zero angle will be chosen to be the case where the line is parallel to the cone axis.

The control angle can be varied - giving wider or narrower cones. *When this angle is zero, i.e., the surface-line and cone-axis are parallel, the cone becomes a ruled-cylinder.* It is in this way that both the ruled cylinder and the cone have the structure $\mathbb{R} \textcircled{w} SO(2)$. Notice that the case where the surface-line has rotated to intersect with the cone-axis perpendicularly, corresponds to the plane with polar coordinates (as opposed to the plane structure given previously in Sect. 10.2).

As stated above, the action of increasing and descreasing the control angle can be viewed as the effect of a higher $SO(2)$ group. We thus obtain the following 3-fold structure:

$$\mathbb{R} \textcircled{w} SO(2) \textcircled{w} SO(2). \tag{10.2}$$

Notice that the additional $SO(2)$ should be considered as controlling the angle of each of the surface lines simultaneously.

Now let us consider the issue of maximization of transfer and recoverability. With respect to transfer, observe that a great benefit of the structure just given is that the organization of the ruled cylinder is transferred onto the cone. With respect to recoverability, notice that the operation of increasing the control angle above zero can be regarded as symmetry-breaking on the ruled cylinder: That is, the formation of the cone from the cylinder breaks the north-south symmetry of the cylinder. This exemplifies our use of control groups as symmetry-breaking operations. In fact, more strongly, the third level in (10.2) is *asymmetry-building*, in accord with the theory in Sect. 2.18.

Observe also that this third level is used as an *external* group. That is, any shape (e.g., a ruled cylinder or a cone) elaborated in that structure, corresponds to the occupancy of only one point in the third level $SO(2)$. Thus, we should regard only the fiber $\mathbb{R} \textcircled{w} SO(2)$ as the primitive and add the action of the control-angle $SO(2)$ when deformations of that primitive are required. Notice that since this $SO(2)$ is symmetry-breaking, then to allow recoverability, the primitive must be chosen to be the ruled cylinder; i.e., in accord with the Asymmetry Principle. This is a primitive in addition to the cross-section cylinder described earlier. Therefore we conclude this:

To maximize transfer and recoverability, two of the standard primitives, the cylinder and cone, must be replaced by a cross-section cylinder and ruled cylinder. The cone must be classified as derived from the ruled cylinder under the external symmetry-breaking action of increasing control angle.

10.5   Level-Discrete Primitives

Surveying the CAD literature on primitives, there is one standard primitive that has not been mentioned in the list in Sect. 10.2. This is the *block*. The list in Sect. 10.2 was of what we call the *level-continuous* primitives (Definition 10.1, p. 231). The block, however, is *level-discrete*; i.e., one level in the wreath representation must be discrete. In fact we are going to argue that there are *three* alternative wreath representations of a block. Each is important because each implies a different action structure to be exploited by a designer, a robot manipulator, navigator, etc.

(1) The cross-section block.
In this first representation, the block is created by taking the cross-section cylinder $SO(2) \textcircled{w} \mathbb{R}$ and discretizing the rotation group $SO(2)$ to its subgroup $\mathbb{Z}_4$. We will assume this block is *primitive* because, to obtain a block with a rectangular cross-section, one simply applies an external deformation group. Thus the primitive block has the structure:

$$\mathbb{Z}_4 \textcircled{w} \mathbb{R}.$$

However, this representation omits one piece of important information: Each face of the block is continuous. To include this information, the full structure becomes this:

$$\mathbb{R} \textcircled{w} \mathbb{Z}_4 \textcircled{w} \mathbb{R}.$$

Notice that the first two levels constitute the structure of the square as described it in Chapter 3. This structure should now be understood in a new light. In the transition from the cross-section cylinder $SO(2) \textcircled{w} \mathbb{R}$ to the cross-section block, the $SO(2)$ group is replaced by $\mathbb{R} \textcircled{w} \mathbb{Z}_4$, not just $\mathbb{Z}_4$. This preserves the continuity of the cross-section.

Definition 10.2. *The transition:*

$$\{e\} \textcircled{w} SO(2) \longrightarrow \mathbb{R} \textcircled{w} \mathbb{Z}_n$$

will be called a **continuity-splitting** *of* $SO(2)$.

Intuitively, the $\{e\}$ on the left can be considered as expanding to become $\mathbb{R}$ on the right, and $SO(2)$ on the left as contracting to become $\mathbb{Z}_n$ on the right. Notice that we are permitting the discrete order n to take on any value, and this allows primitive blocks to have a regular polygonal cross-section of any degree. Throughout the discussion, the case $n = 4$ will frequently be used - simply as the main example.

Finally, we emphasize that this block is called a *cross-section* block because it is the translation of the cross-section $\mathbb{R} \textcircled{w} \mathbb{Z}_n$, as fiber, along its control group $\mathbb{R}$.

(2) The ruled or planar-face block.
In this second representation of a block, a continuity-splitting of $SO(2)$ is performed in the wreath representation $\mathbb{R} \textcircled{w} SO(2)$ of the *ruled cylinder*, thus obtaining the following 3-fold hierarchy:

$$\mathbb{R} \textcircled{w} \mathbb{R} \textcircled{w} \mathbb{Z}_4.$$

Notice that, in this 3-fold wreath hierarchy, the 2-level fiber $\mathbb{R} \textcircled{w} \mathbb{R}$ is a *planar face* of the block. It is sent around onto the other faces by the highest control group $\mathbb{Z}_4$. In contrast, in the previous representation of the block, the faces were not encoded at all. The 2-level fiber was the square cross-section, which was then moved along the block by the highest control group $\mathbb{R}$. Thus two very different generative structures are encoded in these two block representations.

It should be emphasized that the planar-face block is related to the frequent representation of the block as produced by planes as half-spaces. However, in our formulation, there is the additional generative structure that *transfers* the plane around the block. Notice also that in this block representation, the planar fiber $\mathbb{R} \textcircled{w} \mathbb{R}$ is one of our previously given primitives (Sect. 10.2). Thus we can choose to encode or not encode the plane within this block representation.

There are alternative names we can give this block representation to be consistent with the terms used so far: Either we can call it the *ruled* block, to emphasize its relation to the ruled cylinder - this also emphasizes its Level 1 fiber $\mathbb{R}$ - or we can call it the *planar-face* block, to emphasize its 2-level fiber $\mathbb{R} \textcircled{w} \mathbb{R}$, a planar face, which does not exist in the cross-section block.

(3) The cube.
The third representation of a block is non-primtive: It is as a deformation of a cube. In this non-primitive representation, the block has memory of the *reflection* structure of the cube. Thus the cube must be added as the final primitive to our list. Expressed as a wreath c-polycyclic group, the cube has the wreath-polycyclic hyperoctahedral structure

$$\mathbb{Z}_2 \textcircled{w} \mathbb{Z}_3.$$

However, the information that the faces are planes, has not yet been included. Since the plane is a primitive in its own right, we can again either understand the hyperoctahedral structure to be combined with the plane as primitive, or include the plane as part of the description of the cube. In either case, the plane enters as a lower fiber thus:

$$\overbrace{\mathbb{R} \textcircled{w} \mathbb{R}} \textcircled{w} \mathbb{Z}_2 \textcircled{w} \mathbb{Z}_3.$$

Finally a comment about wedges: The wedge is sometimes included as a primitive in the literature. However, since our generative theory maximizes transfer and recoverability, the wedge will be understood to be a planar-face block which has undergone an external symmetry-breaking operation that rotates a face, in much the same way that we defined a cone to be a ruled cylinder with a rotating action on the fiber.

Table 10.1. Classification of primitives by maximizing transfer and recoverability

LEVEL-CONTINUOUS	
Plane	$\mathbb{R} \text{ ⓦ } \mathbb{R}$
Sphere	$SO(2) \text{ ⓦ } SO(2)$
Cross-Section Cylinder	$SO(2) \text{ ⓦ } \mathbb{R}$
Ruled Cylinder	$\mathbb{R} \text{ ⓦ } SO(2)$

LEVEL-DISCRETE	
Cube	$\mathbb{R} \text{ ⓦ } \mathbb{R} \text{ ⓦ } \mathbb{Z}_2 \text{ ⓦ } \mathbb{Z}_3$
Cross-Section Block	$\mathbb{R} \text{ ⓦ } \mathbb{Z}_n \text{ ⓦ } \mathbb{R}$
Ruled or Planar-Face Block	$\mathbb{R} \text{ ⓦ } \mathbb{R} \text{ ⓦ } \mathbb{Z}_n$

10.6 Formulation of Primitives to Maximize Transfer and Recoverability

The purpose of sections 10.2 - 10.5 has been to develop a classification of primitives to maximize transfer and recoverability. The results are given in Table 10.1.

Notice that there is no torus or cone. To maximize transfer and recoverability these two shapes are regarded as the result of applying an external symmetry-breaking action on the sphere and the ruled cylinder respectively. Similarly, the wedge is omitted because it is regarded as the result of applying an external symmetry-breaking action in the planar-face block.

Notice also the important relations in the table: The cross-section and ruled cylinders are the same as each other except that their component groups are reversed in the transfer structure - thus giving a different generative structure. Correspondingly the cross-section block and ruled block are the same as each other except that their second and third factors are reversed. This follows directly from reversal of the two factors in the cross-section cylinder and ruled cylinder, since the two blocks are derived from the two cylinders by continuity splitting.

Notice that the cube is not derivable by continuity splitting, e.g., from the sphere $SO(2) \text{ ⓦ } SO(2)$, although its control structure $\mathbb{Z}_2 \text{ ⓦ } \mathbb{Z}_3$ might at first seem to come from the sphere. The reason is that the $\mathbb{Z}_2$ of the cube is the reflection group, not a rotation group and therefore not derivable from the fiber $SO(2)$ of the sphere; and also the control group $\mathbb{Z}_3$ of the cube, although being a rotation group, is on a diagonal axis, unlike the control rotation group of the sphere.[1]

Finally, it should be emphasized that, by ensuring the maximization of transfer and recoverability, we have provided a description that maximally yields the required generative structures for applications, e.g., in design, navigation, manipulation, assembly, milling, etc.

10.7   Externalization

By far the most profound property of the primitives, as we have described them, is that they are *memoryless with respect to external history*. We propose that this is the deep reason why they are the main primitives in perception and CAD. To explain:

Recall that the Externalization Principle (p. 53) says that any external inference goes back to an object whose trace-structure is given by an iso-regular group (wreath c-polycylic, wreath-isometric). What we see is that each of the primitives, as given in Table 10.1, is an iso-regular group. None involves an external history, e.g., of deformation.

This is crucial because each can therefore be used as a memory store for external actions, i.e., such actions are recoverable. Notice that those actions could not be recoverable if the primitives themselves already contained external action. By making the primitives iso-regular, we make them "clean" as memory stores, and therefore they maximize recoverability.

---

[1] The cube is the hyperoctahedral wreath group of degree 3, and we could substitute the degree 2 version for $\mathbb{Z}_n$ in the cross-section of ruled blocks of $n = 4$. This will be taken for granted without explicitly saying so.

11. Unfolding Groups, I

11.1 Symmetry Group of a Complex Environment

We now turn to the main purpose of this book: the representation of complex shape. Our goal is to develop algebraic methods for fully generating complex structures.

Consider the main problem: In the generative theory, complexity will necessarily be regarded as corresponding to the amount of asymmetry; i.e., the greater the complexity, the greater the asymmetry. Now, start with a very simple object, such as a square, and sequentially add some asymmetries. The symmetry group reduces very quickly to the identity element. Thus a complex structure, e.g., an architectural or mechanical design, a real-world scene, a metalurgical compound, etc., is so asymmetric that its symmetry group is apparently just the identity element. For this reason, most people have regarded group theory as inappropriate for handling complex structures. Yet group theory is the largest and most powerful branch of algebra.

What we will do here is develop a symmetry group for a complex environment. This will be a powerful structure because it will contain all the required information: A complete specification of the structure of the environment, including all the asymmetries, as well as all the generative information. In fact, by using our theory of recoverability, the asymmetries and the generativity are closely related. Thus the addition of complexity, rather than loosing information, in the sense of reducing the symmetry group, will increase the latter.

Some of the practical advantages of the method to be developed will be as follows:

(1) Robot Navigation: By being able to describe a complex scene in terms of a single symmetry group, one will be able to capture the navigational information needed by the robot to move through the environment.

(2) Computer-Aided Design: By being able to describe a complex design in terms of a single symmetry group, one will be able to capture all the generative information needed by the designer to produce the design.

(3) Computer-Aided Manufacturing: By being able to describe a complex product in terms of a single symmetry group, one will be able to capture all the manufacturing information, e.g., NC-milling, needed by a manufacturer to create the product.

(4) Physical sciences: By being able to describe a complex object (e.g., an actual rock, etc.) in terms of a single symmetry group, one will be able to capture all the historical processes that produced the object. In fact, representing this generativity - causal generativity - is we claim the main purpose of science.

(5) Computation: We argue that a complex program can be represented by a single symmetry group, and that this captures both the asymmetries (breakdowns in nesting, etc.,) and the generative structure of the program; in particular, it captures structured-programming and object-oriented issues. Furthermore, since we argue that the mathematics of program structure must ultimately be the same as the mathematics of memory storage, we will be able to assign a single symmetry group to a memory configuration - a group that exactly captures the memory content of the store.

(6) Human perceptual representation: We argue that the human perceptual representation of a complex scene is a single symmetry group, and that this is a group of the type that will be developed here.

11.2   Concatenation as Symmetry-Breaking

The representation of two or more bodies in a space is a fundamental problem in all the physical sciences, computational disciplines, and design disciplines. In the physical sciences, it appears in issues of *causality*; i.e., in the interaction between two bodies. In the computational disciplines such as computer vision, it appears in the representation of real-world scenes. In the design disciplines, it appears as the successive concatenation of bodies in the construction of the compound design object.

Let us use the word *concatenation* about all these situations; i.e., where two bodies, represented seperately are brought together in a compound situation that is itself given a representation. We now propose this:

CONCATENATION AS SYMMETRY-BREAKING. *In the physical sciences, computational disciplines, and design disciplines, concatenation of two objects is organized such that the objects individually have greater symmetry than their compound structure. Thus concatenation is organized as a symmmetry-breaking operation.*

Consider, for example, the cube and cylinder shown in Fig. 11.1. If the cube were alone in space, then the situation would have several symmetries; for example, the reflectional symmetries of the pairs of opposite faces; i.e., the hyperoctahedral structure. If the cylinder were alone in space, then the situation would be symmetric in a somewhat different way; e.g., there would be the continous rotational symmetry $SO(2)$ around the central vertical axis of the cylinder.

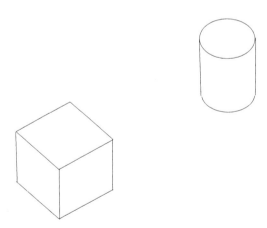

Fig. 11.1. Concatenation breaks symmetry.

However, let us now consider space as containing *both* objects as shown in Fig. 11.1. Several of the symmetries have been lost. For example, whereas the cube had mirror symmetry between the left-most face shown and its opposite, the combined situation does not have this symmetry because the cylinder exists on only one side of the mirror plane involved. Also, the rotational symmetry of the cylinder has been lost, i.e., no axis can be found in this compound situation such that the situation can be continuously rotated onto itself. In conclusion therefore, concatenation is structured as a symmetry-breaking operation.

Now, when one looks at this combined figure one naturally decomposes it into a cube and cylinder. However, there are other decompositions: For example, one could consider a single plane that slices obliquely through both the cube and cylinder simultaneously. This plane decomposes the configuration into two asymmetric parts: (1) the part below the plane and (2) the part above the plane. Then concatenation of these two parts would not be symmetry-breaking. In fact, it might be symmetry-increasing. For example, the combined situation (i.e., Fig. 11.1) has reflection symmetry about the common vertical mirror plane through the cube and cylinder. This symmetry did not exist in the two oblique parts (i.e., prior to the concatenation).

The important question is this: Why does one not define the decomposition and concatenation in the particular way given in the previous paragraph? The answer is crucial: This concatenation operation is non-recoverable. The reason is that it is arbitrary. As we have said in Chapter 1 of Leyton [96], if one tries to recover a past asymmetric structure from a symmetric data set, then one can only be arbitrary, because the relationship between a symmetry and asymmetry is one-many. For exactly this reason, recoverability works only in the opposite direction: From asymmetry back to symmetry.

CONCATENATION AND RECOVERABILITY. *The reason why the physical sciences, computational disciplines, and design disciplines, organize concatenation as a symmetry-breaking operation is because only this will allow recoverability of concatenation.*

11.3 Concatenation as Asymmetry-Building

Describing concatenation as a symmetry-breaking operation has considerable power. However, we argue that much more powerful descriptions are possible in terms of the algebraic method introduced in Sect. 2.18, which we call *asymmetry-building*, and which is the entire basis of our generative theory. In asymmetry-building, one does not loose the initial symmetry ground-state; i.e., there is no actual symmetry-breaking. Instead one creates a group-extension of the ground-state using the symmetry group of the asymmetrizing process as the extending group. In fact, the type of extension used is a wreath product, in which the symmetric ground-state is the fiber, and the control group is the asymmetrizing symmetry group. In this way, the control group transfers the initial symmetry onto the current asymmetric data set. The power of this approach is that it maximizes both transfer and recoverability.

Using this approach, we will now show how concatenation can be described as an asymmetry-building operation. To illustrate, return to Fig. 11.1.

The generative history starts out with the two independent objects, and therefore the symmetry of this starting situation is given by

$$G_{cylinder} \times G_{cube} \tag{11.1}$$

which is the direct product of the groups of the two independent objects.

Warning. *The direct product symbol in (11.1) should not be regarded as representing a direct product between fibers, as previously. It is within a single fiber.*

Now, by the maximization of transfer, the starting group, i.e., the direct product group (11.1), must be transferred onto subsequent states in the generative history, and therefore it must be the fiber of the wreath product in which the control group creates the subsequent generative process. Let us take the control group to be the affine group $AGL(3, \mathbb{R})$ on three-dimensional real space.[1] The full structure, fiber plus control, is therefore the following:

$$[G_{cylinder} \times G_{cube}] \,\textcircled{w}\, AGL(3, \mathbb{R}). \tag{11.2}$$

It is necessary to fix the group representation of this wreath product. First, by our basic theory, the control group must have an asymmetrizing action. Thus proceed like this: The particular fiber-group copy

$$[G_{cylinder} \times G_{cube}]_e$$

corresponding to the identity element e in $AGL(3, \mathbb{R})$, must be the most symmetrical configuration possible. This exists only when the cube and the cylinder are coincident with their symmetry structures maximally aligned. For example, their centers, reflection planes, rotation axes, must be maximally coincident.

Next choose one of the two objects to be a reference object. This will remain fixed at the origin of the coordinate system. Let us choose the cube as the referent.

Given this, now describe the action of the control group $AGL(3, \mathbb{R})$ as providing an affine motion of the cylinder relative to the cube. Each fiber copy

$$[G_{cylinder} \times G_{cube}]_g$$

for some member g of $AGL(3, \mathbb{R})$ is therefore an arrangement of this system. In fact, any fiber copy will be called a **configuration** of the system. For example, Fig. 11.1 shows a configuration.

---

[1] An element of this group is a linear transformation composed with a translation. AGL means Affine General Linear.

It is necessary to understand the action of the affine control group in two respects: First, with respect to the *abstract* group theory, a member a of the affine control group sends one fiber-group copy onto another, like this[2]:

$$[G_{cylinder} \times G_{cube}]_g \quad \xrightarrow{a} \quad [G_{cylinder} \times G_{cube}]_{ag} \ . \tag{11.3}$$

However, with respect to the group *representation* theory, the affine effect is confined to only the cylinder. The best way to understand this is not to consider this as an isolated effect on the cylinder but as an effect on the *configuration* that contains the cylinder and cube. We defined a configuration to be a fiber-group copy $[G_{cylinder} \times G_{cube}]_g$. Therefore, the consequence of the affine action on the cylinder is to alter the configuration of cube and cylinder. This is the real meaning of expression (11.3).

Recall that, in the symmetry-breaking view of the last section, the combined configuration of the cube and cylinder had virtually no symmetry group. In particular, the configuration had almost none of the starting symmetries of the individual cube and cylinder. In contrast, the approach we have taken, using expression (11.2), retains all the information. It codes exactly the visual effect, which is that the configuration is *seen* as containing a cube and a cylinder, and therefore containing both their symmetry groups. Notice that, correspondingly, expression (11.2) acknowledges the presence of these symmetry groups as parts in the whole structure.

One way of understanding expression (11.2) is that it contains *three* symmetries: That of the cube; that of the cylinder; and that of the affine action that relates them.

Most importantly these three symmetry groups appear because a *generative* theory is being given of the compound configuration, and, according to our theory, the generative operations must come from the symmetry groups. Furthermore, since the generative structure is required to be recoverable from the data set, this places strong constraints on how the groups are put together and represented as actions.

Expression (11.2) is the *complete* symmetry group of the situation. It contains an enormous amount of powerful information: It contains all the generative information, as well as the information about the parts, and the information about how the parts are related.

Let us now understand how to add a further object, for example a sphere. First of all, the fiber becomes:

$$G_{sphere} \times G_{cylinder} \times G_{cube}.$$

In such expressions, our rule will be that each object encoded along this sequence provides the reference for its left-subsequence of objects. Thus the

---

[2] In fact, this can be defined in terms of a wreath-direct action as defined in Sect. 13.3. In the present case, the control group would be $AGL(3, \mathbb{R}) \textcircled{w} \{e\}$.

cube is the referent for the cylinder-and-sphere, and the cylinder is the referent for the sphere. Accordingly, there are now two levels of control, each of which is $AGL(3, \mathbb{R})$, and each of which is added via a wreath product. Thus we obtain the 3-fold wreath product:

$$[G_{sphere} \times G_{cylinder} \times G_{cube}] \, \circledW \, AGL(3, \mathbb{R}) \, \circledW \, AGL(3, \mathbb{R}). \qquad (11.4)$$

This is interpreted in the following way: Initially, the three objects (cube, cylinder, sphere) are coincident with their symmetry structures maximally aligned. The higher affine group $AGL(3, \mathbb{R})$ moves the cylinder-sphere pair in relation to the cube. The lower affine group $AGL(3, \mathbb{R})$ moves the sphere in relation to the cylinder.

Any configuration is expressed as a fiber copy

$$[G_{sphere} \times G_{cylinder} \times G_{cube}]_{g_1, g_2}$$

where the indexes g_1 and g_2 come from the two successive affine groups. The index g_2 expresses the relation of the cylinder-sphere pair to the cube, and the index g_1 expresses the relation of the sphere to the cylinder. Notice that any element,

$$[h_1, h_2, h_3]_{g_1, g_2} \qquad (11.5)$$

in this fiber copy, expresses a symmetry relation within the configuration; e.g., h_3 exchanges sides within the cube, while h_2 simultaneously rotates the cylinder about its axis, while h_1 simultaneously rotates the sphere about any one of its axes. We argue that this symmetry exists despite the affine action created by the two control elements g_1 and g_2. If we call the symmetry $[h_1, h_2, h_3]$, a triple isometry of the starting configuration, then the element (11.5) is a control-nested τ-conjugate of the triple isometry.

The above discussion has been illustrating a class of groups we are soon going to propose, which we will call **telescope groups**. This class is part of a still larger class which we will also be proposing, which we will call **unfolding groups**. Unfolding groups will be the most important class of algebraic structures introduced in this book. The reader should now read p. 246 describing unfolding groups, following by p. 247 describing telescope groups; and then return here.

Comments on p. 246 and p. 247.

(1) As stated in Sect. 1.14, a basic principle of our generative theory is exhaustiveness, which means that there are no primitives; i.e., there is generativity all the way down. This means that what have been called primitives on p. 246 are in fact objects that themselves have generative structures, which is what

UNFOLDING GROUPS

Unfolding groups are characterized by the following two properties:

SELECTION: The control group acts *selectively* on only part of its fiber.

MISALIGNMENT: The control group acts by *misalignment.*

The control group will be called the *unfolding control*; and the selected part of the fiber will be called the *unfolding fiber* or *target*. In major classes of unfoldings the above two conditions, selection and misalignment, are achieved in the following way:

(1) There are n objects which have symmetry groups $G_1, \ldots, G_n$ respectively. These will also be called the *primitives*.

(2) One forms the direct product $G_1 \times \ldots \times G_n$, and makes this the fiber group of a wreath product, with control group $C(G)$.

(3) In this wreath product, any fiber-group copy, i.e., any copy of $G_1 \times \ldots \times G_n$, is called an *object-configuration*, or just *configuration*.

(4) Let the fiber-group copy in which the object symmetry groups $G_1, \ldots, G_n$ are maximally aligned with each other, be called the *alignment kernel*. Choose this to be the fiber-group copy corresponding to the identity element of the control group. By abuse of language, the fiber-group itself will often be referred to as the alignment kernel.

(5) Notice that the above structure realizes *transfer* and *recoverability*, in the following way: (a) It is a transfer structure because the control group maps object-configurations $[G_1 \times \ldots \times G_n]_{g_i}$ onto object-configurations $[G_1 \times \ldots \times G_n]_{g_j}$. (b) It achieves recoverability because (at least locally to the identity element), the action of the control group is to pull the objects out of alignment - which creates asymmetrization.

TELESCOPE GROUPS

(1) Conceptually, a *telescope* group is an unfolding group that has an *opening telescope structure,* as follows:

(2) The fiber group is an alignment kernel of n primitives, possibly including repetitions of those primitives. It is written like this $[G_1 \times \ldots \times G_n]_T$, where T denotes "telescope".

(3) The unfolding control group itself is a wreath product of order $n - 1$, and hence, with the alignment kernel, the entire unfolding group is a wreath product of order n.

$$[G_1 \times \ldots \times G_n]_T \ \textcircled{w} \ C(G)_1 \ \textcircled{w} \ \ldots \ \textcircled{w} \ C(G)_{n-1}. \qquad (11.6)$$

(4) The *telescope opening effect* is realized in the following way. Work down the control groups: The highest control group, $C(G)_{n-1}$, moves only the left subsequence $G_1 \times \ldots \times G_{n-1}$ of the alignment kernel - thus moving it out of alignment with G_n. Nevertheless it keeps the alignment of groups $G_1, \ldots, G_{n-1}$ with respect to each other. Similarly, the next lower control group, $C(G)_{n-2}$, moves only the left subsequence $G_1 \times \ldots \times G_{n-2}$, of the alignment kernel - thus moving it out of alignment with G_{n-1}. Nevertheless it keeps the alignment of groups $G_1, \ldots, G_{n-2}$ with respect to each other. And so on down the control groups.

(4) The unfolding is called a *projective (affine, or Euclidean) telescope group* if each of the control groups $C(G)_i$ is a copy of the projective (affine, or Euclidean) group.

the groups $G_1, \ldots, G_n$ actually give. The objects are understood as primitives only in relation to the higher actions provided by the unfolding control groups.

(2) In defining the group representation of the alignment kernel, there can be a number of alternative maximal alignments. For example, a cylinder can maximally align with a cube in three alternative ways, i.e., with its cross-section parallel to either of the three pairs of opposite sides of the cube. Furthermore, by starting with one configuration of maximal alignment, the action of the control group can pull it eventually into another. Nevertheless, the configuration will first travel through intervening states of non-alignment. Thus, the control group will be understood as creating asymmetrization locally to its identity element. This is, in fact, all that is needed.

(3) In expression (11.6), there are n components to the telescope fiber, and $n-1$ components to the control chain. However, even if a Cartesian reference frame were not included in the primitives 1 to n, both chains could be made of equal length n, by adding an extra control component $C(G)_n$ on the right, that gives the relation between the right-most primitive (n) to the implicit Cartesian frame. Nevertheless, in many examples of unfolding groups, the Cartesian frame will be explicitly given as the cube.

(4) We will define three classes of unfolding groups that will be of particular significance in this book. They will be introduced in three successive chapters, starting with the present one:

Chapter 11 :   Telescope Groups
Chapter 12 :   Super-Local Unfoldings
Chapter 13 :   Sub-Local Unfoldings

11.4   Serial-Link Manipulators as Telescope Groups

We will now argue that it is valuable to formulate serial-link manipulators as telescope groups. This formulation will be developed in three stages, each of which will emphasize particular properties:

(Stage 1) In Chapter 6, a serial-link manipulator was formulated as a wreath-isometric group

$$SE(3) \, \textcircled{w} \, \ldots \, \textcircled{w} \, SE(3) \, \textcircled{w} \, SE(3). \tag{11.7}$$

It is important to notice the Cartesian-frame assignment to this group. By Cartesian frame, we mean here the conventional sense of the term. Each level in the above group sequence is assigned its Cartesian frame, and this defines the group representation of that level on Cartesian space. Correspondingly, each link in the manipulator receives one frame, and standardly this is placed at the distal end of the link.

Most crucially, the structure follows our algebraic theory of parent-child hierarchies in Sect. 7.3. Each level in the wreath product is the parent of the level to the left of it, in the sequence.

(Stage 2) We have called the group (11.7), the "full" group of the serial-link manipulator for the following reason: The standard assumption in robotics is that, because the transformations connecting the base frame to the effector frame are special Euclidean transformations, the group linking those two frames is the special Euclidean group. We have argued that this is incorrect because it assumes that the link-structure is frozen. Instead, the structure is semi-rigid and therefore should be given by the much larger group, expression (11.7).

However, we now wish to go further, and add even more algebraic structure. The reason is as follows: The Cartesian frames used in the above discussion are the ordinary ones in the standard literature. The problem with these is that they do not explain why one assigns the frames to the links in the typical way. The placement of an ordinary frame can effectively be arbitrary. However, when one examines the usual attachment of frames one finds that they are assigned to the *symmetry* structure of links. For example, Fig. 11.2 shows a typical link. The $n + 1$ frame is the one assigned to the link, and we see that it is aligned with the symmetry structure of end. Indeed, the frame of the previous link n is also shown, and we see that it also corresponds to symmetries in this link. (This diagram uses the conventional numbering of frames from proximal to distal.)

Chapter 14 will discuss this symmetry structure in much greater detail, but for now, its most significant component will be used - which we argue is the hyperoctahedral wreath group $\mathbb{Z}_2 \textcircled{w} \Sigma_3$ (Sect. 8.12, p. 205). Its fiber-group product expresses the reflection structure associated with the cube.

Its clear therefore, that the assignment of a Cartesian frame to a link uses this symmetry structure. This greatly reduces the arbitrariness of where to assign the Cartesian frame. To code this, we place the hyperoctahedral group $\mathbb{Z}_2 \textcircled{w} \Sigma_3$ as a fiber of the wreath sequence of special Euclidean groups, thus:

$$[\mathbb{Z}_2 \textcircled{w} \Sigma_3] \textcircled{w} SE(3) \textcircled{w} \ldots \textcircled{w} SE(3) \textcircled{w} SE(3). \qquad (11.8)$$

The hyperoctahedral group is interpreted as corresponding to the identity element of each of the control groups, and simultaneously being assigned to the maximal symmetry structure of each of the link-joints.

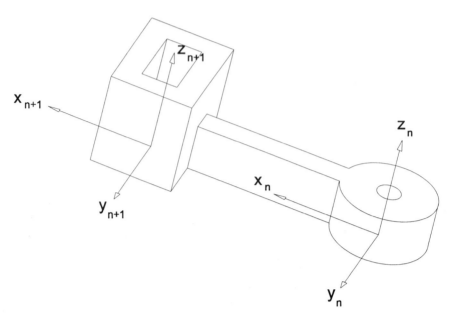

Fig. 11.2. In a robot link, coordinate frames correspond to symmetries.

(Stage 3) This stage provides an additional convenient way of describing manipulators. Here the lowest fiber is no longer a single hyperoctahedral wreath group, i.e., a single Cartesian frame, but the *configuration* of frames across the links.

This approach is offered because we argue that there is a strong similarity between the serial-link manipulator and an ordinary physical telescope. Consider a telescope. It consists of a set of components that are initially in a collapsed form, i.e., they are maximally aligned. When one extends the telescope, one pulls the components out of alignment with each other. We argue that exactly this structure is the basis of assigning frames to a serial-link manipulator. The fact that the frames are related by the special Euclidean group, means this: In the case where one selects the identity element in each of the link groups, the frames are all coincident at the base frame of the manipulator. This is exactly analogous to the telescope in its collapsed form. The pulling out of the manipulator frames in relation to each other corresponds to extending the telescope out from its collapsed form.

To describe such structures, we have created a class of groups which we call telescope groups (p. 247). The telescope group for the serial-link manipulator is this:

$$[[\mathbb{Z}_2 \textcircled{w} \Sigma_3]_1 \times \ldots \times [\mathbb{Z}_2 \textcircled{w} \Sigma_3]_n \times [\mathbb{Z}_2 \textcircled{w} \Sigma_3]_{n+1}]_T$$
$$\textcircled{w} \, SE(3)_1 \, \textcircled{w} \, \ldots \, \textcircled{w} \, SE(3)_n. \tag{11.9}$$

The first line in this expression is the alignment kernel, and the second line is the control group which will unfold the kernel as an opening telescope. The alignment kernel consists of the direct product of $n + 1$ copies of the hyperoctahedral wreath group, one for each each link (indicated by the subscripts 1 to n), and an additional copy (subscript $n + 1$), for the origin frame. The subscripts 1 to $n + 1$ are ordered distal-to-proximal, as in Sect. 6.4, because this will produce the control ordering 1 to n that corresponds to the numbering used in this book for wreath products. Initially, the frames are maximally aligned - i.e., in the collapsed form. The telescope opening effect is realized by the second line, in the following way. Work down the control groups: The highest control group, $SE(3)_n$, moves only the left-subsequence $[\mathbb{Z}_2 \, \textcircled{w} \, \Sigma_3]_1 \times \ldots \times [\mathbb{Z}_2 \, \textcircled{w} \, \Sigma_3]_n$ of the alignment kernel - thus moving it out of alignment with $[\mathbb{Z}_2 \, \textcircled{w} \, \Sigma_3]_{n+1}$. Nevertheless it keeps the alignment of groups $[\mathbb{Z}_2 \, \textcircled{w} \, \Sigma_3]_1, \ldots, [\mathbb{Z}_2 \, \textcircled{w} \, \Sigma_3]_n$ with respect to each other. Similarly, the next lower control group, $SE(3)_{n-1}$, moves only the left subsequence $[\mathbb{Z}_2 \, \textcircled{w} \, \Sigma_3]_1 \times \ldots \times [\mathbb{Z}_2 \, \textcircled{w} \, \Sigma_3]_{n-1}$, of the alignment kernel - thus moving it out of alignment with $[\mathbb{Z}_2 \, \textcircled{w} \, \Sigma_3]_n$. Nevertheless it keeps the alignment of groups $[\mathbb{Z}_2 \, \textcircled{w} \, \Sigma_3]_1, \ldots, [\mathbb{Z}_2 \, \textcircled{w} \, \Sigma_3]_{n-1}$ with respect to each other. And so on down the control groups.

11.5 Constructive Solid Geometry (CSG)

In CAD, solid primitives are standardly combined using Boolean operations, and the theory of combination is called Constructive Solid Geometry (CSG). The purpose of the remainder of this chapter is to **formulate Constructive Solid Geometry group-theoretically**.

This paragraph quickly reviews CSG for readers not familiar with it: The procedure of CSG is illustrated in Fig. 10.1 (p. 230). One starts with the simple primitive solids - shown at the terminal (lowest) nodes of the tree - and successively combines the primitives upward through the tree, until one obtains the object shown as the top node. Each node is a use of one of the three Boolean operations: subtraction, intersection or union. Observe that the object at the top is more complex than any one of the primitives. Essentially, the Boolean operations are being used to *add or remove material*. Notice that, for the procedure to work, the primitives must be sized and positioned in space. Secondly, all objects in the hierarchy must be closed sets in the sense of topology; i.e., they must contain their own limit points - that is, a boundary. Thirdly, the Boolean operations used cannot be the standard weak operations of ordinary set theory, but considerably more powerful operations - called *regularized* Boolean operations - that satisfy the following properties: When a Boolean operation is used to combine the closed set A with the closed set B, the resulting set C must (1) also be closed, (2) preserve the dimension of A and B, and (3) not lead to inhomogeneous objects - i.e., objects with dangling or disconnected parts of lower dimensions. A good introduction to regularized Boolean operations is Mortenson [110] p318-332. For a detailed survey of CSG, the reader should consult Requicha and Rossignac [124]. Throughout this book, regularized Boolean operations will be referred to simply as Boolean operations.

11.6   Boolean Operations as Symmetry-Breaking

According to our theory, the most important fact about Boolean operations in CSG is that are actually *symmetry-breaking*. It is worth considering this carefully, by looking at the three operations as illustrated in Fig. 11.3, to which we will return a number of times in the book. The left-most part of the figure shows two primitive solids - a cube and cylinder - to which the Boolean operations will be applied. The remaining parts of the figure show the actual application of the Boolean operations to the primitives; i.e., respectively:

$$Cube \bigcup Cylinder \quad , \quad Cube - Cylinder \quad , \quad Cube \bigcap Cylinder$$

Each of these will be called a "compound object".

The crucial thing to observe is that the resulting object in each case has much less symmetry than the two independent primitives. The three Boolean operations are each therefore symmetry-breaking. This re-inforces our claim that the design process generally is symmetry-breaking.

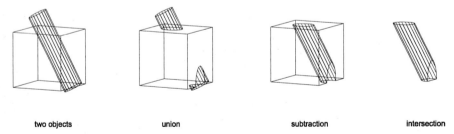

two objects   union   subtraction   intersection

Fig. 11.3. The Boolean operations.

Most importantly therefore, when presented with one of these compound objects as a data-set, the *inference* that a Boolean operation was used in the construction of that object, is an example of the Asymmetry Principle, which says that *recoverability* is possible only if a present asymmetry goes back to a past symmetry ground-state.

Observe that this relates to the fact that the mind chooses to define the compound object as a Boolean combination of highly symmetric primitives.

That is, as pointed out in an argument on p. 242: One could, theoretically, choose to decompose the compound object into primitives that are extremely complex; but one does not do this because this inference would be arbitrary. Only by making the inference go back to more symmetric objects can arbitrariness be eliminated.

Thus we observe that the primitives *maximize symmetry* so that the design process can *maximize generativity and recoverability.*

We can deeply characterize the idea of primitives maximizing symmetry, as follows: Recall the Externalization Principle (Sect. 2.11), which states that any external inference goes back to an iso-regular group, i.e., a wreath c-polycyclic, wreath-isometric group (intuitively described as a control-nested hierarchy of repetitive isometries).

Now, our theory of solid primitives (Chapter 16) says that solid primitives are characterized by having bounding surfaces which are iso-regular groups. In contrast, observe that the bounding surfaces of the compound objects in Fig. 11.3 are not characterized by iso-regular groups. This means that in the backwards generation of a Boolean operation - from the compound object back to the independent primitives - one is going from a non iso-regular group back to iso-regular groups. This exactly conforms to the Externalization Principle.

BOOLEAN OPERATIONS AND THE EXTERNALIZATION PRINCIPLE. *Each Boolean operation conforms to the Externalization Principle. It is an external operation, whose past application is inferred from the compound object by going back to an iso-regular group (a wreath c-polycyclic, wreath-isometric group).*

11.7   Boolean Operations as Telescope Groups

In the previous section, we claimed that Boolean operations are constructed so that they are symmetry-breaking. By symmetry-breaking, we mean that there is a *reduction* in symmetry group. This corresponds to Method 1 for specifying generativity (p. 70). However, fundamental to our theory is the approach we developed called Method 2 (p. 70). Here symmetry-breaking is re-described as an *increase* in symmetry group, i.e., a group extension. The extension is created by a wreath product, thus expressing transfer. The extending group is a symmetry group of the asymmetrizing process. We called this approach *asymmetry-building.*

The asymmetry-building approach supports our goal of maximizing transfer and recoverability. Therefore it is extremely important to formulate Boolean operations in terms of asymmetry-building.

GOAL

Formulate Boolean operations as group extensions via wreath products (transfer) where the extending group is the group of the asymmetrizing process involved in concatenation. This goal is achieved using unfolding groups.

Our solution is to use the particular class of unfolding groups we call *telescope groups* (defined p. 247). Thus, consider two primitives that one wishes to combine via a Boolean operation. Let their symmetry groups be respectively G_1 and G_2. Form the direct product $G_1 \times G_2$ of these symmetry groups. Then wreath super-append the affine group $AGL(3, \mathbb{R})$ as the control group thus:

$$[G_1 \times G_2]_{\mathcal{T}} \; \circledW \; AGL(3, \mathbb{R}).$$

The subscript $\mathcal{T}$ indicates that the direct product is the fiber of a telescope group. This defines the group-representational aspects as follows: First, the fiber-group copy $[G_1 \times G_2]_e$, corresponding to the identity element e in the control group, is the configuration in which the two objects have their symmetry-structures maximally aligned. Next, choose one of the objects (that corresponding to the final member of the direct product), as the referent object, and define $AGL(3, \mathbb{R})$ as moving the other object relative to the referent.

Generally, for an n-fold concatenation of objects, follow the full instructions given on p. 247 for constructing telescope groups.

11.8   Spatial Group Equivalence of Boolean Operations

It is important to observe that, according to our approach, each of the Boolean operations - union, intersection, difference - is represented in the same way. That is, starting with two primitive objects in a fixed spatial and deformational relationship to each other, then, no matter whether one takes the union, intersection, or difference, the resulting object will have the same symmetry group. Thus for example in Fig. 11.3, the union object, intersection object, and difference object, all have the same symmetry group. This is extremely valuable. For example, it solves a particularly important problem in shape-representation: the *completion* of information, e.g., after occlusion, or merging, or machine-milling, etc. Whether one takes union, intersection, or difference, information on parts of the separate primitives is perceptually lost - e.g., in the union operation in Fig. 11.3, one looses the cylinder information within the object since the material from the cube and cylinder is merged; again, in the difference operation in Fig. 11.3, one looses the extension of the negative shape (the cylinder) out past the edges of the cube. The complete information is essentially the same in all three operations and is given by the group structure we have defined. It is well known in computer vision (e.g., occlusion phenomena), or in scientific discovery (e.g., the prediction of complete multiplets of quantum-mechanical particles), etc., that the complete information is more important than the incomplete data set. Our approach shows how to represent the complete information, and indeed makes this information the dominant factor in defining the object. Generally, this can be understood as solving the problem of "Gestalt completion", which is basic to processes of *insight* and *intelligence*.

Given what has just been said, one then asks: How is it possible to distinguish *between* the three particular Boolean operations? The distinction will be provided by the occupancy group $\mathbb{Z}_2$, which we wreath sub-append to the entire structure, i.e., there is copy of the group at each point in the structure. The type of Boolean operation determines the occupancy switch state for each point, in the obvious manner: Union forces all on-switches in the separate objects to be on in the compound object; intersection switches off all on-switches which were not shared in the separate objects, etc.

This is a very convenient formulation, as follows:

TWO-LEVEL DEFINITION OF A BOOLEAN OPERATION. *We divide a Boolean operation into two levels: (1) The spatial level: Here, the three Boolean operations all have the same structure, which is an important requirement for occlusion phenomena, etc. This level is given by a telescope group. (2) The color level: Here, the Boolean operations are distinguished, corresponding to the membership difference. This level is given by the occupancy group.*

12. Unfolding Groups, II

12.1 Importance of Selection in Generativity

Selection is a profound phenomenon in generativity: In many situations, it is necessary for an agent to be able to *select* part of an existing structure and perform generative operations with respect to only that part. Examples are as follows: (1) Navigation: It is often the case that a path is defined as progressing from one of the surfaces of a scene, while ignoring other surfaces, which might be moving independently. For example, a path might be defined with respect to a wall but not an opening door in the wall; or conversely the manipulation of a door-handle might be defined with respect to the opening door rather than the wall. (2) Computer-Aided Design: During the design process, one frequently needs to modify only part of the existing design; e.g., move a block face with respect to the rest of the block, rotate only a subset of nodes in a facial animation. (3) Computer vision: The representation of a scene can often involve the creation of a tree of relationships through the scene, such that any particular branch selectively connects only some of the objects and not others. (4) Science: A causal trajectory might depend on a property that only a specific class of objects in the situation possess, e.g., those that are not electrically neutral. (5) Biology: The limbs on a body emerge at only specific points. (6) Music: Marsella & Schmidt [103] have identified structures in which there is dependency selectively between a part in one hierarchy and a part in another hierarchy.

Because our generative theory requires that transfer is maximized, we want to be able to describe the above selective type of generativity as a process of transferring structure. This in turn will give us access to the mathematics of wreath products. It is not at first easy to see how to do this. Nevertheless, the elaboration of a solution is fundamentally important because so many generative processes have a selective action. There is also another deep reason for developing a solution: We want to be able to give a single symmetry group for an arbitrarily complex object, and any such object inevitably involves selective action.

In this chapter and the next, we develop two alternative solutions, respectively, to this problem: *super-local* and *sub-local unfoldings*. To begin to understand these terms, we introduce the following:

Definition 12.1. *Transfer is* **super-local** *if it is selective (local) and is created by wreath-appending a control group above the existing wreath sequence. Transfer is* **sub-local** *if it is selective (local) and is created by wreath-appending a group below some existing level in the wreath sequence.*

The super-local and sub-local actions to be defined will be called *unfoldings* because they *asymmetrize by misalignment*. That is, they fullfil the basic requirments of unfoldings as defined on page 246.

In order to illustrate the concepts in this and the next chapter, a particular example will be progressively constructed: The structure of a house. This example happens to be in architecture. However, in later chapters, examples will be given in many other areas such as mechanical design, assembly planning, etc.

12.2   Super-Local Unfolding

In this chapter, we will define a type of group structure which has the following action:

SUPER-LOCAL UNFOLDING.   *In super-local unfolding, one applies a higher level control group, which although acting in a control-nested fashion on the entire situation, actually only affects part of the situation, unfolding new structure from that part, by transfer.*

A control group having a super-local unfolding action is labelled by the part X it selects beneath it in the hierarchy, thus:

$$[G_1 \dots G_j] \circledw G_n^X$$

where X is some selection from the group sequence $[G_1 \dots G_j]$. It is crucial to observe that the group G_n^X is added to the sequence $[G_1 \dots G_j]$ via a *wreath product*, and therefore acts by transfer on $[G_1 \dots G_j]$. Thus, like any wreath product, the control group G_n^X sends copies of $[G_1 \dots G_j]$ onto copies of $[G_1 \dots G_j]$ via conjugation by elements $g \in G_n^X$, thus:

$$[G_1 \dots G_j]_{g_1} \xrightarrow{\;g-g^{-1}\;} [G_1 \dots G_j]_{gg_1} \tag{12.1}$$

Fig. 12.1. Two rooms that need to be connected by a door-opening.

In this particular expression, we are assuming that the wreath product is regular, i.e., that the copies of the fiber group $[G_1 \ldots G_j]$ are indexed in the control group G_n^X. For example, in expression (12.1), not only is g in G_n^X, but so is the indexing element g_1.

Now, we have just discussed the *algebraic* action, which is standard for a wreath product. However, let us now consider the group-representational aspect. By this we mean the effect of this action on the fiber sets. As usual, each copy $[G_1 \ldots G_j]_{g_1}$ of the fiber group acts on its own copy F_{g_1} of the fiber set. In super-local unfolding, the crucial thing is that each fiber set is exactly the same as each other fiber set, except for some selected subset. The selected subsets of each of the fiber sets are transformationally related by the control group G_n^X. This will now be illustrated.

12.3   Establishing a Target for Super-Local Unfolding

Any unfolding acts on a target, that part of the overall structure selected for generativity. It is necessary to understand how a target is set up in the full group sequence.

In the present section, this will be illustrated by formalizing a standard procedure in architectural drawing: Fig. 12.1 (p. 259) shows the floor-plan of two rooms that need to be connected by a door-opening. A standard design technique is this:

(1) Draw two *infinite* lines where the door-opening will be, as shown in Fig. 12.2.
(2) Trim the lines, as shown in Fig. 12.3.

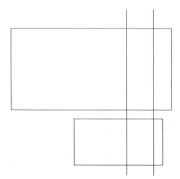

Fig. 12.2. Standardly, the door-opening is first specified with infinite lines.

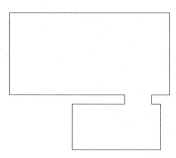

Fig. 12.3. Finally, one trims the lines.

Someone who is not a designer would probably be horrified by the apparent wastefulness of a procedure in which, to draw the two small vertical edges of the door-opening in the final figure, one first draws them as infinite lines. However, this procedure is used in several thousand architecture and design firms throughout the world, every day. Why?

The answer comes from our generative theory. The two vertical door-edges shown in the final figure might be small, but they each are specified by a tracing generator - an infinitessimal translation - that is *repeated* in one direction to produce that small door-edge. The order of information increase, over the course of the procedure, is this: The repetition of a generator defines an infinite line, the group $\mathbb{R}$, and to make the line finite, one then has to specify extra information to switch the repetition *off*. This is therefore a process of *asymmetry-building*.[1]

We argue that this descriptive procedure is general; for example, it is exactly the procedure used by the human perceptual system: i.e., one understands the short segment in terms of its infinite completion (standard Gestalt principle), and sees the segment as the result of surgery performed by the occupancy group. Perception works by the same principle as the design system because both are generative. Indeed both are generative-recoverable. This means that the generative procedure is powerfully constrained by the fact that its recovery is produced by the Asymmetry Principle. Thus, in the forward-time direction, one must start with the infinite line, and then cut it down.

Let us now understand the full group-structure of the final figure. First note that in human perception, the smaller rectangle is referenced to the larger one. This has been well-corroborated in the Gestalt literature (see Leyton [96] for review), where is has been shown that smaller objects in a scene are referred to larger ones. Furthermore, exactly this referential structure is exploited in navigational situations.

In fact, according to our theory, the referential structure must involve at least two stages: Not only is the smaller rectangle referred to the larger one, but the larger one, must in turn be referred to a square because it is seen as a deformed version of a square. This latter stage is an example of external inference (Sect. 2.9), and according to our Externalization Principle, an external inference must lead back to an iso-regular group, which in the present case, must correspond to a square. Simply because it is easier to notate, the group to be used for a square (and Cartesian frame) will be the 2-level group $\mathbb{R}\circledS\mathbb{Z}_4$ based on rotation, rather than the 3-level group $\mathbb{R}\circledS\mathbb{Z}_2\circledS\mathbb{Z}_2$ based on reflection. This will make the group sequences shorter.

According to our theory, the full group structure is this:

---

[1] Recall that asymmetry-building can also occur by wreath *sub*-appending the added group. Here, the added group is the occupancy group $\mathbb{Z}_2$.

$$[[[\mathbb{R}_e \times \mathbb{R}_{r_{90}} \times \mathbb{R}_{r_{180}} \times \mathbb{R}_{r_{270}}] \ \text{Ⓢ}\ \mathbb{Z}_4]_{P_1}]_{\mathcal{U}}$$
$$\text{Ⓦ}\ {}_{baP_1, eaP_1}AGL(2,\mathbb{R})^{aP_1}\ \text{Ⓦ}\ {}_{aP_1}AGL(2,\mathbb{R})^{P_1}\ \text{Ⓦ}\ \mathbb{R}^{ea}\mathbb{R}_{r_{90}} \qquad (12.2)$$

Let us explain this sequence.

The first line gives the alignment kernel, indicated by the subscript $\mathcal{U}$ on the outer brackets. Within this bracket, we will generally have the direct product of the primitives to be used, each bracketed by a symbol P_i labeling the primitive. In the above example, the alignment kernel contains only one primitive P_1, which is the square, given by $[\mathbb{R}_e \times \mathbb{R}_{r_{90}} \times \mathbb{R}_{r_{180}} \times \mathbb{R}_{r_{270}}]$ Ⓢ $\mathbb{Z}_4$. Because there are no other primitives, there is no actual *alignment* of primitives, and therefore no (non-trivial) alignment kernel. Therefore, one could simply use the group given for the square as an *ordinary* fiber. However, in the next chapter, extra primitives will be added, and therefore it is useful to start using the notation of unfolding groups.

The second line in expression (12.2) exhibits the control groups. There are three of them: two affine groups and one translation group. To understand the structure of any unfolding group, one starts with the control-group level which has the primitives P_i as its superscripts. In the present example, this is the higher affine group ${}_{aP_1}AGL(2,\mathbb{R})^{P_1}$.

As a superscript, P_1 represents the square in its starting state, i.e., the copy of the fiber group P_1 corresponding to the identity element in each of the three control groups. This copy is the unit square at the origin of the world frame.

Therefore, the higher affine group ${}_{aP_1}AGL(2,\mathbb{R})^{P_1}$ acts on the unit square P_1 at the origin of the world frame and moves and stretches it to become the major rectangle aP_1 in Fig. 12.1 (p. 259). Thus the label aP_1 is placed as the lower index on the affine group ${}_{aP_1}AGL(2,\mathbb{R})^{P_1}$, indicating that it is the result of the affine action a on the primitive P_1 given in the superscript.

Notice that we have adapted the notation used in Sect. 6.3 for serial-link manipulators, i.e., where a transformation T going from frame i to frame $i-1$ was given as ${}_{i-1}T^i$. On the group ${}_{aP_1}AGL(2,\mathbb{R})^{P_1}$, the upper index P_1 will be called the **input index**, and the lower index aP_1 will be called the **output index**.

Now consider the lower affine group ${}_{baP_1}AGL(2,\mathbb{R})^{aP_1}$ in the above sequence. This is used to move and stretch the major rectangle to produce the minor one. Thus, the input (upper) index on this affine group is aP_1, which was the output index from the previous affine group. There are two output indexes baP_1, eaP_1 on the lower affine group. We will generally work from right to left along indexes, numbering them in that order. The first index eaP_1 is the major rectangle aP_1 multiplied by the identity element e of this second affine group. The second index baP_1 is the first rectangle aP_1 having undergone an affine transformation b, producing the smaller rectangle. Thus the two affine groups produce the two rectangles. (The mathematics of double indexes will be fully explained in the next chapter.)

So far, there was no reason to use the unfolding-group notation of indexes. We could have simply used the notation employed so far in this book:

$$[\mathbb{R} \, \text{ⓢ} \, \mathbb{Z}_4] \, \text{ⓦ} \, AGL(2, \mathbb{R}) \, \text{ⓦ} \, AGL(2, \mathbb{R}). \tag{12.3}$$

However, there is an extra piece of structure that now has to be encompassed: the placing of the two additional lines in Fig. 12.2 that will form the doorway. Let us assume that the designer has created these lines as offsets (parallel displacements) of the right side of the major rectangle, i.e., as *transfers* of the right side. This is a typical trick in CAD.

For this, we use *super-local unfolding*, which is an action that targets only part of an existing structure. Here the target is the right side of the major rectangle. The remainder of the structure is unaltered.

This is expressed as follows: Add the group $\mathbb{R}^{ea\mathbb{R}_{r90}}$, as an extra wreath level above the existing group sequence, as shown in expression (12.2). This is the parallel translation group $\mathbb{R}$ applied to the target indicated by its superscript $ea\mathbb{R}_{r90}$. This target $ea\mathbb{R}_{r90}$ is derived from the component $\mathbb{R}_{r90}$ *within the alignment kernel* (first line), but *after* the action a has been applied from the higher affine group. We deduce this because its two prefixes ea come respectively from the two copies of $AGL(2, \mathbb{R})$ in the above sequence. Therefore, we infer that the target $ea\mathbb{R}_{r90}$ is the right side of the major rectangle. Thus what has been encoded is the fact that each of the two infinite door-lines is a translate of this right side. Note that, according to our theory, the right side is itself an infinite line underlying an occupancy structure. Therefore, the infinite door-lines can truly be understood as translates of the infinite line underlying the right side. Finally, the lines are trimmed by switches in the occupancy groups.

It is now necessary to understand the real meaning of a super-local unfolding, as follows:

12.4 Super-Local Unfolding and Wreath Coverings

We now need to understand what distinguishes super-local unfoldings from other kinds of wreath products. All wreath products are about actions. What distinguishes a super-local unfolding is the type of action.

The super-local unfolding group given in expression (12.2) above is a 5-fold wreath product. The first thing to understand is that, *algebraically*, it works exactly in accord with everything said earlier about n-fold wreath products. In particular, it is structured by control-nested τ-automorphisms. To grasp the full meaning of super-local unfolding, it is necessary to understand what these automorphisms do in this particular example.

Fig. 12.4 shows this group schematically. The five levels in the figure correspond to the five levels of the group. Each node in the figure is a copy of the group on its level in the list on the far left. The diagram is schematic in the sense that, on each level, a set of nodes dominated by a single node above it, should have a cardinality equal to the order of the group of that dominating node. For example, in the third level down, a set of nodes dominated by a single node above should correspond to all the members of the affine group $AGL(2, \mathbb{R})$ rather than just the three nodes shown. Similarly, the set of nodes on the second level down should form a continuum equivalent to the real line $\mathbb{R}$ above it.

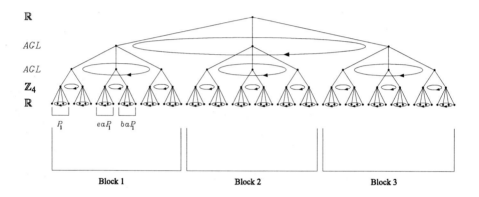

Fig. 12.4. A control-nested τ-automorphism in the group (12.2) on p. 262.

The entire set of circular arrows in the diagram constitutes an example of a control-nested τ-automorphism in the group. Notice that there is no difference between the structure of this diagram and the general schematic diagram given earlier for a control-nested τ-automorphism in an arbitrary wreath product, Fig. 4.4 (p. 124).

Now consider the lowest type of subtree in Fig. 12.4. It consists of four nodes on the bottom level, dominated by a node above it. The four nodes are the fiber copies $\mathbb{R}_e, \mathbb{R}_{r_{90}}, \mathbb{R}_{r_{180}}, \mathbb{R}_{r_{270}}$. The node above is their control group $\mathbb{Z}_4$. Thus the subtree corresponds to the wreath product $[\mathbb{R}_e \times \mathbb{R}_{r_{90}} \times \mathbb{R}_{r_{180}} \times \mathbb{R}_{r_{270}}] \circledS \mathbb{Z}_4$ which is $\mathbb{R} \circledW \mathbb{Z}_4$. Therefore the subtree will be referred to as a $[\mathbb{R} \circledW \mathbb{Z}_4]$-subtree. Clearly, the bottom of the figure is the set of $[\mathbb{R} \circledW \mathbb{Z}_4]$-subtrees.

It is now necessary to understand the indexes on the group in expression (12.2) on p. 262. First, we said that the meaning of P_1 as a *superscript* (i.e., on the upper affine group) is the square primitive located at the origin of the world frame. In Fig. 12.4, the identity element of each group will be taken to be the left-most node in the node set corresponding to the group. Therefore the superscript P_1, the starting state of the square, corresponds to the left most $[\mathbb{R}\text{⑩}\mathbb{Z}_4]$-subtree in Fig. 12.1. Note that this has been marked by P_1 in the diagram (extreme bottom left).

Now, this $[\mathbb{R}\text{⑩}\mathbb{Z}_4]$-subtree is the input index on the upper affine group in expression (12.2) on p. 262. Thus, the affine group sends it to aP_1, as indicated by the output index on that group. aP_1 is the $[\mathbb{R}\text{⑩}\mathbb{Z}_4]$-subtree shown as eaP_1 on the bottom of Fig. 12.4.

Then the lower affine group in expression (12.2) on p. 262 has the effect of sending this subtree to the subtree baP_1 shown at the bottom of Fig. 12.4. Notice that the two subtrees eaP_1 and baP_1 correspond respectively to the larger and smaller rectangles in the architectural plan Fig. 12.1.

Let us now understand the effect of the highest control group $\mathbb{R}$. Observe that the bottom of Fig. 12.4 has been divided into blocks. Notice that each block corresponds to one node on the second level down in the diagram. However, the set of nodes on the second level correspond to the set of members of the group on the top level. Therefore *each block corresponds to one value in the control group* $\mathbb{R}$.

Most crucially, the control group $\mathbb{R}$ has the effect of translating one block onto the next block along the bottom. It is necessary now to understand what effect this has on the geometry.

The effect on the geometry is given by the input index on the highest control group $\mathbb{R}^{ea\mathbb{R}_{r90}}$ in expression (12.2) on p. 262. This index $ea\mathbb{R}_{r90}$ refers to the right side of the major rectangle in the floor-plan. This is the side that will be translated to create the extra two lines as offsets.

The crucial thing to understand now is that this right side, i.e., the input index $ea\mathbb{R}_{r90}$ (on the control group), is actually a node in the hierarchy of nodes in Fig. 12.4. The index $ea\mathbb{R}_{r90}$ is actually the *second* node along the *bottom* of the *subtree labeled* eaP_1. The reader should carefully locate this node before continuing.

Now, the top control group $\mathbb{R}$ targets this node (which the right side of the rectangle) and translates it by some distance in Euclidean space. Therefore, the equivalent node in Block 2 in Fig. 12.4 will represent a translated version of the right side. Furthermore the equivalent node in Block 3 will represent the right side translated even further; and so on.

Note that, according to the notation $\mathbb{R}^{ea\mathbb{R}_{r90}}$, it is only the right side that is translated; i.e., the other three sides of the rectangle are left the same as they are in Block 1.

Now for a crucial question: What do we mean by *targeting* the second node? Algebraically, the targeting does not exist. The control-nested τ-automorphism shown in Fig. 12.4 has a circular arrow for each of the nodes of the hierarchy. Thus, no actual node is privileged. This means for example, that *all* the nodes along the bottom are affected algebraically by the translation on the top. This effect is conjugation. In fact, each circular arrow in the figure represents a conjugation, and these conjugations are perfectly uniform across the whole diagram. Therefore, no privileged nodes exist.

The targeting is therefore not in the abstract algebra of the group. It occurs somewhere else: in the *representation* on Euclidean space.

In order to understand this representation, it is necessary to invoke the concept of a *wreath covering* from p. 133. That is: Let S be a set. A wreath-covering of S is a wreath product together with a set of maps $\phi_{c_i} : F_{c_i} \longrightarrow S$, one for each fiber-set copy F_{c_i}, for all c_i in the control set. The wreath product without the maps is called the *abstract wreath product*, and the maps are called the *covering maps*.

At the time of defining a wreath covering, we had observed that nearly all the wreath products in this book are wreath coverings, because there is some overlap of the sets that one would wish to call the fiber-set copies. For example, in the wreath product of a square, the sides overlap at the corners. This means that the sides are not *independent* sets, as one requires for fiber-set copies. Thus one has to go to sets that cover the sides. The covering sets are independent, and can therefore consititute the fiber copies. As was said on p. 134, in nearly all cases considered in this book, the fact that the wreath products are wreath coverings will be so obvious that this will usually be assumed without explicitly pointing it out.

However, in the case of super-local unfoldings, the concept of a wreath-covering gains a strongly explicit importance. This is because the fiber sets are coincident except at a selected subset, the target. The set coincidence means the necessary specification of a wreath covering.

To understand this, consider our example. Each of the $[\mathbb{R} \textcircled{w} \mathbb{Z}_4]$-subtrees along the bottom of Fig. 12.4 corresponds to a four-sided figure. It is worthwhile re-considering some cases: The subtree marked P_1 is the square at the origin of the world frame, the subtree marked eaP_1 is the larger rectangle, the subtree marked baP_1 is the smaller rectangle, ..., etc. In Block 2, the right-hand side in each of these four-sided figures is moved by a small amount; in Block 3, the right-hand side in each of these four-sided figures is moved by a larger amount; and so on. *However*, the other three sides in each figure are coincident with their corresponding versions in each of the other blocks; i.e., these sides, as sets of points, are not independent. Thus it is necessary to use wreath coverings of the sides. Therefore do this:

Each node along the bottom of Fig. 12.4 represents a copy $\mathbb{R}_{c_i}$ of the fiber group $\mathbb{R}$, where c_1 stands for the multi-index of control sets above the node. Now the fiber-group copy $\mathbb{R}_{c_i}$ acts on its own fiber-set copy F_{c_i} which is a copy of the real line. The copies of the real line are independent (and not in any space). This completely defines the *abstract wreath product* involved, since we now have all the fiber sets involved, and all of the control sets above these (because all the wreath product operations in the hierarchy are regular).

The final stage is to define the effects in the Euclidean plane $\mathbb{R}^2$, on which the designer is working. To do this, one converts this *abstract* wreath product into a wreath covering, by defining a set of maps $\phi_{c_i} : F_{c_i} \longrightarrow \mathbb{R}^2$, one for each fiber-set copy F_{c_i} of the real line, for all c_i. This images the lines in the Euclidean plane.

One can now understand the nature of the index $ea\mathbb{R}_{r90}$ on the highest control group $\mathbb{R}^{ea\mathbb{R}_{r90}}$. We had said that the index refers to the second node along the bottom of the subtree labeled eaP_1 in Fig. 12.4. In fact, the index says this: Consider the covering map for this second node, and the covering map for the corresponding second node in Block 2, and the covering map for the corresponding second node in Block 3, and so on. The *images* of these covering maps are related by the control group $\mathbb{R}$ represented as a horizontal leftward action on the Euclidean plane.

This means that the control group $\mathbb{R}^{ea\mathbb{R}_{r90}}$ has two actions as follows:

(1) Algebraic: This is its conjugation $g - g^{-1}$ action within the *abstract* wreath product, sending *any* node x in Block 1 onto its conjugate $g(x)g^{-1}$ in Block 2 (and that node onto its conjugate in Block 3, and so on). In particular, it sends the "second" node $ea\mathbb{R}_{r90}$ in Block 1 onto its conjugate $g(ea\mathbb{R}_{r90})g^{-1}$ in Block 2 (and that node onto its conjugate in Block 3, and so on).

(2) Representational: This is its action of translating the images of the covering maps defined on each of these conjugate nodes. Thus, let $F_{ea\mathbb{R}_{r90}}$ be the fiber-set copy (real-line copy) corresponding to the "second" node $ea\mathbb{R}_{r90}$, and let $\phi_{ea\mathbb{R}_{r90}}$ be the covering map associated with this node, that is

$$\phi_{ea\mathbb{R}_{r90}} \;\; : \;\; F_{ea\mathbb{R}_{r90}} \longrightarrow \mathbb{R}^2.$$

Then let g be any member of the control group $\mathbb{R}^{ea\mathbb{R}_{r90}}$, that is, a translation. The covering map associated with the conjugate node $g(ea\mathbb{R}_{r90})g^{-1}$ of the node $ea\mathbb{R}_{r90}$ is this:

$$\phi_{g(ea\mathbb{R}_{r90})g^{-1}} \;\; : \;\; F_{g(ea\mathbb{R}_{r90})g^{-1}} \longrightarrow \mathbb{R}^2.$$

The *representational* effect of g is to relate the covering maps; specifically to relate their images thus:

$$g \; : \; \phi_{ea\mathbb{R}_{r90}}\left(F_{ea\mathbb{R}_{r90}}\right) \longrightarrow \phi_{g(ea\mathbb{R}_{r90})g^{-1}}\left(F_{g(ea\mathbb{R}_{r90})g^{-1}}\right) = g[\phi_{ea\mathbb{R}_{r90}}\left(F_{ea\mathbb{R}_{r90}}\right)].$$

In contrast, the effect of g on the images of all other covering maps is to leave their images unchanged. This latter effect can be understood as analogous to the *freezing* effect that can be chosen in advanced drafting systems such as AutoCAD, i.e., freezing an entire structure except for some target to be altered.

12.5   The Symmetry Group of a Complex Object

Now observe this: What we have done in the previous section is to describe the final figure on p. 259 by a *single large symmetry group*, that given by expression (12.2) on p. 262. This single group powerfully encodes the complex structure that a designer, manufacturer, perceiver, navigator, etc., uses in creating, seeing, manipulating, or moving through, that structure. If we had taken the conventional approach used to define the symmetry group of a figure, the group would have consisted of only the identity element - because the operations of creating the rectangles and joining them would have successively broken all the symmetries; i.e., the final organization has no reflectional and rotational symmetries. By capturing the generative operations by group extensions via wreath products, we not only preserve the previous symmetries by transfer, but successively build up recoverable record of the generative procedure, rather than loosing it. This is the true power of our approach.

12.6   Exploitation of Existing Structure

The most important fact about super-local unfolding is its capacity to exploit existing structure. For example, the two vertical lines that appeared in the middle figure on p. 260, were translates of a side of the room.

Clearly, therefore, such groups exemplify our goal of maximizing transfer. Indeed they embody our Fundamental Principle of Cognition (p. 22), which states that *any new structure should be created as the transfer of existing structure.*

In super-local unfolding, one "digs" into an existing structure and finds a part that one can transfer for the creation of the required additional structure. This is used endlessly in computer-aided design - from the design of fuselages for aerospace to the design of heating and ventilation systems for architecture. Furthermore it is the basis of great artworks, such as the symphonies of Beethoven, or the paintings of Raphael (see Leyton [96]). In these works what is ultimately being exploited is the fact that transfer is maximized by the human perceptual systems.

12.7   Cross-Hierarchy in Super-Local Unfolding

As we have said, super-local unfolding "digs into" the existing structure and extracts a part that can be used to create additional structure. In fact, the extracted part can be arbitrarily complex in the following sense:

**CROSS-HIERARCHY ACTION OF SUPER-LOCAL UNFOLD-
ING.**   *An important benefit of super-local unfolding is that it can act on a
selection that cuts across hierarchical boundaries.*

This facility is required an extraordinary number of times in any shape generation. For example, consider computer-aided design or 3D solid modeling.
A situation that occurs many times within a single work session is this: The
designer has created a structure - e.g., a plan of an apartment complex, a
3D model of a biological organ such as a liver, etc. What is required at the
next stage is to act on some selection of parts that cuts across the hierarchy
of that structure. For example, in the plan of an apartment complex, one
might need to select two of the walls in a room in one of the apartments, and
the entire room in another apartment, and an entire apartment on a different floor - and apply a single operation to this selection. Again, in the 3D
modeling of a liver, one might need to select certain mesh faces, on one side,
certain points on another side, and a particular volume within - and apply
a single operation, such as rotation, to this selection. This will give the liver
the subtle deformation needed to make it realistic. Furthermore, this kind of
situation occurs not only in the design disciplines, but also in the physical
sciences: One might need to define a process that targets only certain parts
of a complex molecule - e.g., those parts with negative charge.

What characterizes such situations is that the selection cuts across hierarchical boundaries in two senses: (1) horizontally; i.e., one selects from
different sub-hierarchies, (2) vertically; i.e., one selects from different hierarchical levels.

It is to handle the above range of situations, that we developed the concept
of super-local unfolding. Although, the example in Sect. 12.3 operated on only
a single selected element from the existing structure, one can equally operate
on a cross-hierarchy selection from the structure. For example, consider the
two-room apartment as shown in the final figure on p. 259. Supposing that
we want to simultaneously select the larger room and the bottom wall of the
smaller room, so that a rotation could be applied simultaneously to these.
This is a typical operation used in the architectural work of Eisenman, Gehry,
and Libeskind. What we do is take the group developed so far for that figure,
and we wreath super-append the rotation group $SO(2)$ above this, so that
$SO(2)$ has a non-trivial representation for only the target parts, thus:

$$[[[\mathbb{R}_e \times \mathbb{R}_{r_{90}} \times \mathbb{R}_{r_{180}} \times \mathbb{R}_{r_{270}}] \, \text{Ⓢ} \, \mathbb{Z}_4]_{P_1}]\mathcal{U}$$
$$\text{Ⓦ} \; _{baP_1, eaP_1} AGL(2, \mathbb{R})^{aP_1} \; \text{Ⓦ} \; _{aP_1} AGL(2, \mathbb{R})^{P_1} \; \text{Ⓦ} \; \mathbb{R}^{ea\mathbb{R}_{r_{90}}}$$
$$\text{Ⓦ} \; SO(2)^{\{ba\mathbb{R}_{r_{180}}, \, eaP_1\}}. \tag{12.4}$$

The first two lines of this expression give the existing structure, which was
the previous group. The last line gives the appended $SO(2)$. Its target sites
are eaP_1 which is the larger room, and $ba\mathbb{R}_{r_{180}}$ which is the bottom wall

of the smaller room. It is easy to check that our wreath-covering analysis of super-local unfolding is extendible to this type of situation. Notice that braces "{ }" have been put around the target list, in the super-local unfolding. This will distinguish it from the case without braces, to be discussed later.

What can be observed in this example is the powerful sense in which this type of action is an *asymmetrization by misalignment*. Prior to the $SO(2)$ action, there is a certain amount of symmetry in the room structure, by virtue of the existing hierarchy. That is, the rooms are parallel to each other, and the wall-structure of an individual room has a double-reflectional symmetry. However, after the $SO(2)$ action, the rooms cease to be parallel to each other, and the wall-structure of the smaller room no longer has a double-reflectional symmetry. One can regard this situation as having been created by pulling the major room and bottom wall "out of alignment".

The reader can begin to see, from this example, how our tools can generate and represent an arbitrarily complex shape as a sequence of unfoldings via transfer. As we said before, all the complexities of a structure become entirely comprehensible as the transfers of a minimal set of components in the object's core.

13. Unfolding Groups, III

13.1 Introduction

We now need to invent one final class of unfolding groups, in order to deal with an important remaining type of asymmetry (complexity) that can occur in shape generation. These groups will be called *sub-local* unfoldings because they wreath sub-append some group below a selected component in a level.

The following should be pointed out: We invented the three classes of unfolding groups - telescope, super-local, and sub-local unfolding groups - for the following reason: The recoverability component of the generative theory implies that *plans concerning an object are inferrable from its symmetries*, e.g., plans for designing the object, manufacturing it, manipulating it, navigating with respect to it, etc. This is true whether the object is simple or complex, e.g., a complex environment. Expressing simple objects in terms of symmetry groups is easy. Consequently we had to invent classes of symmetry groups to describe complex objects.

In order to establish these groups, our procedure was this: We worked through every single operation in each of several main CAD, solid modeling, assembly, and animation programs, including AutoCAD 2000, Architectural Desktop, Mechanical Desktop, ProEngineer, 3D Studio Viz, etc., as well as all the major manuals on each of the programs - approximately 15,000 pages of text. Each individual situation was characterized by a group, and a new class of groups was invented for any situation that could not be formalized in terms of any previously created class of groups. Proceeding in this manner, it was eventually found that three classes of groups - telescope, super-local, and sub-local unfolding groups - could handle any newly created situation.

Thus our current assumption is that these groups can handle any complexity in shape generation.[1]

13.2 Symmetry Group of an Apartment

It is now necessary to motivate the final class of unfolding groups, and fully characterize their properties. To do so, we will construct a symmetry group of a seven-room apartment consisting of a central main room and six outer rooms, as shown in Fig. 13.1.

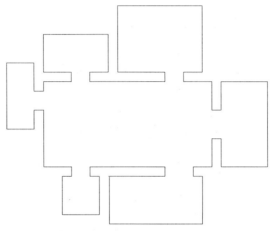

Fig. 13.1. A seven-room apartment.

Within the theory so far given in this book, the symmetry group of this apartment is easy to construct. It is simply this:

$$[\mathbb{R} \textcircled{w} \mathbb{Z}_2 \textcircled{w} \mathbb{Z}_2]$$
$$\textcircled{w} \ AGL(2, \mathbb{R})$$
$$\textcircled{w} \ \ AGL(2, \mathbb{R}). \tag{13.1}$$

Here, the first line is the square, as the hyperoctahedral wreath hyperplane group of degree 2, which can be identified with the Cartesian frame. The control group on the last line, takes the square understood as the World frame and converts it (by affine transformation) into the rectangle constituting the main room. Then the control group on the middle line, takes the main room rectangle and converts it into each of the six outer rooms. To understand this

---

[1] Notice that the term complexity is being used here not in the usual contemporary sense related to instability, but in the sense of amount of structure.

second stage, open up the wreath product in the final line, in its semi-direct product notation. Thus the bottom two lines become:

$$[\ldots \times AGL(2, \mathbb{R}) \times AGL(2, \mathbb{R}) \times AGL(2, \mathbb{R}) \times \ldots]$$
$$ⓈAGL(2, \mathbb{R}). \tag{13.2}$$

The upper line shows the fiber-group product, and now the bottom line exhibits the semi-direct product symbol. The six outer rooms are in fact six occupied fiber-group copies within the fiber-group product. That is, each component of the fiber-group product, can be regarded as taking the main room and transforming it outwards to become one of the outer rooms. Observe that each of the six rooms can therefore be transferred *onto each other* via the control group.

It is worth understanding this structure as analogous to a serial-link manipulator. This is particularly clear in the unopened wreath structure in expression (13.1). One has simply converted the special Euclidean groups $SE(2)$ of the manipulator into the affine groups $AGL(2, \mathbb{R})$ of the room structure. The main room corresponds to a parent link. There is a *single* child link to this parent. This is the affine group in the middle line. Therefore the six outer rooms correspond to six *work-spaces* of that single child robot link. These six work-spaces can be transferred onto each other via the highest control group. Notice, most crucially that, because the parent link has only one child, the manipulator is purely *serial*.

Structures of the form shown in Fig. 13.1 therefore present no problem to the theory we have developed so far. However, let us now make a *local* ammendment. Let us suppose that someone builds an *extension* to the room labeled 5, as shown in Fig. 13.2. The extension is on the bottom right of Room 5.

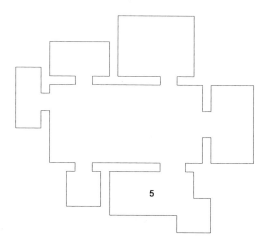

Fig. 13.2. A local "anomaly" - an extension is built onto Room 5.

The problem is that the addition is local to only one of the rooms. This problem occurs in many situations: For example, in architecture a number of design firms (perhaps in different parts of the world) may each be working on the design of a different part of a large building. The decisions are being made independently and are local to the particular needs of the part they are working on. This of course occurs in any large-scale mechanical design - e.g., the design of an airplane. It is also an important problem in defect-crystal physics, where defects tend to be local structures on a crystal.

Let us return to the apartment as now shown in Fig. 13.2 with the "anomalous" extension. There are two solutions to representing this situation: The first is that we can assume that there is the possibility of a corresponding anomaly on each of the other outer rooms. This is reasonable - because at any later time, the owner might wish to make an extension to any of those rooms. Since each room allows for an extension, we therefore loose the sense that an extension is an anomaly. Most crucially, we can mathematically account for the possibility of an extension onto each outer room by wreath sub-appending a further affine group below the sequence of affine groups in (13.1), thus obtaining:

$$\begin{aligned}
&[\mathbb{R} \textcircled{w} \mathbb{Z}_2 \textcircled{w} \mathbb{Z}_2] \\
&\textcircled{w} \ AGL(2, \mathbb{R}) \\
&\textcircled{w} \ AGL(2, \mathbb{R}) \\
&\textcircled{w} \ \ AGL(2, \mathbb{R})
\end{aligned} \tag{13.3}$$

where the group on the second line is the new group. Notice that this structure is still of the *serial-link* type. In robotic terms, there are three successive links: The main room, *one* outer room, and *one* extension. What we see as parallel rooms are merely fiber-group copies arising from a purely serial structure. This means that the rooms on any level can be transferred onto each other.

In the above approach, the anomaly was removed by making a purely serial structure. However, let us consider the following situation in biology: A human body has a head and two arms. Each are extensions of the torso. Certainly the two arms can be mapped onto each other. However, the head cannot be mapped onto either of the two arms. The head therefore constitutes a genuine anomaly. Thus it is necessary to characterize such situations. For this, the final class of groups is needed: sub-local unfoldings. To create such groups, we first need to define a type of wreath product which we will call a *wreath-direct group*.

13.3   Wreath-Direct Groups

Sub-local unfoldings will require us to define the following class of groups:

Definition 13.1. *A group of the form*

$$[H_1 \times H_2] \ \textcircled{w} \ [G_1 \times G_2] \tag{13.4}$$

will be called a **wreath-direct group,** *if (1) the wreath product is regular, i.e., the control group and control set are both $G_1 \times G_2$; (2) if the control group acts on the control set component-wise, i.e., imitating the direct product operation; (3) if the fiber-set is structured as a Cartesian product $F_1 \times F_2$; and (4) the fiber group $H_1 \times H_2$ acts on the fiber-set $F_1 \times F_2$ component-wise.*

As usual, of course, where wreath products are concerned, the term *group* tends to mean *group action*.

A number of different types of examples of wreath-direct groups will occur in sub-local unfoldings. One of the most frequent is this:

$$[P_1 \times P_2] \ \textcircled{w} \ [AGL(2, \mathbb{R}) \times AGL(2, \mathbb{R})] \tag{13.5}$$

where P_1 and P_2 are primitives, and $AGL(2, \mathbb{R})$ are their command groups. We shall now use this example to illustrate Definition 13.4. For this illustration, let the primitive P_1 be the generative action $SO(2)$ for a circle, and the primitive P_2 be the generative action for the square, which will be given simply as $\mathbb{Z}_4$, rather than $\mathbb{R}\textcircled{w}\mathbb{Z}_4$, to avoid including unnecessary details. So the group (13.5) will be taken to be:

$$[SO(2) \times \mathbb{Z}_4] \ \textcircled{w} \ [AGL(2, \mathbb{R}) \times AGL(2, \mathbb{R})]. \tag{13.6}$$

The best way to understand the definition of this group is to go through the 5-stage structure of transfer given in Sect. 3.20.

STAGE 1: Define the Fiber Level and Control Level

The data set is the set of all configurations of a single square and circle. It is decomposed into two levels:

FIBER LEVEL.

An individual fiber is an individual circle-square configuration. Now define:

(1) Fiber set $= S^1 \times S^q$, where S^1 is the unit circle and S^q is the unit square (both of which will be taken here to be point-sets, for ease of exposition).
(2) Fiber group $= SO(2) \times \mathbb{Z}_4$.
(3) Action of fiber group on fiber set is *component-wise*:

$$\begin{cases} [SO(2) \times \mathbb{Z}_4] \ \times \ [S^1 \times S^q] \ \longrightarrow \ [S^1 \times S^q] \\ (\ (s \ , \ z) \ , \ (p,q) \ ) \ \longmapsto \ (sp, zq) \end{cases}$$

CONTROL LEVEL.

This level concerns the mapping of configurations to configurations. The wreath product will be assumed to be regular, i.e., the control set and the control group are the same. Now define:

(1) Control set $= AGL(2, \mathbb{R}) \times AGL(2, \mathbb{R})$.
(2) Control group $= AGL(2, \mathbb{R}) \times AGL(2, \mathbb{R})$.
(3) Action of control group on control set is *component-wise*: That is, given an element (a, b) from the control group, and an element (a_i, b_i) from the control set, the application of the former to the latter is

$$(a, b)(a_i, b_i) = (aa_i, bb_i). \qquad (13.7)$$

Notice that this action is equivalent to multiplication in the direct product $AGL(2, \mathbb{R}) \times AGL(2, \mathbb{R})$.

STAGE 2: Define the Fiber-Group Product and Its Action on the Data Set

Since the wreath product is regular, there are as many copies of the fiber group $SO(2) \times \mathbb{Z}_4$ as there are members (a_i, b_i) in the control group $AGL(2, \mathbb{R}) \times AGL(2, \mathbb{R})$. The copies of the fiber group $SO(2) \times \mathbb{Z}_4$ will be labeled

$$[SO(2) \times \mathbb{Z}_4]_{(a_i, b_i)} \qquad (13.8)$$

for all $(a_i, b_i) \in [AGL(2, \mathbb{R}) \times AGL(2, \mathbb{R})]$. Therefore, the fiber-group product is this:

$$[[SO(2) \times \mathbb{Z}_4]_{(a_1, b_1)} \times \ldots \times [SO(2) \times \mathbb{Z}_4]_{(a_\infty, b_\infty)}]. \qquad (13.9)$$

Now, define the group action of the fiber-group product on the entire data set, as follows: The data set is the Cartesian product of the fiber set $S^1 \times S^q$ (circle × square) and the control set $AGL(2, \mathbb{R}) \times AGL(2, \mathbb{R})$. That is, it consists of all copies

$$[S^1 \times S^q]_{(a_i, b_i)}$$

of the fiber set, for all members (a_i, b_i) of the control set $AGL(2, \mathbb{R}) \times AGL(2, \mathbb{R})$. The fiber-set copies are the circle-square configurations.

The group action of the fiber-group product on the entire data set is defined by making each copy $[SO(2) \times \mathbb{Z}_4]_{(a_i, b_i)}$ of the fiber group act on each copy $[S^1 \times S^q]_{(a_i, b_i)}$ of the fiber set *independently*. The action therefore has the *selective effect*.

STAGE 3: Define the Control Group as an Automorphism Group on the Fiber-Group Product

The fiber-group copies (configurations) will now be related to each other by the control group $AGL(2, \mathbb{R}) \times AGL(2, \mathbb{R})$. This is achieved in the following way: The control group $AGL(2, \mathbb{R}) \times AGL(2, \mathbb{R})$ acts as a τ-automorphism group on the fiber-group product. Thus, given an element (a, b) from the control group, it corresponds to the τ-automorphism $\tau((a, b))$ on the fiber-group product thus:

$$\tau((a, b))[[SO(2) \times \mathbb{Z}_4]_{(a_1, b_1)} \times \ldots \times [SO(2) \times \mathbb{Z}_4]_{(a_\infty, b_\infty)}] \qquad (13.10)$$

which, by applying the action on the indexes, is this

$$[[SO(2) \times \mathbb{Z}_4]_{(a,b)(a_1,b_1)} \times \ldots \times [SO(2) \times \mathbb{Z}_4]_{(a,b)(a_\infty, b_\infty)}] \qquad (13.11)$$

which, by raising the control group action (13.7) from the control set to the fiber-group product, is simply this:

$$[[SO(2) \times \mathbb{Z}_4]_{(aa_1, bb_1)} \times \ldots \times [SO(2) \times \mathbb{Z}_4]_{(aa_\infty, bb_\infty)}]. \qquad (13.12)$$

In other words, the τ-automorphism corresponds to multiplication in the direct product $AGL(2, \mathbb{R}) \times AGL(2, \mathbb{R})$, raised to the indexes.

STAGE 4: Define a Splitting Extension of the Fiber-Group Product by the Control Group

This of course is the same as opening up the wreath product symbol in (13.6), in the semi-direct product notation. Thus, the wreath product is this:

$$\begin{aligned}&[[SO(2) \times \mathbb{Z}_4]_{(a_1, b_1)} \times \ldots \times [SO(2) \times \mathbb{Z}_4]_{(a_\infty, b_\infty)}] \\ &\text{\textcircled{S}}_\tau \ [AGL(2, \mathbb{R}) \times AGL(2, \mathbb{R})]. \end{aligned} \qquad (13.13)$$

STAGE 5: Define the Action of the Wreath Product on the Data Set

Any element of the wreath product group (13.13) is of the form:

$$\langle \ (s_1, z_1), \ldots, (s_\infty, z_\infty) \ | \ (a, b) \ \rangle. \qquad (13.14)$$

An element in the data set is of the form

$$(p, q)_{(a_i, b_i)} \qquad (13.15)$$

i.e., a point p on the circle, and a point q on the square, in the particular circle-square configuration indexed by (a_i, b_i) in the control group.

The action of the group element (13.14) on the data element (13.15) goes in two stages: (1) By the selective effect, one first selects the (a_i, b_i)-indexed component from the *vector* within (13.14). This is the component (s_i, z_i) along this vector, i.e., an element from the fiber-group copy $[SO(2) \times \mathbb{Z}_4]_{(a_i,b_i)}$. It is first applied to the data element $(p, q)_{(a_i,b_i)}$. This results in

$$(s_i p, z_i q)_{(a_i,b_i)}. \tag{13.16}$$

Notice that this action has occurred within the particular circle-square configuration $[S^1 \times S^q]_{(a_i,b_i)}$.

(2) One then applies the control element (a, b) from (13.14) to the result (13.16). This is raised to the indexes of the data point, thus:

$$(s_i p, z_i q)_{(aa_i,bb_i)}.$$

That is, the data point is moved from the circle-square configuration $[S^1 \times S^q]_{(a_i,b_i)}$ to the circle-square configuration $[S^1 \times S^q]_{(aa_i,bb_i)}$.

13.4   Canonical Unfoldings

Let us now go back to the apartment discussed in Sect. 13.2. A problem had arisen because we wished to create an "anomaly" that was particular to Room 5. There were two alternative solutions: (1) The anomaly is transferred as a *possibility* onto each of the other rooms, i.e., it ceases to be an anomaly. (2) The option of transfer is dis-allowed, i.e., the status of anomaly is preserved.

We showed how to represent alternative (1). Now we show how to represent alternative (2). Let us first deal with the seven-room case as shown in Fig. 13.1. Let us also make the assumption that each of the outer rooms is going to be handled independently by a different design company; i.e., each will carry a set of "anomalies" that are particular to that room, i.e., non-transferrable. Previously we gave the group to be this:

$$[\mathbb{R} \ \widehat{w} \ \mathbb{Z}_2 \ \widehat{w} \ \mathbb{Z}_2]$$
$$\widehat{w} \ AGL(2, \mathbb{R})$$
$$\widehat{w} \ AGL(2, \mathbb{R}). \tag{13.17}$$

There is only one component on the second line - it is responsible for all six outer rooms. These are copies within the fiber-group product associated with the second line as fiber. However, we now wish to make the outer rooms more strongly independent within the second line. Thus the group to be constructed is this:

$$[\mathbb{R} \textcircled{w} \mathbb{Z}_2 \textcircled{w} \mathbb{Z}_2]_{\mathcal{U}}$$
$$\textcircled{w}[AGL(2,\mathbb{R}) \times AGL(2,\mathbb{R}) \times AGL(2,\mathbb{R}) \times AGL(2,\mathbb{R}) \times AGL(2,\mathbb{R})$$
$$\times AGL(2,\mathbb{R})]\textcircled{w}AGL(2,\mathbb{R}). \qquad (13.18)$$

The second line is now a direct product of the groups representing the six outer rooms. Notice most crucially that the direct-product symbols are not within a fiber-group product; i.e., they are not between fiber-group *copies*. That is, none of the wreath products in the above expression have been opened up in their semi-direct product notation, which means that we do not actually see any fiber-group copies in the expression. In particular, the second line is a *single* fiber group, and therefore the direct product is a *decomposition of that fiber group*, not a decomposition of the fiber-group product.

The analogy to the robot manipulator is that, whereas the six rooms were previously described as six states of one link, they are now described in terms of six links. Using the more general concept of inheritance, we say that, whereas previously there was only one child, there are now six children.

Now look at the top line. It contains a suffix $\mathcal{U}$, which we now explain:

Definition 13.2. *A wreath product of the form*

$$[\ldots]_{\mathcal{U}} \textcircled{w} G_1 \textcircled{w} G_2 \textcircled{w} \ldots \textcircled{w} G_n$$

will be called a **canonical unfolding.** *The symbol* $[\ldots]_{\mathcal{U}}$ *brackets an alignment kernel consisting of a direct product of all primitives required by all the wreath-direct actions from the control hierarchy.*

Therefore, in the particular case of expression (13.18), the first line $[\mathbb{R} \textcircled{w} \mathbb{Z}_2 \textcircled{w} \mathbb{Z}_2]_{\mathcal{U}}$ means a direct product of six copies of the group $\mathbb{R} \textcircled{w} \mathbb{Z}_2 \textcircled{w} \mathbb{Z}_2$. The action of its control group, on the second line, will be *wreath-direct* on this direct product. That is, the direct product on the second line will act component-wise on the direct product on the first line.

Now let us add indexes to expression (13.18) to show how the rooms are generated:

$$[[\mathbb{R} \textcircled{w} \mathbb{Z}_2 \textcircled{w} \mathbb{Z}_2]_{P_1}]_{\mathcal{U}}$$
$$\textcircled{w} \quad [\quad _{g_1 a P_1}AGL(2,\mathbb{R})^{aP_1} \times {}_{g_2 a P_1}AGL(2,\mathbb{R})^{aP_1} \times {}_{g_3 a P_1}AGL(2,\mathbb{R})^{aP_1}$$
$$\times \ _{g_4 a P_1}AGL(2,\mathbb{R})^{aP_1} \times {}_{g_5 a P_1}AGL(2,\mathbb{R})^{aP_1} \times {}_{g_6 a P_1}AGL(2,\mathbb{R})^{aP_1} \ ]$$
$$\textcircled{w} \ _{aP_1}AGL(2,\mathbb{R})^{P_1} \qquad (13.19)$$

In the top line, the primitive square has now been labeled P_1. Thus, go to the last line. The input (upper) index P_1 on this line represents this primitive in its initial state, where it is the World frame. The output (lower) index aP_1 represents this primitive after the application of the affine transformation a which converts the World frame square into the rectangle constituting the main central room of the apartment. Next, this output index becomes the input to the direct product above it. Most crucially, notice that this index is now the input index for *each* of the six individual components of the direct product. Each component produces as output an index of the form $g_i a P_1$ representing the fact that the component has applied the transformation g_i to the main room rectangle $a P_1$, thus producing one of the outer rooms $g_i a P_1$.

Now let us understand the symmetry effect of an individual element from this group. For this, use the type of diagram introduced in Sect. 4.5 for control-nested τ-automorphisms. The particular diagram needed here is Fig. 13.3. This is a 3-level structure, corresponding to the three levels in expression (13.18). The top node in the diagram is the highest control group $AGL(2, \mathbb{R})$. Each node below this is a *copy* of the 6-fold structure $AGL(2, \mathbb{R}) \times AGL(2, \mathbb{R}) \times AGL(2, \mathbb{R}) \times AGL(2, \mathbb{R}) \times AGL(2, \mathbb{R}) \times AGL(2, \mathbb{R})$. There are as many copies as there are elements in the top node (infinite). Each bottom node is a copy of the alignment kernel, which is the 6-fold direct product of the primitive $\mathbb{R} \ \textcircled{w} \ \mathbb{Z}_2 \ \textcircled{w} \ \mathbb{Z}_2$. Each node above the bottom level dominates an infinite set of nodes on the level below it. The finiteness in the diagram is only schematic.

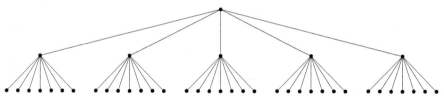

Fig. 13.3. The group action corresponding to the seven-room apartment.

Let us now obtain the form of an element of this group. Given a wreath product $G(F)\textcircled{w}G(C)$, the notation will be this:

$$\langle \, T_1 \, , \, T_2 \, , \, \ldots \, , \, T_n \mid r \, \rangle$$

where each T_i comes from its corresponding copy of $G(F)$, and r comes from $G(C)$. The string $T_1 \, , \, T_2 \, , \, \ldots \, , \, T_n$ will be called the *vector* of the element. We will assume that there are as many elements T_i along the vector as there are elements in $G(C)$; i.e., the wreath product is regular.

Now return to the symmetry group of the apartment, as depicted in Fig. 13.3. An element of the group is a selection of an element from each of the nodes. To understand such an element, let us first consider only the top two levels. These two levels correspond to the wreath product:

$$[AGL(2, \mathbb{R}) \times AGL(2, \mathbb{R}) \times AGL(2, \mathbb{R}) \times AGL(2, \mathbb{R}) \times AGL(2, \mathbb{R})$$
$$\times AGL(2, \mathbb{R})]\textcircled{w}AGL(2, \mathbb{R}). \tag{13.20}$$

We now obtain an element from this group by selecting an element from each of the nodes of the top two levels of Fig. 13.3. Thus an element looks like this:

$$\langle \, (h_1, h_2, \ldots, h_6) \, , \, (g_1, g_2, \ldots, g_6) \, , \, \ldots \, , \, (k_1, k_2, \ldots, k_6) \mid a \, \rangle. \qquad (13.21)$$

The right-most symbol a is an element from the top-level node. Then, to the left of this, is the *vector*. Each of its components, for example $(h_1, h_2, \ldots, h_6)$, will be called a *sub-vector*. A sub-vector is an element from a node on Level 2 of Fig. 13.3. Since such a node is a copy of the 6-fold direct product $AGL(2, \mathbb{R}) \times AGL(2, \mathbb{R}) \times AGL(2, \mathbb{R}) \times AGL(2, \mathbb{R}) \times AGL(2, \mathbb{R}) \times AGL(2, \mathbb{R})$, the sub-vector has six components. There are obviously an infinite number of sub-vectors along the vector in expression (13.21).

Now let us extend expression (13.21) so that we obtain a full group element; i.e., for the entire three levels of Fig. 13.3. To do this, convert each sub-vector in (13.21) into a control element with respect to its own vector, which will be called a sub-sub-vector. That is, obtain

$$\langle \, \langle \, u_1, u_2, u_3, \ldots, u_\infty \mid (h_1, h_2, \ldots, h_6) \, \rangle \, ,$$
$$\langle \, v_1, v_2, v_3, \ldots, v_\infty \mid (g_1, g_2, \ldots, g_6) \, \rangle \, ,$$
$$\ldots \, ,$$
$$\langle \, w_1, w_2, w_3, \ldots, w_\infty \mid (k_1, k_2, \ldots, k_6) \, \rangle \, ,$$
$$\mid a \, \rangle. \qquad (13.22)$$

Notice that this new expression is the same as (13.21) except that each sub-vector, for example, $(h_1, h_2, \ldots, h_6)$ is now extended to its left by a sub-sub-vector, for example, $u_1, u_2, u_3, \ldots, u_\infty$. In a sub-sub-vector, each component, u_i is a selection from a bottom node of Fig. 13.3. That is, it is a selection from a copy of the alignment kernel, which is a 6-fold direct product of $\mathbb{R} \, \textcircled{w} \, \mathbb{Z}_2 \, \textcircled{w} \, \mathbb{Z}_2$. A particular sub-sub-vector $u_1, u_2, u_3, \ldots, u_\infty$ corresponds to a set of bottom nodes dominated by only one node on the second level (i.e., is an element-wise selection from the dominated set). There are as many components to the sub-sub-vector $u_1, u_2, u_3, \ldots, u_\infty$ as there are elements in the 6-fold direct product $AGL(2, \mathbb{R}) \times AGL(2, \mathbb{R}) \times AGL(2, \mathbb{R}) \times AGL(2, \mathbb{R}) \times AGL(2, \mathbb{R}) \times AGL(2, \mathbb{R})$, representing a dominating node.

The action of the group element in expression (13.22) is as follows: Consider a configuration of six outer rooms. The configuration actually corresponds to a bottom node of the diagram. It is a fiber-group copy, and its copy index is given by the succession of nodes upward from it in the diagram. Now apply the group element in expression (13.22). By the selective effect in a wreath product, the configuration will select only those elements in (13.22) to which its index corresponds. Thus on the lowest level, it selects an element from a sub-sub-vector, let's say u_i. Since this is a selection from a copy of the alignment kernel, which is a 6-fold direct product of $\mathbb{R} \, \textcircled{w} \, \mathbb{Z}_2 \, \textcircled{w} \, \mathbb{Z}_2$, the effect of u_i will be to apply translations to each of the sides of the rooms, and a 2-level hierarchy of reflections to those sides. The translations will be individual to each side, and the reflection hierarchy will be individual to each room. Then, also by the selective effect, on the next higher level, one applies the associated sub-vector, which for u_i is $(h_1, h_2, \ldots, h_6)$. This applies an affine transformation h_i to each room individually, relative to the main room. Notice that, in relation to the element u_i, this is using the wreath-direct action. That is, u_i can be regarded as a 6-fold structure, each component of which is personal to each room, and the transformation $(h_1, h_2, \ldots, h_6)$ also acts personally to each room.

Finally, the element a from the top node is applied, and this is an affine transformation of the referent main room. The six rooms will, of course, inherit this transformation, in accord with our algebraic theory of inheritance in Sect. 7.3.

13.5   Incorporating the Symmetry of Referents

Two more pieces of algebraic structure need to be added to the group being studied, in order to complete its symmetry action. The first is this:

The reader will notice that, while the above group allows for internal symmetry transformations of each outer room (translations and reflections of the walls), it does not allow such transformations of the main room. The reason is that, in fact, we do not yet have a main room, as follows:

What we have are six outer rooms, each of which started as coincident with the World frame. They were first transformed to be coincident with what would be the main room. The bottom nodes of Fig. 13.3 allow us to do internal symmetry transformations independently on each of these copies. It is this that tells us that, although they are coincident with what would be the main room, none of them could actually be the main room, since there is no criterion to particularly choose one of them as such, by symmetry. This was not a problem in the purely serial case. In fact, the six rooms in this position must be interpreted as a particular *configuration* of the six *outer* rooms - the particular configuration in which they happen to be coincident with what would be the main room. Furthermore, these internal symmetry transformations, which are applied independently to each of them, are equivalent to the internal symmetry transformations that could be applied independently to them after they have been moved out from this position.

A seventh room is therefore required which will be the main room - and which can receive its own internal symmetry transformations even after the other six rooms have been moved outwards.

The solution therefore is to add an extra copy of the square to the alignment kernel. Thus the kernel becomes the direct product of *seven* copies of the group $\mathbb{R} \ \textcircled{w} \ \mathbb{Z}_2 \ \textcircled{w} \ \mathbb{Z}_2$. Furthermore, this will be *controlled* by an extra component to be added to the 6-fold direct product on the level above; i.e., the middle line in expression (13.18). The component need not be the full affine group because the main room is not going to be moved in relation to itself (only in relation to the World frame which is an affine movement already encoded in the highest control group). Therefore the added component need only be the group $\{e\}$, thus:

$$[\mathbb{R} \,\textcircled{w}\, \mathbb{Z}_2 \,\textcircled{w}\, \mathbb{Z}_2]_{\mathcal{U}}$$
$$\textcircled{w}[AGL(2,\mathbb{R}) \times AGL(2,\mathbb{R}) \times AGL(2,\mathbb{R}) \times AGL(2,\mathbb{R}) \times AGL(2,\mathbb{R})$$
$$\times AGL(2,\mathbb{R}) \times \{e\}]\textcircled{w}AGL(2,\mathbb{R}) \tag{13.23}$$

where the third line shows the added control component. Now, in this expression, we *infer* that there are seven copies of the group $\mathbb{R} \,\textcircled{w}\, \mathbb{Z}_2 \,\textcircled{w}\, \mathbb{Z}_2$ in the alignment kernel, because the suffix $\mathcal{U}$ tells us that the unfolding is canonical (Definition 13.2), and this means that the alignment kernel must contain as many copies of primitives as are required by the *wreath-direct* action from above. Thus we infer the number of copies in the alignment kernel from the number of components in the direct product above - which, in this case, is seven. It should be emphasized that, even though the extra component in the middle-level control group is the trivial $\{e\}$, it nevertheless is explicitly required to make the middle level act on the alignment kernel via a *wreath-direct* action; i.e., the components of the 7-fold middle level act on the components of the 7-fold alignment kernel *component-wise*.

When indexes are put on the groups, expression (13.23) then becomes

$$[[\mathbb{R} \,\textcircled{w}\, \mathbb{Z}_2 \,\textcircled{w}\, \mathbb{Z}_2]_{P_1}]_{\mathcal{U}}$$
$$\textcircled{w} \quad [\quad _{g_1 a P_1}AGL(2,\mathbb{R})^{a P_1} \times \ _{g_2 a P_1}AGL(2,\mathbb{R})^{a P_1} \times \ _{g_3 a P_1}AGL(2,\mathbb{R})^{a P_1}$$
$$\times \ _{g_4 a P_1}AGL(2,\mathbb{R})^{a P_1} \times \ _{g_5 a P_1}AGL(2,\mathbb{R})^{a P_1} \times \ _{g_6 a P_1}AGL(2,\mathbb{R})^{a P_1}$$
$$\times \ _{a P_1}\{e\}^{a P_1} \quad]$$
$$\textcircled{w} \quad _{a P_1}AGL(2,\mathbb{R})^{P_1} \quad . \tag{13.24}$$

Careful consideration reveals that there is one more copy of the Cartesian frame (square) that needs to be added to the alignment kernel. This gives the internal symmetry group of the World frame. The reason this is needed is because, according to our theory, the World symmetry group is actually used in shape generation. For example, it defines the symmetry structure of the starting construction plane in CAD, as given in the theory of construction planes on p. 212.

With this addition to the alignment kernel, we need to add a corresponding component $\{e\}$ onto the highest control group, via direct product, to obtain the requisite *wreath-direct* action. The $\{e\}$ will mean that the World-frame will not actually be moved, but can undergo internal symmetry transformations from its corresponding component in the alignment kernel. The new group is thus:

$[[\mathbb{R} \ \textcircled{w} \ \mathbb{Z}_2 \ \textcircled{w} \ \mathbb{Z}_2]_{P_1}]\mathcal{U}$

$\textcircled{w} \quad [\quad _{g_1 a P_1} AGL(2,\mathbb{R})^{a P_1} \ \times \ _{g_2 a P_1} AGL(2,\mathbb{R})^{a P_1} \ \times \ _{g_3 a P_1} AGL(2,\mathbb{R})^{a P_1}$

$\qquad \times \ _{g_4 a P_1} AGL(2,\mathbb{R})^{a P_1} \ \times \ _{g_5 a P_1} AGL(2,\mathbb{R})^{a P_1} \ \times \ _{g_6 a P_1} AGL(2,\mathbb{R})^{a P_1}$

$\qquad \times \ _{a P_1}\{e\}^{a P_1} \quad]$

$\textcircled{w} \quad _{a P_1} AGL(2,\mathbb{R})^{P_1} \ \times \ _{P_1}\{e\}^{P_1} \quad .$ \hfill (13.25)

The reader will recall that P_1, as an *input index* on a control group, means the primitive P_1 in its initial state. Here P_1 is the square, i.e., the hyperoctahedral wreath group which is the group of the Cartesian frame. Therefore the *input index* P_1 means the Cartesian frame in its initial state, which is the World frame W. Therefore all occurences of the index P_1 in the last line can be written as W.

13.6   Why Internal Symmetry Groups

The reader might wonder why we have been so careful to ensure the presence of an internal symmetry group for each object. The reason is extremely powerful: A basic claim of our system is that *plans are inferred from symmetries.* This comes from our recoverability rules in Chapter 2. Thus the actual use of the rooms - e.g., to draw them, to build them, to navigate them, etc. - is inferrable from their symmetry structures.

13.7   Base and Subsidiary Alignment Kernels

When defining unfolding groups (p. 246), we characterized them by two properties: their effect is *selective,* and they act by *misalignment.* We have seen how canonical unfoldings are selective - they use the mechanism of *wreath-direct* products. What now needs to be understood is the way in which they work by misalignment.

Let us consider expression (13.25). What one can see here is that there is a successive misalignment effect, as follows: In the last line, the two component groups both have input indexes P_1, the World frame. Now the *output* index on the component $\{e\}$ is P_1; that is, there has been no alteration of P_1 via this component. In contrast, the output index on the other component $AGL(2,\mathbb{R})$ is aP_1, which is P_1 having undergone an affine movement a. Therefore the output indexes on the two components indicate that there has been a *misalignment* with respect to the World frame P_1.

Now go to the next level, as shown by the three lines above this. Here the input index, on all of the seven direct-product components, is aP_1. Notice that the output index on the component $\{e\}$ is of course the same as the input index. In contrast, observe that the output indexes on all the other components are different. This means that there has been a misalignment with respect to aP_1.

It can be seen therefore that, as one progresses up the lines in expression (13.25), there is a *successive* misalignment, first with respect to the World frame, and second with respect to the altered World frame, which is the main room.

Notice that alignments are actually represented by common input indexes on the same level. That is, on the first level (i.e., the last line) the common input index P_1 represents the alignment of copies of P_1. On the next level, the common input indexes aP_1 represent the alignment of copies of aP_1.

Also notice the following: In the last line the two input indexes P_1 actually present *eight* copies of P_1. This is easy to infer from the structure of indexes in the entire group. Thus the input indexes on the last line represent the entire eight objects in the alignment kernel. Next, on the three lines above, the seven indexes represent seven aligned copies aP_1. Therefore we can consider this to be a subsidiary alignment kernel. Thus the initial alignment kernel will be referred to as the **base alignment kernel**, and any remaining alignment kernels, on subsequent levels, will be referred to as **subsidiary alignment kernels**. Of course the subsidiary alignment kernels correspond to those subsets of the base alignment kernel where alignments have not yet been broken.

13.8 Cloning

Cloning is a basic operation in object-oriented software. Computer-aided design and solid modeling would be almost impossible without it. Also, cloning is fundamental to prototype-based languages, because, unlike the conventional object-oriented approach, which starts with classes and generates objects by instantiating classes, the prototype approach starts with actual objects and generates further objects by cloning and modification, see Blaschek [11].

Using the concepts developed so far in this chapter, we are now able to give an algebraic representation of cloning. Expression (13.25) illustrates our approach. As was pointed out, one infers from the wreath-direct action in this expression that there are a total of eight copies of the World frame $W = P_1$. The group therefore keeps a record of the number of copies created of an object. We thus offer the following algebraic theory of cloning:

ALGEBRAIC THEORY OF CLONING. *The set of cloned objects is given by the alignment kernel. As each clone is created, extend the alignment* *kernel by its symmetry group, and simultaneously extend a direct product within one of the control levels by a component representing the command group of the cloned object. This is captured by a canonical unfolding group.*

The fact that command groups act by misalignment on clones is important: This feature can clearly be seen in CAD and solid modeling, where clones are initially coincident, and are subsequently moved away from each other and deformed. Thus, generally, successive actions on clones is symmetry-breaking.

13.9   The Inference Structure

The notation in (13.25) is complete in the sense that it allows us to infer the entire structure of the group and its action. The inference is as follows: The symbol $\mathcal{U}$ tells us that the unfolding is canonical, and that there are therefore as many copies of the given primitive as are required by the wreath-direct actions from above it. Thus the required copies must be counted by going through those wreath-direct actions. Start with the last line. Here there are two components each with an input index P_1. Furthermore, the *output* index aP_1, on one of these two components, becomes the input index on the seven components of the direct product on the next level. Therefore, the conclusion is that there are a total of eight copies in the alignment kernel.

13.10   Group Elements

It is now necessary to understand what an element of the group (13.25) looks like. For this, a notation is required that is slightly more explicit than that just given. We must register the fact that the middle level of the group is not a 7-fold direct product but is actually an 8-fold direct product. This is inferrable from the previous notation: The notation lead to the conclusion that the alignment kernel is 8-fold. Therefore, to allow a wreath-direct action from the middle level, there must actually be an eighth component in the middle level. The missing component is simply $_{P_1}\{e\}^{P_1}$ from the final line. This component must be copied into the middle-level direct product. Although the crucial role of this component occurs only in the final line, where it is part of defining the misalignment represented there, the algebraic structure requires that it is carried over into the middle level, where it is simply a "dummy" place-holder allowing the wreath-direct action. Therefore a more explicit notation for the group is:

$$[[\mathbb{R} \;\textcircled{w}\; \mathbb{Z}_2 \;\textcircled{w}\; \mathbb{Z}_2]_{P_1}]_{\mathcal{U}}$$

$$\textcircled{w} \quad [\quad _{g_1 aP_1} AGL(2,\mathbb{R})^{aP_1} \;\times\; _{g_2 aP_1} AGL(2,\mathbb{R})^{aP_1} \;\times\; _{g_3 aP_1} AGL(2,\mathbb{R})^{aP_1}$$

$$\times \; _{g_4 aP_1} AGL(2,\mathbb{R})^{aP_1} \;\times\; _{g_5 aP_1} AGL(2,\mathbb{R})^{aP_1} \;\times\; _{g_6 aP_1} AGL(2,\mathbb{R})^{aP_1}$$

$$\times \; _{aP_1}\{e\}^{aP_1} \qquad \times \; _{P_1}\{e\}^{P_1} \quad]$$

$$\textcircled{w} \quad _{aP_1} AGL(2,\mathbb{R})^{P_1} \;\times\; _{P_1}\{e\}^{P_1} \quad . \tag{13.26}$$

Notice that the last component of the final line has been copied directly above it. The entire direct product on lines 2, 3, and 4, will be called the middle-level direct product.

With this notation, one can more easily understand an element of the group, as follows. Again, let us use Fig. 13.3 (p. 280). One obtains a group element by selecting an element from each of the nodes simultaneously. The top node corresponds to the last line in expression (13.26), and each of the middle-level nodes is a *copy* of the entire the middle-level direct product in expression (13.26). Each bottom node in the graph is a copy of the alignment kernel which is an 8-fold direct product of the primitive $\mathbb{R} \;\textcircled{w}\; \mathbb{Z}_2 \;\textcircled{w}\; \mathbb{Z}_2$.

The process of constructing a group element is the same as given in Sect. 13.4. However, in the current situation, the group element will actually record the misalignments as follows:

First, as before, considering only the top two levels of Fig. 13.3, the group element is structured like this:

$$\langle\; (h_1, h_2, \ldots, h_6, e, e)\;,\; (g_1, g_2, \ldots, g_6, e, e)\;,\; \ldots\;,\; (k_1, k_2, \ldots, k_6, e, e) \mid a, e \;\rangle. \tag{13.27}$$

Obviously, the control element a, e at the far right comes from the final line in (13.26), and each of the sub-vectors comes from a copy of the middle-level direct product. Most importantly, we see the introduction of a number of identity elements, not contained in the previous formulation (13.21) on p. 281. These will be crucial to both contructing the misalignments and allowing internal symmetries of the referents, as will be seen.

Next extending expression (13.27), to get a full group element, one obtains

$$\langle\; \langle\; u_1, u_2, u_3, \ldots, u_\infty | (h_1, h_2, \ldots, h_6, e, e)\;\rangle\;,$$

$$\langle\; v_1, v_2, v_3, \ldots, v_\infty | (g_1, g_2, \ldots, g_6, e, e)\;\rangle\;,$$

$$\ldots\;,$$

$$\langle\; w_1, w_2, w_3, \ldots, w_\infty | (k_1, k_2, \ldots, k_6, e, e)\;\rangle\;,$$

$$\mid a, e \;\rangle. \tag{13.28}$$

where each sub-vector has been extended leftward by an associated sub-subvector, for example, $u_1, u_2, u_3, \ldots, u_\infty$. In this new version, each component, u_i is a selection from a copy of an *8-fold* alignment kernel, rather then a 6-fold kernel as before.

The action of the group element in expression (13.28) is as follows: Any configuration can be considered to be a configuration of eight "rooms", where the eighth is the World frame. The above group operation will be applied to such a configuration.

In particular, let us take the configuration to be the starting one, where all eight rooms are coincident clones of the World frame. This starting configuration (as with any configuration) will correspond to a particular node on the bottom of the tree (in fact, the left-most node, for the starting configuration) - and will select its corresponding elements upward in the full group element. Thus, let us suppose that the configuration corresponds to u_1 in the first sub-sub-vector in expression (13.28), and therefore selects sub-vector $(h_1, h_2, \ldots, h_6, e, e)$ above it.

Now apply the full group element to the starting configuration. The control element a, e, on the last line in expression (13.28), prescribes an affine transformation for these eight clones, in the following way: The e applies to the World frame, which will therefore not move. The a applies to the seven *actual* rooms, and will mean that they move out together from the World frame, to become seven coincident copies of the main room.

Next, apply the associated sub-vector $(h_1, h_2, \ldots, h_6, e, e)$. The six h_i will move six of the rooms to become the outer rooms, and the first e will retain the seventh room as the unaltered main room, and the second e will retain the eighth as the unaltered World frame.

Finally, apply the internal symmetry u_1. Since this is a selection from a copy of the alignment kernel, which is a 8-fold direct product of $\mathbb{R} \, \textcircled{w} \, \mathbb{Z}_2 \, \textcircled{w} \, \mathbb{Z}_2$, the effect of u_1 will be to apply translations to each of the sides of the eight rooms, and a 2-level hierarchy of reflections to those sides. The translations will be individual to each side, and the reflection hierarchy will be individual to each room.

We have just described the application of the group element to the starting configuration. But the same group element can be applied to any configuration, i.e., any other node along the bottom of the figure, and the selective effect will work accordingly. Most crucially, the identity elements e will allow misalignments with respect to the same referents as before (main room and World frame), because the identity elements e occur in the same positions. Furthermore, these identity elements concern only the affine movements on the referents. The referents can still alter *internally* via their associated symmetry groups in the alignment kernel.

13.11   Adding the Anomaly

We are now ready to solve the main issue of this chapter: adding an anomaly. The example being considered is the apartment in Fig. 13.2 where an extension has been built onto Room 5, and we are choosing to regard this as an anomaly.

The techniques developed so far in this chapter were designed to handle this in a natural way. First, for convenience, let us take expression (13.26), for the apartment without the anomaly, and re-write it here, thus:

$$[[\mathbb{R} \; ⓦ \; \mathbb{Z}_2 \; ⓦ \; \mathbb{Z}_2]_{P_1}]_{\mathcal{U}}$$
$$ⓦ \; [\; _{g_1 a P_1} AGL(2,\mathbb{R})^{aP_1} \; \times \; _{g_2 a P_1} AGL(2,\mathbb{R})^{aP_1}$$
$$\times \; _{g_3 a P_1} AGL(2,\mathbb{R})^{aP_1}$$
$$\times \; _{g_4 a P_1} AGL(2,\mathbb{R})^{aP_1}$$
$$\times \; _{g_5 a P_1} AGL(2,\mathbb{R})^{aP_1}$$
$$\times \; _{g_6 a P_1} AGL(2,\mathbb{R})^{aP_1}$$
$$\times \; _{a P_1}\{e\}^{aP_1} \quad \times \; _{P_1}\{e\}^{P_1} \quad]$$
$$ⓦ \; _{a P_1} AGL(2,\mathbb{R})^{P_1} \; \times \; _{P_1}\{e\}^{P_1} \quad . \tag{13.29}$$

Here, for readability the groups for Rooms 3, 4, 5 and 6, have been put on Lines 3, 4, 5 and 6, respectively. These are easily identifiable by the fact that the left-most index g_i identifies Room i. Each of these groups on these lines, generates the associated Room i (output index) from a copy of the main room (input index).

Now, to generate the anomaly, we merely wreath sub-append a component to Room 5 on the Line 5 as follows:

$$[[\mathbb{R} \; ⓦ \; \mathbb{Z}_2 \; ⓦ \; \mathbb{Z}_2]_{P_1}]_{\mathcal{U}}$$
$$ⓦ \; [\; _{g_1 a P_1} AGL(2,\mathbb{R})^{aP_1} \; \times \; _{g_2 a P_1} AGL(2,\mathbb{R})^{aP_1}$$
$$\times \; _{g_3 a P_1} AGL(2,\mathbb{R})^{aP_1}$$
$$\times \; _{g_4 a P_1} AGL(2,\mathbb{R})^{aP_1}$$
$$\times \; [[_{s_1 g_5 a P_1} AGL(2,\mathbb{R})^{g_5 a P_1} \; \times \; _{g_5 a P_1}\{e\}^{g_5 a P_1}] \; ⓦ \; _{g_5 a P_1} AGL(2,\mathbb{R})^{aP_1}]$$
$$\times \; _{g_6 a P_1} AGL(2,\mathbb{R})^{aP_1}$$
$$\times \; _{a P_1}\{e\}^{aP_1} \quad \times \; _{P_1}\{e\}^{P_1} \quad]$$
$$ⓦ \; _{a P_1} AGL(2,\mathbb{R})^{P_1} \; \times \; _{P_1}\{e\}^{P_1} \quad . \tag{13.30}$$

Before explaining the added component, observe that, because it is wreath sub-appended to only one of the direct-product components, it will affect only those parts of the alignment kernel controlled by the direct-product component to which it is sub-appended.

Definition 13.3. *A* **sub-local unfolding group** *is a canonical unfolding group in which one of the control groups in a wreath-direct action, within the group, is wreath sub-appended.*

THEORY OF ANOMALIES. *Anomalies are modeled by sub-local unfolding groups.*

Now let us study expression (13.30). The wreath product symbol $\text{\textcircled{w}}$ on Line 5 indicates that the output objects on the left of $\text{\textcircled{w}}$ are children of the output objects to the right of $\text{\textcircled{w}}$, in accord with our theory of (run-time) link inheritance in Chapter 7. Now let us look at the wreath sub-appended component on Line 5, that is:

$$[{}_{s_1 g_5 a P_1} AGL(2, \mathbb{R})^{g_5 a P_1} \times {}_{g_5 a P_1}\{e\}^{g_5 a P_1}]$$

Here, the left group is ${}_{s_1 g_5 a P_1} AGL(2, \mathbb{R})^{g_5 a P_1}$ which moves a coincident copy of Room 5 out to become the extension room, the anomaly. The right group is ${}_{g_5 a P_1}\{e\}^{g_5 a P_1}$ which merely holds Room 5 where it is. In other words, we have created a *misalignment* of two clones of Room 5. Notice that the input indexes on both the left and right group are the same, that is, $g_5 a P_1$. Furthermore, this is the output index from the group ${}_{g_5 a P_1} AGL(2, \mathbb{R})^{a P_1}$ on the right of the wreath product $\text{\textcircled{w}}$ in Line 5. Therefore, intuitively, the parent-child relationships are as follows: The anomaly is a child of Room 5, and Room 5 is a child of self. However, rigorously, one should express this in terms of the clones involved.

The inference rules tell us that the alignment kernel now has nine clones of the group $\mathbb{R} \text{ \textcircled{w} } \mathbb{Z}_2 \text{ \textcircled{w} } \mathbb{Z}_2$. It is worth the reader tracking down the fate of these copies. Notice in particular that the input index aP_1 on Line 5 now represents two copies, rather than one as before. These are fed out to the two direct-product components that are wreath sub-appended in this line.

Let us give the label G to the wreath product on Line 5; that is:

$$G \quad = \quad [{}_{s_1 g_5 a P_1} AGL(2, \mathbb{R})^{g_5 a P_1} \times {}_{g_5 a P_1}\{e\}^{g_5 a P_1}] \text{ \textcircled{w} } {}_{g_5 a P_1} AGL(2, \mathbb{R})^{a P_1}. \tag{13.31}$$

Notice that an element in this wreath product is of the form:

$$h_5 \quad = \quad \langle\, (b_1, e)\,,\, (b_2, e)\,,\, \ldots\,,\, (b_\infty, e) \mid \tilde{h}_5\,\rangle \tag{13.32}$$

where $\tilde{h}_5$ is selected from the affine control group in G, and each (b_i, e) is from a copy of the direct-product constituting the fiber. The control element $\tilde{h}_5$ acts on two clones within the alignment kernel, clone 5 (which is Room 5) and clone 9, which is the anomaly. It moves them simultaneously out from the main room. Then any fiber element (b_i, e) acts on these two rooms thus: the b_i moves the anomaly out from Room 5, and the e keeps Room 5 unchanged. Because the anomaly is the ninth clone, it will be referred to as Room 9, even though it is actually the eighth room of the eight actual rooms.

Now let us go back to the *entire* group in (13.30). It is necessary to understand how an element in this group acts on configurations. For this, an appropriate modification should first be made to the graph in Fig. 13.3 (p. 280). Consider any tree descending from a middle-level node. Let us call this a *terminal tree*. An example is shown on the left in Fig. 13.4. Next consider any branch in a terminal tree. This is illustrated in the center of Fig. 13.4. In the new situation (i.e., with anomaly), each such branch has to be expanded as shown on the right in Fig. 13.4. This will be called an *expanded branch*. It encompasses the extra structure in Line 5 of the group (13.30). The expanded branch in Fig. 13.4 means this: In Line 5, there is a newly introduced level, the direct product, i.e., the wreath sub-appended level. This corresponds to the newly introduced middle node in the expanded branch. The extra nodes at the bottom arise from extra fiber-group copies, due to the introduction of this middle level.

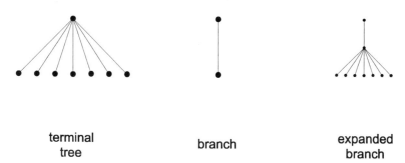

terminal tree   branch   expanded branch

Fig. 13.4. Expansions of the branches of a terminal tree.

Notice that the terminal tree on the left, and any branch as shown in the center, come from the *previous* situation, where there was no anomaly. The expansion on the right is the situation *after* the anomaly is introduced. Furthermore, this expansion is applied to *every* branch in the terminal tree on the far left. This expansion should be understood more fully, as follows:

As was said, the branch in the center of Fig. 13.4 comes from the previous situation of the apartment without the anomaly. Consider the bottom node in the branch. It is a copy of the alignment kernel and therefore has index:

$$(h_1, \ldots, h_4, h_5, h_6, e, e) \tag{13.33}$$

in the node above it. This index represents a configuration of the six outer rooms in relation to the main room. In the new situation (with anomaly), the index from the upper node is exactly the same, except that the fifth entry h_5 now comes from the wreath-product group that constitutes Line 5 in (13.30); that is the group labeled G in expression (13.31). As stated, any element of G is of the form (13.32) above. Therefore, substituting (13.32) in (13.33), one gets:

$$(h_1, \ldots, h_4, \overbrace{\langle\, (b_1, e)\,,\, (b_2, e)\,,\, \ldots\,,\, (b_\infty, e)\mid \tilde{h}_5\, \rangle}^{h_5}, h_6, e, e) \tag{13.34}$$

Most crucially, notice that within the overbrace, there is now a *list* of elements (b_i, e). They correspond to the different alternative misalignments of Room 9 with respect to Room 5. These correspond to the nodes at the bottom of the expanded branch in Fig. 13.4.

We can therefore express the structure in the following way: In the previous situation, the bottom node of a branch (center of Fig. 13.4) corresponded to a single configuration of the outer six rooms. Furthermore, each of the bottom nodes in the *expansion* (right of Fig. 13.4), corresponds to *exactly the same configuration of six outer rooms*. However, each of those bottom nodes have a different relation of Room 9 to Room 5. To repeat: *All the bottom nodes in an expanded branch have the same configuration of the six outer rooms, but a different relation of Room 9 to Room 5.*

Next, let us look at the index structure of a bottom node in the expanded set. Each bottom node is a copy of the alignment kernel. This is a configuration of the nine rooms, including the World frame and Room 9 (the anomaly). The index on a copy of the alignment kernel, is the group element (13.34), except for this: only one of the (b_i, e) must be selected from the list of such elements within the overbrace. In other words, the index is of the form:

$$(h_1, \ldots, h_4, \langle\, (b_i, e)\ |\ \tilde{h}_5\, \rangle, h_6, e, e). \tag{13.35}$$

The elements along this index tell us exactly what to do with each of the nine rooms. Any element of the form h_i tells us to move Room i in relation to the main room by the amount h_i. The element $\langle\, (b_i, e)\ |\ \tilde{h}_5\, \rangle$ tells us to move Rooms 5 and 9 together by amount $\tilde{h}_5$, and then Room 9 by amount b_i. Finally, the two elements e fix the World frame and main room.

Notice that an extra index (a, e) must be added at the end telling us the misalignment of the main room from the World frame. Thus one gets this:

$$[(h_1, \ldots, h_4, \langle\, (b_i, e)\ |\ \tilde{h}_5\, \rangle, h_6, e, e), (a, e)]. \tag{13.36}$$

Therefore, any bottom node is of the form:

$$[[\mathbb{R} \,ⓦ\, \mathbb{Z}_2 \,ⓦ\, \mathbb{Z}_2]\mathcal{U}]_{[(h_1, \ldots, h_4, \langle (b_i, e)|\tilde{h}_5\rangle, h_6, e, e), (a, e)]}$$

where the symbol $\mathcal{U}$ here indicates the 9-fold direct product of the iso-regular group it encloses.

Let us now define an element from the *entire* group in expression (13.30). Again, one obtains such an element by selecting an element from each of the nodes simultaneously. First, as before, considering only the top two levels of the entire graph, we find that the group element is structured like this:

$$\langle\, (h_1, \ldots, h_5, h_6, e, e)\, ,\ (g_1, \ldots, g_5, g_6, e, e)\, ,\ \ldots\, ,\ (k_1, \ldots, k_5, k_6, e, e)\ |\ a, e\, \rangle. \tag{13.37}$$

Although this looks the same as in the previous sections, the *fifth* entry in any sub-vector, for example entry h_5, is now of the wreath form shown in (13.32).

Therefore element (13.37) is really

$$\langle\ (h_1,\ldots,h_4,\overbrace{\langle\ (b_1,e)\ ,\ (b_2,e)\ ,\ \ldots\ ,\ (b_\infty,e)\ |\ \tilde{h}_5\ \rangle}^{h_5},h_6,e,e)\ ,$$

$$(g_1,\ldots,g_4,\overbrace{\langle\ (c_1,e)\ ,\ (c_2,e)\ ,\ \ldots\ ,\ (c_\infty,e)\ |\ \tilde{g}_5\ \rangle}^{g_5},g_6,e,e)\ ,$$

$$\ldots\ ,$$

$$(k_1,\ldots,k_4,\overbrace{\langle\ (d_1,e)\ ,\ (d_2,e)\ ,\ \ldots\ ,\ (d_\infty,e)\ |\ \tilde{k}_5\ \rangle}^{k_5},k_6,e,e)\ ,$$

$$|\ a,e\ \rangle. \tag{13.38}$$

However, for readability, use the shorter notation (13.37). To get a full group element, one extends expression (13.37) to obtain:

$$\langle\ \langle\ u_1,u_2,u_3,\ldots,u_\infty|(h_1,\ldots,h_5,h_6,e,e)\ \rangle\ ,$$

$$\langle\ v_1,v_2,v_3,\ldots,v_\infty|(g_1,\ldots,g_5,g_6,e,e)\ \rangle\ ,$$

$$\ldots\ ,$$

$$\langle\ w_1,w_2,w_3,\ldots,w_\infty|(k_1,\ldots,k_5,k_6,e,e)\ \rangle\ ,$$

$$|\ a,e\ \rangle. \tag{13.39}$$

Finally, one can apply the group element to any configuration. A configuration is a bottom node, and its index is of the form (13.36), which will now be re-written with primes, thus:

$$[(h'_1,\ldots,h'_4,\langle\ (b'_i,e)\ |\ \tilde{h}'_5\ \rangle,h'_6,e,e),(a',e)]. \tag{13.40}$$

The group element is a simultaneous selection from each of the nodes in the graph. When it is applied to a configuration, the configuration selects only those components that correspond to its upward path in the graph. In this way, one configuration is sent to another configuration by a control-nested τ-automorphism.

Most crucially therefore, the group we have constructed acts as a symmetry group on the space of configurations. According to our theory, this symmetry description is basic to any *plans* involving the configuration, i.e., its design, construction, navigation, etc.

13.12 Adding more Primitives

This chapter has so far discussed the situation in which the alignment kernel consists of several clones of only one primitive. However, many situations require more than one primitive. According to our theory, primitives are *isoregular groups* (wreath c-polycyclic, wreath-isometric). One adds primitives

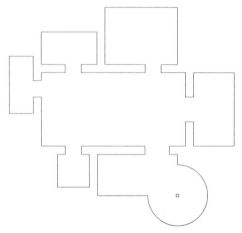

Fig. 13.5. A structure using two iso-regular groups.

by adding iso-regular groups to the alignment kernel. The remaining hierarchy of the group is constructed using the same rules as before.

An example is shown in Fig. 13.5. Here the anomaly has been chosen to be a circle. This means that the alignment kernel contains a second primitive P_2 which labels a circle. The full group of the structure is now:

$$
\begin{aligned}
&[[SO(2)]_{P_2} \times [\mathbb{R} \ \textcircled{w} \ \mathbb{Z}_2 \ \textcircled{w} \ \mathbb{Z}_2]_{P_1}]_{\mathcal{U}} \\
&\textcircled{w} \ \ [\ \ {}_{g_1 a P_1} AGL(2,\mathbb{R})^{aP_1} \times {}_{g_2 a P_1} AGL(2,\mathbb{R})^{aP_1} \\
&\qquad \times {}_{g_3 a P_1} AGL(2,\mathbb{R})^{aP_1} \\
&\qquad \times {}_{g_4 a P_1} AGL(2,\mathbb{R})^{aP_1} \\
&\qquad \times [[{}_{s_1 g_5 a P_2} AGL(2,\mathbb{R})^{g_5 a P_2} \times {}_{g_5 a P_1} \{e\}^{g_5 a P_1}] \\
&\qquad\qquad \textcircled{w} \ {}_{g_5 a P_2}, \ {}_{g_5 a P_1} AGL(2,\mathbb{R})^{aP_2}, \ {}^{aP_1}] \\
&\qquad \times {}_{g_6 a P_1} AGL(2,\mathbb{R})^{aP_1} \\
&\qquad \times {}_{a P_1} \{e\}^{aP_1} \qquad \times {}_{P_1} \{e\}^{P_1} \quad] \\
&\textcircled{w} \ \ {}_{a P_2}, \ {}_{a P_1} AGL(2,\mathbb{R}) \ {}^{P_2}, \ {}^{P_1} \times {}_{P_1} \{e\}^{P_1} \quad .
\end{aligned}
\qquad (13.41)
$$

This is exactly the same as the previous group (13.30) on p. 289, except for quite simple modifications on the first, fifth and final lines, as follows: The modification on the first line is to add the new primitive into the alignment kernel. The symbol $\mathcal{U}$ now means that there will be as many clones of *each* primitive P_i within the kernel as are required by the wreath-direct actions defined from above by the control groups.

Next, go to the last line. Here the only modification is that the input index on the affine group now contains both primitives, and that the output index aP_2, aP_1 is the application of the affine transformation a to this pair of primitives.

The final modification is in Line 5. The input index aP_2, aP_1 on this line is the output index from the bottom line. The affine control group applies g_5 to this pair, and then the two components of the direct-product fiber-group *misalign* this pair, the affine component acting on only the circle and the identity component acting on only the square.[2]

Notice that all other lines in expression (13.41) are the same as those in the previous expression (13.30). Notice also that the double inputs in the index on Line 5 also existed in the previous group, but had to be *inferred* from its direct-product fiber.

One sees therefore that the addition of primitives is easily handled in our system. The underlying reason for this is the power of canonical unfoldings. For example, it was this that allowed the change from the square anomaly to the circle anomaly to be made so easily. The algebraic change very much mirrors the actions that a designer would make in changing his/her design, and the inheritance effects that would propagate as a result of the changes.

13.13   Multi-index Notation

In fact, we actually did not define the meaning of the multi-indexes that appear in the last line and Line 5 of expression (13.41). This is done now:

Notation 13.1 *In canonical unfoldings, multi-indexes will be interpreted as (group) direct products, thus:*

$$_{x,y}G^{u,v} \;=\; {_x}G^u \times {_y}G^v \qquad\qquad for\ x \neq u; y \neq v.$$

For example, the affine component in the last line of (13.41) is:

$$_{aP_2,\,aP_1}AGL(3,\mathbb{R})^{\,P_2,\,P_1} \;=\; {_{aP_2}}AGL(3,\mathbb{R})^{P_2} \times {_{aP_1}}AGL(3,\mathbb{R})^{P_1}.$$

Similarly, on Line 5, the double index on the control group means that this group is a direct product. Observe an implication of this: The wreath product symbol ⓦ on Line 5 relates a direct-product control group and a direct-product fiber group. The consequence is that one can now understand that the ⓦ means a *wreath-direct* action.

---

[2] Note that, throughout this expression, there is no use of the identity group for the circle. This is because there is no *referent* circle *explicitly* used in the design - which means that it is not necessary to hold, within the alignment kernel, a copy of the independent internal symmetries for such a *referent* circle (i.e., only for the circle that is actually explicit). Internal symmetries for a non-explicit copy would be unnecessary because the symmetries serve the role of plans, e.g., for design and navigation, and plans are not needed for non-explicit copies. Note that the cloning structure of the alignment kernel therefore gives us a system for specifying occupancy that is different from our other system using an occupancy group.

On Line 5, the fiber-group direct product can also be reduced using a double index, via the following rule:

Notation 13.2 *In canonical unfoldings, multi-indexes with an* **unchanged coordinate** y *will be interpreted thus:*

$$_{x,y}G^{u,y} \ = \ _xG^u \ \times \ _y\{e\}^y$$

(with the corresponding expression for an unchanged first coordinate).

13.14   Symmetry Streaming

The issue to be examined now applies equally to the first anomaly using the square primitive, and the second anomaly using the circle primitive. Therefore, let us return to the former example.

The issue is this: When one inspects the extension room, one notices that it is not merely derived from a square, e.g., by deformation, but it actually *is* square. This means that it is open to two interpretations, corresponding to the two types of inference defined in Sect. 2.9, external vs. internal inference. All inferences are of one of these two forms, as follows. (1) In external inference, the data set is assumed to contain a record of only one state of a process, and any past state is therefore assumed to be outside what is observable in the data set. With respect to this, note that the Externalization Principle says that any external inference goes back to an iso-regular group. (2) In internal inference, the data set is assumed to contain records of more than one state of a process, and therefore past states are observable in the data set.

Now let us see what the two types of inference do with the square anomaly. (1) In the case of external inference, the square, being an iso-regular group, is assumed to be a starting state. (2) In the case of internal inference, one sees the square as derived by deformation from the rectangular Room 5 to which it is attached.

Both inferences are possible. The advantage of the external inference is that one does not assume an unnecessary history of deformation; i.e. starting with a square primitive, then deforming to produce the main room and Room 5, and then *un*-deforming to produce the square extension room. However, the disadvantage is that one looses the child-parent relationship that the anomaly has with respect to Room 5. In contrast, the advantage of the internal inference is that one retains the child-parent relation to Room 5, but introduces apparently unnecessary deformations.

The problem is resolved by regarding some of the parameters of the affine group as external and some as internal. In accord with the theory of Chapter 2, it is appropriate to regard the deformational parameters as external, and the translational ones as internal. The result is this: The clone of the square primitive, used by the anomaly, was initially situated at the World frame exactly like the others. This clone underwent the translation to the main room position - i.e., maintaining positional alignment - and the further translation to the position of Room 5 - again maintaining positional alignment - however, it did not undergo the deformations that were used to create the main room and Room 5. In this way, the symmetry of the square is protected through the derivational history till it becomes the square of the anomaly. We will call this process **symmetry streaming**; i.e., it streams the symmetry through the derivational history.

In the particular example being studied, the effect of the symmetry streaming is that *alignment* is preserved with respect to position but *not* with respect to deformation. This means that deformational misalignment started earlier than was coded in the previous formulation. In other words, without symmetry streaming, deformational misalignment is delayed so that it can occur together with the positional misalignment.

There is thus a choice in formulation: either one can use the previous formulation, in which case our inference structure is entirely *internal*; or one chooses the new formulation in which the inference structure is *partially external*. The diagram of the apartment is perceptually ambiguous as to which formulation one should choose, and the empirical validity of this ambiguity is easy to demonstrate in psychological experiments. Therefore, rather than regarding the ambiguity as a disadvantage, one should code for both interpretations. The formulation of the first interpretation has been given in expression (13.30) on p. 289. In fact, we need only two modifications of that formulation to give the second interpretation. The latter is given thus:

$$
\begin{aligned}
&[[\mathbb{R} \ \text{\textcircled{w}} \ \mathbb{Z}_2 \ \text{\textcircled{w}} \ \mathbb{Z}_2]_{P_1}]_{\mathcal{U}} \\
&\text{\textcircled{w}} \quad [\quad {}_{g_1 a P_1} AGL(2,\mathbb{R})^{a P_1} \ \times \ {}_{g_2 a P_1} AGL(2,\mathbb{R})^{a P_1} \\
&\qquad \times \ {}_{g_3 a P_1} AGL(2,\mathbb{R})^{a P_1} \\
&\qquad \times \ {}_{g_4 a P_1} AGL(2,\mathbb{R})^{a P_1} \\
&\qquad \times \ [[{}_{s_1 g_9 b P_1} AGL(2,\mathbb{R})^{g_9 b P_1} \ \times \ {}_{g_5 a P_1}\{e\}^{g_5 a P_1}] \\
&\qquad\qquad \text{\textcircled{w}} \ {}_{*g_9 b P_1} \ , \ {}_{g_5 a P_1} AGL(2,\mathbb{R})^{b P_1} \ , \ {}^{a P_1}] \\
&\qquad \times \ {}_{g_6 a P_1} AGL(2,\mathbb{R})^{a P_1} \\
&\qquad \times \ {}_{a P_1}\{e\}^{a P_1} \qquad \times \ {}_{P_1}\{e\}^{P_1} \quad] \\
&\text{\textcircled{w}} \ {}_{*b P_1} \ , \ {}_{a P_1} AGL(2,\mathbb{R})^{P_1} \ , \ {}^{P_1} \ \times \ {}_{P_1}\{e\}^{P_1} \quad .
\end{aligned}
\tag{13.42}
$$

The modifications occur in only the fifth and last lines. In the last line, the affine group now has two input indexes, both P_1. The first output index is aP_1 where a is the translation and deformation needed to produce the main room. The second output index is bP_1 where b is only the *translation* needed to move a copy of the square out to the main room position, *without deformation*. In other words, b is the isometric component of the transformation a used for the rectangle. A star * is put in front of b to indicate this fact. Thus we introduce following notation:

Notation 13.3 *Consider two maximally aligned objects x and y. Then, a starred multi-index of the form:*

$$_{*ax,by}G^{x,y} \;=\; _{*ax}G^x \;\times\; _{by}G^y$$

will mean that operation a is the isometric part of operation b shown in the index immediately to its right. The star will ensure that ax and by are **maximally aligned up to deformation.** *(Notationally, a star will prefix a particular index only when that index is output; i.e., at the point of application of a.)*

To return to the group sequence (13.42). The index aP_1 in the bottom line will eventually become the square anomaly. Thus, the two output indexes bP_2, aP_1 become the two input indexes on Line 5, and the rest of Line 5 is altered accordingly. Notice another use of the star notation in that line. This will help ensure that the anomaly is square. In more general circumstances, the successive use of the star can delay deformation for as long as the designer wishes it to be delayed.

13.15   Complex Shape Generation

Unfolding groups formulate fundamental aspects of our theory of complex shape. For we argue:

> **Complex shape generation proceeds by a series of symmetry-breaking phase-transitions, that occur by** *selective misalignment.* **This is given by a wreath-product hierarchy in which the fiber groups, representing the symmetry ground-states, are the alignment states, and the successive control groups create the selective misalignments by transfer. We call this process,** *unfolding.*

While we have, in the preceding chapters developed the structure of unfolding groups, much more now needs to be said about about how they work in shape-generation. For this we will take the approach of applying them to particular areas such as mechanical and architectural design. In particular, the next chapter on mechanical design is crucial for all readers.

14. Mechanical Design and Manufacturing

14.1 Introduction

The techniques developed in this chapter are relevant to all areas of shape representation from human perception to quantum mechanics, and therefore this chapter should not be omitted by the reader.

For example, a structural factor that is important to all domains is *assembly* - the fitting together of objects. Assembly works by the imposition of constraints, e.g., that the side of one object must mate with the side of another object, or that their axes must align. Such constraints are basic to our world on every level. For example, on the visual level, a garage is attached to a house, and objects are placed on tables, or in chemistry the spin axes of two particles can be aligned. Thus assembly structure should be regarded as basic to perceptual psychology (where it has been completely ignored) as well as physics and chemistry, etc. Assembly is analyzed in the final sections of this chapter. Furthermore, several other sections have this general importance to shape representation.

14.2 Parametric, Feature-Based, Solid Modeling

Mechanical computer-aided design (MCAD) and mechanical computer-aided manufacturing (MCAM) are enormously sophisticated disciplines, capable, for example, of designing an airplane of several million assembled parts.

The major software systems for mechanical design (MCAD) are *parametric, solid-modeling, feature-based* systems. These terms are explained as follows:

(1) *Parametric.* This means that the physical shape of an object is determined by a number of constraints - e.g., a hole appears at the center of a block, or at a specific distance from one edge. If you modify the object, e.g., change the size of the block, the constraints will propagate with the changes, e.g., the hole will appear at the center of the modified block, or at the same specific distance from one of its edges.

(2) *Solid modeling.* This refers to the fact that what is constructed is not a surface model, but a model with material between the surfaces. This is useful for many aspects of MCAD, for example, because the eventual machining of the object will be achieved by the removal of material, e.g., by drilling or milling.

(3) *Feature-based.* This means that the object is designed by adding successive features[1]. Standardly, MCAD programs such as *Pro/Engineer* and *Mechanical Desktop* use three types of features:

> *(a) Sketched features.* These begin with sketching a profile curve on a 2D plane. The sketch is then resolved (i.e., cleaned up); and finally it is swept to create a 3D object.

> *(b) Placed features.* These are standard features such as holes, cuts, pockets, chamfers, rounds, slots, and shafts. They are pre-existing in the software, and therefore the user merely has to instance and place them at run-time. Therefore, the use of these features avoids sketching.

> *(c) Datum features.* Conventionally these are understood not as physical parts of the object, but are used as reference elements in the process of designing the object; e.g., as reference points, axes, and planes.

One often refers to (a) and (b) as *physical* features, and (c) as *non-physical* (reference) features.

14.3   A Generative Theory of Physical Features

The physical features of a part tend to be defined with respect to functions/actions in which they are involved rather than by the geometric structure that they represent. For example, a feature such as a hole is defined

---

[1] Cunningham & Dixon [25], Rossignac [133], Shah [141], Hoffmann & Juan [59], Chen & Hoffmann [20].

by the function of drilling in the manufacturing stage. The term "hole" is therefore used, rather than its geometric term, which would be "cylinder". Other examples of features are cuts, slots, pockets, chamfers, rounds, shafts, etc. Because of the use of functionally-dependent terms rather than geometric terms, there are often considerable problems converting features defined in one phase of the product life-cycle (e.g., design) into features defined in another phase of the product life-cycle (e.g., manufacturing). Problems include the fact that, in the design phase, one tends to use both addition and removal of material, whereas in the machining phase, one tends to use only removal of material. Therefore an object, created in the design stage by addition and removal of material, has to be converted into an object created only by removal, in the manufacturing stage. Because of this, there are considerable difficulties in defining features, as well as recognizing them and converting them.

According to this book, the reason why there are these difficulties is this: The distinction between geometry and function is incorrect. In our generative theory, function and geometry are the same. Shape is simply the frozen representation of action, and the construction of this representation, is achieved by rules for recoverability. These actions include the drawing actions of the designer, the machining actions of the manufacturing process, the manipulating actions of the user, etc.

We argue that the widely-accepted distinction between shape and function is due to the history of mathematics having taken an entirely incorrect approach in which figures are essentially "dead" sets, and geometric properties are memoryless with respect to action, i.e., invariant under groups of actions (see Chapter 22).

In accord with the system developed in the present book, the argument will be that the fundamental problems of mechanical CAD/CAM are solved when one takes the following two-part approach: (1) Representation: Shape must be represented generatively - which means in terms of action/function. (2) Recoverability: The generative representation of shape must be recoverable from data sets.

The method developed in this book, for solving the representational problem, is based on our fundamental claim that shape generation proceeds by group extensions. Applying this approach to the issue of *features*, we say this: Shape-generation proceeds by the successive addition of features. Each feature corresponds to a group. The successive addition of features corresponds to successive group extensions.

$$\textbf{Features} \quad \longrightarrow \quad \textbf{Mathematical Groups } G_i.$$
$$\textbf{Addition of Features} \quad \longrightarrow \quad \textbf{Group Extensions } G_1 ⓔ G_2 ⓔ \ldots ⓔ G_n.$$

Furthermore, we place the additional restriction of maximizing transfer, which means that, in any such extension, $G_1 ⓔ G_2$, the lower group G_1 has a decomposition into fibers, and the upper group G_2 transfers the fibers onto each other; i.e., the extension corresponds to a wreath product.

The issue of recoverability is solved by ensuring that the upper group is symmetry-breaking/asymmetry-building on the fibers of the lower group.

The enormous power of this approach comes when one understands the following:

THE ROLE OF GROUPS.   *The groups G_i in the generative group extensions, correspond simultaneously to:*

(1) **symmetries** *of the object;*

(2) **phases** *of the generative process;*

(3) **ground-states** *of the generative process.*

The correspondence between (1) and (2) is due to the History Symmetrization Principle. This essentially implies one of our basic principles that *symmetries are the channels of action* and therefore that *plans are inferred from the symmetry structure of a situation.* The correspondence between (2) and (3) is due to the fact that, by transfer, i.e., the wreath construction, a phase becomes a fiber on which the next control level acts. The phase thus becomes the symmetry-ground state for the next generative level.

Given this, our overall theory of features can now be stated as follows: We argue that features are generative phases:

$$\text{features} \quad \longleftrightarrow \quad \text{generative phases.}$$

Generative phases are given by the fiber structure of the wreath hierarchy describing the object:

$$\text{phase-structure} \quad \longleftrightarrow \quad \text{fiber-structure.}$$

Notice that, by our theory of grouping (Chapter 5), the groupings in an organization are given by the fibers (left-subsequences) of the wreath hieararchy:

$$\text{fiber-structure} \quad \longleftrightarrow \quad \text{grouping-structure.}$$

This idea is important because features must be perceptually identified by grouping. Therefore, inferentially, we argue that one goes from the symmetry structure to the features, and this proceeds in the following way:

THEORY OF FEATURE-EXTRACTION

$$
\begin{aligned}
\text{symmetry-structure} \quad &\longrightarrow \quad \text{generative-structure} \\
&\longrightarrow \quad \text{phase-structure} \\
&\longrightarrow \quad \text{fiber-structure} \\
&\longrightarrow \quad \text{grouping-structure} \\
&\longrightarrow \quad \text{feature-structure.} \qquad (14.1)
\end{aligned}
$$

We claim that this approach solves several major problems in mechanical CAD/CAM. It explains deeply many aspects of features, their structure, their relation to function, the ways that they can be added to each other, their relation to assembly, etc. In the contemporary literature, all these aspects remain separate, non-systematic and confused. In contrast, in our theory, these different aspects come together in a single unified structure that explains them and shows their powerful logical organization.

14.4   Datum Features

According to the theory to be elaborated, the contemporary distinction between physical and datum features is due to a lack of insight into the generative process. Our claim will be that datum features are really a subset of physical features:

THEORY OF DATUM FEATURES

Since datum features are reference objects for generative phases, they must correspond to phase-transitions. Therefore, by our theory of reference objects (Chapter 8), datum features must be the *symmetry ground-states* of successive phases. The generative process proceeds by breaking the symmetries of the datum features.

Let us contrast our view with the standard view of datum features, which can be represented by the following quotation:

"Datum planes and axes are features used to provide references for other features, like sketching planes, dimensioning references, view references, assembly references, and so on. Datum planes and axes are not physical (solid) parts of the model, but are used to aid in model creation (or, eventually, in an assembly). A datum is a plane (or axis) that extends off to infinity." Toogood [148] p5-2.

It will be useful to go successively through each sentence in the above quotation and to interpret it in terms of our generative theory of shape.

Sentence (1): "Datum planes and axes are features used to provide references."

Interpretation: As said above, in our theory, references are starting points of generative phases, and they are therefore the symmetry ground-states of the successive asymmetry building (symmetry-breaking) processes.

Second part of sentence (1): "references for other features, like sketching planes, dimensioning references, view references, assembly references, and so on."

Interpretation: This quotation shows the wide applicability of datum objects. Nevertheless, it will be seen, throughout this chapter, that our symmetry ground-state theory of datum objects works for each of these areas of applicability, from sketching planes at the beginning of part-design, to constraint-imposition in the final assembly phase.

Sentence (2): "Datum planes and axes are not physical (solid) parts of the model, but are used to aid in model creation (or, eventually, in an assembly)."

Interpretation: The view we take is very different. We argue that the datum features are actually part of the objects geometry. Because they correspond to symmetry states, they correspond to either the fiber group or the control group of the wreath hierarchy describing the geometry of the object. For example, a datum plane corresponding to the side of a block, represents a fiber of the wreath product.

Sentence (3): "A datum is a plane (or axis) that extends off to infinity."

Interpretation: According to our theory, the extension to infinity means that the datum corresponds to the *full* symmetry group of the fiber or control in the wreath product, i.e., before occupancy has been imposed. This supports our view that the full group underlies the organizational (e.g., perceptual structure) of the object. Occupancy is imposed by adding another level to that full symmetry structure but not removing it. This is critical for the perceptual representation of the shape by the user, as well as feature-decomposition in manufacturing. Generally, according to our theory: All objects, fill their entire group. That is, the entire group is relevant in perception, design, manufacturing, etc.

14.5   Parent-Child Structures as Wreath Products

All mechanical CAD, from part-design to assembly, is structured into parent-child hierarchies as the design proceeds. In any major program, such as *Pro/Engineer*, these hierarchies are available to the designer in clearly laid out tables, which the designer can access at any point in time. That is, the designer can click on a feature, and request its position in the parent-child hierarchy, and the program will show the parents and children of the feature both by highlighting these in the model, as well as displaying them as lists in a hierarchy table. This information is crucial for all designers. The reason is that, typically, a designer needs to modify features, and the modification of a feature will, in most cases, seriously affect the children of a feature. It will be seen throughout this chapter that, in accord with our algebraic theory of inheritance in Sect. 7.3, parent-child structures in MCAD are best modeled by *wreath products*.

14.6   Complex Shape Generation

This chapter will develop a theory of part design, assembly and manufacturing, in mechanical CAD/CAM. The argument exemplifies our general theory of *complex* shape-generation: A complex shape is best represented as generated from a hierarchy of symmetry-ground states by a succession of asymmetry-building (symmetry-breaking) phase-transitions that maximize transfer and recoverability via unfolding groups.

Each phase in the design and manufacturing of the object will follow this structure, from the creation of the base feature of a part, to feature attachment, to assembly, to machining.

We will claim that this theory captures the CAD/CAM process in a powerful way that maximizes the goals of the designer - which include the intelligent and insightful organization of the design, as well as the optimal organization of the manufacturing end-point.

14.7   Review of Part Design

Mechanical design proceeds by designing *parts* and then *assembling* them. This section will briefly review the main aspects of mechanical part design, and then, Sect. 14.9 will start to present our own theory of this process.

The design of a mechanical *part* begins with a phase which has major ramifications on all aspects of the design and assembly process. This phase is called *sketching* - which is a rough drawing of the mechanical part - a drawing that will later be *resolved* by the program into an accurate drawing.

The role of sketching is very profound and is not simply that of allowing initial inaccuracy. It is, in fact, a powerful means of allowing *parameters and constraints* to be defined - and these eventually become basic to the assembly-linking and kinematic aspects of mechanical design that distinguish it, for example, from architectural design. The reader should note that sketching and its extraordinary relation to parameters and constraints is basic to major mechanical software packages such as *Pro/Engineer* and *AutoCAD Mechanical Desktop*.

Let us begin by describing and illustrating the role of *sketching*. First, the work-flow sequence in the design of a mechanical part follows these eight stages:

WORK-FLOW IN MECHANICAL PART-DESIGN

(1) Choose a construction plane.
(2) Sketch.
(3) Align.
(4) Dimension: Define parameters.
(5) Resolve.
(6) Edit constraints and parameters.
(7) Resolve again.
(8) Sweep (e.g., extrude, revolve, etc).

This procedure will now be illustrated in order to show the critical role of sketching. For this initial illustration, the simplest possible example will be taken: the drawing of a rectangular block. It is necessary to fully understand this example before going on to complex part design.

(1) Choose a construction plane.
It was seen in Sect. 8.14 (p. 210), that the standard procedure for drawing a block has two phases: The designer traces out one face of the block on a construction plane, and then extrudes (sweeps) this face in the perpendicular direction to produce the block. This procedure is used here in the above work-flow sequence, except for the additional factor that the initial face of the block is merely "sketched".

The procedure is as follows: The first phase is the same as in any 3D solid modeling program: the definition of a construction plane, as was seen in Sect. 8.14. *Pro/Enginneer* visually presents the designer the three Cartesian planes, as shown in Fig. 14.1, and asks the designer to select one of these as the construction plane - called the sketch plane (the term "sketch plane" is also used in *AutoCAD Mechanical Desktop*). Each of the three planes is a datum plane - called *DTM1, DTM2, DTM3*. In our particular example, the assumption will be that the designer chooses $DTM3$ as the construction plane (sketch plane). The program then rotates this plane to become the view plane facing the designer; i.e., identified with the screen itself. The program thus becomes like a drafting program, except that in the next phase, *sketching*, a new set of issues appears.

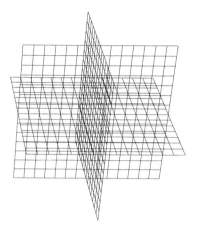

Fig. 14.1. The designer chooses one of the 3 Cartesian planes as a construction plane.

(2) Sketch.
As said above, the designer will sketch the first face of the block on the construction plane (sketch plane). Let us consider how the issue of inaccuracy first appears here. In *standard* 3D modeling, the drawing of the first face of the block is entirely accurate. This is because the designer has explicitly selected the block option on the program, and the program therefore knows that the edges are parallel or perpendicular to each other and to the grid lines of the construction plane. That is, these *constraints* are built into the choice of the block option.

However, in "sketching" in mechanical CAD, one will not explicitly choose the block option. This is because one will want to discover the constraints in the process of design - to drive the subsequent assembly and kinematic aspects.

Therefore, because one has not explicitly selected the block option on the program, one is going *necessarily* to be inaccurate in drawing. The nature of this inaccuracy is as follows: To draw the face of the block, one draws a polyline - consisting of a sequence of four straight lines. These lines are accurately straight, because straight lines are the easiest kind of lines to draw in CAD - one merely selects the endpoints by clicking the cursor at points

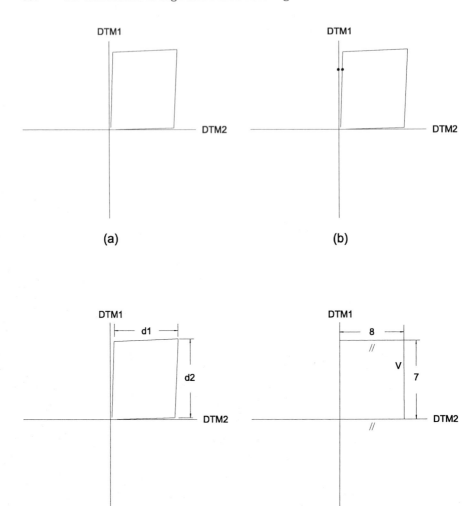

Fig. 14.2. The sketch-resolve sequence in mechanical CAD.

on the screen and the program provides a straight line between any pair of successive endpoints. However, what *is* inaccurate is the designer's placement of the points. Thus one obtains a somewhat skewed rectangle, with sides not really parallel or perpendicular to each other and the grid lines. The situation is therefore as shown in Fig. 14.2a.

Note that the vertical and horizontal *axes* represent datum planes $DTM1$ and $DTM2$ respectively; the sketch plane facing the viewer is $DTM3$.

(3) Align.
In this phase, one starts the procedure by which the constraints of the design will be determined. Specifically, one specifies alignment with the axes of the construction plane. Typically, in this example, one will specify that the left edge of the rectangle is aligned with the vertical datum plane, DTM1, and that the bottom edge of the rectangle is aligned with the horizontal datum plane, DTM2. One specifies alignment by clicking the pair of entities to be aligned; for example, in Fig. 14.2b, the two dots represent the two clicks that select the left line and the vertical axis to be aligned.

(4) Dimension: Define parameters.
Here one chooses certain sides on which one will create dimensions, as shown in Fig. 14.2c. These dimensions will be specified by *symbols* (not numbers!) at this stage, for example, $d1$ and $d2$, as shown (Fig. 14.2c). The dimensions are the *parameters* of the drawing.

(5) Resolve.
This phase is extraordinarily powerful: The program is asked to *interpret* the sketch. The result of this interpretation will be that the program will return a new version of the drawing, as illustrated in Fig. 14.2d, which is a "cleaned up" version of the original sketch. (The term "resolve" is taken from *AutoCAD Mechanical Desktop*; the program *Pro/Engineer* uses the term "regenerate" for this phase.)

Standardly, one considers resolution as a problem in *geometric constraint solving* (Bouma, Chen, Fudos, Hoffmann, & Vermeer [13]). The constraints are of three types:

(a) Geometric constraints.
(b) Alignments.
(c) Dimensioned parameters.

It was seen that type (b) is supplied by the designer in phase (3) above. Furthermore, it was seen that the designer also supplies type (c) - but in a symbolic form; for example, the symbols $d1$, $d2$, in Fig. 14.2c. These symbolic dimensions are now given actual numerical values by the program; for example, $d1 = 8$ and $d2 = 7$, as shown in Fig. 14.2d.

Type (a), the geometric constraints, are added by the program. To illustrate these, observe, in Fig. 14.2d, that the program has added a number of symbols not of the numerical-dimension type. In this example, the symbols are (1) the letter V on the right side, indicating that the computer has interpreted this side as a vertical line, even though it was not really vertical in the original sketch; (2) a parallel-bar symbol on both the top and bottom lines, indicating that the computer has interpreted the top and bottom lines as parallel.

Table 14.1. Geometric constraints and interpretation rules in mechanical CAD

Constraints	Interpretation Rules
HORIZONTAL	Interpret approximately horizontal lines as horizontal.
VERTICAL	Interpret approximately vertical lines as vertical.
PERPENDICULAR	Interpret almost perpendicular lines as perpendicular.
PARALLEL	Interpret approximately parallel lines as parallel.
TANGENT	Interpret approximately tangential line/arc pairs as tangential.
COLINEAR	Interpret approximately colinear lines as colinear.
CONCENTRIC	Interpret approximately concentric arcs as concentric.
CLOSURE	Interpret nearly closed figures as closed.
EQUAL X-VALUE	Interpret almost equal X-values as equal.
EQUAL Y-VALUE	Interpret almost equal X-values as equal.
EQUAL RADIUS	Interpret circles with approximately equal radius as having equal radius.
EQUAL LENGTH	Interpret lines with approximately equal lengths as having equal lengths.
REFLECTION	Interpret approximately reflectional parts as truly reflectional.

The addition of geometric constraints is a product of several rules that the program uses. These are shown in Table 14.1.

(6) Edit constraints and parameters.
Having now been presented with the resolved sketch, the designer can edit the two factors appearing in the resolved sketch: (1) geometric constraints, e.g., the designer can remove the parallelism constraint that appears in Fig. 14.2d; and (2) dimension values, e.g., the designer can change the numerical value of the length of a line.

(7) Resolve again.
The resolve instruction is given again, and the drawing regenerates to exhibit the consequences of the previous constraint editing.

(8) Sweep (e.g., extrude, revolve, etc).
In this final stage the designer sweeps the profile in the perpendicular direction from the construction plane. As in Chapter 18, the term sweep will be used to cover extrusion (translation) and revolving (rotating), as well as more general sweeping, i.e., along an arbitrary smooth path. However, in the initial examples, the sweepings considered here will be pure translations - for ease of exposition.

14.8   Complex Parts

The work-flow for part-design, as listed on page 306 is used to create not just rectangular blocks, but more complex parts. An illustration is given in Fig. 14.3. The top diagram shows the rough sketch made by the designer, and the middle diagram shows the resolution made by the program. The resolution used more interpretation rules and constraints than the resolution of the previous rectangular block. For example, consider the lines A and B in the sketch at the top. As can be seen from the resolution in the middle, these lines have been interpreted not only as vertical but as colinear. Also observe that the rounds (arcs) C and D in the sketch at the top, have been interpreted as having the same radius in the resolution in the middle, even though they were not the same in the sketch at the top. Again, notice that the two protrusions on the left have been interpreted as having the same shape.

Finally, the bottom diagram shows the 3D extrusion.

14.9   A Theory of Resolution

We now begin our theory of resolution, which will be developed over the next few sections. In developing this theory, we will provide an explanation of certain crucial aspects of current resolution programs, but we also argue for a substantial expansion of those programs, to make them much more helpful to both the designer and manufacturer.

To present an initial overview of our argument, recall from p. 309 that resolution involves the incorporation of three types of constraints:

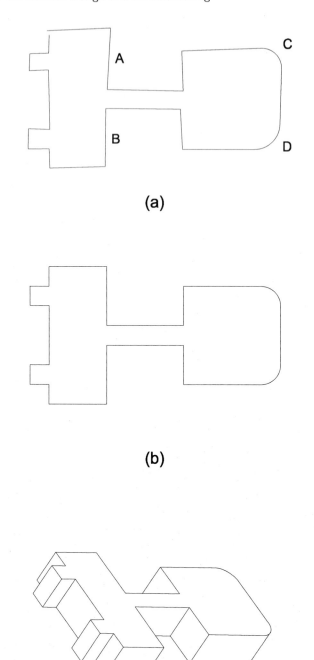

(a)

(b)

(c)

Fig. 14.3. Sketching, resolution, and extrusion for a more complex object.

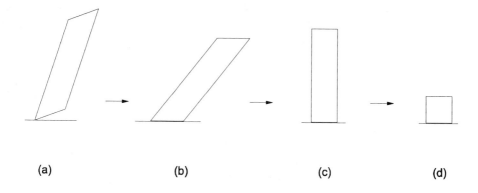

(a)   (b)   (c)   (d)

Fig. 14.4. Subjects were discovered to go through this reference sequence.

(a) Geometric constraints.
(b) Alignments.
(c) Dimensioned parameters.

Each of these constraints will be interpreted in terms of our generative theory of geometry, as follows:

(a) Geometric constraints.

Let us consider again the psychological experiment reported in Chapter 2: We found that subjects, presented with a rotated parallelogram, produced the successive-reference shown in Fig. 14.4.

The theory of what is happening in this example was also presented there, and is complex, involving issues such as the alignment of eigenvectors with symmetry axes, externalization to control-nested hierarchies of isometries, and so on. However, the two most basic factors are these: The subjects are using the Asymmetry Principle, which factorizes asymmetries to obtain symmetries, and also the Symmetry Principle, which preserves symmetries.

One can see that this is exactly what is going on in the resolution in Fig. 14.2. Indeed the transition from Fig. 14.2a, the sketch, to Fig. 14.2d, the resolution, has strong similarities with Fig. 14.4. Why?

The answer comes from examining the list of interpretation rules in Table 14.1. Most crucially, we propose this:

THE MEANING OF GEOMETRIC CONSTRAINTS AND THEIR INTERPRETATION RULES. *The geometric constraints are* *symmetries. The interpretation rules in Table 14.1 are uses of (1) the Asymmetry Principle, i.e., asymmetries in the current configuration are removed leaving symmetries; and (2) the Symmetry Principle, i.e., symmetries in the current configuration are preserved backwards in time. Both give the symmetry ground-states of the generative process.*

Notice also another important similarity between the successive reference example in Fig. 14.4, and the resolution in Fig. 14.2: The symmetries to which one returns are the symmetries of the reference frame. This accords with our theory that a reference frame embodies the symmetry ground-state in the generative process.

Comment 14.1 It is important to understand that, in resolution, the asymmetries that are removed are those that are within the tolerances set by the system or designer. For example, the interpretation of a slanting line as vertical occurs only if the angle of slant lies within a tolerance around the vertical direction. The slanting line is then captured by the vertical constraint and pulled into the vertical position. However, if the line lies outside the tolerance, then it is left alone by this constraint. Thus, it is important to recognize that the rotated parallelogram in Fig. 14.4 is resolved to a square, only if it falls within the associated tolerance. We will soon return to this issue.

(b) Alignments.

Alignments are *indistinguishabilities*, and are therefore symmetries; i.e., ground states. Thus, from now on, they will be classed structurally with geometric constraints. Notice their similar role: as illustrated in Fig. 14.2 alignments are often with respect to reference frames, which we claim are symmetry ground-states. (The main difference between the geometric constraints and the alignments is that the former are discovered by the resolution program and the latter are specified by the designer.)

(c) Dimensioned parameters.

This is the final class of resolution constraints. The first crucial point we wish to make about these constraints, is that, whereas the first two classes of constraint were *symmetry ground-states* of the generative process, this third class are *asymmetry end-states* of the generative process. However, this issue is enormously subtle and requires careful examination as follows.

One needs to distinguish between two types of asymmetries that occur within the sketch, as follows: Comment 14.1 said that resolution to the symmetry ground-states, i.e., to geometric constraints, occurs only if the asymmetries fall within the tolerances of those ground-states. The consequence is that the rotated parallelogram in Fig. 14.4 could be treated in either of two different ways by the resolution program: (1) if it fell within the tolerance, then it would resolved to the symmetry ground-state; or (2) if it did not fall within the tolerance, then it would not be resolved to the symmetry ground-state.

TWO CLASSES OF ASYMMETRIES. *The asymmetries in the configuration fall into two classes: (1)* **Inessential asymmetries** *- those falling within the tolerance of the geometric constraints, and therefore removed from the sketch to obtain the symmetry ground-states expressed by the geometric constraints. (2)* **Essential asymmetries** *- those falling outside the tolerance, and therefore not removed from the sketch.*

However, we argue that, even though the second type are actually not removed from the sketch, they are removed in the *inference process* that produces an intelligent resolved sketch. The reason is that the second class of asymmetries must also be given a generative description, from symmetry ground-states, in order that the resolved sketch is one that provides a useful feature-based structure for the designer and manufacturer. It will take a number of sections to fully grasp this, but we begin as follows:

Conceptually, one should understand the two classes of asymmetries as two levels, one above the other. The inessential asymmetries are removed first, and then the essential asymmetries. The reason is this: The inessential asymmetries are like noise, and their removal can be conceived of as the removal of noise. This removal leaves the resolved sketch. One then has to remove the essential asymmetries in order to give a generative description of the resolved sketch.

As an illustration, consider Fig. 14.5, which is the rotated parallelogram sequence except that it has been preceded by an additional figure, which is the actual sketch drawn by the designer. Notice for example, that the opposite sides are not parallel in this first figure. Now, let us suppose that this non-parallelism falls within the tolerance of the parallelism geometric constraint, i.e., the non-parallelism is a *inessential* asymmetry. It is then removed in going from the first figure to the second, leaving a figure with truly parallel sides. Now suppose that the asymmetries in this second figure do not fall within any of the tolerances of the geometric constraints; i.e., the asymmetries in this figure are *essential* asymmetries. Then the second figure corresponds to the resolved sketch, and the remaining sequence back to the square gives the generative explanation of the resolved sketch.

One can now understand the role of the first phase, removing the inessential asymmetries (i.e., going from the first figure to the second). It was seen that the removal of the inessential asymmetries (non-parallelism) in the first

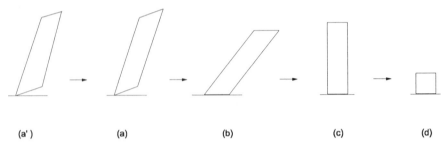

(a')   (a)   (b)   (c)   (d)

Fig. 14.5. Removal of inessential asymmetries followed by essential asymmetries.

figure, produced the symmetries (parallelism) in the second figure. Now, according to the Symmetry Principle, these symmetries must be preserved backwards in time through the generative process; that is, the parallelism must be preserved back to the square. This means that what the removal of the inessential asymmetries did was to provide the symmetries that would be preserved in the generative explanation of the essential asymmetries. This generative explanation is the remaining sequence back to the square.

THEORY OF GEOMETRIC CONSTRAINTS. *Geometric constraints establish the symmetries that are to be input to the Symmetry Principle, in giving a generative description of the resolved sketch.*

Let us now understand the full structure of symmetries and asymmetries in the resolution process. This is shown in Fig. 14.6. On the left, there is the designer's sketch. It consists of three levels: inessential symmetries (sketching noise), essential asymmetries, and some symmetries. In accord with the Asymmetry Principle, both levels of asymmetries will have to be removed leaving symmetries. The two removals are shown by the two successive arrows in the diagram. First, the inessential asymmetries are removed in producing the middle stage, which is the resolved sketch. Then, the essential asymmetries are removed in producing the final stage on the left. Notice therefore that, on the right, all three levels are symmetries.

The diagram shows the uses of the Asymmetry Principle. However, at each stage, one uses also the Symmetry Principle. This acts on the symmetries in a stage, preserving them for the next stage. Thus the Symmetry Principle goes between any two successive blocks that are labeled as symmetries.

Thus each of the phases backwards through the generative sequence uses *both* the two main rules of our generative theory, the Asymmetry and Symmetry Principles. There is no difference between the inference structure that produces the second stage from the first, and the inference structure that produces the third stage from the second.

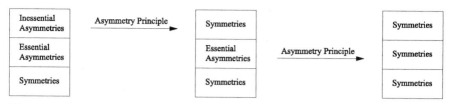

Fig. 14.6. The structure of asymmetries and symmetries in resolution.

Now turn to the dimensioned parameters; i.e., the third class of resolution constraints. These are necessarily asymmetries. To see this consider first, the specification of the dimension parameter of a line, i.e., the specification of its length. It is clear that this gives what we call the occupancy structure of the line, and in particular, the points where the occupancy changes from off to on (at one end of the line) and from on to off (at the other end of the line). These are the asymmetries of the line; i.e., the end-points in the process for generating the line. The same concept applies to angular dimensions: The specification of an angle gives the end-point of a rotation process.

The crucial point is that the dimensioned parameters are a subclass of the *essential* asymmetries. For example, in the first figure in Fig. 14.5, one would not dimension the angle of convergence of the non-parallel opposite sides because one would want this non-parallelism to disappear in the sketch resolution; i.e., in going to the next figure. However, one might wish to dimension the angle of the figure with respect to the horizontal line. This is because the angle lies outside the specified tolerances of the geometric constraints, and will therefore be an essential part of the figure. Notice that this angle is given a generative explanation in the remainder of the sequence.

The dimensioned parameters are in fact a chosen subclass of the essential asymmetries, such that the remainder of the essential asymmetries can be produced from them; i.e., by a geometric constraint solver. We shall call them the **core essential asymmetries.**

In conclusion, the three types of constraints used in resolution have the following roles in our generative theory of shape:

SYMMETRY/ASYMMETRY INTERPRETATION OF CONSTRAINTS.
Geometric and alignment constraints = *symmetries that will be preserved in the generative description of the resolved sketch.* **Dimensioned constraints** = *core essential asymmetries.*

With the description of the constraints in terms of symmetries and asymmetries, it is now possible to turn to the generative description that we will argue is at basis of the resolved sketch.

14.10   A Theory of Sketching

We now give a generative theory of the resolved sketch. The argument is that only a generative description (one that maximizes transfer and recoverability) will contain the information needed for intelligent and insightful design. This generative description must be based on a theory of the designer's sketching process. Thus it is first necessary for us to give a theory of sketching. This will be elaborated over several sections.

The first thing to understand about sketching is that the designer has an envisioned goal object in mind. The goal object is different from the physical sketch that is produced, but the designer creates the sketch in such a way that he/she hopes that the resolution process will result in the desired goal object. Thus the imagined goal object guides the sketch process. It dominates the designer's mind during the entire sketching process.

One must therefore distinguish between the goal object, which is a *mental* construct, and the physical sketch that results from the designer's drawing movements. Based on this distinction, the overall components of the sketch process can be stated thus:

(1) The designer's mental analysis of the mental goal object.
(2) The designer's physical sketch which is guided by the mental analysis.
(3) The computer's resolution of the physical sketch, resulting in a physical realization of the designer's mental goal object.

Thus, since the designer's mental analysis drives the entire process, it is crucial for us to give a mathematical theory of the designer's mental analysis, as follows:

14.11   A Mathematical Theory of the Designer's Mental Analysis

As said earlier, one cannot understand the design process, without first giving a rigorous theory of the designer's mental analysis of the mental goal object. It is the way human beings analyze goal objects that unconsciously drives the way CAD software is structured. Furthermore, if one is to improve this software, one must rigorously understand the designer's mental analysis. Let us proceed:

The mental goal object is a 3D organization which the designer structures in a certain way. In fact, we argue that, since this mental object will itself embody the designer's insight and intelligence, it will maximize *transfer* and *recoverability*. Therefore it will embody our mathematical theory for realizing those aims.

For an illustration, suppose that the 3D mental object which the designer imagines is the object shown in Fig. 14.7a. Chapters 11, 12, and 13 argued that the way to represent a complex object - such that the representation maximizes transfer and recoverability - is to generate it using unfolding groups. Let us show how this is done using this particular example.

First let us consider the issue of the *solidity* of the object: Mechanical CAD is conceptually driven by solid modeling. One tends, for example, to understand a model such as Fig. 14.7a to be solid. Chapter 16 will give our theory of solid structure. That chapter will show that it is possible to correspond the solid primitives with the surface primitives (iso-regular groups) of Chapter 10; i.e., since the latter can be considered to be the bounding surfaces of the solid primitives. Thus for ease of exposition, the present chapter will refer to the surface primitives (iso-regular groups) of Chapter 10, when in fact, we will mean the corresponding solid primitives of Chapter 16.

Let us begin: The basis of an unfolding group is an alignment kernel which contains the Cartesian frame as a primitive. According to our theory, the Cartesian frame contains the hyperoctahedral wreath group,

$$\mathbb{Z}_2 \textcircled{w} \Sigma_3 \;=\; [\mathbb{Z}_2 \times \mathbb{Z}_2 \times \mathbb{Z}_2] \textcircled{s} \Sigma_3.$$

and in particular, the triple reflection group which is the fiber-group product $\mathbb{Z}_2 \times \mathbb{Z}_2 \times \mathbb{Z}_2$ of that wreath product. This structure will act as a control group on the triple of planes which represent the Cartesian system - with the crucial understanding that the planes are *reflection planes*. (This will be discussed in detail in Chapter 16).

Now let us assign this triple-reflection structure to the 3D mental model shown in Fig. 14.7a. The most obvious way to do this is as follows:

First reflection plane: Considering the plan view, shown in Fig. 14.7b, it is obvious that the figure is reflectionally symmetric about the dashed line shown. Therefore, the first reflection plane should be assigned along this axis. Let us call it Datum Plane 1, that is, $DTM1$. To emphasize: datum planes, in our system are reflection structures - in fact, fibers from the hyperoctahedral wreath group.

Second reflection plane: Looking at the 3D object itself, Fig. 14.7c, there are three obvious candidates for the second reflection plane: (1) The bottom plane of the object, as shown. This is a reflection plane of the bottom face of the object as an independent fiber in its own right. (2) The top plane of the object, which is a reflection plane of the top face as an independent fiber. (3) The middle plane shown, i.e., half-way between the bottom and top planes. This is a reflection plane of the entire object. Any one of these three is suitable, and will be called Datum Plane 2, that is, $DTM2$. These are exactly the three choices offered by mechanical CAD programs such as *Pro/Engineer*

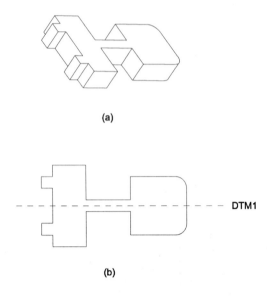

(a)

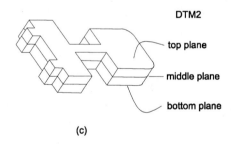

DTM1

(b)

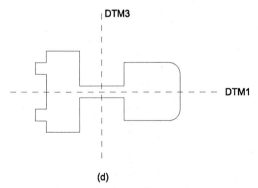

DTM2

top plane

middle plane

bottom plane

(c)

DTM3

DTM1

(d)

Fig. 14.7. Assignment of the triple-reflection structure to an particular mechanical part.

and we explain them here in terms of the symmetry groups involved and, in particular, in terms of the wreath structure of those symmetry groups.

Third reflection plane: This is determined to be perpendicular to the other two planes. Because it must be a reflectional plane, it must lie along one of the faces (fibers) of the object, or one of the reflectional symmetries of the main unfolded primitives. Here, this plane has been chosen to lie along the central reflectional symmetry of the central rectangular component, as shown in Fig. 14.7d, where it is labeled $DTM3$. One could have equally chosen the corresponding reflection plane of either the left main block, or the right main block. The issue is a matter of which block one wishes to unfold the structure from, in the sense of *unfolding groups*. This is because our theory says that the Cartesian frame is the symmetry ground-state of the unfolding process. We will return to the issue of choice later, when discussing the meaning of *intent managers* in mechanical CAD programs.

Now according to our theory, this triple-reflection structure can be regarded as equivalent to the cube primitive. That is, both can be given as the hyper-octahedral wreath hyperplane group of dimension 3,

$$\mathbb{R}^2 \ \textcircled{w} \ \mathbb{Z}_2 \ \textcircled{w} \ \Sigma_3.$$

Note the meaning of using two wreath products in this expression: Each reflection group carries with it *two* translation planes (Sect. 16.5).

In this way, we identify the frame with the cube primitive within the alignment kernel. Most crucially, whether it is understood as the Cartesian frame, or as a cube, it is unfolded from the world coordinate system, which is the first position of this hyperoctahedral wreath structure.

Now let us unfold the entire object shown in Fig. 14.7a. The unfolding will first be described intuitively, and then rigorously: The cube primitive is unfolded by an affine action (translation and stretch) to become the central rectangle shown by the number 1, in Fig. 14.8. From this position, it is then pulled leftward, under an affine action, to become the rectangle indicated by 2 in Fig. 14.8. From this new position, two copies of it are then pulled in vertically opposite directions to become rectangles 3 and 4 in Fig. 14.8. This produces the component grouping 2,3,4 on the left of the figure.

Now for the right side. While there is a leftward movement of the cube primitive from 1 to 2, there is also a simultaneous rightward movement to produce Component 5. The simultaneity really means that the relation between left and right is that of a *direct* product.

Also, moving rightward from 1, there is the *cylinder* primitive. It moves to position 5 (aligned with the moving cube primitive) but then is moved to positions 6 and 7, where it becomes the rounds shown (curved edges). This completes the entire unfolding.

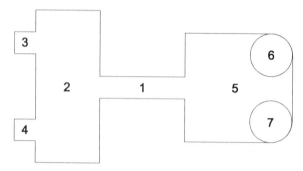

Fig. 14.8. Numbering the unfolded primitives.

Now let us look at the structure rigorously.

(1) The alignment kernel. This contains the cube primitive P_1 and the cylinder primitive P_2. It is therefore written as follows:

$$[[G_{cylinder}]_{P_2} \times [G_{cube}]_{P_1}]\mathcal{U}.$$

Algebraically, the alignment kernel is thus the direct product of a number of clones of each primitive. From a group-representation point of view, all of the clones should be understood as maximally aligned with each other and with the World-frame. The number of clones will be determined by the wreath-direct actions with the control groups. Until the number of clones of each primitive is determined, we shall often speak of *the* cube, and *the* cylinder.

(2) The hierarchy of rectangular components. Let us first consider only the rectangular components in the structure; i.e., those derived from the cube. They have a particular referential order which is given by the following group:

$$[[G_{cylinder}]_{P_2} \times [G_{cube}]_{P_1}]\mathcal{U}$$
$$\textcircled{w} \;_{faP_1} AGL(3, \mathbb{R})^{aP_1}]$$
$$\times [[_{daP_1} AGL(3, \mathbb{R})^{baP_1} \times _{cbaP_1} AGL(3, \mathbb{R})^{baP_1}$$
$$\times _{baP_1} \{e\}^{baP_1}] \textcircled{w} \;_{baP_1} AGL(3, \mathbb{R})^{aP_1}]$$
$$\times _{aP_1} \{e\}^{aP_1}]$$
$$\textcircled{w} \;_{aP_1} AGL(3, \mathbb{R})^{P_1} \times _{P_1} \{e\}^{P_1} \quad . \tag{14.2}$$

Let us go through the above expression from bottom-to-top. In the bottom line, the right-hand group fixes a copy of P_1, the World frame. The purpose of this is to hold a copy of the *internal symmetries* of the construction plane, as is required by the theory of construction planes, summarized on p. 212. The left-hand group also takes the cube primitive P_1 as input, but outputs an affine-altered version of it, aP_1, corresponding to the central rectangular component - numbered 1 in Fig. 14.8.

The output index aP_1 then becomes the input index on each of the three lines above; i.e., at the right-hand end of each of these lines. Each of these lines determines a different fate for aP_1. Thus working upwards through these lines: Line 4 fixes a copy of Component 1 - that copy seen in Fig. 14.8. Next, considering Line 3, the right-most group takes aP_1 and applies an affine transformation b to it, producing the output index baP_1, which represents Component 2 in the plan-view in Fig. 14.8. Then, moving left-ward on this line, there is a wreath-product symbol Ⓦ, the fiber of which is a direct product, each component of which takes baP_1 as input. The first component is the identity group $\{e\}$ which fixes a copy of baP_1, which is Component 2 as seen in Fig. 14.8. The other two groups in the direct product have output $cbaP_1$ and $dbaP_1$ respectively, which correspond to Components 3 and 4 in Fig. 14.8. The direct product shows that these two objects are misalignments with respect to the copy held by the identity component; i.e., there are *symmetry-breakings by misalignment*.

Finally, Line 2 also takes aP_1 as input, but outputs faP_1 which represents Component 5 in the plan-view Fig. 14.8.

Thus, the reader can see from the entire expression (14.2) that there must be six clones of the cube primitive in the alignment kernel.

(3) The addition of rounds 6 and 7. Now add the rounds shown as 6 and 7 in the plan-view, Fig. 14.8. These are understood as coming off the rectangular Component 5. However, Component 5 is understood as coming off Component 1, which is itself derived from the alignment kernel. Thus, the rounds are passed up from the alignment kernel, where they start as clones of the cylinder primitive P_2. Then they are moved to be aligned with Component 1; then moved again to be aligned with Component 5; and finally misaligned to become rounds 6 and 7. The appropriate group description is therefore the following:

$$[[G_{cylinder}]_{P_2} \times [G_{cube}]_{P_1}]u$$
$$Ⓦ \ [[[_{w'vuP_2}AGL(3,\mathbb{R})^{vuP_2} \times {}_{wvuP_2}AGL(3,\mathbb{R})^{vuP_2} \times {}_{faP_1}\{e\}^{faP_1}]$$
$$Ⓦ \ {}_{*vuP_2}, {}_{faP_1}AGL(3,\mathbb{R})^{\ uP_2},{}^{aP_1}]$$
$$\times [[_{daP_1}AGL(3,\mathbb{R})^{baP_1} \times {}_{cbaP_1}AGL(3,\mathbb{R})^{baP_1}$$
$$\times {}_{baP_1}\{e\}^{baP_1}] Ⓦ {}_{baP_1}AGL(3,\mathbb{R})^{aP_1}]$$
$$\times {}_{aP_1}\{e\}^{aP_1}]$$
$$Ⓦ \ {}_{*uP_2}, {}_{aP_1}AGL(3,\mathbb{R})^{\ P_2},{}^{P_1} \times {}_{P_1}\{e\}^{P_1}. \tag{14.3}$$

This is the same as the group given at (14.2), except that the new Line 2 has been inserted above the previous Line 2. This new line is *wreath sub-appended* onto the previous Line 2, and represents the final positioning of the two rounds. The only other changes are the extra cylinder indexes (i.e., based on P_2) that are inserted in two of the remaining lines.

Thus, starting with the bottom line, the affine group now has two input indexes representing the cube and cylinder primitives P_1 and P_2 from the alignment kernel. By Notation 13.3 (p. 298), the affine group here actually means the direct product of affine groups:

$$_{*uP_2\ ,\ aP_1} AGL(3, \mathbb{R})\ ^{P_2\ ,\ P_1} = _{*uP_2} AGL(3, \mathbb{R})^{P_2} \times _{aP_1} AGL(3, \mathbb{R})^{P_1}.$$

The output index uP_2 is prefixed by an star because the u is chosen such that the cylinder remains maximally aligned with the cube up to deformation; that is, u is the non-deformation part of a.

The two lines above the bottom line are the same as before, because they concern only the grouping of rectangular blocks on the left in Fig. 14.8.

Then above this, Line 3 is the same as the previous Line 2 except that it now has cylinder indexes. Notice that the input to this line is the output of the affine component of the bottom line. Also, the double index on the group of this line indicates that it is a direct product, thus:

$$_{*vuP_2\ ,\ faP_1} AGL(3, \mathbb{R})\ ^{uP_2\ ,\ aP_1} = _{*vuP_2} AGL(3, \mathbb{R})^{uP_2} \times _{faP_1} AGL(3, \mathbb{R})^{aP_1}.$$

The effect of this line is to move the cylinder out together with the rectangular component, which becomes Component 5 in Fig. 14.8. As indicated by the star, the cylinder is not deformed but remains maximally aligned with the rectangular component, up to deformation.

Finally, go to the wreath sub-appended Line 2. Working from right to left in Line 2, there is first an identity component which fixes a copy of faP_2, Component 5. The next two affine groups on this line move the two cylinders out from alignment with respect to Component 5, thus becoming the two rounds.

Notice that Line 2 allows us to infer the distribution of clones in Line 3. It tells us that the cylinder index in Line 3 actually corresponds to two clones; and the cube index in Line 3 corresponds to only one clone.

Using this, we can work down the lines in the group and conclude that there are a total of eight clones: six cubes and two cylinders. These all start maximally aligned in the alignment kernel, and are successively pulled out of alignment. Notice, for example, that since, in the bottom line, the right-hand group fixes the World-frame, the left-hand group must take, as input, the five remaining cube clones and the two cylinder clones. It applies the affine transformation a to the five cube clones, and the affine transformation u to the two cylinder clones. The bottom line has therefore broken two alignments, that of the World-frame in relation to all the other clones, and the deformational alignment of the cubes in relation to the cylinders. However, in the output from the bottom line, there is still considerable alignment remaining; i.e., the cubes with respect to each other, and the cylinders with respect to each other. Then, in the successive lines above, all of these alignments will eventually be broken.

If one wishes to compress the notation of the group maximally, then one can collapse all direct products into multi-indexes (except the direct product of wreath products). The result is:

$$
\begin{aligned}
&[[G_{cylinder}]_{P_2} \times [G_{cube}]_{P_1}]_{\mathcal{U}} \\
&\textcircled{w} \; [_{w'vuP_2} \, , \, _{wvuP_2} \, , \, _{faP_1} AGL(3,\mathbb{R})^{vuP_2} \, , \, ^{vuP_2} \, , \, ^{faP_1} \\
&\textcircled{w} \; _{*vuP_2} \, , \, _{faP_1} AGL(3,\mathbb{R})^{\;uP_2} \, , \, ^{aP_1}] \\
&\times [_{dbaP_1} \, , \, _{cbaP_1} \, , \, _{baP_1} AGL(3,\mathbb{R})^{baP_1} \, , \, ^{baP_1} \, , \, ^{baP_1} \\
&\textcircled{w} \; _{baP_1} \, , \, _{aP_1} AGL(3,\mathbb{R})^{aP_1} \, , \, ^{aP_1}] \\
&\textcircled{w} \; _{*uP_2} \, , \, _{aP_1} \, , \, _{P_1} AGL(3,\mathbb{R})^{\;P_2} \, , \, ^{P_1} \, , \, ^{P_1} \; .
\end{aligned}
\tag{14.4}
$$

It is worth examining this carefully because it emphasizes certain features more strongly. Notice for example, the triple of identical indexes in the middle of the third line, indicating the alignment at this stage. Then notice the triple of output indexes coming from this line (far left), showing how the misalignment has been formed with respect to one of the indexes in the triple.

Generally then, the group developed in this section illustrates the main aspect of our theory of complex shape: Complex shape generation proceeds by a series of symmetry-breaking phase-transitions, that occur by *selective misalignment.* This is given by a wreath-product hierarchy in which the fiber groups, representing the symmetry ground-states, are the alignment states, and the successive control groups create the selective misalignments by transfer. We call this process, *unfolding.*

14.12 Constraints and Unfolding

Before continuing with our theory of sketching, let us consider the issue of constraints. This issue arises later - in fact, it occurs several times later. However, according to our theory, it is fundamentally related to the unfolding structure of the object. Since this unfolding structure has just been discussed, it is best to look at the issue of constraints at this point.

Section 14.9 gave a lengthy discussion of constraints and argued that the geometric constraints correspond to the symmetries of the resolved sketch, and the dimension constraints correspond to the (core) asymmetries. It can now be seen that these arise from the unfolding groups. For example, consider again the part we have been discussing in Fig. 14.7.

Symmetries. Observe in Fig. 14.7, that the symmetries consist, for instance, of reflections between the sides of each block. These clearly come from the $\mathbb{Z}_2$ components within the alignment kernel. They would correspond to the geometric constraints. Notice this important fact: Whereas, in standard constraint-solving programs, such symmetries have a relatively "flat" structure, in our theory, they are placed hierarchically within an unfolding group, and thus have substantial relations to each other. For example, the reflectional symmetries within a block are related hiearchically to each other, and this hierarchy is itself related to the reflectional hierarchy in another block through the unfolding structure. The relations building the hierarchies are the other type of constraints (asymmetries - as described below).

However, we are not arguing against constraint-solvers. In fact, we believe that they are deeply important, and have contributed much to mechanical CAD. We are asking only that they be expanded by the generative approach elaborated in this book - i.e., based on maximization of transfer and recoverability, and the associated wreath/symmetry group formulation.

Asymmetries. The asymmetries of the part are the factors that *relate the symmetries* to each other. For example, a block is itself a hierarchy of symmetries and the relation of one block to another block within the part is given by the unfolding group; i.e., the unfolding group gives the relation between hierarchies of symmetries. This relation is *misalignment* which is the type of asymmetry embodied in unfolding. Now the distances between blocks are given by dimensions, e.g., between their centers. Since distances between blocks are produced by the unfolding control groups, dimensions relating blocks give information about the unfolding control groups.

In fact, dimensions must account for all asymmetries, not just unfolding asymmetries. They must account for the lengths of sides, which are occupancy asymmetries in the translational symmetries of individual sides.

14.13   Theory of the Sketch Plane

The previous section discussed constraints again because of the close relation between constraints and unfolding. However, it is now necessary to return to the issue of sketching. This will collapse the unfolding structure to a 2D plane, and it is necessary to understand how this is done.

Recall that we have divided the sketch process into three stages:

(1) The designer's mental analysis of the mental goal object.
(2) The designer's physical sketch which is guided by the mental analysis.
(3) The computer's resolution of the physical sketch, resulting in a physical realization of the designer's mental goal object.

The first stage was dealt with above, where we said that the designer's mental analysis of the mental goal object, is to generate it using unfolding groups. Let us now go onto the second stage.

The designer's physical sketch of the mental goal object uses a 2D construction plane - usually called the sketch plane in mechanical CAD. It is now necessary to develop a theory of how the designer uses the construction plane. For this we will incorporate the theory of construction planes developed in Sect. 8.14. Again, for ease of exposition, assume that the sweeping structure is given by a pure translation; however, the theory can easily be extended to revolved structures and general sweepings. The point is that all cases are, according to our theory, *control-nested hierarchies of symmetries*.

The main points of our theory of construction planes were as follows: A construction plane is a reflectional $\mathbb{Z}_2$ fiber taken from the hyperoctahedral wreath group defining the Cartesian frame. Now, because the plane is a *reflection* plane, any drawing made on the plane is necessarily reflectionally symmetric about the plane. In mechanical CAD, the designer will make a drawing on the plane, and then sweep it in the perpendicular direction. Most crucially, this creates a structure of nested control (a wreath product) in which the drawing is the fiber, and the sweeping is the control. The importance of identifying this structure of nested control, is that the reflectional symmetry of the 2D drawing is not lost in the final 3D model: This is because the 2D drawing is a *fiber* of the 3D model. That is, according to our theory, the power of a structure of nested control is that the asymmetrizing process is added onto the symmetry of the fiber via a wreath product, and therefore both the symmetry group of the fiber, and the symmetry group of the asymmetrizing process are present in the final structure.

With this in mind, we now propose the main aspects that determine the choice of construction plane:

CHOICE OF CONSTRUCTION PLANE.   *The designer must select* **two symmetry groups** *in the goal object:*

(1) A triple reflection structure $\mathbb{Z}_2 \times \mathbb{Z}_2 \times \mathbb{Z}_2$ within the object.
(2) A translation symmetry $\mathbb{R}$ of the object.

The construction plane *will be the particular $\mathbb{Z}_2$ component in (1) that* **maximizes**, *perpendicular to itself, the translational symmetry $\mathbb{R}$ in (2).*

The maximization clause just given leads to the following:

Definition 14.1. *We will speak of the $(\mathbb{Z}_2, \mathbb{R})$ choice for construction plane.*

The maximization clause is crucial, and to illustrate it, return to the example shown in Fig. 14.7a (p. 320). Recall that the three lower parts of this figure show the choice of the three hyperoctahedral $\mathbb{Z}_2$ fibers - which, according to our theory, provide the three datum planes $DTM1$, $DTM2$, and $DTM3$. The logic behind this selection was given on page 319.

One of these planes will be the candidate for the construction plane. Which one? In order to decide this, look at the translation structure in the perpendicular direction to the plane. First, consider $DTM1$, shown in Fig. 14.7b. The translation direction would be vertical in the diagram. Notice that the translational symmetry in this direction is not optimal for a number of reasons: For example, the line on the left side zigzags up the figure, i.e., destroys translational symmetry vertically. Furthermore, the fillets on the right also destroy vertical translational symmetry.

In contrast, consider $DTM2$, shown in Fig. 14.7c. This is either the bottom, middle or top plane, because each of these is a reflectional symmetry of the object or one of its fibers, as noted on page 319. The crucial thing to observe is that either one of these reflection planes *maximizes translational symmetry* in its perpendicular direction. In particular, there are only straight parallel lines in that perpendicular direction.

Now, consider the final case $DTM3$, shown in Fig. 14.7d. Here the translational movement perpendicular to $DTM3$ would have to be horizontal. Clearly, translational symmetry does not exist in the horizontal direction.

Thus, in conclusion, one sees that the $\mathbb{Z}_2$ fiber, $DTM2$, maximizes translational symmetry in its perpendicular direction. Therefore it is this fiber, from the hyperoctahedral group, that should be chosen as the construction plane.

Comment 14.2 It is important to understand here that we have not even begun to consider the actual sweeping of the construction plane done by the *computer program*. What has been examined here is the far more critical thing of the *mental analysis* that the designer must make of the model so as to choose the construction plane on which to do the drawing. That is, the designer takes the mental model and identifies in it the *two symmetry groups* defined above - in order to discover an optimal plane on which to do the sketch.

Comment 14.3 For ease of exposition, the preceding sections have referred to the sweeping control group as a translational symmetry. In fact, the more general case is given in Chapter 18 in our theory of sweep representations. The sweeping group is a continuous isometry group (translation or rotation) that can have external action that deforms it. The theory presented above still holds for this more general structure: The designer must identify the sweeping group as a *symmetry* group in the 3D model.

14.14   Solidity

The solidity of the model can now be considered. In Chapter 16, the solid structure is given by the wreath c-polycyclic group $\mathbb{R}Ⓦ\mathbb{R}Ⓦ\mathbb{R}$, which is the successive hierarchical sweeping of the three translational dimensions of ordinary space. We argue that this is related to the construction plane as follows:

THE FUNDAMENTAL ALIGNMENT.   *After making the $(\mathbb{Z}_2, \mathbb{R})$ choice (Definition 14.1) to determine the construction plane, one then makes the following correspondence between the construction choice $(\mathbb{Z}_2, \mathbb{R})$ and the solid structure $\mathbb{R} Ⓦ \mathbb{R} Ⓦ \mathbb{R}$:*

$\mathbb{Z}_2 \longleftrightarrow$ *first two levels* $\mathbb{R} Ⓦ \mathbb{R}$ *in* $\mathbb{R} Ⓦ \mathbb{R} Ⓦ \mathbb{R}$

$\mathbb{R} \longleftrightarrow$ *final level* $\mathbb{R}$ *in* $\mathbb{R} Ⓦ \mathbb{R} Ⓦ \mathbb{R}$.

This correspondence will be called the **fundamental alignment** *of a solid model.*

The fundamental alignment is a deep and very powerful concept that brings together much of our theory of solid modeling. It is not simply an issue of bringing together the construction plane and the solid structure, but it involves this: Recall that the construction plane actually comes from the hyperoctahedral wreath hyperplane group. This group is essentially the surface that *distinguishes* the cube from the other primitives. In Chapter 16, a distinguishing surface is called a *surface kernel*. A surface kernel breaks the symmetry of the infinite solid structure, the wreath c-polycyclic group $\mathbb{R} Ⓦ \mathbb{R} Ⓦ \mathbb{R}$. What the fundamental alignment does is tell us how the two components - the surface kernel and solid structure - are aligned when this symmetry-breaking occurs. The components of this symmetry-breaking are not arbitrarily aligned. They are aligned in the manner described by the fundamental alignment. This phenomenon is profound in its own right - and affects all of design. However, still more profound is the underlying basis of the phenomenon: What justifies the fundamental alignment is the History Minimization Principle, which implies that, although we have symmetry-breaking, it must be the minimal that can occur. Hence, even the two groups that break each other's symmetry, must preserve as much of each other's symmetry as possible.

14.15   A Comment on Resolution

In the above sections, we have discussed the mental analysis that the designer makes in the production of the 3D base part. With the completion of his/her mental analysis, the designer then draws a 2D sketch on the sketch-plane chosen by this analysis. This collapses the 3D unfolding structure onto a 2D unfolding structure - with no real loss of information because of the translational symmetry in the third dimension.

Resolution can then begin. The main factors in our theory of resolution have been given in the previous sections. In particular, according to our theory: *Resolution is the attempt to recover the designer's unfolding structure of the object.* For example, as was argued in Sect. 14.12, the geometric constraints correspond to the symmetries in the alignment kernel and the dimension constraints give information about the unfolding control groups.

At first, the reader might question our theory of resolution as the recovery of the designer's unfolding, accusing this theory of attempting a "psychic mind-reading". However, in mechanical CAD, *design intent* controls everything - e.g., the managing of constraints and feature relationships, etc. Major programs such as *Pro/Engineer* contain powerful facilities called "intent managers", which will be discussed later. Any system within the computer, that can rigorously "read the mind" of the designer from the designers actions, is strongly in keeping with the way in which modern CAD programs are set up. We are suggesting that this "mind reading" should be formulated as a *recovery problem on shape.* CAD intent-management and resolution are among our main motivations in creating a generative theory of shape that maximizes *recoverability.*

14.16   Adding Features

The above sections described the creation of what is generally called a *base* part. Standardly, in mechanical CAD, the designer proceeds by first creating a base part, and then adding *features* such as holes, slots, etc., to this base. Usually features correspond to machining operations in the manufacturing process; for example, a hole corresponds to the use of a drill to remove material.

Let us therefore take the base part constructed in the previous sections, and add to it some standard features: three holes (left) and a slot (right) in Fig. 14.9. Notice the following: These features could not have been created in the preceding design phase because they destroy the translational symmetry in the sweeping direction (vertical). It was for this reason that they had to be inserted after the sweeping phase. This again reinforces our claim that the design (and manufacturing) process is one of symmetry-breaking/asymmetry-building.

It is now necessary to understand how the addition of these features is represented in terms of our generative theory of shape. There are two methods, within the theory, of representing this structure: (1) super-local unfolding,

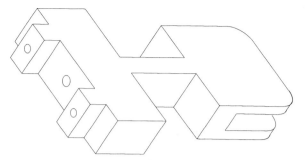

Fig. 14.9. The addition of three holes (left) and a slot (right)

and (2) sub-local unfolding. The two representations describe two different structures as follows:

(1) Super-local unfolding.

In the first of these methods, we take the unfolding group defining the base part, i.e., as given in expression (14.3) on p. 323, and we wreath super-append control groups that each unfold the extra primitives corresponding to the new features. For example, consider the central hole in Fig. 14.9. This is a cylinder unfolded from the alignment kernel, via an affine control group that is super-appended to expression (14.3), and acts (like any control group) on its entire fiber, but acts non-trivially on only the cylinder from the alignment kernel; i.e., it keeps the remaining structure frozen. Indeed, a number of design programs such as *AutoCAD* allow the designer to explicitly freeze the structure while such operations take place. According to our theory, one super-appends a control group for each of the features, and does so in the order in which the features are created.

The advantage of super-local representation is that the wreath hierarchy corresponds to the order of the design decisions. This is not a necessary part of our theory. The reader will recall that the true power of the wreath hierar-chy is that it corresponds to the hierarchy of *control* rather than the order of generation; for example, in a wreath polycyclic group, the canonical plan cor-responds to nested do-while loops in a computer program - and this nesting corresponds to the wreath hierachy, whereas the execution point oscillates between the levels in the hierarchy. A simple example is the standard gener-ation of the square: One draws a side (lower level), then executes a rotation (upper level), then draws the next side (lower level), then executes another rotation (upper level), etc.

Thus, it is not necessary to want the wreath hierarchy to correspond to the generation order. However, super-local unfolding has the advantage that this correspondence is the natural interpretation.

It is important to observe however that, in this representation, the movement of the hole is in complete disregard to the unfolding structure of the base part. One can think of the base part as a single solid of one uniform material with no internal boundaries, and the hole as having been moved with respect to this uniform material.

There is however a very different representation, to which we now turn.

(2) Sub-local unfolding.

Using constraints in the attachment of features is a fundamental part of design, assembly, and manufacturing. Such constraints should be modeled in the most powerful and informative way. We propose that these constraints are best modeled by *sub-local unfolding.*

First observe that constraints in feature-attachment use the unfolding hierarchy within the pre-existing structure - which is the base part in the current situation.

Let us consider the particular example of Fig. 14.9 (p. 331). Observe the following obvious visual facts: The two side-holes are each centered on their respective small side-blocks. The intervening hole is centered on the entire part. The slot is centered on its own block. These are constraints that one might wish to keep in the entire design. In fact, the advantage of parametric feature-based design is exactly that it allows one to define constraints that will hold despite modification of the design at later stages. Thus, one could have the following constraints for the four features added in Fig. 14.9:

> *Constraint 1:* Assume that the middle hole will eventually contain an axle around which the object will rotate in the final assembly. Thus require the constraint that this hole runs centrally through the thin neck of the object.
>
> *Constraint 2:* The slot feature is central to the end-block which contains it, even if the block is widened or moved in later phases of the design.
>
> *Constraints 3 and 4:* Assume that the two side-holes are actually screw-holes that will attach the small side-blocks to the larger left block. Thus, one would wish to keep the constraint that they are *centered* on the small side-blocks throughout the design process, despite the fact that the dimensions of the part might be altered at some point.

Observe that these constraints use the unfolding structure of the base part; i.e., expression (14.3). For example, the fact that the side-holes are aligned with the center of the small side-blocks, means that they are constrained - within the parent-child unfolding hierarchy of the base part - such that they are aligned with the side-blocks as children of the large parent block to which the small blocks are attached.

In this method of representation, the features are placed directly within the parent-child structure described by the unfolding group of the base part, expression (14.3). In this way, one realizes Constraints 1-4 (given above) in a very natural way, as follows.

Constraint 1: Here the central hole is unfolded as a cylinder together with the block defining the central neck. The constraint embodies the fact that the two are aligned - i.e., they have coincident symmetry structures. By the Symmetry Principle, which states that symmetries are preserved backwards through the generative sequence, this alignment is preserved backwards, which means that it comes from the alignment of the cylinder and block in the alignment kernel. Thus the constraint is expressed in terms of the unfolding structure from the alignment kernel.

One should be aware of the following: This hole, being an axle hole, runs through the entire part. At first this might seem to be a problem because it runs through subsidiary blocks such as the large block at either end, even though these are unfolded later than the central neck. However, this is not a problem. First, one must understand that a hole is really a solid. For example, *Pro/Engineer* lists it as a solid, and manuals for *Pro/Engineer* explain this by the fact that a hole corresponds to the solid of material which is to be removed. Furthermore, from a theoretical point of view, a hole is a solid in the sense of constructive solid geometry.

Thus the hole is a solid that is unfolded from the alignment kernel together with the solid neck with which it is aligned. Now, the cylinder defining the hole does not require its end-planes - i.e., the cylinder can be considered to be infinite. The ends of the hole become defined when the two large blocks are added at each end of the neck. In other words, the end-planes of the cylinder are given indirectly as the boundary planes of the two large end-blocks. This is a reasonable description.

Constraint 2: The slot is unfolded as a block together with the block in which it is located - the right block. The constraint specifies that the slot is vertically central to the right block. This is a symmetry that is preserved from the alignment kernel. However, the slot is non-central to the right block in the horizontal direction - which means that symmetry has been broken in the horizontal direction. Thus the slot has been unfolded one step further, i.e., with respect to the right block, since misalignment is created by unfolding. This is achieved by wreath sub-appending the additional unfolding group below that of the right block. Thus the slot becomes a child of the right block.

Constraint 3 and 4: This is the constraint that the two side-holes are aligned with the centers of the two small blocks on the left end. This constraint is realized by making the two hole cylinders follow the same unfolding history as the two small blocks in which they are located. In other words, the original alignment of cylinder with block, within the alignment kernel, is preserved through the unfolding process, first to the central neck, then leftward to the large block, and then splitting into two copies (i.e., a direct product). Therefore Constraints 3 and 4 are actually described as memory of the original alignment in the alignment kernel. This is the advantage of the symmetry-breaking nature of our generative theory: One does not need to create these constraints as extra events - they already exist in the origin of the object. Only the loss of these constraints would have to be represented - i.e., as symmetry-breaking.

In this section, it was assumed that the feature constraints were *center-based* with respect to the unfolded primitives. A frequent alternative, in mechanical CAD, is to base a feature (and its constraint) on one of the *faces* of the unfolded primitive. Here, the same theory applies except that alignments are preserved with respect to faces as *fiber groups*. This will be studied later in our theory of assembly.

14.17   Model Structure

There are substantial differences between the way we specify the model structure and the way in which current CAD programs specify this structure. According to our theory, the model is represented by its unfolding group. This expresses the structure maximally in terms of transfer and recoverability - which gives the designer the most important information concerning the model structure. In contrast, consider the representation that would be given in a major program such as *Pro/Engineering-* where the representation is called the *model tree*. For example, the model tree of the part we have been constructing would be this:

1. Datum Plane
2. Datum Plane
3. Datum Plane
4. Coordinate System
5. Base Part
6. Hole
7. Hole
8. Hole
9. Slot

It is important to notice a number of things about this structure: First, item 5, the Base Part is not given any structure. This is because there is no real structural analysis of this item by the program. While the sketch-resolution performs a constraint-solving analysis on the part, the constraints - geometric and dimensional - are not given any organization in the powerful sense defined in the previous sections where the geometric constraints are elements of the alignment kernel, and the dimension constraints are elements of the unfolding control groups and the occupancy groups. It is only when one understands these constraints to be elements from an unfolding group, that one obtains a organization for the base part. We argue that this organization is essential for all later aspects of the design and manufacturing. For example, it is basic for understanding how the added features are related to the base part - i.e., into whose components they are to be placed - and for understanding design intent and design reformulation. This is discussed in the next section. Furthermore, the unfolding group is basic for understanding the intelligent structuring of manufacturing operations. Also observe that, according to our theory, the lower Items 1 to 4 - the datum planes and coordinate system - are part of the unfolding structure, i.e., they are symmetries of that structure, and should be understood as such in order to obtain more intelligent design.

14.18   Intent Manager

Major programs such as *Pro/Engineer* now have what is called an *intent manager*. The principle issue here is that a design can be produced in a number of different ways, and therefore be structurally defined in a number of different ways. Thus for example, the constraints will change with different ways of defining the structures.

Now, our generative theory provides a particularly powerful way of enumerating these alternative ways. Consider for example the part we were designing in Fig. 14.7, p. 320. In our generative representation of the part, we chose the Cartesian frame as centered on the neck of the part, as shown in the bottom diagram on p. 320. However, we could have equally chosen the center of either of the two main side blocks, or the fiber faces. Our theory exactly predicts the alternative ways in which the part could have been generated because the unfolding comes from the alignment kernel, which contains the Cartesian frame, and the Cartesian frame is a symmetry structure. Thus the alternative methods of generating the figure can be produced from only the alternative symmetry structures, which are limited, and easy to enumerate.

For each alternative, the unfolding group changes. For example, if one chooses the generative origin to be the reflection structure at the center of the main left block, then the neck is placed on the wreath level below this, and the right block is placed on the level below that. This means that the right block is now two wreath levels below the left block. However, if one chooses the generative origin to be the neck (as we did before), then the right and left block are on the same level.

Notice how the issue of constraints naturally arises here. According to our theory, the geometric constraints are symmetries from the alignment kernel and the dimension constraints are asymmetries from the unfolding control groups (and occupancy groups). It can be seen therefore that, with the change of unfolding groups, one gets a change in the geometric and dimension constraints.

14.19   Intent Managers: Gestalt Principles

A powerful function of some intent managers, such as that in *Pro/Engineer*, is that they can work "online" with the designer's sketching actions, to assign geometric and dimensioning constraints to the sketched elements as the designer proceeds with sketching. For example, as the designer draws a new line, the intent manager will try to judge whether it should be regarded as parallel to an existing line; and, if the verdict is affirmative, will assign a parallelism constraint, which will then be explicitly shown on the screen.

Thus the intent manager literally acts like a gestalt perceiver, i.e., like the human visual system. We therefore propose that an intent manager should incorporate our theory of gestalt perception; i.e., the theory of grouping in Chapter 5 and its extension to unfolding groups (Chapters 11, 12, and 13).

14.20   Slicing as Unfolding

Slicing an object is an essential part of most design. For example, in mechanical design, one needs to slice through mechanisms to show how the pieces interlock; and in architectural design, one needs to take horizontal slices through a massing study in order to generate the floorplates. Slicing is also basic in 3D modeling programs for chemical structure, geological data, etc.

Although, to our knowledge, the following fact has never been pointed out, the most important thing to recognize about slicing is that it is a symmetry-breaking operation. Almost inevitably, an object after it has been sliced is more asymmetric than before slicing. If this were not the case, one would have created the second more symmetric state *first* - because it is more easily generated.

What is the appropriate mathematical description of slicing? We argue that it is unfolding groups. To understand this, let us begin by considering an example in detail: slicing a cylinder with a plane, as shown in Fig. 14.10.

The first thing to observe is that each half of the cylinder is more asymmetric than the initial unbroken cylinder. Thus, as we have said: Slicing tends to break symmetries. Notice that one obtains an individual part by regarding

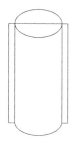

Fig. 14.10. Slicing tends to create asymmetrical parts.

the slice-plane as a (one-sided) occupancy boundary. One can cycle between the two parts by simply reversing the on-off sides of the plane. The cycle is given by $\mathbb{Z}_2$ understood as a color group.

The second thing to observe is that not all slicings produce the same amount of asymmetry. In Fig. 14.11, the slice plane is off-center, and therefore produces two unequal parts. Furthermore, each part is more asymmetric than a part in Fig. 14.10. Even further, if the plane sliced the cylinder obliquely, then the parts would be more asymmetric still.

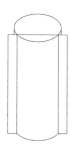

Fig. 14.11. The asymmetry of slicing can be increased further.

We argue that the way to capture this information is by using unfolding groups. Create an unfolding group in which the alignment kernel contains both the plane and cylinder; i.e., their symmetries are maximally aligned. Then unfold this using the affine group. Group theoretically, the structure is this:

$$[[\mathbb{R}^2]_{P_2} \times [SO(2) \, ⓦ \, \mathbb{R}]_{P_1}]]_{\mathcal{U}} \, ⓦ \, _{aP_2} \, , \, _{P_1} AGL(3, \mathbb{R}) \, ^{P_2 \, , \, P_1}.$$

The primitive $[SO(2) \, ⓦ \, \mathbb{R}]_{P_1}$ is the cylinder, and the primitive $[\mathbb{R}^2]_{P_2}$ is the plane. The affine group $_{aP_2} \, , \, _{P_1} AGL(3, \mathbb{R}) \, ^{P_2 \, , \, P_1}$ is shown as unfolding the plane with respect to the cylinder. Notice that, in this particular example, the group sequence happens to be the same as that of a telescope unfolding - but this is only a coincidence.

The successively greater asymmetrizations described above are given by the levels of structural allowability within the affine group, as follows: Recall that structural allowability (Sect. 2.17) depends on the extent to which the actions preserve the symmetry structure of the configuration on which one is acting. Thus, horizontal or vertical translations of the plane with respect to the cylinder, will preserve some of the symmetries of the initial objects. However, rotation of the plane about some non-vertical axis will destroy those symmetries. Thus one can hierarchically rank the structural allowability of operations within $AGL(3, \mathbb{R})$, and this will correspond to the degree of asymmetry of the associated slicings.

14.21   Assembly: Symmetry-Breaking Theory

Mechanical CAD involves two successive stages: (1) creating the parts, each of which is produced in its own separate file; and (2) assembling the parts together. The second phase is achieved by viewing the parts simultaneously on the screen, and successively fitting them together. The process is called *virtual assembly*. Throughout this chapter, the term assembly will usually mean the process of virtual assembly in MCAD, although much of what we say is generalizable to all assembly situations.

A number of powerful approaches have been developed to represent the assembly process; for example, using degree-of-freedom analysis, e.g., Miura & Ikeuchi [109], or the group-intersection approach of Liu [99], Popplestone, Liu, & Weiss [120], and Liu & Popplestone [101]. We will propose here a different approach based on unfolding groups. Each of the alternative approaches is valuable because it offers a different conceptual analysis of the problem of assembly. Our approach emphasizes the generative structure of the assembly, and therefore accords strongly with the perceptual and design aspects.

It is well established that assembly takes place by enforcing a succession of constraints that are alignments. To illustrate, consider a typical assembly situation that can occur in mechanical design: Fig. 14.12c shows a cylinder that needs to be inserted in the hole in the cube. This assembly task is solved by establishing two successive constraints:

Constraint 1: The axis of the cylinder must be aligned with the axis of the hole. This constraint is achieved by moving the cylinder from its position in Fig. 14.12c to its position in Fig. 14.12b.

Constraint 2: The center of the cylinder should be aligned with the center of the hole (understood as a complete cylinder). This constraint is achieved by translating the cylinder down along its common axis with the hole in Fig. 14.12b, until the centers of the cylinder and hole are aligned. The result is the assembled configuration Fig. 14.12c.

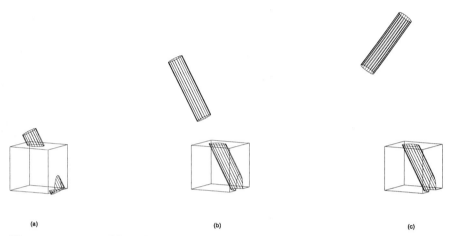

(a)   (b)   (c)

Fig. 14.12. From (c) to (a): Assembling a cylinder and cube.

The important thing to observe is that both **Constraint 1** and **Constraint 2** are alignments, and therefore the successive imposition of the constraints is a successive imposition of alignments.

Our theory of assembly planning can now be stated: It is given on p. 340, which the reader should read carefully before continuing.

14.22   Unfolding Groups, Boolean Operations, and Assembly

The issue of alignment in assembly planning can be further clarified by using unfolding groups to shed light on the difference between ordinary constructive solid modeling and assembly. Let us look at how to generate the cube and cylinder shown in Fig. 14.13. First let us consider how to generate the configuration using constructive solid modeling (CSG), which is the method that uses Boolean operations. In this method, one simply takes the Boolean union of the cube and the cylinder. Within our theory of CSG, this is achieved by an unfolding group, where one starts with the cube and cylinder with their

SYMMETRY-BREAKING THEORY OF ASSEMBLY PLANNING

(1) Because, assembly takes place by creating successive alignments, and because unfolding takes place by successive mis-alignment, assembly can be described as *reverse unfolding*.

(2) One should plan assemblies using two successive unfolding stages, thus:

$$\text{alignment kernel} \overset{unfolding}{\longrightarrow} \text{assembled state}$$

$$\overset{unfolding}{\longrightarrow} \text{dis-assembled state}$$

(3) The two stages successively break alignments in the alignment kernel. The alignments broken at the second stage are what people call the "assembly constraints". Notice that this means that *an assembly constraint is the alignment of two unfolded versions of the same fiber from the alignment kernel.*

(4) The action of assembling is the reverse of the second phase. What is called the "imposition of constraints" is actually the *recovery* of a particular subset of alignments in the alignment kernel.

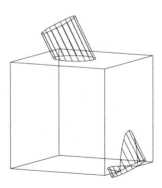

Fig. 14.13. A cylinder/cube configuration that can be described either by CSG or assembly.

symmetries maximally aligned. One then introduces the affine group as a control group that creates the required symmetry-breaking.

One must distinguish however between a CSG representation of this figure, and an assembly representation. Most crucially, whereas in the CSG representation, one uses only one copy of the cylinder, in the assembly representation, one requires *two* copies of the cylinder, one representing the hole and the other representing the solid that will exist in the hole. This conforms to part (3) of our theory of assembly on p. 340 - that an assembly constraint is two coincident fibers transferred up from the alignment kernel.

This is reinforced by what one needs to do to generate the configuration on a computer, using a full solid modeling program[2]. To draw the figure as a CSG configuration, one (1) creates a cube and a cylinder, (2) moves the cylinder to the required position/orientation, and (3) performs a Boolean *union* of the cylinder and cube. However, to draw the figure as an assembly configuration, one needs to create two coincident copies of the cylinder. One of them will be Boolean *subtracted* from the cube, creating the hole. The other will be left coincident with the hole. Inexperienced students in CAD often do not anticipate requiring the second copy - not realizing that the first copy will "disappear" in the creation of the hole.

Let us now use unfolding groups to contrast the CSG representation with the assembly representation. The CSG structure is given by this group:

$$[[G_{cylinder}]_{P_2} \times [G_{cube}]_{P_1}]_{\mathcal{U}}$$
$$\textcircled{W}\ _{aP_2\ ,\ P_1}\ AGL(3,\mathbb{R})\ ^{P_2\ ,\ P_1} \qquad (14.5)$$

Notice that there are only *two* input (upper) indexes on the bottom line: one copy of the cube P_1 and only one copy of the cylinder P_2. Because they are brought from the alignment kernel, their symmetries are maximally aligned; i.e., their reflection planes are coincident, etc. The affine group on the bottom line then misaligns the cube and cylinder by creating the angle between the cylinder and the cube. This is indicated by the fact that the lower indexes are P_1 showing that the cube is unchanged, and aP_2 showing that the cylinder has undergone the affine transformation a.

In contrast, the assembly structure is given by this group:

$$[[G_{cylinder}]_{P_2} \times [G_{cube}]_{P_1}]_{\mathcal{U}}$$
$$\textcircled{W}\ _{aP_2\ ,\ aP_2\ ,\ P_1}\ AGL(3,\mathbb{R})\ ^{P_2\ ,\ P_2\ ,\ P_1} \qquad (14.6)$$

Notice that there are now *three* indexes entering the bottom line - a single copy of the cube P_1 and two copies of the cylinder P_2. The output (lower) indexes in the bottom line show that the cube has remained unaltered, but the two cylinders have undergone the *same* transformation a. This means that the two cylinders are still aligned with each other, despite being at an angle to the cube.

---

[2] By a "full" solid modeling program, we mean an advanced animation program like *3D Studio Max* or *AutoCAD Architectural Desktop* which begin with 3D solid primitives, rather than a mechanical program like *Pro/Engineer* which begins with 2D sketching. We believe that a "full" solid modeling program best corresponds to the designer's conceptual design phase that precedes the sketching phase that begins a mechanical program.

The most crucial thing about this expression is that it embodies our theory that the progression from left to right in Fig. 14.12 (p. 339) is preceded by a previous phase, which goes from the alignment kernel to Fig. 14.12a. This preceding phase is the first phase in the two-phase unfolding in

$$\textbf{alignment kernel} \quad \overset{unfolding}{\longrightarrow} \quad \textbf{assembled state} \quad \overset{unfolding}{\longrightarrow}$$
$$\textbf{dis-assembled state}$$

which was given in part (2) of our theory of assembly on p. 340. The first phase, while destroying some alignments from the alignment kernel, does not destroy all of them. In the second phase (dis-assembly), further alignments will be destroyed. However, most crucial in our theory - these latter alignments nevertheless came up from the original alignment kernel.

Let us therefore move on to the second phase, i.e., dis-assembly. This corresponds to the sequence of stages from Fig. 14.12a to Fig. 14.12c (p. 339). Since this is a further phase of unfolding, one must add further unfolding control groups to expression (14.6) in order to create the further mis-alignments. The full group is this:

$$
\begin{aligned}
&[[G_{cylinder}]_{P_2} \times [G_{cube}]_{P_1}]_{\mathcal{U}} \\
&\textcircled{w}_{\, cbaP_2} \; SE(3) \;^{baP_2} \\
&\textcircled{w}_{\, baP_2 \, , \, aP_2} \; \textbf{Trans}_z \;^{aP_2 \, , \, aP_2} \\
&\textcircled{w}_{\, aP_2 \, , \, aP_2 \, , \, P_1} \; AGL(3, \mathbb{R}) \;^{P_2 \, , \, P_2 \, , \, P_1}
\end{aligned}
\qquad (14.7)
$$

This is expression (14.6) with two extra lines (Lines 2 and 3) giving the breaking of the two assembly constraints. First consider Line 3: The input (upper) indexes are the two aligned cylinders brought up from Line 4. The control group $\textbf{Trans}_z$ on Line 3 breaks **Constraint 2** (page 338) by translating the second cylinder along its z-axis out of the hole. This translation is indicated by the b on one of the output (lower) indexes on Line 3. The other index (the hole) has remained untouched.

The actual mis-alignment created by this translation group is as follows: According to our theory of solid cross-section cylinders, in Sect. 16.10 (p. 413), any such cylinder is swept from a circle in a starting plane which can be considered to be the reflection mid-plane of the object. Let us call this reflection plane the generating plane. Returning to the cube and cylinder example, we conclude that the translation action has mis-aligned the generating plane of the solid cylinder with respect to the generating plane of the cylindrical hole.

Now go on to Line 2 in expression (14.7): This creates the final dis-assembly phase Fig. 14.12b to 14.12c, by breaking **Constraint 1** (page 338). The group used here is the special Euclidean group $SE(3)$ applied to the solid cylinder - as shown on Line 2. (For readability, various identity components have been omitted from Lines 2 and 3, but can easily be inferred, in accord with our theory of cannonical unfoldings.)

The important point is that expression (14.7) contains the entire information of assembly and dis-assembly: The bottom line describes the assembled configuration shown in Fig. 14.12a. The line above this produces the first stage of dis-assembly, shown in Fig. 14.12b. The line above this produces the last dis-assembly stage shown in Fig. 14.12c.

Movable parts. Suppose the solid cylinder in the above example is an axel that will be rotating within the hole in the functioning assembly. In our system, this is represented very easily. The solid cylinder is rotating with respect to the hole cylinder. This means that the two cylinders - initially aligned because they were brought up together from the alignment kernel - undergo rotational mis-alignment. However, this type of mis-alignment does not break the assembly structure, and therefore is on the wreath control level above (i.e., line below) the dis-assembly groups in expression (14.7); that is, we insert the rotation group $SO(2)$ between the third and fourth lines in that expression. Notice that this copy of $SO(2)$ is the circular symmetry contained within the generative structure of the cylinder:

$$[G_{cylinder}]_{P_2} \quad = \quad SO(2) \textcircled{w} \mathbb{R}.$$

This structure $[G_{cylinder}]_{P_2}$ is in the alignment kernel in expression (14.7). Therefore $SO(2)$, used as a *control* group to actually rotate the cylinder in the hole, is pulled up from the cylinder's generative structure $SO(2) \textcircled{w} \mathbb{R}$ within the alignment kernel. Similarly, the dis-assembly control group $\mathbf{Trans}_z$ originated as the $\mathbb{R}$ component from this $SO(2) \textcircled{w} \mathbb{R}$ structure within the alignment kernel.

14.23   The Designer's Conceptual Planning

Before continuing with our theory of assembly, the reader might wonder at the validity of regarding the assembled state as generatively preceding the dis-assembled state, as in part (2) of our theory on p. 340 - and furthermore, of regarding the final assembly as *recovered*, as in part (4) in the theory. After all, mechanical CAD, with major programs such as *Pro/Engineer*, begins with the sketching of parts, each within their separate files, and only subsequently creates assembly by bringing the parts together.

However, although the use of the software begins with part-design, no mechanical designer begins the design process here. The process begins with a long conceptual phase, in which the designer's first drawings, usually on paper, are of the *assembled* object. Only when this is done, is the designer ready to draw the actual independent parts with the software. The reason is simple: Parts must actually match: The fact that a cylinder is supposed to fit in a hole is not suddenly discovered later in the design process, after the cylinder and hole have been created in independent part files. They fit because, in the initial conceptual phase, the designer drew the cylinder actually in the hole. This is crucial because, from this, the designer will know, from the beginning, that the radii of the cylinder and hole have to be the same. Thus, in the next phase, when the designer draws the cylinder and hole separately, in different part files, he will use the same radius values in the two parts. He will know to do this only because the alignment was present in the initial design drawings in the conceptual phase. Thus, later still, when he fits the part files together into an assembly, the imposition of the alignment constraints merely *recovers* what was present at the beginning.

It is surprising that major mechanical programs do not incorporate an initial conceptual phase in the software, i.e., allowing the designer to do all this work within the program, rather than before-hand. The conceptual phase drives the entire design process. Observe that such a phase is incorporated in the major architectural program, *AutoCAD Architectural Desktop*. Anyone familiar with this program knows the powerful advantages of having this phase as part of the software.

An important thing should be observed: The conceptual phase in *AutoCAD Architectural Desktop* is a full 3D solid modeling program. By full, we mean that it is 3D from the beginning, and works by bringing together *solid* primitives, rather than starting with 2D line sketching as in standard mechanical CAD. We have worked extensively with both types of programs and found that the conceptual phase in the former type is enormously valuable exactly because it accomplishes what one needs to do in any case before sketching in mechanical CAD.

Most crucially, the conceptual phase accords with the psychological design process. We have said before that the unfolding of solid primitives from an alignment kernel into a complex object corresponds to the process of solid modeling in a major 3D animation program such as *3D Studio Max/Viz*. For example, one can understand the menu of standard primitives in such a program, as corresponding to the alignment kernel of an unfolding group, and the parameter window of each primitive, as well as the parent/child model hierarchy, as corresponding to the unfolding group structure.

Accomplishing this true type of 3D solid modeling corresponds to the first phase of the two-phase unfolding in our theory of assembly:

$$\textbf{alignment kernel} \quad \xrightarrow{unfolding} \quad \textbf{assembled state} \quad \xrightarrow{unfolding}$$
$$\textbf{dis-assembled state.}$$

Any mechanical designer must actually go through this in the conceptual phase, whether the software allows this or not. The subsequent sketching of individual parts - each in its own separate part file - is really the *dis-assembly* of the object, i.e., corresponding to the second phase in our theory. Then the later assembly of the independent part-structures, into one object, is really the *recovery* of the original assembly in the initial conceptual phase.

14.24 Holes through Several Layers

Let us consider the following frequently encountered situation in MCAD: a multi-layered structure requiring a hole through all the layers. A typical example is where an axel has to go through a pulley wheel, and through washers on either side of that wheel, and through bushings on either side of those washers, etc. The wheel, the washers, and the bushings, will all have a hole of the same radius. To assemble them from a dis-assembled state, one will typically use the constraint of aligning the holes[3]. Another situation is illustrated in Fig. 14.14a where a bolt will eventually bind together a number of flat plates, e.g., of different materials. Each of the plates will have the same radius hole. To assemble them from a dis-assembled state shown in Fig. 14.14b, one will use the same constraint - that the holes in the different plates must be aligned.

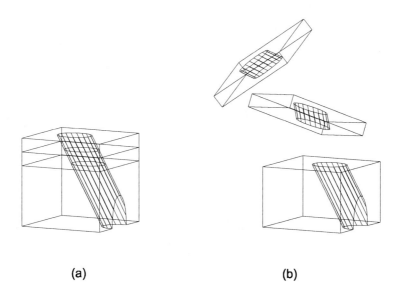

(a)   (b)

Fig. 14.14. A hole through layered objects.

---

[3] There will be two constraints (1) the hole-alignment constraint just mentioned, and (2) a face-mating constraint of the type discussed later in Sect. 14.26.

The issue to be considered now is this: Ignoring the bolt which will eventually go through the hole, we seem to need only one cylinder to go through all the layers, to create the hole. This is not correct.

To understand this, it is useful again to consider what one does in a full solid-modeling program. One first creates the assembled structure shown in Fig. 14.14a. However, if one used only a single cylinder for the hole, and subtracted it from the set of layers, then one would not be able to separate the layers, because the set of layers would be a parent to the single hole cylinder. Therefore, in a solid-modeling program, one needs to create as many coincident copies of the hole-cylinder as there are layers. Each cylinder-copy is subtracted from one layer. Then, when the layers are dis-assembled (moved away from each other) each layer moves away with its own hole, as shown in Fig. 14.14b. For re-assembly, one can then exploit the constraint of aligning the hole-cylinders.

For exactly the same reasons, our generative theory requires us to pull up from the alignment kernel, as many coincident copies of the hole-cylinder as there are layers. Dis-assembly will then mis-align these copies, and re-assembly will then *recover* their original alignment.

14.25   Analogy with Quantum Mechanics

The idea of creating coincident copies, which are then separated, is remarkably analogous to the splitting of degeneracy in quantum mechanics: A degenerate state has multiply coincident eigenvalues; i.e., where linearly independent eigenstates have the same eigenvalue. The splitting of this degeneracy, by a perturbation force, pushes the coincident eigenvalues apart. Thus, linearly independent eigenstates, which previously had the same eigenvalue, now have different eigenvalues under the perturbing action.

14.26   Fiber-Relative Actions

In this section, we are going to introduce two operations which together have the power to generate a wide range of assembly configurations and their associated constraints.

A number of assembly constraints (e.g., mating constraints) are specified as relative to particular surfaces. To handle this, it is first necessary to state part of our theory of solid structure as presented in Chapter 16. To illustrate this statement, let us give the structure of the *solid cross-section cylinder* as presented in Sect. 16.10 (p. 413). The structure is shown in Fig. 14.15

which is explained as follows: In Fig. 14.15a, there is the alignment kernel which consists of the straight line and the circle maximally aligned - i.e., these are the primitive lines from which the primitive solids are generated, as follows: From the straight line one generates the infinite solid structure of 3D space as the wreath c-polycyclic group $\mathbb{R} \, \textcircled{w} \, \mathbb{R} \, \textcircled{w} \, \mathbb{R}$, where the first two factors $\mathbb{R} \, \textcircled{w} \, \mathbb{R}$ generate the horizontal plane, and the last factor $\mathbb{R}$ translates the plane in the vertical direction. This gives the stack of planes shown in Fig. 14.15b. According to our theory, this is the solid structure common to all solid primitives.

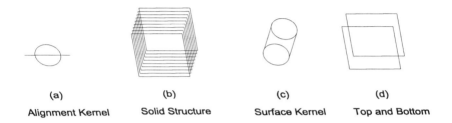

(a)	(b)	(c)	(d)
Alignment Kernel	Solid Structure	Surface Kernel	Top and Bottom

Fig. 14.15. The structure of a solid cross-section cylinder.

Then, in Fig. 14.15c, one unfolds the infinite cylindrical surface from the circle primitive. This cuts down the infinite symmetry of the solid structure of the previous figure. We call this cutting surface, the *surface kernel* of the primitive. A surface kernel is what distinguishes a particular solid primitive from all the other solid primitives. Finally, in Fig. 14.15d, one generates the two bounding planes of the cylinder (top and bottom): these break the infinite symmetry of the cylindrical surface. These two planes are respectively a half unit above and a half unit below the starting horizontal plane; they are related by the reflection group $\mathbb{Z}_2$. The full group of the solid cross-section cylinder is

$$
\begin{aligned}
&[[SO(2)]_{P_2} \times [\mathbb{R}]_{P_1}]_{\mathcal{U}} && \text{alignment kernel} \\
&\textcircled{w} \, [\mathbb{R} \, \textcircled{w} \, \mathbb{R}]^{P_1} && \text{solid structure} \\
&\textcircled{w} \, \mathbb{R}^{P_2} && \text{surface kernel} \\
&\textcircled{w} \, \mathbb{Z}_2^{[\mathbb{R} \, \textcircled{w} \, \mathbb{R}]_{1/2}} && \text{top and bottom planes} && (14.8)
\end{aligned}
$$

where the four successive lines correspond to the four successive parts of Fig. 14.15, and each of the unfoldings is super-local.

Notice in particular how Line 4 of expression (14.8), works: The input index consists of the horizontal Cartesian plane $[\mathbb{R} \, \textcircled{w} \, \mathbb{R}]$ which has undergone a translation by 1/2 vertically. This translation was provided by the last $\mathbb{R}$ of the Line 2. This translated plane is the top plane of the cylinder. The control group on Line 4 makes also a reflectional copy of this plane, producing the bottom plane.

Now, using this structure let us consider how to express the mating constraint in Fig. 14.16. The constraint is that the top of the lower cylinder mates with the bottom of the upper cylinder.

Fig. 14.16. Two cylinders mated at a plane.

What we observe is that the contact structure here can be described by a reflection operation; i.e., the material above the contact surface is reflectionally symmetric to the material below the surface. This fact can be exploited very simply in generating the configuration shown: That is, the generation starts with two coincident copies of the cylinder (pulled up from the alignment kernel of solid primitives); then one of the two coincident cylinders is reflected about the top plane understood as a reflection plane. This creates the configuration shown. The only thing that is necessary to account for is the reflection plane in the top plane. This is easy enough. Recall that, according to expression (14.8), the top plane started out as the horizontal Cartesian plane. This is coincident with the reflection plane of the entire cylinder - i.e., the reflection plane given in the aligned hyperoctahedral wreath structure which would also be in the alignment kernel of solid primitives. Therefore, we merely demand that, in translating the horizontal mid-plane to become the top plane, one translates with it a copy of the hyperoctahedral wreath structure. Then use the translated horizontal reflection plane to produce the structure shown in Fig. 14.16. This accords completely with the psychological theory of Cartesian frames presented in Chapter 8, where it was argued that the human visual system "snaps" a Cartesian *triple-reflection* structure to every flat surface of an object, e.g., recall Fig. 8.13 (p. 208).

We shall call the operation of reflecting an object about the reflection plane coincident with one of its flat faces, a **face-relative reflection**.

This type of operation can be combined with another type of operation, which we will now define, to produce a wide range of contact constraints. This other class of operations will be called **sub-primitive operations**.

Sub-primitive operations again rely on the components of the solid structure, as defined for example by expression (14.8) in the case of the solid cross-section cylinder. This book so far has tended to apply unfolding operations to primtives P_i as a whole. However, there is no problem applying them to parts of primitives. In particular, it will be useful to apply them to surfaces in the primitive. According to the theory in Chapter 16, the surfaces of a typical solid primitive are (1) the surface kernel - which is the surface that distinguishes the particular primitive from the other primitives, and (2) the top and bottom planes. These surfaces will be indicated by the letters

$$S = \text{Surface Kernel}$$
$$T = \text{Top Plane}$$
$$B = \text{Bottom Plane.} \qquad (14.9)$$

When wishing to explicitly refer to these surfaces in a primitive P_i, we shall write $P_i(S, T, B)$. A control group can act specifically on one or more of these surfaces. For example, the notation:

$$_{P_i(aS,T,B)}\ AGL(3, \mathbb{R})\ ^{P_i(S,T,B)}$$

means that an affine transformation a has been applied to the surface kernel S but the top and bottom planes have not been altered. For example, a cylinder has been widened without altering the top and bottom.

An operation of this kind will be called a sub-primitive operation.

We shall now argue that sub-primitive operations are a powerful means of expressing alignment in assembly. Let us illustrate this by first generating a bolt. Start with two coincident cylinders P_1 brought up from the alignment kernel - these are shown in Fig. 14.17a which represents two coincident cylinders. One of the cylinders, Cylinder1, will be left untouched. The other cylinder, Cylinder2, will undergo a sub-primitive operation, narrowing its surface kernel and translating its bottom plane downwards. The result is shown in Fig. 14.17b.

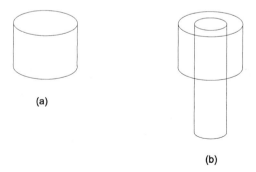

(a)

(b)

Fig. 14.17. Example of sub-primitive unfolding.

Notice that, because the top plane has been left untouched, the top planes of the two cylinders have remained aligned, whereas the bottom planes are no longer aligned, and the surface kernels are now of different widths. The sub-primitive operation that was used on Cylinder 2, was

$$P_1(aS,T,bB) \; AGL(3,\mathbb{R}) \; ^{P_1(S,T,B)}$$

where a is the contraction on the surface kernel S, and b is the downward translation on the bottom plane B. To understand this as an unfolding (mis-alignment) control group, write simply this:

$$P_1(aS,T,bB) \, , \, P_1(S,T,B) \; AGL(3,\mathbb{R}) \; ^{P_1(S,T,B) \, , \, P_1(S,T,B)}.$$

Notice how much information this expression gives us: The input (upper) indexes tell us that all surfaces of the two cylinders are aligned. The output (lower) indexes tell us that the top planes remain aligned whereas the surface kernels have been mis-aligned and the bottom planes have been mis-aligned.

Using the Boolean operations, different structures can now be created from this. Thus, if one uses Boolean union of the two cylinders, one gets a bolt. Alternatively, let us duplicate Cylinder2 - that is, add another P_1 index in the above expression and make it undergo the same sub-primitive operation as Cylinder2. Then let us subtract this second copy of Cylinder2 from Cylinder1, producing a hole in Cylinder1. The first copy of Cylinder2 now sits in the hole in Cylinder1 with the top faces aligned. This expresses another common assembly situation.

Notice that the two constraints in this structure - the alignment of the top faces, and the axial alignment - came up from the alignment kernel. That is, in generating this structure, we have gone through phase 1 of the symmetry-breaking theory of assembly on p. 340:

$$\textbf{alignment kernel} \quad \overset{unfolding}{\longrightarrow} \quad \textbf{assembled state} \quad \overset{unfolding}{\longrightarrow}$$
$$\textbf{dis-assembled state.}$$

The second phase, dis-assembly, will then translate one of the cylinders away from the other along their common z-axis, breaking the alignment of their top surfaces, and will then apply an arbitrary Euclidean movement to that moved cylinder, breaking the axial alignment. These two actions (z-translation and arbitrary Euclidean movement) are represented in exactly the same way as in expression (14.7) on p. 342 for the dis-assembly of the cube and cylinder; that is, merely add the translation group $\textbf{Trans}_z$, and the special Euclidean group $SE(3)$, sub-locally to the sub-primitive operation defined above (which gave phase 1).

Sub-primitive operations are powerful also when they are combined with the other type of operation defined in this section: face-relative reflections. Let us

Fig. 14.18. Mating = combination of sub-primitive unfolding and face-relative reflection.

generate the situation in which two different size cylinders are axially aligned and in face contact, as shown in Fig. 14.18.

Again start with two coincident cylinders from the alignment kernel Fig. 14.17a, and apply the appropriate sub-primitive operation to Cylinder2, to obtain Fig. 14.17b. Then simply apply a face-relative reflection of Cylinder2, using the top face of Cylinder1. This generates Fig. 14.18. Again, notice that the alignment (mating) of the two faces came up from the alignment kernel. That is, we have gone through phase 1 of the sequence

$$\textbf{alignment kernel} \quad \overset{unfolding}{\longrightarrow} \quad \textbf{assembled state} \quad \overset{unfolding}{\longrightarrow}$$
$$\textbf{dis-assembled state.}$$

The second phase, dissembly, will then move the two cylinders apart, breaking the remaining alignment of the two faces, and then breaking their axial alignment. Again, this is achieved merely by sub-locally adding the relevant translation group $\textbf{Trans}_z$, and the special Euclidean group $SE(3)$, to break the respective constraints.

It is important to note that the two objects in Fig. 14.18 need not have been both cylinders. For example, exactly the same generation procedure works for a cylinder and block; i.e., one starts with a maximally aligned cylinder and block from the alignment kernel, then applies the appropriate sub-primitive operation adjusting there relative sizes, but leaving their top faces aligned, and finally one applies a face-relative reflection, flipping one about the top face of the other.

To demonstrate further the remarkable versatility one gets by combining the face-relative reflections with sub-primitive operations, let us generate a bolt placed in a block, as shown in Fig. 14.19. The procedure is easy:

(1) Start with an alignment kernel of a block-primitive and cylinder-primitive, i.e., with their symmetries maximally aligned.

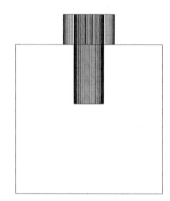

Fig. 14.19. A bolt and block.

(2) Apply a sub-primitive operation to the cylinder to contract it while maintaining the alignment of its top face with the top face of the block. The result is shown in Fig. 14.20a.

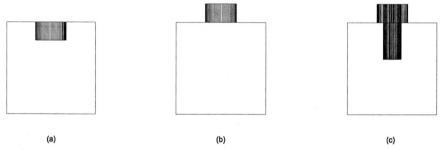

| (a) | (b) | (c) |

Fig. 14.20. Unfolding the bolt and block.

(3) Apply a face-relative reflection of the cylinder about its top face. It now sits on top of the block as shown in Fig. 14.20b.

(4) Create a copy of this cylinder[4].

(5) Apply a sub-primitive operation to the copy, contracting the radius of its surface kernel, lowering its bottom face, but maintaining the alignment of its top face[5]. The result is shown in Fig. 14.20c. The first and second cylinder now form the bolt.

(6) Create a copy of the second cylinder and subtract it from the block. This creates the hole in the block.

---

[4] As is standard in our system, creating a copy of some unfolded primitive means unfolding it together with the latter, so that it has the same generation history as the latter and occurs together with the latter as an index.

[5] Top and bottom will now refer to viewer's top and bottom.

The assembled structure in Fig. 14.19 is now complete. There are two assembly constraints: (1) The bottom face of the bolt head is aligned with the top face of the block, and (2) the bolt is axially aligned with the hole. To break these constraints, i.e., remove the bolt from the hole, insert the groups **Trans**$_z$ and $SE(3)$ as usual.

14.27   The Full Group of the Robot Serial-Link Manipulator

We will now continue our theory of mechanical assembly by giving the full group of a serial-link manipulator.

In Chapter 6, when giving the "full group" of a robot serial-link manipulator, we meant this: In standard group-theoretic representations of such manipulators, one collapses the transformations across the successive joints to obtain a single transformation from base to effector, and thus defines the group of such transformations as $SE(3)$. In contrast, in the "full group" approach that is taken in this book, we construct a group which represents the fact that the links form a control-nested hierarchy of action spaces, and we therefore argue that the group from base to effector is not $SE(3)$, but the much more complicated $SE(3) \, ⓦ \, SE(3) \, ⓦ \, \dots \, ⓦ \, SE(3)$.

In the current section, this notion of a full group will be extended still further. By "full", we will now mean, not only that the group involves the wreath structure just defined - which expresses the *reference-frame* organization of the link hierarchy - but we also include in the group the *geometric* shape of the link hierarchy. This is particularly important when it comes to expressing the workspace structure of the robot - for example, it gives a powerful algebraic representation of configuration space, Minkowski sums, and the sweeping structure involved.

First, we propose the following: The Denavit-Hartenberg representation of the serial-link structure, i.e., as a succession of Cartesian frames outward from the base, should be re-interpreted as a *symmetry-breaking structure* using *unfolding groups*.

Let us show why this is the case: First of all, it is clear that the Cartesian frames assigned along the manipulator links, always coincide with the symmetries of the links: i.e., given a coordinate frame, one of the axes coincides with the rotation axis in a revolute joint, or the translation axis of a prismatic joint, and all pairs of coordinate axes coincide with the reflectional symmetries of the link structure (usually its fibers). This is illustrated in Fig. 14.21, which is a link connecting a revolute and prismatic joint; the assignment of Cartesian frames shown is the standard assignment, and clearly corresponds to the translational, rotational, and reflectional symmetries.

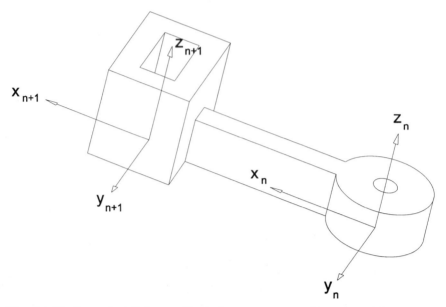

Fig. 14.21. In a robot link, coordinate frames correspond to symmetries.

Now, in the Denavit-Hartenberg [26] representation of the serial-link structure, one "pushes" the reference frames successively outward from the base. According to our theory, this means that what one is actually doing is creating a successive *mis-alignment* of the symmetry structures; i.e., one is *unfolding* the joint symmetry groups outward from initial alignment in the base. Therefore, one can regard the Denavit-Hartenberg representation as corresponding to an unfolding group, with an alignment kernel centered in the base.

The approach just defined is the key to understanding how to obtain the full group of the geometry of the robot manipulator. One unfolds not only the symmetries represented by the frames, but the symmetries of each of the solid primitives represented by the joints and connecting links. To do this, all of the concepts developed in our theory of assembly will be used, as follows:

Let us discuss an example: a robot arm consisting of a base and one link. This is shown in a dis-assembled state in Fig. 14.22. Begin by considering the assembly task. This is solved by establishing two successive constraints:

Constraint 1: The circular surface of the hole in the proximal joint of the link must be aligned with the circular surface of the drum on the top of the base, as indicated in Fig. 14.23.

Constraint 2: The flat bottom surface of the solid cylinder defining the proximal joint of the link must mate with the upper surface of the round at the end of the base, as shown in Fig. 14.24.

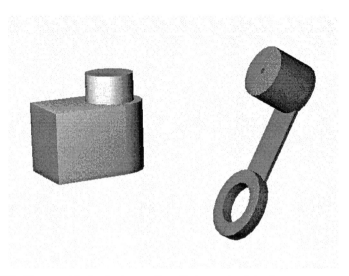

Fig. 14.22. Unassembled robot arm (two link manipulator).

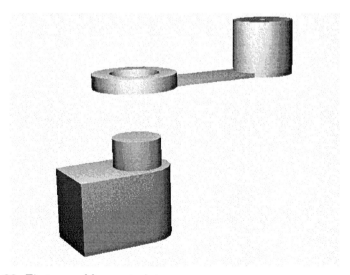

Fig. 14.23. First assembly constraint.

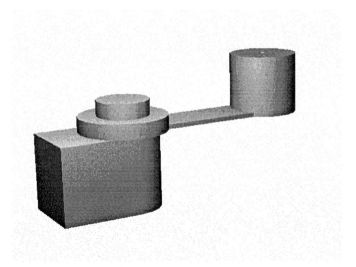

Fig. 14.24. Second assembly constraint.

With this in mind, the unfolding of the entire structure then becomes easy: Simply unfold the *assembled* arm Fig. 14.24 from left to right in the figure by sub-local unfolding, starting with an alignment kernel of only two primitives, a block and a cylinder. These two primitives start out at the left end of the base, where their symmetries are maximally aligned. This is the starting configuration of the shape. All parts of the arm are successively pulled leftwards from these two primitives by cloning and mis-alignment. The above two constraints are two alignments that have not been broken in this process, as follows: Constraint 1 expresses the fact that there are two cylinders at the joint whose surface kernels are coincident - the drum cylinder and hole cylinder. Constraint 2 expresses the fact that the solid joint cylinder at the proximal end of the link, is generated from the cylindrical round below it by a sub-primitive affine transformation that preserves the top face of the round, followed by a face-relative reflection with respect to that top face.

Finally, in this unfolding group, the following three groups are inserted between the base and the link parts: the rotation group $SO(2)$ to create rotation between the base and link; and the groups **Trans**$_z$ and $SE(3)$ to successively break the two assembly constraints. Section 14.22 showed how to insert these three groups. In conclusion:

FULL GROUP OF THE ROBOT ARM. *We have given a full group of the robot arm, expressing its shape, kinematics and dis-assembly, by unfolding, i.e., by a succession of symmetry-breaking operations via mis-alignment.*

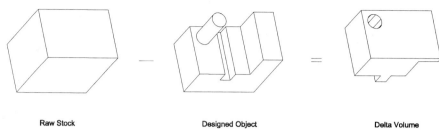

| Raw Stock | Designed Object | Delta Volume |

Fig. 14.25. The raw stock minus the designed object gives the delta volume.

14.28 Machining

When the design phase (CAD) is completed, one can begin the manufacturing phase (CAM). The CAD model of a part is sent to a CAM file for machining. Computer aided process planning (CAPP) is the means by which the CAD model can be used to generate a sequence of instructions to physically create the part. In machining, the part is produced by taking the designed shape for the part and subtracting it from the raw stock (usually a rectangular block). The remaining shape is called the *delta volume*. This is the shape that the machining process must remove from the raw stock. Fig. 14.25 illustrates the fact that the raw stock minus the designed object gives the delta volume.

The delta volume will be removed incrementally from the raw stock by a moving cutter. Therefore, the cutter actually *generates* the delta volume. This means that, once again, shape must be treated generatively. This is a *recovery* problem because the generative operations must be inferred from the delta volume which is the required goal-state of the generative process. This means that the situation exactly accords with our generative-recoverable theory of shape.

The primitives that must be unfolded through the delta volume are the shapes of the cutters. The most frequent cutter shape is the cylinder, shown in Fig. 14.26. As it rotates, it removes material from the stock. This type of cutter produces three basic types of *machined features*: (1) the cylindrical hole, (2) the slot, and (3) the pocket. These correspond respectively to the three shapes shown in Fig. 14.26, Fig. 14.27, and Fig. 14.28.

Let us express these three basic machined features in terms of our generative theory. First consider the cutting structure that produces the hole: The cutter itself is a cylinder, $SO(2) \circledW \mathbb{R}$. This is rotated with respect to itself, and therefore the rotation is the use of $SO(2)$ as a control group with respect to the cylinderical cutter as fiber, thus:

$$\overbrace{[SO(2) \circledW \mathbb{R}]}^{cutter} \circledW \overbrace{SO(2)}^{rotation} .$$

Now, also include the fact that the cutter can approach or withdraw from the raw stock, along its z axis. Since the cutter is rotating much faster than

Rotating Cutter

Fig. 14.26. A rotating cylindrical cutter.

its approach-withdrawal translation, the translation acts as a control group which sets the height of the rotation. Therefore, this translation factor is wreath-appended as an extra control group above the previous expression. That is, the resulting structure is this:

The Cutting Structure of a Hole

$$[SO(2) \ⓦ \ \mathbb{R}] \ ⓦ \ [SO(2) \ ⓦ \ \mathbb{R}]. \tag{14.10}$$

Notice that this consists of two cylinders $[SO(2) \ ⓦ \ \mathbb{R}]$, one placed above the other in a wreath relation. That is, the symmetry group of the cylindrical cutter is used as the group of its motion, and the latter acts on the former as a wreath control group. This reinforces our claim that plans come from symmetries. In fact, it is the symmetries that provide the structurally most allowable actions on an organization.

Notice, carefully, why the relation between the upper cylinder and lower cylinder is a wreath product. The upper cylinder of movements makes copies of the lower cylinder, one for each position and orientation of the latter.

Notice the power of the structure just given. It is actually a 4-fold wreath product, and this entire structure is needed to fully describe the situation. For example, the cutter cylinder is described *internally* as the wreath product $SO(2) \ ⓦ \ \mathbb{R}$, which means that it is the *transfer* of the cross section through its height. This is its sweeping structure. However, this sweeping structure is itself *transferred* through a higher level sweeping structure which represents the motion of the cylinder. That is, there is *transfer of transfer.*

Our theory of aesthetics says that aesthetics is the maximization of transfer. The above structure shows that the choice of cutter and motion is dictated by aesthetics, i.e., maximizing transfer. We argue that this is a general principle in engineering and science.

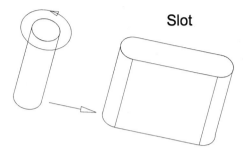

Fig. 14.27. A machined slot.

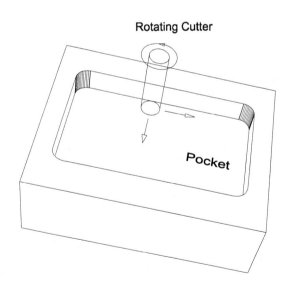

Fig. 14.28. A machined pocket.

Given the above structure, it is a simple matter now to give the cutting structure of the *slot*, Fig. 14.27. One merely wreath super-appends, to the previous structure, a control group giving one-dimensional horizontal translation of the cutter, thus:

The Cutting Structure of a Slot

$$[SO(2) \textcircled{w} \mathbb{R}] \textcircled{w} [SO(2) \textcircled{w} \mathbb{R}] \textcircled{w} \mathbb{R}. \qquad (14.11)$$

Notice that this translational action is one of the symmetries of the flat end-face of the cutter. This is part of the solid structure of the cutter, and it appears in our full symmetry group of the solid cross-section cylinder in expression (14.8) on p. 347, and Fig. 14.15, p. 347. This is fully dealt with in the chapter on solid structure.

Finally, it is easy to give the expression of the *pocket*, Fig. 14.28. Merely wreath super-append an extra translation group to the previous expression, thus:

The Cutting Structure of a Pocket

$$[SO(2) \textcircled{w} \mathbb{R}] \textcircled{w} [SO(2) \textcircled{w} \mathbb{R}] \textcircled{w} [\mathbb{R} \textcircled{w} \mathbb{R}]. \qquad (14.12)$$

Notice that the pocket perimeter is given by the occupancy structure of the final double-translation group.

Now, as said above, the problem of machining is essentially that of generating the delta volume in terms of motions of the cutters used. Furthermore, this generative structure must be recovered from the delta volume. Notice that this implies that the CAD/CAM system must decompose the delta volume in terms of machined features (e.g., hole, slot, and pocket), because, according to our theory, the machined features are *phases* in the generative history - in fact, wreath fibers in the full group. Current research on CAD/CAM integration views the problem of inferring features as that of a feature-recognition task, and feature-recognition is the subject of a major research effort in computer-aided process-planning[6].

A major problem is that of *intersecting features*; for example, slots can cross and overlap each other, e.g., as can be seen in the part shown in Fig. 14.29, which is similar to one of the benchmark parts (Han Part 1) used in the 1997 special ASME panel on feature recognition[7].

A major way of dealing with this problem is to use the observation that the machining operations, producing the features, should leave *traces* on the part, even when the features overlap or cross. This is the basis of what is called the

---

[6] Kyprianou [81], Henderson & Anderson [56], Choi, Barash & Anderson [23], Falcidieno & Giannini [32], [33], Joshi & Chang [68], Kim [75]; see also p1019-20 of Lee & Kim [82] for a classification of the methods developed.

[7] See the special issue of the journal *Computer Aided Design* Vol 30, No.13.

hint-based approach to feature-recognition[8]. Hints are partial surfaces, which, when subjected to a completion procedure, produce the maximally extended removal volume compatible with the hint. For example, in Fig. 14.29, the large diagonal slot has a number of incomplete faces. These faces *hint* at the slot, which, on completion, will itself become the trace of the cutter. In other words, the *boundary of the cutter movement* will actually be a completed version of these incomplete surfaces.

The reader should notice the enormous similarity between hint-completion and Gestalt perception in human psychology - which involves the human perceptual system conjecturing completed structures (Gestalts) despite missing parts. In our book, Leyton [96], we gave a theory of human Gestalt perception, arguing that any grouping is a history. This exactly accords with the view in MCAD that completions come from the *traces* of the machine cutters. In Chapters 2 and 6 of Leyton [96], we gave a theory of the *inference of traces*.

Now let us return to the present book - to Chapter 5, which gives a theory of Gestalt grouping. According to that theory, a grouping corresponds to a fiber (left-subsequence) in a wreath hiearchy. The link to generative history comes from our argument that the phase-structure of a generative history corresponds to the fiber structure of a wreath product - the reason being that the generative history is structured by maximizing transfer; i.e., a phase of the generative history becomes the symmetry ground-state for the next higher control group. Now it is clear that machined features correspond to phases in the generative history of the part:

$$\text{machined features} \longleftrightarrow \text{phase-structure.}$$

However, our theory of grouping claims:

$$\text{phase-structure} \longleftrightarrow \text{fiber structure} \longleftrightarrow \text{grouping structure.}$$

This means that the machined features correspond to the fiber structure which itself corresponds to the grouping structure.

One sees therefore that the recovery of the machined features (e.g., slots) is given by the recovery of groupings. Let us see how this is done. First consider an individual machined feature (e.g., a slot). This is a *trace*, a phase of *internal* structure (Sect. 2.9). Now the interference of another machined feature, (e.g., a second overlapping slot), creates *external* structure, in the form of breaks in the occupancy structure. That is, the second machined feature breaks the symmetry of the first machined feature, by introducing those external asymmetries. Therefore, the recovery of the first machined feature as a complete grouping (trace) means removing the external asymmetries, leaving the original symmetries of that trace structure.

This is exactly what one can see in the example shown in Fig. 14.30, which uses the hint-based method developed by Han, Regli & Brooks [48].

---

[8] Vandenbrande [149], Vandenbrande & Requicha [150], Han & Requicha [49], Gupta, Nau, Regli, & Zhang [?], Regli [122], Han, Regli & Brooks [48].

This figure gives the machined features inferred from Fig. 14.29. Seven slots and two holes are inferred. In Fig. 14.29, the surfaces of these features were incomplete; in Fig. 14.30 they are complete, and give the boundary structures of the machined features. According to our theory, the completions are *symmetries* inferred from the asymmetries in Fig. 14.29. The symmetries are given by the mathematical groups:

$$
\begin{aligned}
\text{Hole} &= [SO(2) \circledW \mathbb{R}] \circledW [SO(2) \circledW \mathbb{R}] \\
\text{Slot} &= [SO(2) \circledW \mathbb{R}] \circledW [SO(2) \circledW \mathbb{R}] \circledW \mathbb{R} \\
\text{Pocket} &= [SO(2) \circledW \mathbb{R}] \circledW [SO(2) \circledW \mathbb{R}] \circledW [\mathbb{R} \circledW \mathbb{R}].
\end{aligned}
\tag{14.13}
$$

Notice that, in accord with our theory, this fully defines the grouping structure.

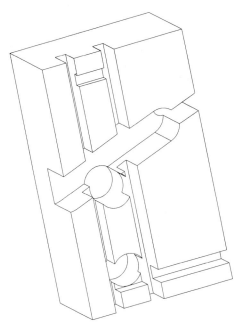

Fig. 14.29. Part similar to Han Part 1 (see, Han, Regli & Brooks [48]).

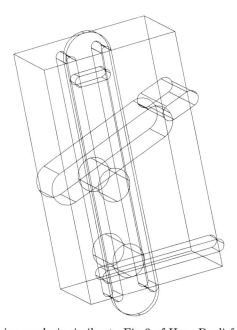

Fig. 14.30. Machining analysis similar to Fig 3 of Han, Regli & Brooks [48].

15. A Mathematical Theory of Architecture

15.1 Introduction

Our mathematical theory of architecture is a means of describing the entire complex structure of a building (from the large-scale massing volumes down to the heating and ventilation systems) by a single symmetry group. This single group is required for all planning with respect to the building, because, by our theory of recoverability, plans come from the symmetries of a structure.

To explain further: Since almost all of the environment is artificially constructed, the rigorous study of architecture is essential not just for the understanding of design but for the study of perception, motor control, navigation and robotics.

We argue that all plans used with respect to a building - e.g., plans for designing the building and plans for navigating through the building - come from the same generative description of the building. Therefore, we must ask: What is the mathematical structure of a building such that this structure facilitates the user's plans with respect to the building?

The answer, we propose, is a generative theory of the building such that the representation *maximizes transfer and recoverability*. This maximization is essential for any intelligent use of the building.

Our theory of recoverability says that all plans that maximize intelligence come from a particular kind of symmetry group - a wreath product (transfer structure) in which each transfer level is the symmetry group of an asymmetrizing process. For complex structures, such a group is an unfolding group.

Comment 15.1 *Substantial parts of our theory of mechanical design are relevant to this theory of architecture but will not be repeated here, to save space. We will therefore assume that the reader has thoroughly absorbed Chapter 14 on mechanical design and regards most of that chapter as part of the present chapter.*

15.2 The Design Process

Essentially, the design process in architecture follows the successive stages[1] shown in Table 15.1 p. 369. The reader should go through the table before continuing.

Two of our basic proposals will be as follows:

BASIC PROPOSALS.
1. The standard architectural design process (Table 15.1) moves forward by successive symmetry-breaking, or more deeply, asymmetry-building.
2. The process is a sequence of unfoldings - in the strict mathematical sense of our unfolding groups.

The approach we develop gives considerable insight into the nature of architectural structure. For example, it will be seen that the *mass groupings* in a building are in fact unfoldings. Furthermore, it will also be seen that the slicings of a building - to produce floor plates - uses exactly the same mathematics - that of unfoldings. The difference is merely one of dimension. Thus, all phases of the design process are formulated in terms of unfolding groups.

15.3 Massing Studies

Let us start at the top of Table 15.1, with the first phase of the design process.

It is generally accepted that architectural design begins with a *massing study*. This should be strongly distinguished from architectural drafting which

---

[1] I am extremely grateful to Dhanraj Vishwanath for his enormously informative explanations of current architectural practice. As an experienced architect, he made himself constantly available for highly detailed expositions on all the systems (engineering, etc.) involved in the architectural process. Also of considerable help was the outstanding software documentation of *AutoCAD Architectural Desktop*, which is the most advanced architectural design program currently available.

is a late phase in the design process and which is concerned with producing precise 2D blue-prints for communicating with construction personel. The purpose of a massing study is very different: to define the main distinctive structure of the building. It can be characterized by three features: (1) It is inherently *3-dimensional* despite being represented by 2D drawings. (2) It is *conceptual* - defining the large-scale structure that characterizes the building both for the designer and the future viewer. (3) It is *exploratory*, allowing the architect to try to develop and discover different solutions.

15.4 Mass Elements

A massing study proceeds by creating, deforming, and combining, *mass elements*; i.e., solid primitives. Chapter 16 will give our theory of solid structure. That chapter will show that it is possible to correspond the solid primitives with the surface primitives (iso-regular groups) of Chapter 10; i.e., since the latter can be considered to be the bounding surfaces of the solid primitives. Thus for ease of exposition, the present chapter will refer to the surface primitives (iso-regular groups) of Chapter 10, when in fact, we will mean the corresponding solid primitives of Chapter 16.

In fact, we call the surface primitives of Chapter 10, the *surface kernels*, of the corresponding solid primitives. Surface kernels are the surface components that distinguish one solid primitive from another. The full specification of the surface kernels (surface primitives) was presented in Table 10.1 (p. 237). In our theory of solid structure, each solid primitive essentially consists of two components: (1) the infinite solid component, which is the wreath c-polycyclic translation group $\mathbb{R} \ \textcircled{w} \ \mathbb{R} \ \textcircled{w} \ \mathbb{R}$, and (2) a surface kernel which breaks the symmetry of the solid component in a way that is particular to that primitive alone.

15.5 The Hierarchy of Mass Groups

Generally, a mass element can undergo two types of operation (1) deformation, and/or (2) combination with another element; i.e., the operations of constructive solid modeling (CSG). Our theory of CSG has been given in the chapters on unfolding groups.

In architecture one considers mass elements to be combined into *mass groups*. These mass groups are themselves *grouped into higher groupings*, and so on. Our main claim will be this:

ARCHITECTURAL MASS GROUPS.
(1) A mass group is the unfolding of an alignment kernel; i.e., there is a correspondence between mass groups and alignment kernels.

$$mass\ groups\ \leftrightarrow\ alignment\ kernels$$

(2) The alignment kernels are themselves unfolded hierarchically from a base alignment kernel up through subsidiary alignment kernels.
(3) A single symmetry group gives the entire unfolding structure of the building.
(4) This maximizes transfer and recoverability.

To understand the above proposals, and to appreciate the deep power of our generative theory to reveal the structural relationships in a massing study, the reader should go carefully through the worked example in the next section.

15.6   Symmetry Group of a Massing Structure

Let us systematically analyze a particular example, a massing study of an office building. Two views of this study are shown in Fig. 15.1. The reader should note that the study is inherently 3D, as are all massing studies, despite being given by 2D images.

In our discussion, the masses will be labeled by numbers, as shown in the plan in Fig. 15.2. Conventionally plans are 2D structures that come later in the design process. However, a plan has been given here merely to show the labeling of the masses by numbers.

Let us first give an intuitive description of the massing structure of the building. A much more powerful description will be given in the mathematical discussion that follows this, because, by the use of wreath products and direct products, we will really be able to understand how the masses are grouped.

However, to begin intuitively: Observe that, because the largest mass is mass 1, shown in the plan, the eye takes this to be the mass to which the others are referred. All the masses are clustered around this main mass, and therefore one can regard the entire structure as a mass grouping. However, within this, there are subsidiary mass groups:

(1) Clearly, masses 2, 3, and 5, form a mass group that goes off the main mass 1.

(2) Furthermore, it is obvious from the massing study Fig. 15.1, that the large cylindrical tower with its dome forms a mass group. The dome is conceptually part of a sphere which is aligned with the cylinder as shown in the wire-frame view in Fig. 15.3. In the massing views shown in Fig. 15.1, the

Table 15.1. The architectural design process.

I. Conceptual Design.
    1. The massing study.
    2. Slicing the massing study to create floorplates.
    3. Space Planning.
II. Design Development.
    1. Choice of materials.
    2. Doors and windows.
    3. Structural column grid.
    4. Ceiling grid.
    5. Stairs.
    6. Shafts.
    7. Roof.
    8. Development of accuracy.

III. Construction Documentation.
    1. Creating elevations and sections.
    2. Dimensioning.

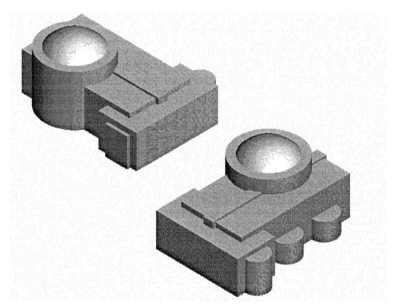

Fig. 15.1. Two views of a massing study.

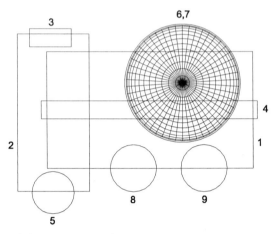

Fig. 15.2. Plan of the massing study.

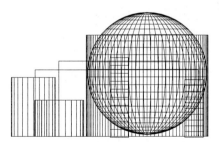

Fig. 15.3. Wireframe side-view of the massing study.

dome is that part of the sphere which is visible above the cylinder. The complete tower is therefore the Boolean sum of the cylinder and sphere masses.

(3) In addition, there are some individual elements that come off the main mass: (a) The thin rectangular spine, mass 4 in the plan; and (b) the two cylindrical lobby entrances, shown as mass 8 and 9 in the plan.

Having described the massing structure intuitively, we shall now understand it much more deeply using the generative theory of shape. The structure will be constructed in stages:

(1) **The alignment kernel.** This contains three primitives: the cube P_1, the cylinder P_2, and the sphere P_3. It is therefore written as follows:

$$[[G_{sphere}]_{P_3} \times [G_{cylinder}]_{P_2} \times [G_{cube}]_{P_1}]_{\mathcal{U}}.$$

Algebraically, the alignment kernel is thus the direct product of a number of clones of each primitive. From a group-representation point of view, all of the clones should be understood as maximally aligned with each other and with the World-frame. The number of clones will be determined by the wreath-direct actions with the control groups. Until we determine the number of clones of each primitive, we shall often speak of *the* cube, *the* cylinder and *the* sphere.

(2) **The hierarchy of rectangular masses.** Let us first consider only the rectangular masses in the structure. They have a particular referential order which is given by the following group:

$$
\begin{aligned}
&[[G_{sphere}]_{P_3} \times [G_{cylinder}]_{P_2} \times [G_{cube}]_{P_1}]_{\mathcal{U}} \\
&\text{\textcircled{w}}\ [_{daP_1} AGL(3,\mathbb{R})^{aP_1} \\
&\times\ [[_{cbaP_1} AGL(3,\mathbb{R})^{baP_1} \times\ _{baP_1}\{e\}^{baP_1}]\ \text{\textcircled{w}}\ _{baP_1} AGL(3,\mathbb{R})^{aP_1}] \\
&\times\ _{aP_1}\{e\}^{aP_1}] \\
&\text{\textcircled{w}}\ _{aP_1} AGL(3,\mathbb{R})^{P_1} \times\ _{P_1}\{e\}^{P_1}\quad. \qquad\qquad (15.1)
\end{aligned}
$$

Let us go through this expression from bottom-to-top. In the bottom line the right-hand group fixes a copy of P_1, the World frame. The purpose of this is to hold a copy of the *internal symmetries* of the construction plane, as is required by the theory of construction planes, summarized on p. 212. The left-hand group also takes the cube primitive P_1 as input, but outputs an affine-altered version of it, aP_1, corresponding to the main rectangular mass - numbered 1 in Fig. 15.2.

The output index aP_1 then becomes the input index on each of the three lines above; i.e., at the right-hand end of each of these lines. Each of these lines determines a different fate for aP_1. Thus working upwards through these lines:

Line 4 fixes a copy of mass 1 - that copy seen in Fig. 15.2. Next, considering Line 3, the right-most group takes aP_1 and applies an affine transformation b to it, producing the output index baP_1, which represents mass 2 in the plan-view in Fig. 15.2. Then, moving left-ward on this line, there is a wreath-product symbol ⓦ, the fiber of which is a direct product, each component of which takes baP_1 as input. The first component is the identity group $\{e\}$ which fixes a copy of baP_1, which is mass 2 as seen in Fig. 15.2. The other group in the direct product has output $cbaP_1$ which corresponds to mass 3 in Fig. 15.2. The direct product shows that this object is a misalignment with respect to the copy held by the identity component; i.e., we have *symmetry-breaking by misalignment*.

Finally, Line 2 also takes aP_1 as input, but outputs daP_1 which represents mass 4 in the plan-view Fig. 15.2.

Thus, the reader can see from the entire expression (15.1) that there must be five clones of the cube primitive in the alignment kernel.

(3) The addition of mass 5. We now add the cylinder shown as mass 5 on the plan-view, Fig. 15.2. This is understood as coming off the rectangular mass 2. However, mass 2 is understood as coming off mass 1, which is itself derived from the alignment kernel. Thus, the cylinder comes from the alignment kernel and is moved to be aligned with mass 1, and moved again to be aligned with mass 2, and finally misaligned to become mass 5. The appropriate group description is therefore the following:

$$
\begin{aligned}
&[[G_{sphere}]_{P_3} \times [G_{cylinder}]_{P_2} \times [G_{cube}]_{P_1}]\mathcal{U} \\
&ⓦ \; [_{daP_1} AGL(3,\mathbb{R})^{aP_1} \\
&\times \; [[_{wvuP_2} AGL(3,\mathbb{R})^{vuP_2} \times \; _{cbaP_1} AGL(3,\mathbb{R})^{baP_1} \times \; _{baP_1}\{e\}^{baP_1}] \\
&ⓦ \; _{*vuP_2 \, , \, baP_1} AGL(3,\mathbb{R})^{\; uP_2 \, , \, aP_1}] \\
&\times \; _{aP_1}\{e\}^{aP_1}] \\
&ⓦ \; _{*uP_2 \, , \, aP_1} AGL(3,\mathbb{R})^{\; P_2 \, , \, P_1} \times \; _{P_1}\{e\}^{P_1} \quad .
\end{aligned}
\tag{15.2}
$$

This is the same as the group given before, except that a cylinder mass has been added into the direct product in Line 3, where it is the left-most component. For notational reasons the control group has been moved down a line. A cylinder index (based on P_2) has been placed on this control group and that on the bottom line.

Thus, starting with the bottom line, the affine group now has two input indexes representing the cube and cylinder primitives P_1 and P_2 from the alignment kernel. By Notation 13.3 (p. 298), the affine group here actually means the direct product of affine groups:

$$
_{*uP_2 \, , \, aP_1} AGL(3,\mathbb{R})^{\; P_2 \, , \, P_1} \; = \; _{*uP_2} AGL(3,\mathbb{R})^{P_2} \times \; _{aP_1} AGL(3,\mathbb{R})^{P_1}.
$$

The output index uP_2 is prefixed by an star because the u is chosen such that the cylinder remains maximally aligned with the cube up to deformation; that is, u is the non-deformation part of a.

Then, two lines above this, Line 4 takes as input the output of the affine component of the bottom line. Also, the double index on the group of this line indicates that it is a direct product, thus:

$$_{*vuP_2} ,\ _{faP_1} AGL(3, \mathbb{R})\ ^{uP_2 ,\ aP_1} = {}_{*vuP_2} AGL(3, \mathbb{R})^{uP_2} \times {}_{faP_1} AGL(3, \mathbb{R})^{aP_1}.$$

The effect of this line is to move the cylinder out together with the rectangular mass, which becomes mass 2 in Fig. 15.2. As indicated by the star, the cylinder is not deformed but remains maximally aligned with the rectangular mass, up to deformation.

Next, go upward one line to Line 3. Working from right to left, there is first an identity component which fixes a copy of aP_2, mass 2. The next affine group on this line moves the rectangle out from alignment with respect to mass 2, to become Mass 3; and the next affine group moves the cylinder out from alignment with respect to mass 2, to become the entrance mass 5.

Notice that Line 3 allows us to infer the distribution of clones in Line 4. It tells us that the cube index in Line 4 actually corresponds to two clones; and the cylinder index in Line 4 corresponds to only one clone.

Finally, Line 2, which produces mass 4, is the same as before.

(4) The mass grouping (6,7). The tower shown as (6,7) in the plan consists of the sphere and cylinder. This is unfolded in parallel with the mass grouping (2,3,5). Therefore, it will be in direct-product relation to the latter. However, one also needs to bring in the sphere P_3 from the alignment kernel so that this can happen. The resulting group is:

$$
\begin{aligned}
&[[G_{sphere}]_{P_3} \times [G_{cylinder}]_{P_2} \times [G_{cube}]_{P_1}]_\mathcal{U} \\
&\text{\textcircled{w}}\ \ [_{daP_1} AGL(3, \mathbb{R})^{aP_1} \\
&\times\ [[_{wvuP_2} AGL(3, \mathbb{R})^{vuP_2} \times {}_{cbaP_1} AGL(3, \mathbb{R})^{baP_1} \times {}_{baP_1}\{e\}^{baP_1}] \\
&\text{\textcircled{w}}\ \ _{*vuP_2} ,\ _{baP_1} AGL(3, \mathbb{R})\ ^{uP_2 ,\ aP_1}] \\
&\times\ _{muP_3} ,\ _{muP_2} AGL(3, \mathbb{R})\ ^{uP_3 ,\ uP_2} \\
&\times\ _{aP_1}\{e\}^{aP_1}] \\
&\text{\textcircled{w}}\ \ _{*uP_3} ,\ _{*uP_2} ,\ _{aP_1} AGL(3, \mathbb{R})\ ^{P_3 ,\ P_2 ,\ P_1} \times {}_{P_1}\{e\}^{P_1} \quad .
\end{aligned}
\tag{15.3}
$$

Notice that this is the same as the previous expression (15.2), except for two changes: An extra index has been added to the last line, indicating the unfolding of the sphere primitive P_3. And one extra line has been inserted, Line 5. The inputs to this line are the (sphere, cylinder) outputs from the bottom line. This produces the new mass grouping (6,7). Notice that this line is in parallel with the mass grouping (2,3,5) and also in parallel with mass 4.

Consequently, there are the three direct product symbols shown down the left side of the entire group-sequence.

(5) The addition of lobby-entrances 8 and 9. Finally, we add the cylindrical lobby entrances 8 and 9 (in plan-view Fig. 15.2). These come off the main mass 1, and are therefore parallel to the spine mass 4, and to the mass grouping (2,3,5), and also to the tower mass grouping (6,7). The full group is therefore given thus:

$$[[G_{sphere}]_{P_3} \times [G_{cylinder}]_{P_2} \times [G_{cube}]_{P_1}]_{\mathcal{U}}$$
$$Ⓦ \; [_{daP_1} AGL(3,\mathbb{R})^{aP_1}$$
$$\times \; [[_{wvuP_2} AGL(3,\mathbb{R})^{vuP_2} \times {}_{cbaP_1} AGL(3,\mathbb{R})^{baP_1} \times {}_{baP_1} \{e\}^{baP_1}]$$
$$Ⓦ \; {}_{*vuP_2} , {}_{baP_1} AGL(3,\mathbb{R})^{\; uP_2} , {}^{aP_1}]$$
$$\times \; {}_{huP_2} AGL(3,\mathbb{R})^{uP_2} \times {}_{guP_2} AGL(3,\mathbb{R})^{uP_2}$$
$$\times \; {}_{muP_3} , {}_{muP_2} AGL(3,\mathbb{R})^{\; uP_3} , {}^{uP_2}$$
$$\times \; {}_{aP_1} \{e\}^{aP_1}]$$
$$Ⓦ \; {}_{*uP_3} , {}_{*uP_2} , {}_{aP_1} AGL(3,\mathbb{R})^{\; P_3} , {}^{P_2} , {}^{P_1} \times {}_{P_1} \{e\}^{P_1} \quad . \qquad (15.4)$$

This is the same as the previous expression in (15.4) except that a new Line 5 has been inserted, unfolding the two cylinder lobbies. This of course does not merely alter the algebraic structure by this one insertion, but adds two extra cylinder clones to the alignment kernel, and to each cylinder index below Line 5.

15.7   Massing Structure and Generativity

The previous section began by giving an *intuitive* description of the massing structure of the example building. This description gave a list of mass groupings in the building. However, this description, which is typical of architectural dialogue, does not give what we regard as the most important information in a massing structure: the *generative* organization. Most importantly, it is only by considering generativity, that the deep relationships between the masses become evident.

Thus, after the intuitive description, we gave a description based on our generative theory of shape. In this, the massing structure is truly revealed. According to our theory, any mass grouping comes from its own *alignment kernel* - recall the distinction, in Sect. 13.7, between base and subsidiary alignment kernels. Indeed there is a correspondence between the mass groupings and the alignment kernels in the generative structure. Furthermore, a mass grouping is the *unfolding* of the associated kernel. These kernels are themselves unfolded ultimately from the base kernel.

This description reveals other powerful structural factors. The generative hierarchy contains direct products and wreath products. These show the complex and subtle relationships between the masses. Masses and mass groupings that are linked by direct products are unfolded parallel to each other in the full structure. Masses and mass groupings that are linked by wreath products are unfolded in a control-nested fashion, and are therefore referentially related, in accord with our algebraic theory of parent-child hierarchies. The entire unfolding structure then becomes a configuration of direct and wreath products that tell us exactly the unfolding relationships within the mass structure.

15.8 Slicing the Massing Study to Create Floorplates

With the massing structure completed, the next phase in the architectural design process is to *slice* through the 3D structure to obtain the floorplates. Floorplates give the perimeter geometry of the building. Therefore, what one is really slicing are the boundary surfaces of the solid model.

In Sect. 14.20 (starting p. 336), we showed how slicing can be expressed by unfolding. Now apply this theory to architecture, as follows: To obtain the slicing that produces floorplates from the 3D model, take one of the planes from the cube primitive within the alignment kernel. This primitive always exists in the alignment kernel either explicitly or as its surrogate, the gravitational frame. The cube is given in Table 10.1 (p. 237) as the hyperoctahedral wreath hyperplane group consisting of a control group $\mathbb{Z}_2 \textcircled{w} \Sigma_3$ acting on the fiber-group product which consists of six copies of the plane - the six sides of the cube. The particular copy of the plane that will be selected for slicing, is the bottom plane. This becomes the ground plane of the building. It is then unfolded upwards through the building to become the successive floorplates.

Now the following important distinction needs to be made: The plane which became the bottom plane of the cube was actually produced earlier in the generation sequence: Recalling Fig. 14.15 (p. 347), it is the first plane to appear in the creation of the infinite solid, i.e., the plane is the first two levels $\mathbb{R} \textcircled{w} \mathbb{R}$ of the solid as a 3-fold wreath c-polycyclic group $\mathbb{R} \textcircled{w} \mathbb{R} \textcircled{w} \mathbb{R}$. Prior to the introduction of the cube, this plane has already been swept upward via the third level of $\mathbb{R} \textcircled{w} \mathbb{R} \textcircled{w} \mathbb{R}$ because this was how the infinite solid was created. Thus, it seems that, to maximize transfer, one does not need to sweep this plane upward again to create the floor plates; i.e., one can regard the floor plates as particular *selected* members of the original continuum of parallel planes that constitute the solid.

However, the floor planes have a different role: Being slice structures, they are actually *boundary* surfaces. This means that conceptually one should think of them as transferred versions of the bottom plane of the cube as a boundary structure - i.e., their existence occurs only after the establishment of the cube not before. Remember: The cube is a boundary structure because it is introduced as cutting the previously established infinite solid. The floor planes, as slice structures, are conceptually transfers of this cutting structure.

The way we handle this is to use the theory of *super-local* unfolding. The reader should think of the situation as being exactly analogous to the example given in Sect. 12.3 - particularly the figures on p. 259 - where a door opening was created by two slice-lines which were transferred versions of the sides of one of the rooms. That previous example is simply a version which is one dimension lower than the example now being studied.

Most crucially therefore, the group generating the slice planes is placed on a control level *higher* than the group creating the massing structure, and unfolds only part of that lower structure. That is, we have this:

$$K \ \textcircled{w} \ G_m \ \textcircled{w} \ G_s^{a...bP}$$

where K is the alignment kernel, G_m is the unfolding group of the *massing* structure, and G_s is the unfolding group of the *slice* structure. The input index $a...bP$ on G_s is the plane P taken from the bottom plane of the cube, but after P has undergone some of the unfolding operations $a...b$ from the lower unfolding group G_m. Notice that this is therefore a *super-local* action.

Some additional comments: The fact that we defined the gravitational coordinate system as algebraically equivalent to the cube primitive, is particularly relevant here: The unfolded slice plane literally represents the horizontal plane with respect to gravity; i.e., floors must *necessarily* be gravitationally horizontal.

This fact becomes particularly conspicuous in deformed structures. For example consider Gallery 304 shown in the center of Gehry's Guggenheim Museum in Fig. 15.4. This gallery is a mass that is deformed upwards, as can be seen by its upper roof-line, and yet its floor must be horizontal. Notice that this floor is actually coincident with the third-floor slice-plane shown in the rectangular structures on the right. Thus the plane has been unfolded through the deformed structure and ignores that deformation. This illustrates the power of super-local unfolding to ignore chosen aspects of a structure - as discussed in Sect. 12.7.

Let us now turn to the issue of regularity and irregularity in the spaces between slice-planes. It is often the case that floors are spaced at equal intervals upward through the building, in which case the affine action that moves the slice plane through the building will be an occupied subset of the infinite

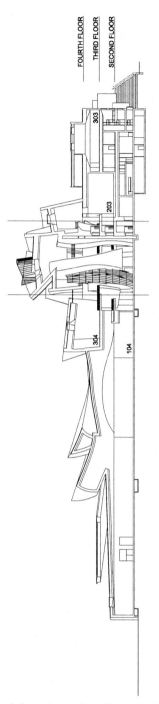

Fig. 15.4. Frank Gehry and Associates, Inc: Guggenheim Museum at Bilbao.

cyclic group $\mathbb{Z}$. However, in some buildings, certain floors can be of different spacings; for example, in an apartment building it is often the case that the ground floor has a greater ceiling height than the other floors. This can be modeled as a *deviation* from the regular action of $\mathbb{Z}$ established by the other floors; i.e., an additional layer of external history, in accord with our Externalization Principle which says that any external history goes back to an internal structure that is a control-nested hierarchy of repetitive isometries. Here the internal structure is the regular floor structure $\mathbb{Z}$, and the local deviation becomes the fiber to that group.

15.9   Space Planning: Unfolding of Space Volumes

The next phase in the design process is *space planning*. We shall formalize this using a two-stage cycle.

Space-planning begins by considering the volumes between the floorplates. There is no architectural term for these volumes, and the term "floor volume" is ambiguous because it could be confused with the floor-slab volume, which will be needed later. So the French word *étage* will be used, thus:

ÉTAGE VOLUME.   *The volume between two successive floorplates will be called the* **étage volume**.

To make the discussion easier to follow, let us assume that the étage volumes are rectangular. It becomes an simple matter to extend the discussion to non-rectangular étage volumes.

It is necessary to consider the étage volumes as entities in their own right, as follows: Consider how these volumes were derived. The first design stage created the massing volumes that will form the basis of the entire design process. The second stage unfolded the slice-planes through the massing volumes, creating the floorplates. Clearly, one can define the étage volumes between the floorplates merely as *parts* of the original volume. Most crucially, in this description, they are relics of the original volume, in the same sense that the slices of a cake are relics of the complete cake that existed before the slicing operation.

However, in space planning, one needs to incorporate also a different description of the étage volumes as follows: These *part* volumes should be redescribed as *whole* volumes, in their own right. By this we mean that each étage volume should now correspond to a *whole* primitive in the alignment kernel.

When the étage volume is rectangular, the particular primitive from which it is derived can be regarded either as a cube or a primitive block. There are some advantages to using the latter alternative, because the CAD architect will often define aspects of the étage volume by an upward sweeping structure. This again is strongly evident in a major CAD program such as *AutoCAD Architectural Desktop*.

Therefore, to allow this to occur, the block primitive must be added to the alignment kernel. Table 10.1 (p. 237) of primitives gives two alternative versions of the block primitive: the cross-section block or ruled block. In the current situation, the cross-section block is the appropriate description because it is generated by taking a cross section $\mathbb{R} \circledw Z_4$, and sweeping it along the perpendicular translation direction $\mathbb{R}$, which acts as a control group moving the cross-section as a fiber: Thus the cross-section block has this structure:

$$\mathbb{R} \circledw Z_4 \circledw \mathbb{R}$$

where the upper $\mathbb{R}$ is the sweeping control group. Now, this translation control group follows the same direction as the translation movement that pulled the slice-planes through the building to produce the floorplates. Thus the structure of the cross-section block accords with the control action that has already been established in the symmetry structure of the floorplate arrangement.

To conclude: To generate an étage volume, as a conceptual entity in its own right, load a primitive cross-section block into the alignment kernel.

An additional factor needs to be specified: The block primitive is a cube, whereas the étage volumes are rectangular; i.e., the floors break the symmetry of the cube. Therefore, unfold the étage volumes not only using the floorplate translation group, but also using the appropriate stretch transformation from the remainder of the affine group.

In this section, we have defined two stages, thus:

INDIRECT/DIRECT UNFOLDING CYCLE.
(1) Indirect unfolding: *Slice through a whole volume that corresponds to an unfolded primitive, thereby producing part-volumes.*
(2) Direct unfolding: *Then unfold these part-volumes directly from their own primitives; i.e., redescribing them as whole-volumes.*

The Indirect/Direct Unfolding Cycle will be used again and again at ever smaller scales, as follows.

15.10   Space Planning: Unfolding the Boundary and Void Spaces

The next stage in space planning is to define the boundary of the étage volumes as having *thickness*; i.e., they are actual floors and ceilings. The reason for defining these thicknesses is that, in space planning, the designer begins to incorporate factors that will be important to the main engineering systems: load-bearing factors (structural systems), and the heating and ventilation ducts (building systems), etc.

We shall formalize this stage using the Indirect/Direct Unfolding Cycle, as follows: Take any whole-volume coming out of the previous stage, i.e., any étage volume - a volume between the successive floorplates - and do this:

(1) **Indirect Unfolding:** Slice the étage volume into four parallel part-volumes: The floor-slab; the spatial void above (through which people will move); the ceiling slab; and the above-ceiling clearance. Notice that this involves unfolding slice-planes through the étage volume.

(2) **Direct Unfolding:** Take the part-volumes just produced by indirect unfolding, and directly unfold them from their own primitives in the alignment kernel.

15.11   Unfolding the Room Volumes

Let us regard the next stage of space planning as the partitioning of the étage into individual room volumes. Again, carry this out by another use of the Indirect/Direct Unfolding Cycle. However, now use vertical slice planes:

(1) **Indirect Unfolding:** Slice the étage using *vertical* slice-planes. The resulting part-volumes will be the room volumes.

(2) **Direct Unfolding:** Take the part-volumes just produced by indirect unfolding, and directly unfold them from their own primitives in the alignment kernel; i.e., the rooms will be described as whole-volumes in their own right.

Notice that this is the first time that the plan structure of an étage has appeared.

15.12   Unfolding the Wall Structure

The walls are vertical slab boundary volumes to the rooms, and are unfolded using the same cycle. Conceptually, one can regard this phase as the vertical equivalent to the unfolding of the floor-slabs as boundary volumes of the étage.

Comment 15.2 *Looking back over the entire space planning process, it can be seen that we have simply used the Indirect/Direct Unfolding Cycle repeatedly at ever smaller scales. Thus, the cycle was used to create the étage volumes from the mass volumes; then to create the boundary volumes (floor and ceiling slabs) of the étage volumes; then vertically to create the room volumes (and hence the first plan); and finally to create the boundary volumes (walls) of the room volumes.*

15.13 Complex Slicing

The approach we have defined allows for complex slicing. Consider for example the slicing of an étage into a complex room-structure. Fig. 15.5 shows a slicing which is not simply a sequence of planes across the étage. Consider, for instance, the room labeled A. Its right and bottom wall were created together as a single polyline - that shown on the far right of the figure. A polyline in CAD is a continuous set of line and arc segments, drawn sequentially from one end to the other. At first, this approach might seem very different from the scenario given above. But close examination reveals that it is not, as follows:

In drawing a polyline, the computer prompts one with a series of questions starting with: "Line or circular arc?" With the selection of a line, one chooses the direction together with the start and end point. With the selection of an arc, one chooses the center together with the start and end point. At each end point, the computer offers again the option of drawing another line or arc. Note also that in modern architectural CAD (e.g., AutoCAD's *Architectural Desktop*), one has the option of drawing a *wall* directly in this manner; i.e., a polyline with *height* in the third dimension. If this facility is not available, one standardly sweeps the polyline vertically to create the wall height.

Now, close examination of what has just been described reveals that it conforms to the method we described for slicing by unfolding, as follows: As was said, in drawing the polyline wall, the computer first prompts one with the question "Line or arc?". We argue that the choice of line, is equivalent to bringing a plane up from the alignment kernel. The plane will have a start and finish point, i.e., an occupancy specification on its infinite extent, but this was the case with our previous scenario. Thus, as in our basic scenario, one is unfolding an infinite plane from the alignment kernel to slice the space. A sequence of planes along the polyline is simply a sequence of such slicings.

Supposing instead that one chooses the "arc" alternative, at some step along the creation of a polyline wall. Although we did not previously handle

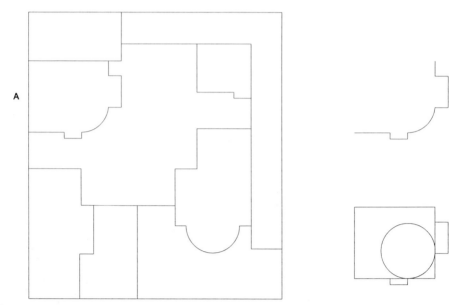

Fig. 15.5. Polyline-to-Volume Conversion.

slicings by non-planes, the principles are exactly the same. The slice-objects are pulled up from the alignment kernel. For an arc, the slice object is a cylindrical surface; and again the start and end points merely specify occupancy.

The wall polyline is therefore simply a set of slice planes and cylinder surfaces unfolded sequentially from the alignment kernel.

This stage in fact corresponds formally to the first stage in the Indirect/Direct Unfolding Cycle; i.e., the creation of part-volumes by the unfolding of slice objects from the alignment kernel. Thus, for example, the room volume shown at A in Fig. 15.5, has been created by a slicing produced by the wall-polyline that was discussed. The room volume is a part volume of the étage - a part volume that has been created by the slicing.

In the second stage of the Indirect/Direct Unfolding Cycle, one redescribes this part volume by directly unfolding whole volumes as primitives from the alignment kernel. For this room, the unfolding is shown in the bottom-right diagram in Fig. 15.5. Notice that the previous slice planes are actually boundaries of the rectangular whole-volumes. This is exactly as it should be: These slice planes in fact came from the edge-planes of the cube in the alignment kernel, and the cube is the primitive now used to create the whole-volume. Similarly, the circular arc was previously created by using exactly the same cylinder as the cylinder that is now brought up as a whole-volume.

One final thing should be noticed: The second stage described here - i.e., the direct unfolding of a room volume - is an example of *sub-local unfolding* (Chapter 13). For example, the diagram given here for unfolding a single room - the bottom right diagram in Fig. 15.5 - is strongly reminiscent of the type of diagram produced when we unfolded the eight-room house in Fig. 13.5 (p. 294). Observe for example, in the room example being described here - bottom right diagram Fig. 15.5 - the circle is clearly seen as a *misalignment* from the center of the main rectangle. The same is true of the minor rectangles. An arbitrarly complex *room* can thus be generated by sub-local unfolding in the same way as an arbitrarily complex *house* was generated previously.

This strongly attests to the power of our theory to provide a single generative method for a large variety of shape situations.

15.14   Design Development Phase

The previous section essentially ended the conceptual development phase, and we now come onto design development. There are a number of purposes of design development:

1. Choice of materials.
2. Doors and windows.
3. Structural column grid.
4. Ceiling grid.
5. Stairs.
6. Shafts.
7. Roof.
8. Development of accuracy.

15.15   Choice of Materials

So far the space-division plans of the building have been purely *geometric*. In such a plan, there has been no distinction between the space within a wall and the space outside it. In the design-development phase, one has to choose different materials for these spaces. For example, a masonary material is chosen for the space within a wall, and one implicitly acknowledges the fact that this is not the material, air, that is outside the wall.

Since our theory is *generative*, we want to be able to describe the choice of materials generatively. Furthermore, since the theory ultimately rests on identifying the symmetry structure of a situation, we propose to do this by using color groups. Recall the fundamental reason why the Russians developed color groups: Enormous benefits had been gained by describing the *geometric* properties of crystals in terms of symmetry and therefore group theory. However, this approach seemed unable to handle non-geometric properties, which were usually referred to as *material properties*. It was in order to overcome this barrier, that the Russians, Heesch (in the 1920's) and later Belov and Shubnikov, developed the concept of color groups. A color group consists of transformations that are non-geometric, but instead *change material.*

Let us apply this concept to architecture. Consider the boundary between a stone wall and air. First ignore the materials involved. The wall and air are separated by a plane which reflects one side onto the other. Now consider the material properties. This reflection operation has to change material from stone to air. The material change is not a geometric operation, but a color one. (The term "color" is the technical term used for any property that is not geometric.)

The incorporation of color groups into our theory is an easy matter. Notice that the "symmetry" in the color group actually generates the asymmetry involved in the distinguishability between colors/materials. This accords exactly with our theory of recoverability which states that the generative operations produce asymmetries, which are encompassed within a higher order symmetry group.

15.16   Doors and Windows

As can be seen from Fig. 15.6, doors and windows are almost always *arrangements of rectangular blocks.* A door consists of a door panel (which is a rectangular block), a frame (built of rectangular blocks), and door-stops (which are rectangular blocks). A window consists of a sheet of glass (which is a rectangular block), a frame (which is built of rectangular blocks), and sash components (which are rectangular blocks) [2].

It should be clear to the reader by now that the natural way to generate these two structures is by *sub-local unfolding* (Chapter 13).

Since the placement of doors and windows is being given as an unfolding process, this means that their placement is being represented in terms of creating misalignment from initial alignment. This accords perfectly with architectural CAD. For example, in the placement of doors and windows in AutoCAD's *Architectural Desktop*, the program prompts the designer with the option "Automatic Offset/Center", which means the positioning of the

---

[2] Occasionally, doors and windows can have non-rectangular elements - and these can be easily handled within our system, i.e., by pulling additional primitives from the alignment kernel. However, it is almost certainly the case that the room in which the reader is sitting now has doors and windows that constructed only from rectangular blocks; and therefore, for ease of discussion, we shall assume that we are dealing with this typical case.

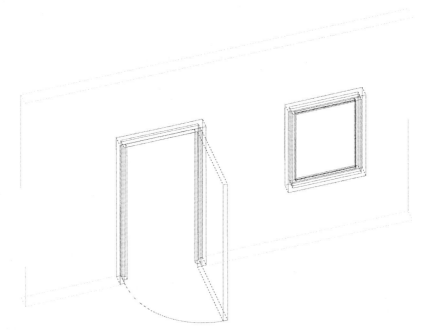

Fig. 15.6. A door and window consist of blocks.

door or window at either an offset distance from one end of the wall, or at the center of the wall. Clearly, the placement of the door or window at the center of the wall means the maximal alignment of their symmetry structure: Then, at a later stage, the designer has the option of moving the door or window, thus creating misalignment from the center - i.e., *unfolding* as we have defined it. The same is true of the offset option. The placement of the door or window at the end of the wall actually means the alignment of the boundary fiber (edge face) of the door or window with the the boundary fiber (edge face) of the wall; i.e., there is maximal alignment of the $\mathbb{R}^2$ symmetry structures of their boundary fibers. The theory of this kind of structure was given by us in the sections on *assembly* in Chapter 14 on mechanical CAD. Finally, the creation of an offset distance becomes the creation of misalignment of these symmetry structures.

The Within-Wall Constraint.
Doors and windows are constrained to lie within walls. Therefore, in unfolding them from the wall structures, one must use only that subgroup of the affine group that keeps them within the wall; e.g., involving only the two-dimensional translation group corresponding to the length and height of the wall. This is an easy matter to specify because the subgroup is based on the symmetries of the wall pulled up from the alignment kernel.

The Influence of the Gravitational Frame.
The fact that doors and windows tend to be built of rectangular blocks attests to the over-riding importance of the gravitational frame. We have argued that the gravitational frame is equivalent to the cube primitive, and the cube primitive is the generative starting state of the blocks unfolded from the alignment kernel. In other words the blocks in a door and window are memory of the gravitational frame.

The Symmetry-Breaking Effect of Doors and Windows.
One can see that there is a successive diminishing of scale starting with the building masses, then going to the étage volumes, then the rooms, the walls, and now the doors and windows. This is a natural process of symmetry breaking. Furthermore, each successive level can be moved within the higher level preceding it, making use of the symmetry of the higher object. For example, the étage volumes can be moved within the mass volumes; the doors and windows can be moved within the wall volumes, etc.

This is the very deep form of symmetry-breaking described in this book. Consider for example a wall before the addition of doors. The wall has translational symmetry along it until the wall's edges. Then, when one adds the door, this translational symmetry is broken. Observe, most importantly, that, in the design process, one moves the door along the wall until one finds the right position. This uses the symmetry of the wall. Thus the very symmetry that is broken in the wall, is used by the object (the door) that breaks that symmetry. This is an essential part of our system. Symmetries are not actually lost in our system: one adds asymmetries by control-nesting symmetries within each other.

15.17   Structural Column Grid

A grid of columns is one of the fundamental structures of architecture: It is the basic means by which the weight is supported in a building - from ancient temples to modern libraries, banks, and airports. One starts by defining a regular grid of columns through the building so that it will support the higher structures; i.e., so that it satisfies the structural engineering analysis. One then might move some of the columns out of regular alignment, to satisfy other constraints such as the building functions, skin geometry, etc, but always with the ultimate goal of maintaining weight support. Modern engineering allows the columns to be moved further from regularity than classical architecture, but nevertheless the design proceeds by *starting* with the regular grid, and then creating deviations from that regularity. For example, this is the way that the menus are standardly set up in architectural CAD; that is, the first menu allows one to choose the spacing of a regular grid, and then one can choose subsequently to move selected columns out of regularity. This means that the design of the column grid proceeds by a process of symmetry-breaking, which is exactly what our theory predicts.

Let us begin by looking at the first stage: The formation of the regular column grid.

COLUMN GRIDS.   *A column grid is an example of what we defined in Sect. 5.4 as a* **generative crystallographic group.**

Recall that a generative crystallographic group was defined as an iso-regular group (wreath c-polycyclic, wreath-isometric) of the form

$$G \;=\; H \; \circledw \; [\mathbb{Z} \; \circledw \; \mathbb{Z} \; \circledw \; \ldots \; \circledw \; \mathbb{Z}]. \tag{15.5}$$

Note that the component H, being iso-regular, has itself a wreath c-polycyclic wreath-isometric decomposition $G_1 \circledw G_2 \circledw \ldots \circledw G_n$ (Definition 5.1 p. 140).

Apply this structure to the column grid, as follows: The column grid has a generative crystallographic group, in which the grid group is the control group, and the column group is the fiber. The grid group is $\mathbb{Z} \circledw \mathbb{Z}$, that is, the movement of a line of columns through the discrete set of intervals in the perpendicular direction. The column group is given by Table 10.1 p. 237 as $SO(2) \circledw \mathbb{R}$ if the column is understood as a cross-section cylinder, and $\mathbb{R} \circledw SO(2)$ if the column is understood as a ruled cylinder. Without loss of generality, assume that the cylinder is the cross-section type. The total group of the column grid is therefore

$$[SO(2) \; \circledw \; \mathbb{R}] \; \circledw \; [\mathbb{Z} \; \circledw \; \mathbb{Z}].$$

Both the control translation-group and the fiber column-group are pulled up from the alignment kernel. The translation group is a discrete subgroup of the continuous translation group $\mathbb{R} \circledw \mathbb{R}$ that forms the bottom plane of the cube primitive; and the column group is the cylinder primitive in the alignment kernel.

Now, to handle the deviation of columns from the regular grid points, merely wreath-append the same bottom continuous plane $\mathbb{R} \circledw \mathbb{R}$, of the cube, below the control group $\mathbb{Z} \circledw \mathbb{Z}$ but above the column $SO(2) \circledw \mathbb{R}$, thus:

$$\underbrace{[SO(2) \; \circledw \; \mathbb{R}]}_{column} \; \circledw \; \underbrace{[\mathbb{R} \; \circledw \; \mathbb{R}]}_{deviation} \; \circledw \; \underbrace{[\mathbb{Z} \; \circledw \; \mathbb{Z}]}_{grid}$$

To understand this consider the wreath product symbol just below the grid. For this wreath product, the control group is the grid, and the fiber is everything below, which means that a plane of deviation exists at each grid point. Furthermore, because the deviation plane is the control group of the column below it (as shown in the above sequence), there is a deviation of the column within the plane located at the grid point. The structure just given therefore beautifully describes the column grid, with its allowed deviation. Note that, because the levels, in the above expression, are entirely linked by wreath products, the structure is entirely described as a hierarchy of transfer, thus satisfying the goal of maximization of transfer.

15.18   Ceiling Grid

The ceiling grid can be specified mathematically in the same way as the column grid. The repetitive placing of lighting fixtures, which is typical in a ceiling-grid situation (e.g., in a restaurant), can again be given as a generative crystallographic structure. A lighting fixture is given by an unfolding group. This group is then put below the grid group in the wreath hierarchy.

15.19   Stairs

As can be seen from Fig. 15.7, stairs are a repetitive use of a horizontal block. We now argue that they are best understood in terms of *transfer* and *unfolding*.

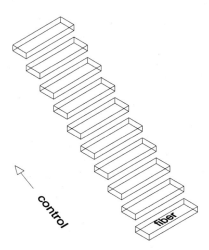

Fig. 15.7. Stairs are a wreath product.

First observe that, functionally, a person walking up a stair transfers the same cycle of motion from one step to the next. That is, there is a transfer structure in the very motion of the person. Correspondingly, the steps form a transfer structure. Each step is a block, which can be assumed to be an affine transformation of the cross-section block $\mathbb{R} \,\text{\textcircled{w}}\, \mathbb{Z}_4 \,\text{\textcircled{w}}\, \mathbb{R}$ in Table 10.1, p. 237. The block is transferred by the cyclic group $\mathbb{Z}$ along to the successive positions. This gives a nested hierarchy of control and an obvious wreath product, with the block as the fiber and $\mathbb{Z}$ as the control group.

It is insightful now to consider this as an unfolding structure. Notice first that the purpose of the stair is to allow the person to move upwards from the floor. This is actually basic to the unfolding structure, as follows: Observe that the primitive block, from which an individual step is formed, is within the alignment kernel. Most crucially, it is pulled from the kernel at the same time as the block defining the floor; that is, regard the step-block as initially *aligned* with the floor-block. Then, in creating the individual steps away from the floor, pull the step-block out of alignment with the floor-block; i.e., by the successive action of the cyclic group. In this way, the successive transfer structure in the stairs breaks the initial symmetry of the alignment.

This illustrates again our general method of describing symmetry-breaking by a control group that is itself a symmetry group. In the present case, the control group is $\mathbb{Z}$. It is this that creates the misalignment and hence the symmetry-breaking. Nevertheless, it also describes the symmetry and structure of transfer across the set of stairs.

Observe that, notationally, this means that there will be two input indexes on $\mathbb{Z}$ in the unfolding control group: the floor-block and the step-block; i.e., both being aligned transforms of the primitive block from the alignment kernel. In the output index, the floor-block will be untouched and the step-block will be duplicated. Thus, the two indexes correspond to a wreath-direct action in which the control group is $\mathbb{Z} \times \{e\}$.

15.20 Shafts

The insertion of shafts is an important factor at the design development phase. Shafts are required for

(1) Plumbing
(2) Stairs
(3) Elevators
(4) HVAC Systems (Heating, Ventilation, and Air-Conditioning).

Shafts are vertical rectangular blocks that slice upward through the floors. By now, the reader can easily predict that we will handle them as unfolded from the cross-section block primitive in the alignment kernel.

15.21   Roof

On a typical large office building, the roof is a horizontal plate, and can therefore be handled mathematically in the same way that the floor slabs were handled; i.e., unfolded together with the initial floor-slab block and pulled upward through the building parallel with the successive floor slabs till it reaches the required height.

On a conventional house, where the roof is given by a reflectional pair of sloped planes, follow exactly the same procedure, except that an additional two stages are needed. (1) To obtain the sloping of the roof, simply rotate the roof slab to the required slope angle - we shall represent this action as we have typically done, using the overall $ALG(3, \mathbb{R})$ affine group. (2) To obtain the reflectional pair of roofs, simply add the reflection group $\mathbb{Z}_2$.

We propose that the most elegant way to express this is to have the reflection group above the rotation group in the wreath hierachy

$$ALG(3, \mathbb{R}) \text{ⓦ} \mathbb{Z}_2.$$

Use the diagonal elements of the wreath product for sides of equal slope, and the non-diagonal elements for sides of unequal slope.

15.22   Development of Accuracy

Whereas the documents in the conceptual phase were not accurate, one of the main goals of the current phase is the development of documents that are truly accurate (to within the demanding tolerance required by engineering, building codes, etc). In Leyton [88], we developed an approach to the description of accuracy, using the concept of nested control. Let us illustrate this with decimal numbers, which can be extended to any desired accuracy.

In decimal numbers, any digit along the number is to base 10, and can therefore be represented by the cyclic group $\mathbb{Z}_{10}$. We shall now show that a decimal can be represented by a *wreath product* of this form:

$$\ldots \text{ⓦ} \mathbb{Z}_{10} \text{ⓦ} \mathbb{Z}_{10} \text{ⓦ} \mathbb{Z}_{10} \text{ⓦ} \mathbb{Z}_{10}$$

where the right-to-left order represents the left-to-right number order in the decimal. Let us take any example of $\mathbb{Z}_{10}$ along this sequence. The selection of a number n from this group means the selection of a particular fiber copy from the $\mathbb{Z}_{10}$ immediately below it. This fiber copy is itelf divided into 10 units. The selection of a particular member of this group, then selects a particular fiber copy of 10 units below this, etc. It is easy to check that all aspects of a wreath product are satisfied by this structure. Most crucially, the control group on any level, maps the copies on the next level below onto each other. Fig. 15.8

illustrates this: The unit numbers shown, $0, 1, 2, 3, \ldots 10$, can be considered to be a translation group that slides the individual blocks, existing *between* the unit numbers, onto each other. The unit numbers therefore act as a control group permuting the blocks as fiber-copies.

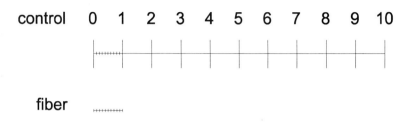

Fig. 15.8. Decimals as a wreath product.

Most crucially each fiber is a refinement in accuracy of the control group. In this way, the issue of *accuracy* has been defined as an issue of *transfer*.

15.23   Construction Documents

Two of the main aspects of construction documents are (1) the creation of sections and elevations, and (2) dimensioning. Other aspects, such as annotation and scheduling, can be seen as supporting aspects of these and the previous phases.

Dimensioning was discussed in the theory of mechanical design (Chapter 14). Therefore the only major topic left to handle is sections and elevations.

15.24   Sections and Elevations

The creation of a section is the creation of a *slicing*, and we have already established a theory of slicing. Furthermore, we have already encountered

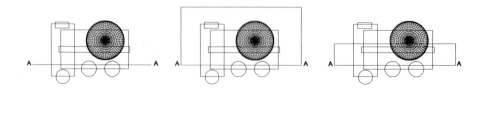

a   b   c

Fig. 15.9. Sections of a plan view of Fig. 15.1.

a major example of architectural slicing in the concept development phase, when the massing study was sliced to produce the floor-plates. In this previous case, the slice was produced by horizontal planes, unfolded from the alignment kernel - in fact, from the bottom face of the primitive cube in the kernel. In contrast, in the case of a building *section*, the slicing is produced by a vertical plane - which therefore must be unfolded from a side face of the primitive cube in the alignment kernel. As an example, in the plan view Fig. 15.9a, the line AA corresponds to a vertical plane slicing through the building. It has been unfolded from the alignment kernel.

Now, when one examines the concept of slicing in architecture, one realizes that, although a slice plane is the most obvious object with which one deals, it is more accurate to say that one deals with a *slice volume*. Fig. 15.9b shows an obvious volume associated with the same slice plane. The section view of the building will be northwards into this volume. The reason for understanding the slice as defining a volume is that the architect often wants the section to include only part of the building beyond the slice plane. This is shown in Fig. 15.9c, where the slice volume - which will still be viewed northward from the slice plane AA - will encompass only the part of the building; and therefore only this part will be visible in the three-dimensional northward view. For example, in *AutoCAD Architectural Desktop*, the definition of a section line automatically defines a section volume which the designer can then modify as required.

The use of a slice volume can easily be encompassed by our unfolding theory: One merely unfolds the entire cube primitive from the alignment kernel (with affine modification) rather than just one of the cube's faces.

A crucial aspect of the theory is that this slicing creates a further level of asymmetrization. For example, observe that the slicing cuts through the set of floors, i.e., destroying the translational symmetry across the floors. Generally because the slicing can cut through the existing structures without accordance to the shape and positioning of those structures, the slicing is a significant addition of asymmetrization.

As an example, consider again Fig. 15.4 (p. 377) showing a section through Frank Gehry's Guggenheim Museum. Consider the spaces in the left half of the building as well as the top center. These are deformed versions of blocks unfolded from the alignment kernel. Their deformation has introduced some level of asymmetrization. However, the vertical slicing now creates an additional level of asymmetrization of those volumes.

Finally let us turn to *elevations*. One can consider elevations to be a somewhat weakened form of slicing. In fact, in a highly advanced program such *AutoCAD Architectural Desktop*, sections and elevations are treated by many of the same menus. For example, both involve the creation of the same type of volumes that can be modified in the same way.

15.25   Conclusion

This chapter has shown that all stages of the architectural design process can be insightfully described as the successive use of unfolding groups. This is valuable not just for understanding the design process generally, but for understanding how computer vision should represent building environments, and for understanding how navigational robotics should deduce motion and manipulation plans in such environments.

An unfolding group is an elegant means of realizing our two primary principles of generativity: the maximization of transfer and recoverability. Unfolding maximizes transfer by expressing the successive stages of generativity as wreath expansions of the previous stages, such that the increasing complexity of the structure is expressed by transferring targeted aspects of the previous stages; i.e., by sub-local or super-local unfolding. The procedure maximizes recoverability by expressing the successive stages as misalignments of the previous stages - ultimately traceable to misalignments of the limited set of primitives constituting the alignment kernel. These misalignments always have a symmetry-breaking (asymmetry-building) effect - which is the primary condition required for recoverability according to our Asymmetry Principle. One also sees the enormous economy of this representation as the system continually uses and re-uses preceding stages, which are themselves used and re-used versions of the limited structural elements in the alignment kernel. This enormous economy embodies our claim that aesthetics is the same thing as intelligence.

15.26   Summary of a Mathematical Theory of Architecture

According to our approach, the mathematical theory of architecture is a means of describing the entire complex structure of a building by a single symmetry group. This group is required for all plans with respect to the building, because, by our theory of recoverability, plans come from the symmetries of a structure. The group is constructed by the following means:

CONCEPTUAL DESIGN.

Massing Study.
A massing study is a misaligned transfer of the alignment kernel. This algebraic structure expresses the *hierarchical* organization of the mass groupings with a level of rigor that has never been achieved before in architecture.

Slicing the Massing Study to Create Floorplates.
Slicing is expressed as the unfolding of the ground-plane fiber of the cube primitive from the alignment kernel. The fiber is unfolded upward through the building to create the floorplates.

Space Planning
The successive stages of space planning - the production of étage volumes, the boundary and void spaces, the room volumes, etc., are all achieved by the repeated use of the Indirect/Direct Unfolding Cycle, which has two phases: (1) *Indirect phase*: One first slices through a whole volume that corresponds to an unfolded primitive - thereby producing part volumes. (2) *Direct phase*: One then unfolds these part volumes directly from their own primitives; i.e., redescribing them as whole volumes.

We also saw that *complex slicing*, e.g., slicing a complex room structure via a polyline in plan view, is really the unfolding of successive planes and cylinders from the alignment kernel.

DESIGN DEVELOPMENT PHASE.

Choice of Materials.
This is achieved generatively by expressing the configuration of materials via color groups.

Doors and Windows.
Doors and windows are expressed as arrangements of blocks unfolded from the alignment kernel. They are appended to the overall structure by sub-local unfolding.

Structural Column Grid.
The structural column grid is an example of what we call *generative crystallography*. The grid placement is the wreath-lattice control group, and the column is a fiber group sub-appended to that wreath structure.

Ceiling Grid.
An exactly analogous structure expresses the ceiling grid.

Stairs.
Stairs are also best understood in terms of transfer and unfolding. The step is an affine transform of the cube primitive, and the sequence of steps is a translation control group.

Shafts.
Shafts are rectangular blocks that slice upward through the floors, and again are produced by unfolding the block primitive from the alignment kernel.

Roof.
The flat roof is unfolded in the same way as the foor slabs. The standard roof of a house - i.e., a reflectional pair of sloped planes - is given by a wreath product: the control group is the reflection group, and the fiber group is the affine group responsible for rotation of the pair of unfolded planes.

The Development of Accuracy.
We gave a theory of accuracy in which *refinement was expressed as transfer*.

CONSTRUCTION DOCUMENTS.

Sections and Elevations.
Sections and elevations are really slice volumes that are unfolded from the alignment kernel.

16. Solid Structure

16.1 Introduction

This chapter develops a theory of solid structure. Indeed, since most of the concepts of this book are involved in this theory, one can regard most of the analysis so far as an elaboration of that theory. All that in fact remains is to define, in detail, the nature of solid primitives - which is the main topic of this chapter. The theory of modeling, in the previous chapters - e.g., the theory of CSG, construction planes, constraints, assembly, etc. - provide the rest.

3D solid modeling is a true miracle of computer science. It is now basic to most of CAD, where once 2D drafting drove the entire process. When CAD was essentially 2D drafting, it looked very much like design before the use of computers - plans or blueprints were drawn and dimensioned; the actions of the designer produced lines and arcs. In 3D solid modeling, the actions of the designer produce 3D solids which imitate the desired goal object. Thus the designer begins with the representational language of the goal object and this allows the first design phase to be strongly conceptual - which is a massive stimulus to creativity.

16.2 The Solid Primitives

In solid-modeling, the generation of the initial 3D model standardly begins by choosing and combining 3D solid primitives. Currently, there is no systematic analysis and classification of these primitives. We argue that the absence of such an analysis severely holds back understanding in nearly all aspects of CAD. For example, without a real understanding of solid primitives, one cannot understand the relation of primitives to construction planes and other datum features, and one cannot understand the nature of assembly.

The purpose of most of this chapter is to develop a systematic theory of solid primitives. What we will do is develop *symmetry groups* for the solid primitives - symmetry groups that contain enormous amounts of information that we argue is exploited in all solid modeling. These groups are used in other parts of the book to explain powerful aspects of mechanical engineering, assembly, and spline structure - aspects which could not have otherwise been explained. We have left the detailed discussion of the primitives till late in the book, i.e., till this chapter, exactly because this level of detail might have been a barrier to the reader continuing.

16.3   The Solid n-Cube

The solid n-cube is basic to our physical world. As such it is a fundamental structure occurring in all the physical, computational, and design sciences. For example, it is basic in determining most of the crystallographic hierarchy of point groups, basic to solid-state physics, basic to computer storage media, basic to 3D graphics where solid n-cubes constitute the parameter domains of splines, etc.

The first half of this chapter will develop a symmetry group of the solid n-cube. It is currently assumed, in the research literature, that the group of the n-cube is reasonably trivial. We will argue that it is not. To construct this group one needs almost every aspect of our generative theory of shape: nested control, symmetry-breaking, memory-inference, super-local action, and so on. The very group theory is determined by issues of generativity and recoverability.

16.4   The Hyperoctahedral Wreath Hyperplane Group

Definition 16.1. *When the hyperoctahedral group is expressed in its wreath-decomposed form thus*

$$\mathbb{Z}_2 \ \text{Ⓦ} \ \Sigma_n$$

it will be called the **hyperoctahedral wreath group.**

We now come to one of the most important groups in our generative theory of shape. It explains much in human perception, mechanical and architectural CAD, etc.

Definition 16.2. *When the hyperoctahedral wreath group is wreath sub-appended by the translation hyperplane $\mathbb{R}^{n-1}$ group, the resulting group*

$$\mathbb{R}^{n-1} \text{ ⓦ } \mathbb{Z}_2 \text{ ⓦ } \Sigma_n$$

will be called the **hyperoctahedral wreath hyperplane group** *HWH(n).*

As usual, throughout this book, we will use the term "hyperplane" for any translated version of a hyperplane.

Definition 16.3. *The* **degree** n *of the hyperoctahedral wreath hyperplane group HWH(n), will be the degree n of the hyperoctahedral component - which is also the dimension of the space $\mathbb{R}^n$ of action. The figure corresponding to the hyperoctahedral wreath hyperplane group will be called the n-***cube***. The $\mathbb{R}^{n-1}$ fiber in the hyperoctahedral wreath hyperplane group will be called a* **face** *of the cube, and will be assumed to be parallel to one of the coordinate hyperplanes in $\mathbb{R}^n$, called the* **corresponding coordinate hyperplane.**

Degree 1.

In this case, the hyperoctahedral wreath hyperplane group is *HWH(1)*, which is

$$\mathbb{R}^0 \text{ ⓦ } \mathbb{Z}_2 \text{ ⓦ } \Sigma_1.$$

The space in which the group acts is $\mathbb{R}^1$, the line. To see what the corresponding cube looks like, work from left-to-right up the hierarchy. First, a face of the cube is $\mathbb{R}^0$, which is a point. Next, the intermediate control group $\mathbb{Z}_2$ makes a reflectional copy of the point about the origin. Finally, the highest control group Σ_1 is trivial, and therefore no more structure is added. The 1-cube therefore consists of two points on either side of the origin (in one-dimensional space).

Degree 2.

In this case, the hyperoctahedral wreath hyperplane group is *HWH(2)*, which is

$$\mathbb{R}^1 \text{ ⓦ } \mathbb{Z}_2 \text{ ⓦ } \Sigma_2.$$

The space in which the group acts is $\mathbb{R}^2$, the plane. To see what the corresponding cube looks like, work from left-to-right up the hierarchy. First, a face of the cube is $\mathbb{R}^1$, which is a straight line. Next, the intermediate control group $\mathbb{Z}_2$ makes a reflectional copy of the line about its corresponding coordinate axis. Finally, the highest control group Σ_2 makes two copies of the line-pair just generated: one about one coordinate axis and one about the other. The 2-cube is therefore a square (i.e., with infinite "wires" as sides - as we have standardly described it).

Degree 3.
In this case, the hyperoctahedral wreath hyperplane group is $HWH(3)$, which is

$$\mathbb{R}^2 \ \circled{w} \ \mathbb{Z}_2 \ \circled{w} \ \Sigma_3.$$

The space in which the group acts is $\mathbb{R}^3$. To see what the corresponding cube looks like, work from left-to-right up the hierarchy. First, a face of the cube is $\mathbb{R}^2$, which is a plane. Next, the intermediate control group $\mathbb{Z}_2$ makes a reflectional copy of the plane about its corresponding coordinate plane. Finally, the highest control group Σ_2 makes three copies of the plane-pair just generated: one pair about each of the coordinate planes. The 3-cube is therefore the ordinary cube (i.e., with infinite planes as faces - as we have standardly described it).

Degree n.
In this case, the hyperoctahedral wreath hyperplane group is $HWH(n)$, which is

$$\mathbb{R}^{n-1} \ \circled{w} \ \mathbb{Z}_2 \ \circled{w} \ \Sigma_n.$$

The space in which the group acts is $\mathbb{R}^n$. To see what the corresponding cube looks like, work from left-to-right up the hierarchy. First, a face of the cube is $\mathbb{R}^{n-1}$, which is a hyperplane. Next, the intermediate control group $\mathbb{Z}_2$ makes a reflectional copy of the face about its corresponding coordinate hyperplane. Finally, the highest control group Σ_n makes n copies of the face-pair just generated: one pair about each of the coordinate hyperplanes. The n-cube therefore has n pairs of hyperplanes as faces.

16.5   Cubes as Cartesian Frames

The content of this section is extraordinarly profound. Simple as it is, this book demonstrates that it explains truly fundamental issues in perception, navigation, and CAD - e.g., the structure of the human visual system, and mechanical engineering design.

We have often argued that there is a strong relationship between the cube and the Cartesian reference frame. Indeed, our generative theory has often taken the cube and the Cartesian frame as *the same thing*, both with respect to their internal generation and with respect to their use in generating other shapes. Our theory regards them as essentially the same *group*, i.e., the hyperoctahedral wreath hyperplane group, i.e., the wreath-decomposed hyperoctahedral group wreath sub-appended by the translation hyperplane:

$$\mathbb{R}^{n-1} \ \circled{w} \ \mathbb{Z}_2 \ \circled{w} \ \Sigma_n$$

In fact, the Cartesian frame can be regarded as the cube with its faces coincident with the coordinate planes; i.e., a cube of width zero, but with faces still of infinite extent as with a cube - the latter's finiteness being given by occupancy.

This approach is very natural, as can be seen in the case where $n = 3$. Here the fiber structure corresponds to the six standard orthographic projection planes; i.e., when one projects (orthographically) an object from one of its sides, one standardly uses a projection plane in one of six positions: "top", "bottom", "left", "right", "front", "back". They are positioned exactly at the six faces of some imaginary cube. These six projection planes are therefore related by exactly the same symmetries as the faces of a cube.

When considering the 3D Cartesian frame in the *conventional* description of *three* coordinate planes (located at the origin), we will name the three planes:

(1) View Plane (or Image Plane).

(2) Vertical Travel Plane (i.e., direction of movement).

(3) Ground Plane.

Notice however, when the Cartesian frame is understood as the hyperoctahedral wreath hyperplane group, there will actually be six coordinate planes, in three pairs - i.e., there are two coincident xy planes, two coincident yz planes, and two coincident xz planes. The two members of each pair are reflectionally symmetric to each other.

With this in mind, let us now rigorously define the relation between the Cartesian frame and the cube. The intuitive idea behind our approach will be to regard the cube as an "expanded" version of the Cartesian frame - that is, the cube will be regarded as a Cartesian frame where each pair of coincident planes has been been pulled apart by a parallel distance from the origin along the axis perpendicular to the pair of planes. Since the cube is symmetrical, the parallel movement of each of the pair of planes along its own axis, will be the same for each pair. Therefore the amount of movement can be given by a value on $\mathbb{R}$ which will be understood as a translation group - the translation value will be the distance a plane has moved from the origin. This translation group $\mathbb{R}$ will be wreath super-appended to the hyperoctahedral wreath hyperplane group, thus:

$$\mathbb{R}^{n-1} \textcircled{w} \mathbb{Z}_2 \textcircled{w} \Sigma_n \textcircled{w} \mathbb{R}.$$

To understand what is going on here, consider the newly appended control group $\mathbb{R}$. Its fiber group is $\mathbb{R}^{n-1} \textcircled{w} \mathbb{Z}_2 \textcircled{w} \Sigma_n$, and any of the fiber copies

$$[\mathbb{R}^{n-1} \textcircled{w} \mathbb{Z}_2 \textcircled{w} \Sigma_n]_g$$

is really a cube of a width $2g$. Notice that the subscript g on the fiber copy is, of course, a member of the control group $\mathbb{R}$. Thus there is one cube for each member of the control group. Negative values of g will produce what one can regard as negative cubes - i.e., holes.

Two cubes will be particularly important in this book: The cube corresponding to the identity element e in the control group, will be called the **Cartesian frame**. This is the cube of size 0. The cube corresponding to the translation distance $1/2$ in the control group, will be called the **unit cube**, because it is of width 1. In fact, in most cases, $1/2$ will be written for g, even when a unit cube is not actually needed.

Depending on context, we will sometimes regard the Cartesian frame as possessing n pairs of distinct hyperplanes (i.e., a total of $2n$ hyperplanes) and sometimes regard the two hyperplanes within each pair as indistinct - thus reducing the total number of hyperplanes to n. Most crucially, in the latter case, the hyperplanes can be identified with the mirror hyperplanes themselves. Furthemore, when these mirror hyperplanes are moved out to become the faces of cube (of non-zero width), they then express the reflectional symmetry of each individual face in its own right - and thus express the reflectional symmetry of the face as a fiber within the cube.

To illustrate, consider the crucial case of 3D. Let us consider first the Cartesian frame with the two planes within each pair as indistinct - thus reducing the total number of planes to three. The planes can be identified with the mirror planes themselves. Furthemore, when these mirror planes are moved out to become the faces of cube (of non-zero width), they then express the reflectional symmetry of each individual face in its own right - and thus express the reflectional symmetry of the face as a fiber within the cube.

16.6   The Symmetry Group of the Solid n-Cube

We want now to develop a symmetry group for the *solid* n-cube. According to the generative theory, this is a fundamentally important group to all of human perception, 3D solid modeling, and physical structure in the sciences. Up to now, we have been discussing the n-cube as an empty box. Its generative structure was defined as being a wreath product in which the control group is a reflection structure given by the hyperoctahedral group $\mathbb{Z}_2 \ⓦ \ \Sigma_n$, and the fiber is given by the translation hyperplane representing a face of the cube. The resulting structure was this:

$$\mathbb{R}^{n-1} \ⓦ \ \mathbb{Z}_2 \ⓦ \ \Sigma_n$$

which we call the *hyperoctahedral wreath hyperplane group HWH(n)*. Examples, of this group for dimensions 1, 2, and 3, were given in Sect. 16.4 (p. 398). In the case of dimension 1, the empty n-cube is the empty interval, i.e., the set of two end-points of the interval. In the case of dimension 2, the empty n-cube is the empty square, i.e., the square as a line drawing. In the case of dimension 3, the empty n-cube is the empty box. The goal now is to fill these structures - i.e., so that they are *solid*.

The description will be a generative description, because in this book, structural descriptions are generative descriptions. Furthermore, the description must be dictated by our two fundamental criteria for generativity: the maximization of transfer, and the maximization of recoverability. We begin by using the second criterion: recoverability. To fulfill recoverability, one must make sure that the generative structure satisfies the Asymmetry Principle, which means that the generative history of the solid n-cube must be *symmetry-breaking* forward in time. Some consideration reveals that there are two alternative methods that ensure this. These methods are the reverses of each other:

(1) The first method takes infinite swept space and breaks its symmetry using the boundary of a cube.

(2) The second method takes the boundary of a cube and breaks its symmetry using infinite swept space.

Both methods are very important, and as shown elsewhere in the book, they are the basis of different generative techniques; e.g., in CAD. For example, we argue that method (1) is basic to splines, and method (2) is basic to continuous slicing structures that appear for instance in mechanical engineering.

The reader might wonder why two methods, that are reverses of each other, fulfill the recoverability condition of our theory. The reason comes from the discussion in Sect. 11.2 and is as follows: When bringing two symmetric objects together, there is usually a reduction in symmetry. Thus, if the two objects are generated *sequentially*, then, which-ever order is chosen, the symmetry of the first object will be destroyed by the second object.

This section will develop method (1). The other method, which is based on slicing, was described in Sect. 14.20.

The first method of ensuring symmetry-breaking is this: Observe that the *solid* in an n-cube has translational symmetry up to its boundaries, which breaks this translational symmetry. Thus the generative history can be understood to be one in which the full translational symmetry once existed, but was cut down by introducing the bounding box. That is, start with the full translation group of space $\mathbb{R}^n$, and cut it down with the hyperoctahedral wreath structure $\mathbb{R}^{n-1} \textcircled{w} \mathbb{Z}_2 \textcircled{w} \Sigma_n$, which will generate the boundaries.

Since the full translation group $\mathbb{R}^n$ is the preceeding symmetry state, and the asymmetrization process is produced by the introduction of the hyperoctahedral structure, our standard technique requires us to make the translation group the fiber and the hyperoctahedral structure the control group. That is, we obtain a wreath product something like this:

$$\overbrace{\mathbb{R}^n}^{Translation} \qquad \textcircled{w} \qquad \overbrace{\mathbb{R}^{n-1} \textcircled{w} \mathbb{Z}_2 \textcircled{w} \Sigma_n}^{Hyperoctahedral} \qquad (16.1)$$

where the first overbrace gives the translation fiber, and the second overbrace gives the hyperoctahedral control group. The idea is that the hyperoctahedral component breaks the symmetry of the translation component.

However, (16.1) is only a first guess at the structure that is required. To fully create the structure, one must employ also the other of our fundamental criteria: the maximization of transfer. To do this, observe that the hyperplane component $\mathbb{R}^{n-1}$, within the hyperoctahedral part, is a subgroup of the translation part to the far left. This is a non-trivial observation as follows:

Let us consider the case of a 3-dimensional cube. The idea to be used is this: The side of the cube is coincident with a planar slice of 3-space. Thus the generative procedure should go like this: The previous translation phase starts with a plane, and generates all of space by translating that plane along its perpendicular axis. This generates all of space as a stack of planes. Then take one of the planes, and apply the hyperoctahedral group to it, producing the six sides of the cube. If the plane taken from the stack is the one at distance $1/2$ along the translation axis, then the six planes will produce the unit cube.

Generally therefore, in the n-dimensional case, start off *not* with the ordinary translation group $\mathbb{R}^n$, but the wreath-reconstituted translation group $\mathbb{R}^{n-1} \ \textcircled{w} \ \mathbb{R}$. This is the fiber hyperplane $\mathbb{R}^{n-1}$ pulled through n-space and therefore structuring n-space as a stack of parallel hyperplanes. Then, in the next phase, the hyperoctahedral group takes the particular hyperplane that is at distance $1/2$ along the translation axis, and produces the $2n$ sides of the n-cube. The resulting structure is this:

$$\overbrace{\mathbb{R}^{n-1} \ \textcircled{w} \ \mathbb{R}}^{Translation} \quad \textcircled{w} \quad \overbrace{\mathbb{R}^{n-1} \ \textcircled{w} \ \mathbb{Z}_2 \ \textcircled{w} \ \Sigma_n}^{Hyperoctahedral} \tag{16.2}$$

Inspecting this expression carefully, one can see that both the translation component and the hyperoctahedral component act on on a hyperplane $\mathbb{R}^{n-1}$. The translation component translates the hyperplane through space, and the hyperoctahedral component reflects the hyperplane to become the faces of the cube.

All that remains to do is to specify the fact that the hyperplane selected by the hyperoctahedral component is the same as that *already* used by translation component; in other words, there is *transfer*. The way this will be done is as follows. First remove the hyperplane from the hyperoctahedral component thus:

$$\overbrace{\mathbb{R}^{n-1} \ \textcircled{w} \ \mathbb{R}}^{Translation} \quad \textcircled{w} \quad \overbrace{\mathbb{Z}_2 \ \textcircled{w} \ \Sigma_n}^{Hyperoctahedral} \tag{16.3}$$

This is because the hyperoctahedral component uses the same hyperplane as that shown at the beginning of the sequence. However, the other thing that we must do is indicate that the hyperoctahedral component uses only one particular fiber copy of the hyperplane, the copy at translation distance $1/2$. That is, the translation component has generated all the copies of the hyperplane as fibers of its own wreath product; and then the hyperoctahedral group must select only one of these fibers, that at translation distance $1/2$. This will be indicated in the following way:

$$\overbrace{\mathbb{R}^{n-1} \textcircled{w} \mathbb{R}}^{Translation} \quad \textcircled{w} \quad [\overbrace{\mathbb{Z}_2 \textcircled{w} \Sigma_n}^{Hyperoctahedral}]^{\mathbb{R}^{n-1}_{1/2}} \tag{16.4}$$

The superscript at the far right is on the hyperoctaheral component. It indicates what that component will specifically act on. The superscript is $\mathbb{R}^{n-1}_{1/2}$ which is the hyperplane $\mathbb{R}^{n-1}$ indexed by $1/2$, indicating that it is the copy of the hyperplane at translation distance $1/2$. This is the only fiber copy selected from the wreath translation structure below. This is an example of what we call a *super-local* action. In such an action, one applies a higher level control group, which although acting in a control-nested fashion on the lower structure, actually only affects part of the lower structure - that referenced in the superscript. Super-local action was discussed at length in Chapter 12.

The crucial point is that, because the hyperoctahedral component uses a hyperplane that is a fiber in the pervious translation component, it maximally exploits existing structure rather than creating new structure - which is a basic principle of our generative theory, embodied in the maximization of transfer.

Definition 16.4. *The* **solid hyperoctahedral wreath hyperplane group** *S-HWH(n) is the group*

$$\overbrace{\mathbb{R}^{n-1} \textcircled{w} \mathbb{R}}^{Translation} \quad \textcircled{w} \quad [\overbrace{\mathbb{Z}_2 \textcircled{w} \Sigma_n}^{Hyperoctahedral}]^{\mathbb{R}^{n-1}_{1/2}}$$

where the translation component generates infinite n-space as a stack of parallel hyperplanes, and the hyperoctahedral component takes one of those hyperplanes $\mathbb{R}^{n-1}_{1/2}$ and generates from it 2n hyperplanes that cut down the previous translation group.

Example 16.1. Let us compare the specific cases for dimension n, 3, 2, and 1. At this stage, the overbraces used previously, will be removed:

Dimension n:

$$\mathbb{R}^{n-1} \textcircled{w} \mathbb{R} \quad \textcircled{w} \quad [\mathbb{Z}_2 \textcircled{w} \Sigma_n]^{\mathbb{R}^{n-1}_{1/2}} \tag{16.5}$$

Dimension 3

$$\mathbb{R}^2 \textcircled{w} \mathbb{R} \quad \textcircled{w} \quad [\mathbb{Z}_2 \textcircled{w} \Sigma_3]^{\mathbb{R}^2_{1/2}} \tag{16.6}$$

Dimension 2

$$\mathbb{R} \textcircled{w} \mathbb{R} \textcircled{w} \quad [\, \mathbb{Z}_2 \textcircled{w} \Sigma_2 \,]^{\mathbb{R}_{1/2}} \tag{16.7}$$

Dimension 1

$$\mathbb{R}^0 \textcircled{w} \mathbb{R} \textcircled{w} \quad [\, \mathbb{Z}_2 \textcircled{w} \Sigma_1 \,]^{\mathbb{R}^0_{1/2}} \tag{16.8}$$

Using the above groups as a basis, one can now move onto constructing the group of a solid n-cube. The construction procedure will be illustrated with the case of the solid 3-cube. This is based on expression (16.6) above. Now, a bounding face of the solid 3-cube is actually a solid 2-cube, i.e., a *solid* square, and is therefore based on expression (16.7). This is an infinite plane which is cut by lines, as hyperplanes. However, each line is really a solid 1-cube; i.e., a *solid* interval, which is based on expression (16.8). This is an infinite line cut by points, as hyperplanes, which are solid 0-cubes.

We conclude therefore that each successive solid face downward through the ordinary solid 3-cube corresponds to the solid hyperoctahedral wreath hyperplane group of one dimension lower, and so on downward. Thus we can capture the entire structure of all faces downwards merely by recursively substituting solid hyperoctahedral wreath hyperplane groups downwards. However, this must be done so as to use pre-existing structure in the hierarchy. This problem will be solved shortly.

Observe also that, in order to further increase transfer, the hyperplane in the translation component, should be re-constituted as a translation hierarchy of sub-hyperplanes all the way down, i.e., as a wreath c-polycyclic group $\mathbb{R} \textcircled{w} \mathbb{R} \textcircled{w} \dots \textcircled{w} \mathbb{R}$. The fibers of this recursion are then transferred in the recursion defined in the previous paragraph.

Using these two recursions, the full symmetry group for the solid 3-cube can now be constructed. This will be done by generating first an edge of the cube, then a face of the cube, and finally, the full cube itself.

3-Cube: Solid Edge

$$\mathbb{R}^0 \textcircled{w} \mathbb{R} \textcircled{w} \mathbb{R} \textcircled{w} \mathbb{R}$$
$$\textcircled{w} [\, \mathbb{Z}_2 \textcircled{w} \Sigma_1 \,]^{\mathbb{R}^0_{1/2,1/2,1/2}} \;\;. \tag{16.9}$$

Here the top line explicity begins the sequence with the point $\mathbb{R}^0$. In our theory, all generative sequences begin with the point, but we often do not mention this.

Now, following the point in the top line of the above expression is the wreath re-constitution of $\mathbb{R}^3$ as the wreath c-polycyclic group $\mathbb{R} \textcircled{w} \mathbb{R} \textcircled{w} \mathbb{R}$. In the second line of the expression, the superscript $\mathbb{R}^0_{1/2,1/2,1/2}$ indicates that we have selected, from the wreath product on the first line, the point fiber-copy corresponding to the triple index $1/2, 1/2, 1/2$ in the triple wreath c-polycyclic group $\mathbb{R} \textcircled{w} \mathbb{R} \textcircled{w} \mathbb{R}$. This point then enters the hyperoctahedral group $\mathbb{Z}_2 \textcircled{w} \Sigma_1$ on that line, and is reflected to become the other endpoint on the particular edge bisected by the single reflection group involved. Thus the total resulting structure is 3-space (top line) with a particular pair of end edge points suspended at some perpendicular distance from the origin. The line interval between the endpoints is filled - i.e., we have a solid interval - because the entire space is filled with lines by the preceding $\mathbb{R}^0 \textcircled{w} \mathbb{R} \textcircled{w} \mathbb{R} \textcircled{w} \mathbb{R}$ structure. The newly generated endpoints select a specific line and cut it down - in accord with the symmetry-breaking requirement in the generative theory.

3-Cube: Solid Face

$$\mathbb{R}^0 \textcircled{w} \mathbb{R} \textcircled{w} \mathbb{R} \textcircled{w} \mathbb{R}$$

$$\textcircled{w} \; [\, \mathbb{Z}_2 \textcircled{w} \Sigma_1]^{\mathbb{R}^0_{1/2,1/2,1/2}}$$

$$\textcircled{w} \; [\, \mathbb{Z}_2 \textcircled{w} \Sigma_2]^{[\mathbb{R}^0 \textcircled{w} \mathbb{R}]_{1/2,1/2}} \;\; \textcircled{w} \;\; [\, \mathbb{Z}_2 \textcircled{w} \Sigma_1]^{\mathbb{R}^0_{1/2,1/2,1/2}} \qquad . \qquad (16.10)$$

The first two lines are the entire previous expression (16.9) for the solid edge of the cube. The third line will take that edge and reflect it sufficient times to create a square. It does this in the following way: In its superscript $\mathbb{R}^0 \textcircled{w} \mathbb{R}_{1/2,1/2} \;\; \textcircled{w} \; [\, \mathbb{Z}_2 \textcircled{w} \Sigma_1]^{\mathbb{R}^0_{1/2,1/2,1/2}}$, it takes the infinite line $\mathbb{R}^0 \textcircled{w} \mathbb{R}_{1/2,1/2}$ along that edge. Notice that this is a fiber-copy of a line generated by the wreath c-polycyclic group on the top line; i.e., it is selected from the structure already generated. Furthermore, in this subscript, the component $[\, \mathbb{Z}_2 \textcircled{w} \Sigma_1]^{\mathbb{R}^0_{1/2,1/2,1/2}}$ selects, in addition, the two previous endpoints generated on the line above - that cut down that infinite line. It should be recognized that the entire superscript on the final line, is a selection of previously generated structure from the previous lines - i.e., most crucially one of the fibers from the previous two lines.

Then this superscript - which represents the infinite line cut down by its endpoints - enters the hyperoctahedral group shown (on the final line) which reflects it sufficient times to become a square. The square is filled because the entire space is filled by infinite wreath planes by the preceding $\mathbb{R}^0 \textcircled{w} \mathbb{R} \textcircled{w} \mathbb{R} \textcircled{w} \mathbb{R}$. The square selects a particular infinite plane to cut it down - in accord with the symmetry-breaking requirement in our generative theory.

3-Cube: Entire Solid

$$\mathbb{R}^0 \text{\textcircled{w}} \mathbb{R} \text{\textcircled{w}} \mathbb{R} \text{\textcircled{w}} \mathbb{R}$$

$$\text{\textcircled{w}} [\mathbb{Z}_2 \text{\textcircled{w}} \Sigma_1]^{\mathbb{R}^0_{1/2,1/2,1/2}}$$

$$\text{\textcircled{w}} [\mathbb{Z}_2 \text{\textcircled{w}} \Sigma_2]^{[\mathbb{R}^0 \text{\textcircled{w}} \mathbb{R}]_{1/2,1/2}} \text{\textcircled{w}} [\mathbb{Z}_2 \text{\textcircled{w}} \Sigma_1]^{\mathbb{R}^0_{1/2,1/2,1/2}}$$

$$\text{\textcircled{w}} [\mathbb{Z}_2 \text{\textcircled{w}} \Sigma_3]^{[\mathbb{R}^0 \text{\textcircled{w}} \mathbb{R} \text{\textcircled{w}} \mathbb{R}]_{1/2}} \text{\textcircled{w}} [\mathbb{Z}_2 \text{\textcircled{w}} \Sigma_2]^{[\mathbb{R}^0 \text{\textcircled{w}} \mathbb{R}]_{1/2,1/2}} \text{\textcircled{w}} [\mathbb{Z}_2 \text{\textcircled{w}} \Sigma_1]^{\mathbb{R}^0_{1/2,1/2,1/2}} . \tag{16.11}$$

The first three lines of this expression are the entire previous expression
(16.10) for a face of the cube. The final line follows the principles set out
earlier: Select a previously generated boundary together with the infinite
hyperplane that it cuts down. Then feed this into its corresponding hyperoctahedral group. This process, is repeated recursively downward through the
superscripts. The consequence is the solid 3-cube.

SOLID n-CUBE

The symmetry group of the solid n-cube is this:

$$\mathbb{R}^0 \text{\textcircled{w}} \mathbb{R} \text{\textcircled{w}} \mathbb{R} \text{\textcircled{w}} \mathbb{R} \text{\textcircled{w}} \ldots$$

$$\text{\textcircled{w}} [\mathbb{Z}_2 \text{\textcircled{w}} \Sigma_1]^{\mathbb{R}^0_{1/2,1/2,1/2,\ldots}}$$

$$\text{\textcircled{w}} [\mathbb{Z}_2 \text{\textcircled{w}} \Sigma_2]^{[\mathbb{R}^0 \text{\textcircled{w}} \mathbb{R}]_{1/2,1/2,\ldots}} \text{\textcircled{w}} [\mathbb{Z}_2 \text{\textcircled{w}} \Sigma_1]^{\mathbb{R}^0_{1/2,1/2,1/2,\ldots}}$$

$$\text{\textcircled{w}} [\mathbb{Z}_2 \text{\textcircled{w}} \Sigma_3]^{[\mathbb{R}^0 \text{\textcircled{w}} \mathbb{R} \text{\textcircled{w}} \mathbb{R}]_{1/2,\ldots}} \text{\textcircled{w}} [\mathbb{Z}_2 \text{\textcircled{w}} \Sigma_2]^{[\mathbb{R}^0 \text{\textcircled{w}} \mathbb{R}]_{1/2,1/2,\ldots}} \text{\textcircled{w}} [\mathbb{Z}_2 \text{\textcircled{w}} \Sigma_1]^{\mathbb{R}^0_{1/2,1/2,1/2,\ldots}}$$

$$\ldots \tag{16.12}$$

**where each of the dot sequences represent a completion up
to n-fold recursion. The group will be called the** *recursive
solid hyperoctahedral wreath hyperplane group, RS-HWH(n).*

The reader might wonder why we choose to give such a long name to
the group: recursive solid hyperoctahedral wreath hyperplane group. The
reason is that this name identifies the main stages by which the group was
constructed, thus:

(1) First we started with the *hyperoctahedral wreath group* - which is the
reflection structure. It was argued that this is the single most important
structure of the human visual system - and it corresponds to what is probably
the most fundamental determining physical organization on this planet - the
reflection structure arising from the gravitational system. Furthermore, most
crucially for us, it is the most significant structure of the alignment kernel,

and the structure with respect to which all other primitives are aligned in the kernel.

(2) Then, we wreath sub-appended the translation hyperplane group, i.e., making the reflection structure act on hyperplanes. This produced the *hyperoctahedral wreath hyperplane group, HWH(n)*.

(3) Then we added the solid component. This was expressed as an infinite solid whose symmetry was broken by the *HWH(n)* group. The resulting group was called the *solid hyperoctahedral wreath hyperplane group, S-HWH(n)*.

(4) Finally, in order to gain the hierarchy of solid faces within the *S-HWH(n)* group, we substituted that group within itself recursively for descending values of n - while making sure that transfer was maximized. The result was the final group, the *recursive solid hyperoctahedral wreath hyperplane group, RS-HWH(n)*.

In most contexts, the particular structural considerations of the context will require us to mention only part of this full group. When this happens, there will tend to be an order of priority of what is mentioned - and this is determined exactly by the derivation order just described. Thus, typically, the structure that will almost always be mentioned is the starting one: the hyperoctahedral wreath group. Then the second most likely structure to be mentioned is the appended hyperplane group, and so on through the above derivation order. In some contexts, however, such as the theory of splines, all the structure will be needed.

16.7   Solid Interval and Solid Square

Besides giving the structure of the general solid n-cube, the last section also gave the particular example of the solid 3-cube. This book also requires the filled interval - which will be called the *solid interval* - and the filled square - which will be called the *solid square*. In fact, most of applied mathematics involves the solid interval $[a, b]$, which acts as the most frequently used domain of a function.

Solid Interval

This is the group *RS-HWH(1)*, which is given thus:

$$\mathbb{R}^0 \textcircled{w} \mathbb{R}$$

$$\textcircled{w} \, [\, \mathbb{Z}_2 \textcircled{w} \, \Sigma_1]^{\mathbb{R}^0_{1/2}} \tag{16.13}$$

The two lines of this expression correspond to the first two lines of the general formula (16.12) - where the n-fold wreath c-polycyclic translation component in the first line of the general formula is now 1-fold, and the index on the second line is ammended appropriately.

Let us carefully understand what is going on in this group. According to our theory, the interval is an infinite straight line which has been cut down by two endpoints. Line 1 in the above expression is the initial infinite straight line. Line 2 generates the endpoints, in this way: The index is the starting point - the origin translated to position $1/2$ along the line. This translated point comes from the first line of the expression and is a particular fiber copy from the structure generated by the wreath product on that line. Line 2 uses this super-locally, as the input index. The control group $\mathbb{Z}_2 \textcircled{w} \Sigma_1$ is the hyperoctahedral wreath group of order 1. Its action is as follows: It takes the point given by the input index, and first applies the fiber $\mathbb{Z}_2$ - which produces the reflectionally symmetric endpoint. It then applies the permutation group Σ_1 which is trivial. This generates the required pair of endpoints.

It is important that the entire expression (16.13) is a symmetry-breaking structure: Start with the infinite straight line and break its translational symmetry by introducing the endpoints. Most crucially this is all done by transfer. The first symmetry-breaking endpoint is created as a translation-transfer of the origin (first line) and the second symmetry-breaking endpoint is created by the reflection-transfer (second line).

Solid Square

This group is *RS-HWH(2)*, which is given thus:

$$\mathbb{R}^0 \textcircled{w} \mathbb{R} \textcircled{w} \mathbb{R}$$

$$\textcircled{w} \, [\, \mathbb{Z}_2 \textcircled{w} \, \Sigma_1]^{\mathbb{R}^0_{1/2,1/2}}$$

$$\textcircled{w} \, [\, \mathbb{Z}_2 \textcircled{w} \, \Sigma_2]^{[\mathbb{R}^0 \textcircled{w} \mathbb{R}]_{1/2}} \textcircled{w} \, [\, \mathbb{Z}_2 \textcircled{w} \, \Sigma_1]^{\mathbb{R}^0_{1/2,1/2}} \, . \tag{16.14}$$

The three lines of this expression correspond to the first three lines of the general formula (16.12), where the n-fold wreath c-polycyclic translation component in the first line of the general formula is now 2-fold, and the index on the subsequent lines is ammended appropriately. To understand surface splines, it is necessary to fully understand the three lines of (16.14), as follows:

Line 1: This gives the initial infinite structure, i.e., solidity. The required solid structure is a filled plane, which is the wreath c-polycyclic translation group of order 2, that is, $\mathbb{R} \textcircled{w} \mathbb{R}$. Lines 2 and 3, then generate the square boundary as follows:

Line 2: This generates the endpoints of the first side of the square, in the following way: The 3-fold index $\mathbb{R}^0_{1/2,1/2}$ on Line 2 is a fiber taken from the 3-fold wreath product $\mathbb{R}^0 \textcircled{w} \mathbb{R} \textcircled{w} \mathbb{R}$ on Line 1. The index represents the origin $\mathbb{R}^0$ after it has been translated to position $1/2$ along the first generated straight line, which is itself moved a parallel distance of $1/2$ from the origin. Because this doubly-translated point is selected from the entire set of fiber points on Line 1, it is being used super-locally as the input index in Line 2. The control group $\mathbb{Z}_2 \textcircled{w} \Sigma_1$ on Line 2 is the hyperoctahedral wreath group of degree 1. It creates the opposite endpoint on that side of the square. Notice that by this stage, the side is a filled interval because it is on an infinite straight line that already exists on Line 1.

Line 3: This takes the side created on Lines 1 and 2, and reflects it sufficient times to create a square boundary. It does this in the following way: In its superscript $\mathbb{R}^0 \textcircled{w} \mathbb{R}_{1/2} \textcircled{w} [\mathbb{Z}_2 \textcircled{w} \Sigma_1]^{\mathbb{R}^0_{1/2,1/2}}$, it takes the infinite line $\mathbb{R}^0 \textcircled{w} \mathbb{R}_{1/2}$ along that side. Notice that this is a fiber-copy of a line generated by the wreath c-polycyclic group on Line 1; i.e., it is selected from the structure already generated. Furthermore, in this subscript, the component $[\mathbb{Z}_2 \textcircled{w} \Sigma_1]^{\mathbb{R}^0_{1/2,1/2}}$ selects, in addition, the two previous endpoints generated on Line 2 - which cut down that infinite line. It should be recognized that the entire superscript on the final line, is a selection of previously generated structure from the previous lines - i.e., most crucially one of the fibers from Lines 1 and 2.

Then this superscript - which represents the infinite line cut down by its endpoints - enters the hyperoctahedral group (on Line 3) which reflects it sufficient times to become a square. The square is filled because the entire 2D parameter space is filled by the wreath c-polycyclic structure $\mathbb{R}^0 \textcircled{w} \mathbb{R} \textcircled{w} \mathbb{R}$ on Line 1. The square boundary generated on Lines 2 and 3, cuts this down - in accord with the symmetry-breaking requirement in our generative theory.

16.8 The Other Solid Primitives

We now come onto the other solid primitives besides the cube. Chapter 10 derived the set of 3D primitive surfaces that maximize transfer and recoverability. These primitives are given again here in Table 16.1 (p. 412). Our concern now is to use these surfaces as a basis for developing their *solid* counterparts. Of course, the first primitive in Table 16.1, the plane, has no solid counterpart and will therefore be omitted. Also, the solid version of the cube in Table 16.1 has already been derived in the present chapter as the group *RS-HWH(3)*.

Table 16.1. The 3D *surface* primitives established in Chapter 10

LEVEL-CONTINUOUS

Plane	$\mathbb{R} \circledW \mathbb{R}$
Sphere	$SO(2) \circledW SO(2)$
Cross-Section Cylinder	$SO(2) \circledW \mathbb{R}$
Ruled Cylinder	$\mathbb{R} \circledW SO(2)$

LEVEL-DISCRETE

Cube	$\mathbb{R} \circledW \mathbb{R} \circledW Z_2 \circledW Z_3$
Cross-Section Block	$\mathbb{R} \circledW Z_n \circledW \mathbb{R}$
Ruled or Planar-Face Block	$\mathbb{R} \circledW \mathbb{R} \circledW Z_n$

Some comments should be made before beginning:

(1) We shall see that all solid primitives contain the corresponding primitive surface that was developed previously. This surface is important and will be called the *surface kernel*. The surface kernel is different from the alignment kernel because the latter occurs at the lowest level of the group wreath hierarchy. In contrast the surface kernel does not - for a very important reason: it breaks the symmetry of the infinite solid which exists at the lower level.

SURFACE KERNEL. *The surface kernel of a solid primitive is obtained from the group of the solid primitive by omitting the space translation group expressing the solid component as a wreath c-polycyclic group, and omitting also any bounding hyperplanes that make the primitive of finite length. Conceptually, the surface kernel is the first structure to break the infinite symmetry of the translational solid component.*

(2) All the solid primitives to be developed here have the same solid component - i.e., the triple wreath translation group $\mathbb{R} \circledW \mathbb{R} \circledW \mathbb{R}$. This is the commonality that exists before the surface primitive enters and differentiates one solid from another.

(3) We shall see that the solid versions of the level-continuous primitives in Table 16.1 will all have alignment kernels. This is because they are unfolded from both a circle and a straight line. In contrast, the solid versions of the level-discrete surfaces will not require alignment kernels, because they are made entirely out of straight lines.

16.9 The Solid Sphere

According to Table 16.1, the sphere has surface structure, $SO(2) \, \textcircled{w} \, SO(2)$. We now propose its solid structure:

$$[[SO(2)]_{P_2} \times [\mathbb{R}]_{P_1}]_{\mathcal{U}}$$
$$\textcircled{w} \, [\mathbb{R} \, \textcircled{w} \, \mathbb{R}]^{P_1}$$
$$\textcircled{w} \, SO(2)^{P_2} \qquad\qquad (16.15)$$

To understand this expression, let us go through it line-by-line, starting from the top:

Line 1: The first line is the alignment kernel, consisting of the straight line primitive $P_1 = \mathbb{R}$, and the circle primitive $P_2 = SO(2)$. Being in the kernel, their symmetries must be maximally aligned, which means either that (1) the straight line is in the plane of the circle, and is one of its diameters, or (2) the straight line is the rotation axis of the circle. Assume the *first* alternative. Furthermore, without loss of generality, assume that the straight line is in the ground plane. (This will lead to a sphere where the north and south poles are in the horizontal plane.)

Line 2: Here one generates the entire 3D space as $\mathbb{R} \, \textcircled{w} \, \mathbb{R} \, \textcircled{w} \, \mathbb{R}$. The lowest fiber $\mathbb{R}$ of this triple is already in the alignment kernel, and therefore we simply enter it as the index P_1 at the right end of the second line. Thus the group $\mathbb{R} \, \textcircled{w} \, \mathbb{R}$ on the second line is the upper two control groups of the triple $\mathbb{R} \, \textcircled{w} \, \mathbb{R} \, \textcircled{w} \, \mathbb{R}$. From a group-representation point of view, assume that the final $\mathbb{R}$ is vertical, and the preceding two $\mathbb{R}$ are both horizontal.

Line 3: This line corresponds to the expression $SO(2) \, \textcircled{w} \, SO(2)$ in Table 16.1 for the sphere; i.e., the surface kernel. Because the fiber $SO(2)$ is already in the alignment kernel as the primitive P_2, it is necessary only to give it as the index P_2 on the $SO(2)$ component of the third line, thus bringing it up super-locally.

16.10 The Solid Cross-Section Cylinder

According to Table 16.1, the cross-section cylinder has surface structure, $SO(2) \, \textcircled{w} \, \mathbb{R}$. This is an infinite structure. The cross-section cylinder will now be given as a solid. Furthermore, this solid will be finite, as used, for example, in a 3D modeling program. The structure we propose is this:

$$[[SO(2)]_{P_2} \times [\mathbb{R}]_{P_1}]_{\mathcal{U}}$$
$$ⓦ [\mathbb{R} \, ⓦ \, \mathbb{R}]^{P_1}$$
$$ⓦ \mathbb{R}^{P_2}$$
$$ⓦ \, \mathbb{Z}_2^{[\mathbb{R} \, ⓦ \, \mathbb{R}]_{1/2}}$$

(16.16)

To explain, let us go successively down the lines in this expression. Without loss of generality, assume that the cylinder axis is vertical.

Line 1: This is the alignment kernel, consisting of the straight line primitive $P_1 = \mathbb{R}$, and the circle primitive $P_2 = SO(2)$. Being in the kernel, their symmetries must be maximally aligned, which means either that (1) the straight line is in the plane of the circle, and is one of its diameters, or (2) the straight line is the rotation axis of the circle. Assume the *first* alternative. This is shown in Fig. 16.1a.

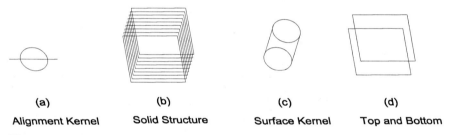

(a)	(b)	(c)	(d)
Alignment Kernel	**Solid Structure**	**Surface Kernel**	**Top and Bottom**

Fig. 16.1. The structure of a solid cross-section cylinder.

Line 2: Here one generates the entire 3D space as $\mathbb{R} \, ⓦ \, \mathbb{R} \, ⓦ \, \mathbb{R}$. The same discussion applies here as that given in Line 2 for the sphere (Sect. 16.9). The result is shown in Fig. 16.1b.

Line 3: This line corresponds to the expression $SO(2) \, ⓦ \, \mathbb{R}$ in Table 16.1 for this cylinder. Because, the $SO(2)$ component is already in the alignment kernel as the primitive P_2, we have only to give it here as the index P_2 on the $\mathbb{R}$ component, thus bringing it up super-locally. To emphasize, there is this equivalence:

Surface kernel:
$$SO(2) \, ⓦ \, \mathbb{R} \quad \longleftrightarrow \quad ⓦ \, \mathbb{R}^{P_2}$$

The surface kernel is illustrated in Fig. 16.1c.

Line 4: This line creates the bottom and top planes of the cylinder; i.e., making the cylinder finite. The index $[\mathbb{R} \,\textcircled{w}\, \mathbb{R}]_{1/2}$ is the plane positioned at parallel distance $1/2$ above the ground plane. This plane has already been generated in the preceding structure. Specifically, this index $[\mathbb{R} \,\textcircled{w}\, \mathbb{R}]_{1/2}$ is really a fiber copy within the triple $\mathbb{R} \,\textcircled{w}\, \mathbb{R} \,\textcircled{w}\, \mathbb{R}$. The subscript $1/2$ indicates the chosen member from the third group of the triple.

Notice carefully how this is all given in the above expression. First of all the index $[\mathbb{R} \,\textcircled{w}\, \mathbb{R}]_{1/2}$ is really $[P_1 \,\textcircled{w}\, \mathbb{R}]_{1/2}$. That is, the $P_1 = \mathbb{R}$ comes from the alignment kernel, and the other $\mathbb{R}$ appears at the beginning of the second line. Then the index $1/2$ is the position of this fiber in the second $\mathbb{R}$ on the second line (which is the third member of the triple $\mathbb{R} \,\textcircled{w}\, \mathbb{R} \,\textcircled{w}\, \mathbb{R}$).

Finally, the main group on the fourth line is $\mathbb{Z}_2$ which is understood as reflection about the ground plane. This produces a reflectional copy of the $1/2$ bounding plane at $-1/2$. This is illustrated in Fig. 16.1d.

16.11   The Solid Ruled Cylinder

According to Table 16.1, the ruled cylinder has surface structure, $\mathbb{R} \,\textcircled{w}\, SO(2)$. We now propose its solid structure to be:

$$[[SO(2)]_{P_2} \times [\mathbb{R}]_{P_1}]_{\mathcal{U}}$$
$$\textcircled{w}\, [\mathbb{R} \,\textcircled{w}\, \mathbb{R}]^{P_1}$$
$$\textcircled{w}\, P_2^{[1/2]_{P_1} \,\textcircled{w}\, e \,\textcircled{w}\, \mathbb{R}}$$
$$\textcircled{w}\, \mathbb{Z}_2^{[\mathbb{R} \,\textcircled{w}\, \mathbb{R}]_{1/2}} \tag{16.17}$$

To explain, let us go successively down the lines in this expression. Without loss of generality, assume that the cylinder is vertical.

Lines 1 & 2: These are the same as for the cross-section cylinder discussed above (Sect. 16.10).

Line 3: This line corresponds to the expression $\mathbb{R} \,\textcircled{w}\, SO(2)$ given in Table 16.1 for the ruled cylinder. To understand this correspondence, note first that the ruled cylinder is a vertical straight line (fiber) that is rotated around a horizontal circle (control). First consider the index

$$[1/2]_{P_1} \,\textcircled{w}\, e \,\textcircled{w}\, \mathbb{R}$$

in the third line of (16.17). This represents the straight line. The index is a subset of the triple $\mathbb{R} \,\textcircled{w}\, \mathbb{R} \,\textcircled{w}\, \mathbb{R}$, the first member of which is P_1. Thus the index $1/2$ is the specific point $1/2$ in the primitive line P_1. The second member e of the index is the identity element of the second member of the triple and therefore does nothing to that point. The third member of the index is $\mathbb{R}$ which acts in the perpendicular direction, and thus produces a vertical straight line from the point. Finally on the third line, the group P_2 is the primitive $SO(2)$, which is now used as a control group rotating the staight line given in the index. Again, the third line gives this correspondence:

Surface kernel:

$$\mathbb{R} \, ⓦ \, SO(2) \qquad \longleftrightarrow \qquad ⓦ \, P_2^{[1/2]_{P_1}} \, ⓦ \, e \, ⓦ \, \mathbb{R}$$

Line 4: This is the same as for the cross-section cylinder, and provides the cutting planes of the cylinder.

16.12   The Solid Cross-Section Block

According to Table 16.1, the cross-section block has surface structure, $\mathbb{R} \, ⓦ \, \mathbb{Z}_n \, ⓦ \, \mathbb{R}$. We now propose its solid structure to be:

$$\mathbb{R}^0 \, ⓦ \, \mathbb{R} \, ⓦ \, \mathbb{R} \, ⓦ \, \mathbb{R}$$
$$ⓦ \, [\mathbb{Z}_n \, ⓦ \, \mathbb{R}]^{[\mathbb{R}^0 \, ⓦ \, \mathbb{R}]_{1/2,0}}$$
$$ⓦ \, \mathbb{Z}_2^{[\mathbb{R}^0 \, ⓦ \, \mathbb{R} \, ⓦ \, \mathbb{R}]_{1/2}} \tag{16.18}$$

Again, let us go downward through these lines in sequence:

Line 1: The first line is the translation group $\mathbb{R}^3$ of infinite solid space, wreath re-constituted as a wreath c-polycyclic group, as occurs in all the solids. The starting point $\mathbb{R}^0$ is being explicitly mentioned, as was done with the solid cube.

Line 2: The upper index on the second line is a particular straight line selected from this previous structure - it is the straight line at perpendicular distance $1/2$ from the origin, within the first generated 2D plane. Notice how this information is given: This index $[\mathbb{R}^0 \, ⓦ \, \mathbb{R}]_{1/2,0}$ is of length four, and is a fiber taken from the 4-fold wreath product on the first line.

Finally, the control group $\mathbb{Z}_n \, ⓦ \, \mathbb{R}$ on this line takes this index as fiber, and first applies the group $\mathbb{Z}_n$ to it - generating the regular n polygon, in the first generated plane - and then the higher $\mathbb{R}$ sweeps the polygon out in the perpendicular direction.

Notice that Line 2 gives the surface kernel for the solid; that is, there is this correspondence:

Surface kernel:

$$\mathbb{R} \,\textcircled{w}\, \mathbb{Z}_n \,\textcircled{w}\, \mathbb{R} \quad \longleftrightarrow \quad \textcircled{w}\, [\mathbb{Z}_n \,\textcircled{w}\, \mathbb{R}]^{[\mathbb{R}^0 \,\textcircled{w}\, \mathbb{R}]_{1/2,0}}$$

(the starting point has been included on the right side).

Line 3: This again is the pair of reflectional cutting planes that make the block of unit length.

16.13   The Solid Ruled or Planar Block

According to Table 16.1, the ruled or planar block has surface structure, $\mathbb{R} \,\textcircled{w}\, \mathbb{R} \,\textcircled{w}\, \mathbb{Z}_n$. Its solid structure is:

$$\mathbb{R}^0 \,\textcircled{w}\, \mathbb{R} \,\textcircled{w}\, \mathbb{R} \,\textcircled{w}\, \mathbb{R}$$
$$\textcircled{w}\, [\mathbb{R} \,\textcircled{w}\, \mathbb{Z}_n]^{[\mathbb{R}^0 \,\textcircled{w}\, \mathbb{R}]_{1/2,0}}$$
$$\textcircled{w}\, \mathbb{Z}_2^{[\mathbb{R}^0 \,\textcircled{w}\, \mathbb{R} \,\textcircled{w}\, \mathbb{R}]_{1/2}} \qquad (16.19)$$

which is exactly the same as the structure of the solid cross-section block, except that there is an order-reversal of $\mathbb{Z}_n$ and $\mathbb{R}$ within the control group of the second line. Notice most crucially that the index on the second line is the same as before - and this means that the same straight line is selected from the first generated plane. The reversal within the control group means that the straight line is now swept to create a plane, which is then rotated - rather than rotated first to create a polygon which is then swept (as in the cross-section block).

16.14   The Full Set of Solid Primitives

Table 16.2 gives the entire set of solid primitives we have proposed. Notice the following features of this table:

(1) The level-continuous primitives all have an alignment kernel, and the level-discrete primitives do not. The alignment kernel for the level-continuous

primitives are all the same - the straight line and the circle. These are the two continuous roots of the Externalization Principle.

(2) The level-discrete primitives all begin with $\mathbb{R}^0 \,\text{ⓦ}\, \mathbb{R} \,\text{ⓦ}\, \mathbb{R} \,\text{ⓦ}\, \mathbb{R}$, which is the infinite solid structure expressed as a wreath c-polycyclic group. In fact, this structure is also contained in each of the continuous level primitives.

(3) For readability, the starting point $\mathbb{R}^0$ has been omitted in the level-continuous primitives. However, it should be placed before the entire alignment kernel.

(4) The cube is the only recursive structure.

The following comment is crucial:

COMPLEXITY OF THE SOLID SYMMETRY GROUPS. *The symmetry groups for the solid primitives are much more complicated than has been previously suspected - e.g., in mathematical crystallography. These complex groups give crucial information not just for the physical sciences (such as solid-state physics and electromagnetism), but are essential for understanding mechanical CAD (e.g., the choice of construction plane), as well as assembly planning (e.g., the contact structure), and also spline manipulation (e.g., the effects of tensor control on parametric sub-surfaces).*

The reader might wonder why objects called primitives are actually complex. The reason is this: According to our theory, the shape of an object is its history. We have seen that a *solid* primitive possesses quite a substantial history. In fact, it is historically reducible down to more basic primitives - the straight line and the circle. What the solid primitives therefore represent is an intermediate level of complexity (history!!) - between (1) the root primitives (line and circle) and (2) the arbitary solids which are made by deforming and Boolean-combining the solid primitives.

Because the groups of the solid primitives are complex, we shall often use the surface kernel as a short-hand expression for these objects. This short-hand is adequate because, by the above theory, the actual solid component is the same in all the primitives - i.e., it is the wreath c-polycyclic translation group. The surface kernel therefore acts as a signature of the solid.

16.15   Externalization in the Solid Primitives

Now observe that the symmetry-breaking step of adding the surface kernel to the infinite solid space, is an *external* transformation. By this we mean that the inference, made backward in time from the final solid primitive,

Table 16.2. The 3D *solid* primitives

LEVEL-CONTINUOUS

Sphere	$[[SO(2)]_{P_2} \times [\mathbb{R}]_{P_1}]_\mathcal{U}$ ⓦ $[\mathbb{R}$ ⓦ $\mathbb{R}]^{P_1}$ ⓦ $SO(2)^{P_2}$
Cross-Section Cylinder	$[[SO(2)]_{P_2} \times [\mathbb{R}]_{P_1}]_\mathcal{U}$ ⓦ $[\mathbb{R}$ ⓦ $\mathbb{R}]^{P_1}$ ⓦ $\mathbb{R}^{P_2}$ ⓦ $\mathbb{Z}_2^{[\mathbb{R} \; ⓦ \; \mathbb{R}]_{1/2}}$
Ruled Cylinder	$[[SO(2)]_{P_2} \times [\mathbb{R}]_{P_1}]_\mathcal{U}$ ⓦ $[\mathbb{R}$ ⓦ $\mathbb{R}]^{P_1}$ ⓦ $P_2^{[1/2]_{P_1}}$ ⓦ e ⓦ $\mathbb{R}$ ⓦ $\mathbb{Z}_2^{[\mathbb{R} \; ⓦ \; \mathbb{R}]_{1/2}}$

LEVEL-DISCRETE

Cube	$\mathbb{R}^0$ ⓦ $\mathbb{R}$ ⓦ $\mathbb{R}$ ⓦ $\mathbb{R}$ ⓦ $[\mathbb{Z}_2$ ⓦ $\Sigma_1]^{\mathbb{R}_{1/2,1/2,1/2}^0}$ ⓦ $[\mathbb{Z}_2$ ⓦ $\Sigma_2]^{[\mathbb{R}^0 \; ⓦ \; \mathbb{R}]_{1/2,1/2}}$ ⓦ $[\mathbb{Z}_2$ ⓦ $\Sigma_1]^{\mathbb{R}_{1/2,1/2,1/2}^0}$ ⓦ $[\mathbb{Z}_2$ ⓦ $\Sigma_3]^{[\mathbb{R}^0 \; ⓦ \; \mathbb{R} \; ⓦ \; \mathbb{R}]_{1/2}}$ ⓦ $[\mathbb{Z}_2$ ⓦ $\Sigma_2]^{[\mathbb{R}^0 \; ⓦ \; \mathbb{R}]_{1/2,1/2}}$ ⓦ $[\mathbb{Z}_2$ ⓦ $\Sigma_1]^{\mathbb{R}_{1/2,1/2,1/2}^0}$
Cross-Section Block	$\mathbb{R}^0$ ⓦ $\mathbb{R}$ ⓦ $\mathbb{R}$ ⓦ $\mathbb{R}$ ⓦ $[\mathbb{Z}_n$ ⓦ $\mathbb{R}]^{[\mathbb{R}^0 \; ⓦ \; \mathbb{R}]_{1/2,0}}$ ⓦ $\mathbb{Z}_2^{[\mathbb{R}^0 \; ⓦ \; \mathbb{R} \; ⓦ \; \mathbb{R}]_{1/2}}$
Ruled or Planar-Face Block	$\mathbb{R}^0$ ⓦ $\mathbb{R}$ ⓦ $\mathbb{R}$ ⓦ $\mathbb{R}$ ⓦ $[\mathbb{R}$ ⓦ $\mathbb{Z}_n]^{[\mathbb{R}^0 \; ⓦ \; \mathbb{R}]_{1/2,0}}$ ⓦ $\mathbb{Z}_2^{[\mathbb{R}^0 \; ⓦ \; \mathbb{R} \; ⓦ \; \mathbb{R}]_{1/2}}$

back to the infinite solid space, is an external inference (Sect. 2.9). Now the Externalization Principle says that any external inference goes back to a control-nested hierarchy of repetitive isometries (a wreath isometric, wreath c-polycyclic group). This is exactly what happens here. The removal of the symmetry-breaking step, backwards in time, leads back to the infinite solid, which is a control-nested hierarchy of repetitive isometries. Thus the Externalization Princple is fulfilled.

Observe also the following: The asymmetrizing object, the surface kernel, is also a control-nested hierarchy of repetitive isometries. This means that the initial control-nested hierarchy of repetitive isometries (the infinite solid) has its symmetry broken, forward in time, by another control-nested hierarchy of repetitive isometries (the surface kernel). This is not a problem. Our theory of asymmetrization accounts for this. In particular the reader should recall our discussion in Sect. 11.2 and observe that Fig. 11.1 (p. 241) is also an example of symmetry-breaking by the concatenation of two objects, both of which are control-nested hierarchies of repetitive isometries. Note that the Boolean combination of any pair of primitive surfaces is also an example of symmetry-breaking by the concatenation of two objects, both of which are control-nested hierarchies of repetitive isometries; see Sect. 11.6 and Fig. 11.3 (p. 252).

16.16   The Unfolding Group of a Solid

Conventionally, a solid model is constructed as a Boolean combination of solid primitives. According to our theory, Boolean combination is realized by unfolding groups. In the case of a solid, we will see that the unfolding groups are more complex than the examples considered previously. Our goal of maximizing transfer will force the alignment kernel to have a transfer structure of four levels. That is, the alignment kernel is a 4-fold wreath product. This is described in Table 16.3, which will now be explained.

The alignment kernel gives the set of primitives that will be used in the particular solid which is to be constructed. Suppose $G_{primitive}$ is the symmetry group of one of those solid primitives. This means that $G_{primitive}$ is one of the groups in Table 16.2 (not Table 16.3!!!). What the four levels in the new Table 16.3 do is to take $G_{primitive}$ and successively factor out from it all those levels which it has in common with the other primitives that are to be used in the particular solid. The succesive factorizations correspond to the successive four levels in Table 16.3, as follows:

Level 1
First, $G_{primitive}$ has the starting point $\mathbb{R}^0$ in common with all the other primitives. So $\mathbb{R}^0$ is factored out and moved to the front of the alignment kernel.

Level 2
Next, $G_{primitive}$ has the first translation level $\mathbb{R}$ in common with all the other primitives, because this is the first translation level of the solid structure. So $\mathbb{R}$ is factored out and moved to the front of the alignment kernel, but just above the first level $\mathbb{R}^0$. Notice that $\mathbb{R}$ is the continuous translation root of externalization (Sect. 2.16). However, parallel with this root is the other continuous root $SO(2)$, if it occurs in any of the solid primtives used (e.g., if the solid cylinder is one of the primitives to be used in the solid). Therefore, form, at this level, an alignment kernel of the two continuous roots, thus:

$$[[SO(2)]_{P_2} \times [\mathbb{R}]_{P_1}]_{\mathcal{U}}. \tag{16.21}$$

Table 16.3. The alignment kernel of a solid object.

THE ALIGNMENT KERNEL IN SOLID MODELING

The alignment kernel for a solid object is a 4-fold wreath product, whose successive levels upward are:

(Level 1) The starting point

$$\mathbb{R}^0.$$

(Level 2) The alignment of the continuous roots:

$$[[SO(2)]_{P_2} \times [\mathbb{R}]_{P_1}]_\mathcal{U}.$$

(Level 3) The wreath unfolding of the translation root through space; i.e., the group

$$[\mathbb{R} \ \textcircled{w} \ \mathbb{R}]^{P_1}.$$

(Note that this completes the infinite solid structure; i.e. the wreath c-polycyclic translation group $\mathbb{R}^0 \ \textcircled{w} \ \mathbb{R} \ \textcircled{w} \ \mathbb{R} \ \textcircled{w} \ \mathbb{R}$.)

(Level 4) The direct product of the remainders of each of the symmetry groups of the solid primitives involved:

$$G_{K_n} \times \ldots \times G_{K_2} \times G_{K_1}.$$

Essentially, this last expression is the direct product of the surface kernels, $G_{K_1}, G_{K_2}, \ldots, G_{K_n}$. However, for any cylinder or block, the G_{K_i} will also include the bounding planes.

The total alignment kernel is therefore

$$\mathbb{R}^0$$
$$\textcircled{w} \ [[SO(2)]_{P_2} \times [\mathbb{R}]_{P_1}]_\mathcal{U}$$
$$\textcircled{w} \ [\mathbb{R} \ \textcircled{w} \ \mathbb{R}]^{P_1}$$
$$\textcircled{w} \ [G_{K_n} \times \ldots \times G_{K_2} \times G_{K_1}] \tag{16.20}$$

Level 3

Next, all primitives have in common the remainder of the translation hierarchy $\mathbb{R} \ \textcircled{w} \ \mathbb{R} \ \textcircled{w} \ \mathbb{R}$ that defines the infinite solid. The first $\mathbb{R}$ has already been factored downward as the continuous root P_1 in Level 2. So now unfold this first $\mathbb{R} = P_1$ via the remaining two translation levels in $\mathbb{R} \ \textcircled{w} \ \mathbb{R} \ \textcircled{w} \ \mathbb{R}$, in this way:

$$[\mathbb{R} \ \textcircled{w} \ \mathbb{R}]^{P_1}.$$

Notice that P_1 originates from expression (16.21) on the previous level (Level 2). Notice that Level 2 and 3 have together generated the complete infinite unbroken solid, $\mathbb{R} \ \textcircled{w} \ \mathbb{R} \ \textcircled{w} \ \mathbb{R}$.

Level 4

Finally, there is what is left of $G_{primitive}$ after all these previous factorizations. Mainly what is left of $G_{primitive}$ is its surface kernel. This is what is not in common with the other primitives - since each primitive has its own surface kernel. So, on this new level, take the direct product of those components of what is not in common between the primitives - essentially their surface kernels. The direct product is expressed as

$$G_{K_n} \times \ldots \times G_{K_2} \times G_{K_1}.$$

where one can think of the $G_{K_1}, G_{K_2}, \ldots, G_{K_n}$ as essentially the alignment kernels.

At this stage the reader should read through Table 16.3 to fully grasp what has been said above.

Having now constructed the alignment kernel of the solid object, let us turn to the *unfolding* of the alignment kernel. Suppose we want to unfold the primitive $G_{primitive}$ from the kernel. Again, $G_{primitive}$ is one of the groups in Table 16.2 (not Table 16.3!!!). What unfolding it means is that it will become an input index for a control group above the alignment kernel. To unfold $G_{primitive}$ from the kernel, one has to "compile" it from the kernel, as follows: One has to move left-to-right along the kernel and select those factors that comprise $G_{primitive}$ as it is defined in Tables 16.2 and 16.3. In compiling it, one must use the same group product operations that exist between the components as they occurred in the alignment kernel.

Notation 16.1 As a short hand, we often express the alignment kernel as the direct product of the solid primitive groups, e.g.,

$$G_{sphere} \times G_{cylinder} \times \ldots \times G_{cube}.$$

However, the reader should understand that this is an abuse of notation. The correct group is the hierarchical 4-level *wreath* product we defined for the solid alignment kernel. However, the *direct* product notation shown here can be regarded as being Level 4, that is, "essentially" the direct product of the surface kernels.

17. Wreath Formulation of Splines

17.1 The Goal of This Chapter

According to our generative theory, a shape is a frozen machine, which can be converted into an active machine using the recovery rules, so that the agent can self-substitute into the machine and thereby use the shape for plans; e.g., for design, manufacturing, manipulation, navigation, etc.

This chapter will elaborate a theory of splines as machines in which there are two levels, each of which is a machine. From higher to lower, the levels are:

Machine 2: In this level, the designer controls the shape of the spline - in fact, splines were invented for exactly this type of control.

Machine 1: A spline can be described *internally* as a space of actions, i.e., as a machine. For example, the spline surface is described as a surface of motion; e.g., in ordinary navigation, or NC machining where the surface has to be milled. This is also fundamental to the use of splines in CAD-based robot kinematics and animation, where trajectories are set up as interpolations between points or key-frames; see Ge & Ravani [39], Jüttler [69], Jüttler & Wagner [70], Kim, Kim, & Shin [74], Park & Ravani [116], Pletinckx [119], Schoemake [136], Röschel [129], Zefran & Kumar [158].

Hierarchically, Machine 2 is above Machine 1, and shapes the action structure of the latter. Thus we will talk of spline-shaping as *machine-shaping*.

GOAL

Formulate a spline as a single machine that contains Machine 1 (actions internal to the spline) and Machine 2 (actions that shape the spline) *as well as* the structural relationship between these two machines. That structural relationship is best expressed by the following wreath product of machines:

$$\text{Spline} \quad = \quad \text{Machine 1} \quad ⓦ \quad \text{Machine 2}.$$

This means that the action of Machine 2 on Machine 1 is one of transfer; i.e., spline-shaping will be understood algebraically by τ-automorphisms of groups. Notice that the agent can self-substitute into either machine. Thus, when self-substituting into Machine 2, the agent could be a designer manipulating the spline shape, and when self-substituting into Machine 1, the agent could be a kinematic arm moving along the spline.

This chapter will give a generative theory of splines that accords with our fundamental principles: the maximization of transfer and recoverability. Thus, in particular, Machine 1, the internal structure of the spline, will itself be decomposed into a hierarchy of transfer. Indeed, this hierachy will give an exhaustive generative account of the internal structure of splines. Furthermore, any level within this exhaustive hierarchy will be recoverable. The importance of this is that it will explain different aspects in the use of spline shape, for example in Mechanical CAD. It will be seen that the different levels are in fact exploited at different times by agents - e.g., in the choice of a construction plane to initiate the spline.

Let us now see how to proceed. A spline is a *parameterized n-surface*. Such a surface is a differentiable map from a solid n-cube (in n-space) into a higher dimensional space. For example, consider the 2-dimensional case. The domain of the parametrized surface is shown in Fig. 17.1. It is the solid 2-cube, i.e., a filled square, in $\mathbb{R}^2$. Fig. 17.2 shows the *codomain* for the parametrized surface, i.e., the image of the square from Fig. 17.1, under the differentiable map. Intuitively, one can think of the codomain as a "deformed" version of the "flat" domain. The deformed version exists in 3-space.

Obviously, this example can be expressed in the following typical way from differential geometry:

$$s : \begin{cases} [0,1] \times [0,1] & \longrightarrow & \mathbb{R}^3 \\ \\ (u,v) & \longmapsto & s(u,v). \end{cases} \tag{17.1}$$

Here the domain is the (filled) 2-cube, expressed as the ordinary Cartesian product of two unit intervals, that is, $[0,1] \times [0,1]$. This cube is called the

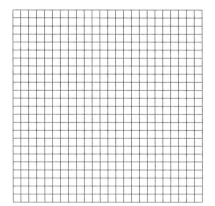

Fig. 17.1. The domain of a parametrized 2-surface.

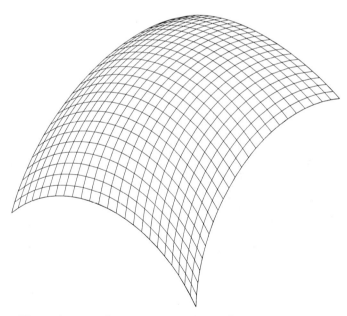

Fig. 17.2. The codomain of a parameterized 2-surface.

parameter domain of the spline. Because the codomain is an image of this cube, it is really a *patch*. However, we will use the standard abuse of terminology, and call it a surface, rather than merely a patch.

This example illustrates the standard way in which the domain, the solid n-cube, is described in the literature: Simply as the set-theoretic Cartesian product of n copies of the unit interval which is itself understood only set-theoretically.

The first step in expressing splines as machines is to convert the ordinary set-theoretic description of the spline parameter domain, i.e., as $[0,1] \times [0,1] \times \ldots \times [0,1]$, into an algebraic structure, in fact, into the recursive solid hyperoctahedral wreath hyperplane group $RS\text{-}HWH(n)$

$$\mathbb{R}^0 \circledw \mathbb{R} \circledw \mathbb{R} \circledw \mathbb{R} \circledw \ldots$$

$$\circledw \; [\; \mathbb{Z}_2 \circledw \Sigma_1]^{\mathbb{R}^0_{1/2,1/2,1/2,\ldots}}$$

$$\circledw \; [\; \mathbb{Z}_2 \circledw \Sigma_2]^{[\mathbb{R}^0 \circledw \mathbb{R}]_{1/2,1/2,\ldots}} \quad \circledw \; [\; \mathbb{Z}_2 \circledw \Sigma_1]^{\mathbb{R}^0_{1/2,1/2,1/2,\ldots}}$$

$$\circledw [\mathbb{Z}_2 \circledw \Sigma_3]^{[\mathbb{R}^0 \circledw \mathbb{R} \circledw \mathbb{R}]_{1/2,\ldots}} \circledw [\mathbb{Z}_2 \circledw \Sigma_2]^{[\mathbb{R}^0 \circledw \mathbb{R}]_{1/2,1/2,\ldots}} \circledw [\; \mathbb{Z}_2 \circledw \Sigma_1]^{\mathbb{R}^0_{1/2,1/2,1/2,\ldots}}$$

$$\ldots \tag{17.2}$$

Recall from section 16.6 (p. 402), the main facts about this group. (1) The group realizes the Asymmetry Principle, by starting with n-space as an infinite solid, and then cutting it down using hyperplanes as faces to produce the n-cube. The hyperplane faces are related by the hyperoctahedral group. (2) Our expression of infinite n-space is as the sweeping of the hyperplane $\mathbb{R}^{n-1}$ via the perpendicular action $\mathbb{R}$. It is this hyperplane that is then used for the faces transferred by the hyperoctahedral group. (3) However, the hyperplane faces are themselves solid $n-1$ cubes, which have solid $n-2$ cubes as faces; and so on downwards. Therefore, we recursively substitute solid cubes for successively descending values of n. (4) Furthermore, transfer is maximized by wreath re-constituting the original infinite solid n-space as a wreath c-polycyclic translation group - i.e., a hierarchical sweeping of hyperplanes of descending dimensions - and we transfer these descending hyperplanes into the descending orders of hyperoctahedral components.[1]

---

[1]   Throughout, this chapter there will be a simple abuse of notation: When describing the surface as a function of the form shown in (17.1), we will assume that the parameter domain is not centered at the origin - i.e., it is as indicated $[0,1] \times \ldots \times [0,1]$, with one corner at the origin. However, when describing the same parameter domain group-theoretically, we shall assume that the domain is centered at the origin, with its hyperoctahedral reflectional symmetries aligned with the coordinate planes. We could, using an extra group action, simply move this out from the origin to be coincident with $[0,1] \times \ldots \times [0,1]$, but this would be inappropriate: The conventional parameter domain $[0,1] \times \ldots \times [0,1]$ is always expressed as an arbitrary choice in CAD; whereas our group theoretic choice is non-arbitrary.

Table 17.1. The wreath formulation of splines.

STRATEGY FOR EXPRESSING SPLINES AS MACHINES

(1) Convert the ordinary set-theoretic description of the spline parameter domain (i.e., as $[0,1] \times [0,1] \times \ldots \times [0,1]$), into a group, in fact, into the recursive solid hyperoctahedral wreath hyperplane group $RS\text{-}HWH(n)$

$$\mathbb{R}^0 \ \textcircled{w} \ \mathbb{R} \ \textcircled{w} \ \mathbb{R} \ \textcircled{w} \ \mathbb{R} \ \textcircled{w} \ \ldots$$

$$\textcircled{w} \ [\ \mathbb{Z}_2 \ \textcircled{w} \ \Sigma_1]^{\mathbb{R}^0_{1/2,1/2,1/2,\ldots}}$$

$$\textcircled{w} \ [\ \mathbb{Z}_2 \ \textcircled{w} \ \Sigma_2]^{[\mathbb{R}^0 \ \textcircled{w} \ \mathbb{R}]_{1/2,1/2,\ldots}} \ \textcircled{w} \ [\ \mathbb{Z}_2 \ \textcircled{w} \ \Sigma_1]^{\mathbb{R}^0_{1/2,1/2,1/2,\ldots}}$$

$$\textcircled{w}[\mathbb{Z}_2 \textcircled{w} \Sigma_3]^{[\mathbb{R}^0 \textcircled{w} \mathbb{R} \textcircled{w} \mathbb{R}]_{1/2,\ldots}} \textcircled{w}[\mathbb{Z}_2 \textcircled{w} \Sigma_2]^{[\mathbb{R}^0 \textcircled{w} \mathbb{R}]_{1/2,1/2,\ldots}} \textcircled{w}[\mathbb{Z}_2 \textcircled{w} \Sigma_1]^{\mathbb{R}^0_{1/2,1/2,1/2,\ldots}}$$

$$\ldots \tag{17.3}$$

(2) Form the set of control tensors into a group $\mathcal{T}$.

(3) Form the infinite group direct product of copies of the recursive solid hyperoctahedral wreath hyperplane group $RS\text{-}HWH(n)$, with one copy for each member of the tensor group $\mathcal{T}$.

(4) Let the tensor group $\mathcal{T}$ be the control group of a wreath product where the fiber-group product is the infinite direct product just defined.

(5) Each spline will correspond to one of the copies in the fiber-group product.

(6) The shaping action that deforms splines onto each other will correspond to the wreath automorphic action that sends copies of the fiber group $RS\text{-}HWH(n)$ onto each other.

Let us now move onto the shaping operations on splines. A spline shape is controlled by a finite set of vectors; these include various position control vectors, tangent vectors, twist vectors, etc. This set of vectors become the entries of a matrix. The matrix is an example of a tensor, i.e., a multi-linear form. One deforms the surface by changing the entries of the tensor.

Let us now see how to realize our goal of expressing splines as machines. The strategy we shall take is presented in Table 17.1 (p. 427). Stage (1) on this table has already been discussed above. The reader should now go through the remainder of the table.

To illustrate the table, return to Figs 17.1 and 17.2 (p. 425). Fig. 17.1 represents the parameter domain for the 2D case. In stage (1) of the table, this is expressed as the *recursive solid hyperoctahedral wreath hyperplane group RS-HWH(2)*. Fig. 17.2 is a spline using one of the control tensors. In stage (2) the set of control tensors is formed into a group $\mathcal{T}$. Then, stage (3) forms the infinite group direct product of copies of the first diagram (Fig. 17.1), with one copy for each member of the tensor group $\mathcal{T}$. (4) Then let the tensor group $\mathcal{T}$ be the control group of a wreath product where the fiber-group product is the infinite direct product just defined. (5) Each spline, like Fig. 17.2, will correspond to one of the copies in the fiber-group product. (6) The shaping action that deforms splines onto each other will correspond to the wreath automorphic action that sends copies of the first figure (Fig. 17.1) onto each other. This illustrates a fundamental principle of our approach:

SPLINE-SHAPING AS AUTOMORPHIC ACTIONS ON GROUPS. *Our theory of splines will formulate the action of deforming splines onto each other as an automorphic action on groups.*

17.2   Curves as Machines

The remainder of this chapter goes systematically up dimension, from 1 to 3 (curves to 3D solids) showing how splines can be usefully formulated in terms of our generative theory of shape. Start first with parameterized curves:

Parameterized curves are used prolifically in CAD. They are usually described as functions that send the unit interval $[0,1]$ into $\mathbb{R}^3$, as follows:

$$c: \begin{cases} [0,1] & \longrightarrow & \mathbb{R}^3 \\ \\ t & \longmapsto & c(t) \end{cases}$$

It is well-accepted in CAD that parameterized curves are useful because they can be regarded as representing the *locus of a moving point* (e.g., Mortenson [110]). According to the theory being developed in this book, such curves can represent the *trajectory of a machine* for which an agent can *self-substitute* in some plan, e.g., as the servo system of a plotter in a drawing process, an electron-beam deflection system in a vector display, or a robot moving through an environment where the curve is the shape of a wall or hillside.

To go from the set of points on a curve, to an actual machine that will trace out the set, one needs our rules for the inference of generative history.

Now the machine, which corresponds to the parametrized curve, is an example of Machine 1, given in Sect. 17.1. It represents the first level of control in our discussion of such curves. In contrast, the designer's control is given by the level above this, Machine 2. Thus for example, supposing that the plan represented by the parametrized curve is one that an *end-user*, e.g., robot, will use as it moves along the edge of a shape. Then the *design control*, which manipulates that curve, is the ability to alter that plan, e.g., deform and move it through space.

As said earlier, these two levels of control, i.e., end-user and designer, will rigorously form a wreath product. Conceptually, the designer can therefore be viewed as *transferring* the trajectories of the end-user around the environment. The end-user is control nested within the designer.

Now let us turn to the algebraic description of the curve as a machine - a machine whose state-space will be shaped by the designer. According to our theory, the parameter domain will be given by the recursive solid hyperoctahedral wreath hyperplane group, which in the 1D case must be *RS-HWH(1)*. This was presented in Sect. 16.7, and is given again here:

$$\mathbb{R}^0 \ \textcircled{w} \ \mathbb{R}$$
$$\textcircled{w} \ [\ \mathbb{Z}_2 \ \textcircled{w} \ \Sigma_1]^{\mathbb{R}^0_{1/2}} \tag{17.4}$$

The two lines of this expression correspond to the first two lines of the general formula (17.3) - where the n-fold wreath c-polycyclic translation component in the first line of the general formula is now 1-fold, and the index on the second line is ammended appropriately.

To understand this structure, the reader should review the discusion concerning the solid interval on p. 409 - 410.

17.3   Cubic Hermite Curves

This section considers a particular well-known example of parameterized curves: the cubic Hermite curves. When looking later at surfaces and 3D solids, we will again choose the Hermite formulation. However, this is not really a restriction. It will be obvious that exactly the same argument applies to any other of the main spline types - Bézier, B-spline, NURBS, etc.

This section will show that the cubic Hermite curve can easily be expressed in terms of our generative theory.

Usually, the control of a cubic Hermite curve-segment is determined by the positions of its endpoints and the tangent vectors at those points. If one simply calls the curve $c(t)$, then the standard endpoints are $c(0)$ and $c(1)$, and the corresponding tangent vectors, at the endpoints, are $c'(0)$ and $c'(1)$ (recall the footnote on p. 426). The *geometric form* of the curve is as follows:

$$c(t) \ = \ c(0)\mathfrak{h}_1(t) + c(1)\mathfrak{h}_2(t) + c'(0)\mathfrak{h}_3(t) + c'(1)\mathfrak{h}_4(t). \tag{17.5}$$

which should be read as an ordinary linear combination; i.e., a sum of basis vectors $\mathfrak{h}_i$ pre-multiplied by coefficients. The four coefficients are the curve endpoints and end-tangents: $c(0)$, $c(1)$, $c'(0)$, $c'(1)$. The four basis vectors $\mathfrak{h}_i$, in this linear combination, are the four *Hermite basis functions*. These are the four cubic equations $\mathfrak{h}_1 = 2t^3 - 3t^2 + 1$; $\mathfrak{h}_2 = -2t^3 + 3t^2$; $\mathfrak{h}_3 = t^3 - 2t^2 + t$; $\mathfrak{h}_4 = t^3 - t^2$.

To obtain the other cubic Hermite curves, hold the four basis functions $\mathfrak{h}_i$ fixed and change the four coefficients, $c(0)$, $c(1)$, $c'(0)$ and $c'(1)$, as one does in any linear combination. Now these four coefficients are each vectors in $\mathbb{R}^3$, since the curve is a map into $\mathbb{R}^3$. So the four coefficients give together $4 \times 3 = 12$ variables. Therefore, they define a 12-parameter group $\mathcal{H}$ which is isomorphic to $\mathbb{R}^{12}$. Let this be the control group $\mathcal{H}$ of our wreath product, i.e., Machine 2. The fiber is Machine 1, and this is *RS-HWH(1)*, which was given in expression (17.4). Therefore the full wreath product describing cubic Hermite curves is:

$$\mathbb{R}^0 \ \textcircled{w} \ \mathbb{R}\textcircled{w} \ [\ \mathbb{Z}_2 \ \textcircled{w} \ \Sigma_1]^{\mathbb{R}^0_{1/2}} \ \textcircled{w} \ \mathcal{H} \tag{17.6}$$

Notice that, although $\mathcal{H}$ is a 12-parameter group, it is infinite. Thus there are an infinite number of copies of the fiber $\mathbb{R}^0 \ \textcircled{w} \ \mathbb{R}\textcircled{w} \ [\ \mathbb{Z}_2 \ \textcircled{w} \ \Sigma_1]^{\mathbb{R}^0_{1/2}}$. Each copy gives one of the cubic Hermite curves. The control group thereby elaborates all the cubic Hermite curves as the fiber-group copies. It deforms and moves them onto each other in 3-space.

The reader should grasp the following important fact: The fiber-group product is the direct product of all the cubic Hermite curves! This means that the control group $\mathcal{H}$, the designer control, acts as an automorphism group on the entire system of curves. In other words, we have expressed *spline shaping as automorphic action on a group* - which is an extremely powerful concept. Notice of course that the automorphic action is the τ-action (or $\hat{\tau}$-action) on the fiber-group product.

Most crucially, we can move between equivalent points on the curves by using this automorphic (conjugacy) operation of the control group. That is, consider a fixed point $\mathbb{R}^0_u$, and its image point $c(\mathbb{R}^0_u)$ on the curve c. Then one can move to its equivalent point $\hat{c}(\mathbb{R}^0_u)$ on another curve $\hat{c}$ by conjugation. Thus, the structure of one curve is *transferred* onto the structure of another curve. That is, the fiber group can be transferred from one curve to another.

As an example, fix the two endpoints $c(0)$ and $c(1)$ in $\mathbb{R}^3$. This has fixed 6 of the parameters of the control group $\mathcal{H}$. Fix also the *directions* of the end-vectors $c'(0)$ and $c'(1)$, but not their magnitudes. This has fixed an extra 4 of the parameters of $\mathcal{H}$. Thus only two parameters are left - the magnitudes of the end-vectors. Varying through these two parameters elaborates a family of curves where all the members of the family share the same endpoints, and the same normalized end-vectors. This is illustrated in Fig. 17.3.

What then do these curves represent? Each curve represents a machine. These machines are *deformed* onto each other. The action of deformation is the control group.

Fig. 17.3. Some Hermite splines sharing the same end-points, but with coincident tangent vectors of different magnitudes.

It is worthwhile comparing the case of cubic Hermite curves with the wreath product we gave for the square:

$$\mathbb{R} \, \textcircled{w} \, \mathbb{Z}_4.$$

First replace the unbroken infinite description $\mathbb{R}$ given for the side of a square with the group $RS\text{-}HWH(1)$, which expresses the side as a finite segment. Thus the above expression of a square is now changed to:

$$\mathbb{R}^0 \, \textcircled{w} \, \mathbb{R}\textcircled{w} \, [\, \mathbb{Z}_2 \, \textcircled{w} \, \Sigma_1]^{\mathbb{R}^0_{1/2}} \, \textcircled{w} \, \mathbb{Z}_4.$$

This settles the occupancy issue.

Now let us compare the case of the square with that of cubic Hermite curves. In the case of a square, there were four "curves," the four copies of $\mathbb{R}$. The control group was $\mathbb{Z}_4$ considered as a rotation group. The four copies of $RS\text{-}HWH(1)$ remain rigidly straight under the action of the control group. In the Hermite case, there are an infinite number of copies of $RS\text{-}HWH(1)$, and the action of the control group $\mathcal{H}$ deforms them and moves them through space. The elegance of the wreath product analysis is that it shows that, despite these differences, the two situations have very deep similarities: Both are organized as translation-structures along lines, i.e., translation-machines, such that these machines are *transferred* from one line to another by a higher level group into which the machines are control-nested.

Let us now return fully to the cubic Hermite group:

$$\mathbb{R}^0 \, \textcircled{w} \, \mathbb{R}\textcircled{w} \, [\, \mathbb{Z}_2 \, \textcircled{w} \, \Sigma_1]^{\mathbb{R}^0_{1/2}} \, \textcircled{w} \, \mathcal{H} \quad . \tag{17.7}$$

It is crucial to understand that the group $\mathcal{H}$ controls the *image* of its fiber $\mathbb{R}^0 \, \textcircled{w} \, \mathbb{R}\textcircled{w} \, [\, \mathbb{Z}_2 \, \textcircled{w} \, \Sigma_1]^{\mathbb{R}^0_{1/2}}$. It does so via the equation (17.5). The clearest way to understand this control is to notice that equation (17.5) has the form of a 1-tensor action. Any tensor, being a multi-linear form, can be expressed as a matrix. Thus, the 1-tensor, which here is the 1×4 matrix $(c(0), c(1), c'(0), c'(1))$, is applied to the 4×1 vector $(\mathfrak{h}_1, \mathfrak{h}_2, \mathfrak{h}_3, \mathfrak{h}_4)$ by ordinary matrix multiplication thus:

$$(\; c(0), \;\; c(1), \;\; c(3), \;\; c(4) \;) \begin{pmatrix} \mathfrak{h}_1 \\ \mathfrak{h}_1 \\ \mathfrak{h}_1 \\ \mathfrak{h}_1 \end{pmatrix}$$

to produce the linear combination in equation 17.5. The control group $\mathcal{H}$ is loaded into the 1-tensor on the left.

TENSOR EQUATIONS AS WREATH PRODUCTS. *The above tensor equation prescribes the way in which $\mathcal{H}$ acts as a control group with respect to the fiber group $\mathbb{R}^0 \; \textcircled{w} \; \mathbb{R}\textcircled{w} \; [\; \mathbb{Z}_2 \; \textcircled{w} \; \Sigma_1]^{\mathbb{R}^0_{1/2}}$ within the wreath product given in (17.7).*

17.4 Parametrized Surfaces as Machines

Now go up one dimension and deal with parameterized surfaces. Again these are used prolifically in CAD, where they usually are described as functions that send the unit square interval $[0,1] \times [0,1]$ differentiably into $\mathbb{R}^3$, thus:

$$s : \begin{cases} [0,1] \times [0,1] & \longrightarrow & \mathbb{R}^3 \\ \\ (u,v) & \longmapsto & s(u,v) \; . \end{cases} \tag{17.8}$$

In many applications, it is useful to consider a parameterized surface as a 1-parameter family of curves that completely decomposes the surface. For example, if one fixes a value v' in the parameter v, then any such curve on the surface has the form $s_{v'}(u)$. These curves are called *iso-parametric curves*. Clearly, this expresses the surface in terms of the wreath c-polycyclic translation group $\mathbb{R} \; \textcircled{w} \; \mathbb{R}$ which accords exactly with what is predicted by our theory - giving the surface a generative description that maximizes transfer.

Observe now that this wreath c-polycyclic group is what we have called the "solid" structure - which is infinite. We need to create a symmetry-breaking of this structure using the square boundary, i.e., a 2D cube - and this means incorporating the hyperoctahedral group of degree 2. Furthermore, each side of the boundary is a 1D cube, i.e., an interval, and this has a boundary - the two endpoints - which are themselves related by the hyperoctahedral group of degree 1. Thus there is a recursive downward structure of hyperoctahedral groups and matching wreath c-polycyclic translation groups. This means, of course, that the full structure of the parameter domain - i.e., the filled square - is given by the recursive solid hyperoctahedral wreath hyperplane group, which in the 2D case must be *RS-HWH(2)*. This group was given in Sect. 16.7 as:

$$\mathbb{R}^0 \circledW \mathbb{R} \circledW \mathbb{R}$$

$$\circledW [\, \mathbb{Z}_2 \circledW \Sigma_1]^{\mathbb{R}^0_{1/2,1/2}}$$

$$\circledW [\, \mathbb{Z}_2 \circledW \Sigma_2]^{[\mathbb{R}^0 \,\circledW\, \mathbb{R}]_{1/2}} \circledW [\, \mathbb{Z}_2 \circledW \Sigma_1]^{\mathbb{R}^0_{1/2,1/2}} \quad . \tag{17.9}$$

To understand our discussion of surface splines, the reader should first thoroughly review our discussion of the solid square p. 410 - 411.

Finally, observe that what has been constructed here is Machine 1 in the 2D case. It is now necessary to add Machine 2 which is the shaping of the 2D surface. This is illustrated in the next section with bicubic Hermite surfaces.

17.5   Bicubic Hermite Surfaces

Now consider the surface corresponding to the cubic Hermite curve: The bicubic Hermite surface. This will be formulated in terms of our generative theory.

First some terminology and notation: Using the conventional description of a parameterized surface

$$s : \begin{cases} [0,1] \times [0,1] & \longrightarrow \quad \mathbb{R}^3 \\[2mm] (u,v) & \longmapsto \quad s(u,v) \end{cases} \tag{17.10}$$

we have the following notation for the partial derivatives:

$$\frac{\partial s(u,v)}{\partial u} = s^u(u,v) \qquad , \qquad \frac{\partial s(u,v)}{\partial v} = s^u(u,v)$$

$$\frac{\partial^2 s(u,v)}{\partial u \partial v} = s^{uv}(u,v) = s^{vu}(u,v) = \frac{\partial^2 s(u,v)}{\partial v \partial u}.$$

It is well-known that the shape and position of a bicubic Hermite surface is fully controlled by 16 vectors, which can be taken to be:

4 position vectors: the position vectors of the four corners: $s(0,0)$, $s(0,1)$, $s(1,0)$, $s(1,1)$.

8 tangent vectors: the end-tangents to each of the four boundary curves, that is, the vectors $s^u(0,0)$, $s^u(0,1)$, $s^u(1,0)$, $s^u(1,1)$, $s^v(0,0)$, $s^v(0,1)$, $s^v(1,0)$, $s^v(1,1)$.

4 twist vectors: the twist vectors at the four corners - these are the second order derivatives $s^{uv}(0,0)$, $s^{uv}(0,1)$, $s^{uv}(1,0)$, $s^{uv}(1,1)$.

A convenient way to conceptualize this is as follows: Essentially the patch is being controlled by the four boundary curves - the images of the four sides of a square. These four curves are cubic Hermite curves. Now, according to Sect. 17.3, a single cubic Hermite is controlled by *four vectors*: the position vectors of the two end-points, and the two end-tangents. Thus, having four boundary curves to control, and four control vectors for each, there are a total of $4 \times 4 = 16$ control vectors. However, because the end-points of each boundary curve coincide with the end-points of the adjacent two curves around the boundary, there are a total of only four independent end-points. This reduces the number of control vectors from 16 to 12. These are the first 12 vectors in the above list.

Four extra control vectors then have to be added to gain full control over the shape of the patch, these are the four twist vectors above.

The full set of 16 vectors can be inserted as entries in a matrix thus:

$$H = \begin{pmatrix} s(0,0) & s(0,1) & s^v(0,0) & s^v(0,1) \\ s(1,0) & s(1,1) & s^v(1,0) & s^v(1,1) \\ s^u(0,0) & s^u(0,1) & s^{uv}(0,0) & s^{uv}(0,1) \\ s^u(1,0) & s^u(1,1) & s^{uv}(1,0) & s^{uv}(1,1) \end{pmatrix}$$

Notice that the matrix divides simply into four quandrants: (1) The upper-left quadrant consists of the four end-points; (2) the lower-left quadrant consists of the end-tangents to the two boundary u-curves ; (3) the upper-right quadrant consists of the end-tangents to the two boundary v-curves; and (4) the lower-right quadrant consists of the four twist vectors.

The most important thing for us to understand is that this matrix controls various aspects of the image of the recursive solid hyperoctahedral wreath hyperplane group *RS-HWH(2)* which is given again here:

$$\mathbb{R}^0 \text{ⓦ} \mathbb{R} \text{ⓦ} \mathbb{R}$$

$$\text{ⓦ} \left[\mathbb{Z}_2 \text{ⓦ} \Sigma_1 \right]^{\mathbb{R}^0_{1/2,1/2}}$$

$$\text{ⓦ} \left[\mathbb{Z}_2 \text{ⓦ} \Sigma_2 \right]^{[\mathbb{R}^0 \text{ⓦ} \mathbb{R}]_{1/2}} \text{ⓦ} \left[\mathbb{Z}_2 \text{ⓦ} \Sigma_1 \right]^{\mathbb{R}^0_{1/2,1/2}}. \tag{17.11}$$

The control is as follows: Let us suppose that the u parameter is the lower $\mathbb{R}$ in Line 1 of the group, and the v parameter is the upper $\mathbb{R}$ in Line 1. Let us now go through the four quadrants of the matrix.

Top left quadrant: The bottom right entry of this quadrant corresponds to the index $\mathbb{R}^0_{1/2,1/2}$ on Line 2 of the group (17.11). This point is the first distinguished corner of the square. The entry above this in the matrix is the reflectional image of this corner as generated by the control group on Line 2 of the group. The first column of the quadrant will control the additional two corners generated by the hyperoctahedral structure in Line 3 of the group.

Bottom left quadrant: The second column of this quadrant controls the straight line that goes through the first pair of corners just mentioned. The first column controls the straight line that goes through the second pair of corners just mentioned. This is the first pair of reflectional lines generated by the $\mathbb{Z}_2$ group in Line 3 of the above group. Notice that this group produces the pair of lines from one of the lines - that given by the index $[\mathbb{R}^0 \textcircled{w} \mathbb{R}]_{1/2}$ on Line 3 of the group.

Upper right quadrant: This quadrant of the matrix concerns the other pair of reflectional lines that bound the square. This pair is generated by the Σ_2 component in Line 3 of the group, acting on its $\mathbb{Z}_2$ fiber which generated the first pair of reflectional boundary lines.

Lower right quadrant: This quadrant controls the relationships between the two directions in the wreath c-polycyclic translation group.

Notice that what has been meant by control, in the discussion of the quadrants, has been control of the *images* of the various components of the group. The matrix therefore corresponds to Machine 2 which is the designer's control over spline shape. The input group to this machine is the group $\mathcal{H}_S$ of matrices H. Thus, we take the group $RS\text{-}HWH(2)$, given in expression (17.11) as Machine 1, and the group $\mathcal{H}_S$ as Machine 2, and form their wreath product:

$$\mathbb{R}^0 \textcircled{w} \mathbb{R} \textcircled{w} \mathbb{R}$$
$$\textcircled{w} \, [\, \mathbb{Z}_2 \textcircled{w} \Sigma_1]^{\mathbb{R}^0_{1/2,1/2}}$$
$$\textcircled{w} \, [\, \mathbb{Z}_2 \textcircled{w} \Sigma_2]^{[\mathbb{R}^0 \textcircled{w} \mathbb{R}]_{1/2}} \textcircled{w} \, [\, \mathbb{Z}_2 \textcircled{w} \Sigma_1]^{\mathbb{R}^0_{1/2,1/2}}$$
$$\textcircled{w} \, \mathcal{H}_S \quad . \tag{17.12}$$

It is crucial to understand that the final line controls the *image* of the previous lines. It does so in the following way:

Whereas the equation for the cubic Hermite *curve* was a *linear* equation, i.e., a 1-tensor action:

$$c(t) \; = \; c(0)\mathfrak{h_1}(t) + c(1)\mathfrak{h_2}(t) + c'(0)\mathfrak{h_3}(t) + c'(1)\mathfrak{h_4}(t). \tag{17.13}$$

the equation for the bicubic Hermite *surface* is the *bilinear* equation

$$s(u,v) = [\mathfrak{h_1}(u), \mathfrak{h_2}(u), \mathfrak{h_3}(u), \mathfrak{h_4}(u)] H [\mathfrak{h_1}(v), \mathfrak{h_2}(v), \mathfrak{h_3}(v), \mathfrak{h_4}(v)]^T \tag{17.14}$$

i.e., in the form of a 2-tensor action, where H, in the center, is the matrix given above, and the symbols $\mathfrak{h}_i$ represent the *same* Hermite basis functions as in curve equation (17.13).

To elaborate the set of bicubic Hermite surfaces, hold fixed the four basis functions $\mathfrak{h}_i$ on either side of the matrix H in equation (17.14), and vary the 16 vector-entries of H. Note that a vector-entry is a 3-vector in (x, y, z)-space, and is therefore changed by altering any one of its (x, y, z) coordinates. Thus H is a $4 \times 4 \times 3 = 48$ matrix of independent coefficents. That is, the alteration of any one of the 48 coefficients will change the surface. Therefore H defines a 48-parameter group which we called $\mathcal{H}_S$ which is isomorphic to $\mathbb{R}^{48}$.

TENSOR EQUATIONS AS WREATH PRODUCTS. *The tensor equation in (17.14) prescribes the way in which $\mathcal{H}_S$ acts as a control group with respect to the fiber group RS-HWH(2) within the wreath product given in (17.12).*

Note that, although $\mathcal{H}_S$ is a 48-parameter group, it is infinite. Thus there are an infinite number of copies of the fiber group *RS-HWH(2)*. Each copy gives one of the bicubic Hermite surfaces. The control group thereby elaborates all the surfaces as the fiber-group copies. It deforms and moves them onto each other in 3-space.

Again, the reader should grasp the following important fact: The fiber-group product is the direct product of all the bicubic Hermite surfaces! This means that the control group $\mathcal{H}_S$, the designer control, acts as an automorphism group on the entire system of surfaces. In other words, we have once again expressed *spline shaping as automorphic action on a group*. Notice of course that the automorphic action is the τ-action (or $\hat{\tau}$-action) on the fiber-group product.

Most crucially, one can move between equivalent points on the surfaces by using this automorphic (conjugacy) operation of the control group. That is, consider a fixed point $\mathbb{R}^0_{u,v}$ in the wreath c-polycyclic fiber $\mathbb{R}^0 \textcircled{w} \mathbb{R} \textcircled{w} \mathbb{R}$ and its image point $s(\mathbb{R}^0_{u,v})$ on the surface s. Then one can move to its equivalent point $\hat{s}(\mathbb{R}^0_{u,v})$ on another surface $\hat{s}$ by conjugation. Thus, the structure of one surface is *transferred* onto the structure of another surface. That is, the fiber group is transferred from one surface to another.

17.6   Parametrized 3-Solids as Machines

Now go up one dimension and deal with parameterized three-dimensional solids. Again these are used prolifically in CAD, where they usually are described as functions that send the unit three-dimensional interval

$[0, 1] \times [0, 1] \times [0, 1]$ into $\mathbb{R}^3$, as follows:

$$s : \begin{cases} [0, 1] \times [0, 1] \times [0, 1] & \longrightarrow & \mathbb{R}^3 \\ \\ (u, v, w) & \longmapsto & s(u, v, w). \end{cases} \qquad (17.15)$$

In many applications it is useful to break this structure into iso-parametric substructures: Thus, for a fixed value w' of w, there is the iso-surface $s_{w'}(u, v)$ obtained by varying u and v. The solid is then obtained by sweeping this surface along the w parameter; thus giving a 1-parameter set of iso-surfaces. Similarly, given any one of these iso-surfaces $s_{w'}(u, v)$, fixing a value v' for v, produces an iso-line $s_{v'w'}(u)$ within that surface. The iso-surface is then obtained by sweeping the iso-line along the v parameter. Clearly, this expresses the entire 3-solid in terms of the wreath c-polycyclic translation group $\mathbb{R} \otimes \mathbb{R} \otimes \mathbb{R}$ which accords exactly with what is predicted by our theory - giving the surface a generative description that maximizes transfer.

Observe now that this wreath c-polycyclic group is what we have called the "solid" structure - which is infinite. We need to create a symmetry-breaking of this structure using the cube boundary, i.e., a solid 3-cube; and this means incorporating the hyperoctahedral group of degree 3. Each face of the 3-cube is a solid 2-cube, i.e., a solid square; and this has a boundary - the four edges - which are themselves related by the hyperoctahedral group of degree 2. Furthermore, each edge of the square is a solid 1-cube, i.e., a solid interval; and this has a boundary - the two endpoints - which are themselves related by the hyperoctahedral group of degree 1. Thus there is a recursive downward structure of hyperoctahedral groups and matching wreath c-polycyclic translation groups. This means, of course, that the full structure of the parameter domain, i.e., the solid 3-cube, is given by the recursive solid hyperoctahedral wreath hyperplane group, which in the 3D case is *RS-HWH(3)*. Using the general formula for *RS-HWH(n)* in (17.3), one can see that *RS-HWH(3)* must be:

$\mathbb{R}^0 \otimes \mathbb{R} \otimes \mathbb{R} \otimes \mathbb{R}$

$\otimes [\mathbb{Z}_2 \otimes \Sigma_1]^{\mathbb{R}^0_{1/2,1/2,1/2}}$

$\otimes [\mathbb{Z}_2 \otimes \Sigma_2]^{[\mathbb{R}^0 \otimes \mathbb{R}]_{1/2,1/2}} \otimes [\mathbb{Z}_2 \otimes \Sigma_1]^{\mathbb{R}^0_{1/2,1/2,1/2}}$

$\otimes [\mathbb{Z}_2 \otimes \Sigma_3]^{[\mathbb{R}^0 \otimes \mathbb{R} \otimes \mathbb{R}]_{1/2}} \otimes [\mathbb{Z}_2 \otimes \Sigma_2]^{[\mathbb{R}^0 \otimes \mathbb{R}]_{1/2,1/2}} \otimes [\mathbb{Z}_2 \otimes \Sigma_1]^{\mathbb{R}^0_{1/2,1/2,1/2}}$

$$(17.16)$$

This structure was fully explained on pages 406 to 408, by building it up from solid edge to solid face to entire solid cube.

Finally, observe that what has been constructed here is Machine 1 in the 3D case. It is now necessary to add Machine 2 which is the shaping of the solid. This will be illustrated in the next section with tricubic Hermite solids.

17.7   Tricubic Hermite Solid

This section discusses the tricubic Hermite solid, and shows that it can easily be expressed in terms of our generative theory.

The shape and position of a tricubic Hermite solid is fully controlled by 64 vectors. These vectors can be easily understood by considering a single corner. At a corner, one has the following eight vectors:

1 position vector: this gives the corner position.

3 tangent vectors: each is the end-tangent to one of the three boundary curves meeting at the corner.

3 twist vectors: each is a twist vector for one of the three curves meeting at the corner.

1 derivative-of-twist vector: the third-order mixed derivative at the corner.

Since there are eight corners, there are a total of $8 \times 8 = 64$ control vectors. These can be placed in a $4 \times 4 \times 4$ matrix. The structure of this matrix can be understood as follows:

Consider the matrix as a cube. Each face of the cube is a 4×4 matrix. Pick the front face, and fill it with the 16 entries corresponding to the matrix H given in Sect. 17.5 for the bicubic Hermite surface on variables u and v. As observed before, this 4×4 matrix (face) divides simply into four quandrants: (1) The upper-left quadrant consists of four point vectors; (2 and 3) the lower-left and upper-right quadrants consist of the tangent vectors; and (4) the lower-right quadrant consists of the four twist vectors.

Having completed the front face, now rotate the matrix cube about the top left corner, till one obtains the top face, which is again a 4×4 matrix. Fill this with the corresponding entries for the variables u and w. Perform the rotation once more till the left face is obtained, which is again a 4×4 matrix. Fill this with the corresponding entries for the variables v and w. The rest of the matrix is easy to fill from this. The full $4 \times 4 \times 4$ matrix itself is obviously divided into 3D quadrants each of which is a $2 \times 2 \times 2$ block. The front top quadrant is the set of 8 corner points of the solid. Its diagonally opposite quadrant (bottom back quadrant) is the set of 8 third-order mixed derivatives. Each remaining quadrant is either a set of tangent vectors, or a set of twist vectors - easily computed from the matrix faces given above.

The entire matrix will be denoted by $\bar{H}$.

The most important thing for us to understand is that this matrix controls various aspects of the image of the recursive solid hyperoctahedral wreath hyperplane group $RS\text{-}HWH(3)$ given in 17.16. The way to see this is to follow the argument we gave in the 2D case, starting p. 434, since this corresponds to significant sections of the 3D case. The remainder of the 3D case can be easily deduced from that previous discussion.

It is important to understand that the controlling role of the matrix is the control of the *images* of the *RS-HWH(3)* group which acts as the fiber. That is, the matrix corresponds to Machine 2 which is the designer's control over spline shape. The input group to this machine is the group $\mathcal{H}_S$ of matrices $\bar{H}$. Thus, we take the group *RS-HWH(3)*, given in expression (17.16) as Machine 1, and the group $\mathcal{H}_S$ as Machine 2, and form their wreath product:

$$\mathbb{R}^0 \text{ⓌⒻ} \mathbb{R} \text{Ⓦ} \mathbb{R} \text{Ⓦ} \mathbb{R}$$

$$\text{Ⓦ} \, [\, \mathbb{Z}_2 \text{ Ⓦ } \Sigma_1]^{\mathbb{R}^0_{1/2,1/2,1/2}}$$

$$\text{Ⓦ} \, [\, \mathbb{Z}_2 \text{ Ⓦ } \Sigma_2]^{[\mathbb{R}^0 \text{ Ⓦ } \mathbb{R}]_{1/2,1/2}} \quad \text{Ⓦ} \, [\, \mathbb{Z}_2 \text{ Ⓦ } \Sigma_1]^{\mathbb{R}^0_{1/2,1/2,1/2}}$$

$$\text{Ⓦ}[\, \mathbb{Z}_2 \text{ Ⓦ } \Sigma_3]^{[\mathbb{R}^0 \text{Ⓦ} \mathbb{R} \text{ Ⓦ } \mathbb{R}]_{1/2}} \text{Ⓦ}[\mathbb{Z}_2 \text{ Ⓦ } \Sigma_2]^{[\mathbb{R}^0 \text{ Ⓦ } \mathbb{R}]_{1/2,1/2}} \quad \text{Ⓦ} \, [\, \mathbb{Z}_2 \text{ Ⓦ } \Sigma_1]^{\mathbb{R}^0_{1/2,1/2,1/2}}$$

$$\text{Ⓦ} \, \mathcal{H}_S \quad . \tag{17.17}$$

It is crucial to understand that the final line controls the *image* of the previous lines. It does so in the following way:

Whereas the equation for the cubic Hermite *curve* was a *linear* equation, i.e., a 1-tensor action,

$$c(t) \; = \; c(0)\mathfrak{h}_1(t) + c(1)\mathfrak{h}_2(t) + c'(0)\mathfrak{h}_3(t) + c'(1)\mathfrak{h}_4(t). \tag{17.18}$$

and the equation for the bicubic Hermite *surface* was the *bilinear* equation, i.e., a 2-tensor action,

$$s(u,v) = [\mathfrak{h}_1(u), \mathfrak{h}_2(u), \mathfrak{h}_3(u), \mathfrak{h}_4(u)]H[\mathfrak{h}_1(v), \mathfrak{h}_2(v), \mathfrak{h}_3(v), \mathfrak{h}_4(v)]^T \tag{17.19}$$

the equation for the tricubic Hermite *solid* is obviously the *trilinear* equation

$$\begin{aligned} s(u,v,w) = \bar{H}\{ & [\mathfrak{h}_1(u), \mathfrak{h}_2(u), \mathfrak{h}_3(u), \mathfrak{h}_4(u)], \\ & [\mathfrak{h}_1(v), \mathfrak{h}_2(v), \mathfrak{h}_3(v), \mathfrak{h}_4(v)], \\ & [\mathfrak{h}_1(w), \mathfrak{h}_2(w), \mathfrak{h}_3(w), \mathfrak{h}_4(w)]\} \quad . \end{aligned} \tag{17.20}$$

In other words, the action of $\bar{H}$ is in the form of a 3-tensor, taking the three arguments shown in the successive three lines. Thus one applies $\bar{H}$ as a matrix first to the 4-vector $[\mathfrak{h}_1(u), \mathfrak{h}_2(u), \mathfrak{h}_3(u), \mathfrak{h}_4(u)]$ of Hermite basis functions on u, which collapses the matrix to a 4×4 matrix; one then applies this collapsed matrix to the 4-vector $[\mathfrak{h}_1(v), \mathfrak{h}_2(v), \mathfrak{h}_3(v), \mathfrak{h}_4(v)]$ of Hermite basis functions on v, which collapses the matrix to a 4×1 matrix; and finally one applies this further collapsed matrix to the 4-vector $[\mathfrak{h}_1(w), \mathfrak{h}_2(w), \mathfrak{h}_3(w), \mathfrak{h}_4(w)]$ of Hermite basis functions on w, which yields a 1×1 matrix. This result is:

$$s(u,v,w) = \sum_{i=0}^{4}\sum_{j=0}^{4}\sum_{k=0}^{4} p_{ijk} \, \mathfrak{h}_i(u)\mathfrak{h}_j(v)\mathfrak{h}_k(w). \tag{17.21}$$

To elaborate the set of tricubic Hermite solids, hold fixed the basis functions $\mathfrak{h}_i$, and vary the 64 vector-entries of the matrix $\bar{H}$. Note that a vector-entry is a 3-vector in (x, y, z)-space, and is therefore changed by altering any one of its (x, y, z) coordinates. Thus $\bar{H}$ is actually a $4 \times 4 \times 4 \times 3 = 192$ matrix of independent coefficents. That is, the alteration of any one of the 192 coefficients will change the solid. $\bar{H}$ therefore defines a 192-parameter group $\bar{\mathcal{H}}_S$ which is isomorphic to $\mathbb{R}^{192}$. Let this group be the control group $\bar{\mathcal{H}}_S$ of the wreath product given in (17.17).

TENSOR EQUATIONS AS WREATH PRODUCTS.   *The tensor equation in (17.20) prescribes the way in which $\bar{\mathcal{H}}_S$ acts as a control group with respect to the fiber group RS-HWH(3) within the wreath product given in (17.17).*

Note that, although $\bar{\mathcal{H}}_S$ is a 192-parameter group, it is infinite. Thus there are an infinite number of copies of the fiber group *RS-HWH(3)*. Each copy gives one of the tricubic Hermite solids. The control group thereby elaborates all the solids as the fiber-group copies. It deforms and moves them onto each other in 3-space.

Again, the reader should grasp the following important fact: The fiber-group product is the direct product of all the tricubic Hermite solids! This means that the control group $\bar{\mathcal{H}}_S$, the designer control, acts as an automorphism group on the entire system of solids. In other words, we have once again expressed *spline shaping as automorphic action on a group*. Notice of course that the automorphic action is the τ-action (or $\hat{\tau}$-action) on the fiber-group product.

Most crucially, one can move between equivalent points in the solids by using this automorphic (conjugacy) operation of the control group. That is, consider a fixed point $\mathbb{R}^0_{u,v,w}$ in the wreath c-polycyclic fiber $\mathbb{R}^0 \textcircled{w} \mathbb{R} \textcircled{w} \mathbb{R} \textcircled{w} \mathbb{R}$ and its image point $s(\mathbb{R}^0_{u,v,w})$ in the solid s. Then one can move to its equivalent point $\hat{s}(\mathbb{R}^0_{u,v,w})$ in another solid $\hat{s}$ by conjugation. Thus, the structure of one solid is *transferred* onto the structure of another solid. That is, the fiber group can be transferred from one solid to another.

17.8   Final Comment

Although only Hermite cubic structures have been used as illustrations in this chapter, it is clear that exactly the same formulation works for the other main splines - Bézier, B-splines, Nurbs, etc. The reason is as follows: First the fiber group in all cases is the same: it is *RS-HWH(n)* of the appropriate degree. Second, in all cases, the control group is a matrix group whose action on images of the fiber group is prescribed by a tensor equation. Third, in all cases, the fiber-group product represents the product of all the splines in that class. Fourth, in all cases, the spline-shaping is given by the τ-action (or $\hat{\tau}$-action) on the fiber-group product. This will deform one spline (in the class) onto another. Furthermore, it does this by *transfer*, i.e., transferring the fiber group from one spline to another.

18. Wreath Formulation of Sweep Representations

18.1 Sweep Representations

Nearly all the representations developed in this book are sweep representations - the movement of some object (e.g., a curve) along a path. The reason for this is that, according to our theory, generativity is maximized by making the representation a wreath c-polycyclic group; i.e., a group that can be decomposed as a wreath product of c-cyclic groups. Each c-cyclic group represents the *time* parameter - i.e., the parameter along which generativity takes place. In a wreath c-polycyclic group, each level represents the time parameter, and generativity is maximized by maximizing transfer, i.e., control-nesting the time parameter. One can say therefore that the entire object is *swept out*; i.e., each level sweeps out the previous level - *all the way down.*

Most of the discussion in this book has therefore been about representations that are sweep representations *all the way down.* However, this chapter is going to discuss the specific class of situations that are referred to in the graphics literature as sweep representations. Such representations are explicitly offered as a choice on most CAD and 3D modeling programs. Usually, the designer is presented a menu in which he or she can select a *profile* curve and a *path*, along which the profile is to be swept; e.g., as shown in Fig. 18.1a (p. 445). Specific examples of this include an *extrusion*, which is usually defined to be the case where the path is a straight line (Fig. 18.1b), and a revolution, which is usually defined to be the case where the path is a circle (Fig. 18.1c). In many software programs, these latter two types are separated from the general sweep option because they are used so frequently. However, in this book, the term "sweep representation" will also include these two options. Also, in the research literature, the profile object can be a higher-dimensional object such as a surface or a solid. Our discussion will also apply to these cases, and the term "profile" will be retained for the higher-dimensional object being swept.

There are a number of set-theoretic and algebraic methods of representing sweep structures in the literature, e.g., equations of affine motion, Minkowski sums. We shall argue this:

SWEEP REPRESENTATIONS

Sweep representations are inherently wreath products in which the profile corresponds to the fiber and the path corresponds to the control; i.e., sweep representations are attempts to define shape as structures of *transfer:*

$$\text{Sweep} \ = \ \text{Profile} \ \textcircled{w} \ \text{Path} \ .$$

In fact, we will, in addition, decompose the fiber and the control themselves into wreath products, because the resulting structure will best explain the special cases discussed in the literature. Our basic claim will, in fact, be this: Sweep representations have the following fundamental structure:

$$(G_{profile} \ \textcircled{w} \ Ext_{profile}) \ \textcircled{w} \ (G_{path} \ \textcircled{w} \ Ext_{path}). \tag{18.1}$$

To understand this structure observe that there is a central wreath symbol $\textcircled{w}$, and the sequence divides naturally into two components, one to the left and one to the right of this central symbol. The part to the left describes the profile, and the part to the right describes the path. However, observe that the two components have *exactly the same structure*. In fact, each is a 2-fold wreath product. Thus the entire structure is an action of a 2-fold wreath product (the right) on a 2-fold wreath product (the left) via an intervening wreath product (the center).

Now let us look at this structure more deeply. Either component, the left or right, has a structure that is illustrated using Fig. 18.2. Purely for the sake of illustration the structure will be assumed to be a curve. According to our theory, this structure has two kinds of asymmetry: (1) External asymmetry in which the structure is seen as a deformed version of a more symmetric object. (2) Internal asymmetry which is a set of distinguishabilities between parts (e.g., points) within the object and which implies the trace structure of the object (recall Sect. 2.9). Thus, in case of the curve shown, the external asymmetry implies that the object is a deformed version of a straight line; and the internal asymmetry implies that the object is the trace of a point. Notice that this description conforms to the Externalization Principle, which states that any external inference goes back to an internal structure that is an iso-regular group; i.e., a control-nested hierarchy of repetitive isometries. In this example, the external inference (removal of deformation) goes back to a straight line whose internal structure is a control-nested hierarchy (only one level) of repetitive translations.

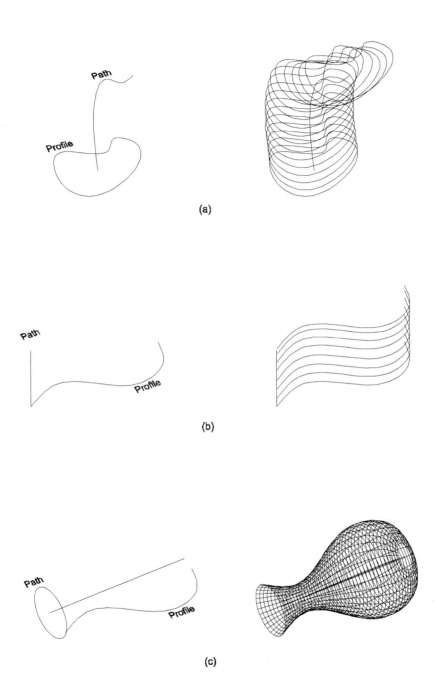

Fig. 18.1. The sweeping of a profile along a path.

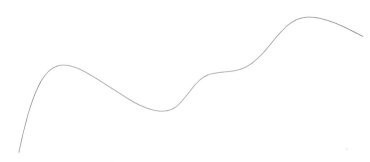

Fig. 18.2. A curve contains both external and internal asymmetry.

Let us return to the general case shown at (18.1) above. Consider the left half $(G_{profile} \text{ⓦ} Ext_{profile})$. This should be interpreted as follows: $Ext_{profile}$ is the group describing the external (deformation) structure of the profile; and $G_{profile}$ is the group describing the internal structure. In accord with the Externalizaiton Principle, $G_{profile}$ is an iso-regular group. When $Ext_{profile}$ is factored from $(G_{profile} \text{ⓦ} Ext_{profile})$, we are left with $G_{profile}$ the wreath-isometric trace structure.

Now, the right half of (18.1), which describes the path, has exactly the same structure; i.e., an external group Ext_{path} which deforms the path, and an internal group G_{path} which describes the trace structure when that deformation has been removed. Once again, the internal structure G_{path} is an iso-regular group.

Note the following: Either the external group $Ext_{profile}$ or Ext_{path} can be a tensor control group as described in our wreath formulation of splines. For example, suppose that Fig. 18.2 is a profile curve, and it is given by a cubic Hermite curve. In this case the group $Ext_{profile}$ would be $\mathcal{H}$ given in Sect. 17.3, and the group $G_{profile}$ would be $\mathbb{R}$. Thus the group of the curve would be:

$$G_{profile} \text{ⓦ} Ext_{profile} \;=\; \mathbb{R} \text{ⓦ} \mathcal{H}. \tag{18.2}$$

This however assumes that the curve is infinite. If the curve has endpoints, as it inevitably does, then the infinite structure $\mathbb{R}$ must be cut down by endpoints, thus giving us the filled interval - what we have called the solid interval. We have argued that the group for this is

$$RS\text{-}HWH(1) \;=\; \mathbb{R}^0 \text{ⓦ} \mathbb{R}\text{ⓦ} [\, \mathbb{Z}_2 \text{ⓦ} \Sigma_1]^{\mathbb{R}^0_{1/2}} \tag{18.3}$$

which was explained in Sect. 16.7 (p. 409). Thus in the case where the profile is given by a cubic Hermite curve, its group would be the group we gave (p. 430) for the cubic Hermite curve:

$$G_{profile} \text{ⓦ} Ext_{profile} \;=\; \mathbb{R}^0 \text{ⓦ} \mathbb{R}\text{ⓦ} [\, \mathbb{Z}_2 \text{ⓦ} \Sigma_1]^{\mathbb{R}^0_{1/2}} \text{ⓦ} \mathcal{H}. \tag{18.4}$$

One should carefully understand how the right-hand side partitions into the internal component $G_{profile}$ and external component $Ext_{profile}$ shown on the left. Recall the addition of endpoints is an *external* action (see Sect. 16.15, p. 418). Therefore, the endpoint component $[\ \mathbb{Z}_2\ \textcircled{w}\ \Sigma_1]^{\mathbb{R}^0_{1/2}}$ in the group *RS-HWH(1)* , is an external group, and must therefore be included in the external component $Ext_{profile}$ of the profile, rather than the internal component $G_{profile}$. Thus the partitioning of the profile into internal and external components would be as follows:

$$G_{profile}\ \textcircled{w}\ Ext_{profile}\ =\ \overbrace{\mathbb{R}^0\ \textcircled{w}\ \mathbb{R}}^{G_{profile}}\ \textcircled{w}\ \overbrace{[\ \mathbb{Z}_2\ \textcircled{w}\ \Sigma_1]^{\mathbb{R}^0_{1/2}}\ \textcircled{w}\ \mathcal{H}}^{Ext_{profile}}. \qquad (18.5)$$

Having shown how the endpoint factor is incorporated, we will, for ease of exposition, ignore the endpoint factor - in the remainder of the chapter. Thus the external component will be considered to consist only of deformation.

Now let us return to the entire structure given in (18.1). The power of this full structure is as follows: (1) The profile and path are described as having the same structure. (2) They are placed in a control-nested relation - the central wreath product. (3) The special classes of sweep structure are explicitly expressed in this structure, as we shall now demonstrate.

Case 1: The External Group Ext_{path} is trivial. When Ext_{path} is trivial, the path is given only by its internal group G_{path}, which is a control-nested hierarchy of repetitive isometries; i.e., a wreath c-polycyclic wreath-isometric group. Fig. 18.1b shows an example where G_{path} is simply the translation group $\mathbb{R}$; and Fig. 18.1c shows an example where G_{path} is simply the rotation group $SO(2)$. A further condition constraining these two examples shown is that the deformation of the profile is constant, i.e., only one element of $Ext_{profile}$ is being used in each example. This restriction can be lifted, as will be illustrated later.

Case 2: The External Group $Ext_{profile}$ is trivial. In this case, where there is no external group $Ext_{profile}$ for the profile, the profile is purely an iso-regular group. However, it can be moved along a path which is a deformed structure. An example is the case of ruled surfaces, where the profile is a straight line, and the path is a deformed curve. An example is shown in Fig. 18.3. Ruled surfaces have many applications in CAD and robot kinematics. A valuable pair of articles analyzing such structures in terms of line geometries is Pottmann, Peternell & Ravani [121], and Peternell, Pottmann & Ravani [118].

We are considering here the case where $Ext_{profile}$ is trivial. We can partition this case as follows: Since $G_{profile}$ has to be a control-nested hierarchy of repetitive isometries, consider first the case where the hierarchy has only one level. Within this case, there are only two types (1) where $G_{profile}$ is the

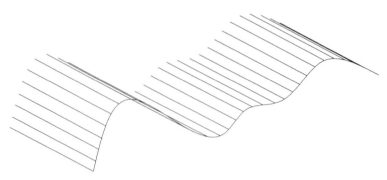

Fig. 18.3. A ruled surface.

1-parameter translation group $\mathbb{R}$, which is the case of ruled surfaces, and (2) where $G_{profile}$ is the 1-parameter rotation group $SO(2)$. In the latter case the swept structure is a tube.

Now let us consider the case where $G_{profile}$ is a two level hierarchy. If both levels are $\mathbb{R}$ then the profile is a plane, and the resulting swept structure is related to developable surfaces. If the two levels are different groups then one must be $\mathbb{R}$ and the other $SO(2)$. In this situation, we can get a cylinder. The resulting swept structure is a moving cylinder, and this structure appears often in CAD/CAM as the trajectory of a cylindrical cutting tool of a milling machine (e.g., Jüttler & Wagner [71]).

Case 3: Both $Ext_{profile}$ **and** Ext_{path} **are non-trivial.** As an example of this case, consider what is called the *synchronous sweep* illustrated in Fig. 18.4, and analyzed by Choi & Lee [22]. The reader can see that the profile deforms as it moved along the path. In Choi and Lee's scheme, the profile is contained in a plane. They control the successive deformation of the profile by a correction calculation to the shape of the profile.

This is interpreted in our scheme as follows: The un-deformed path is the straight line; i.e., the internal group of the path is $\mathbb{R}$. This is deformed by Ext_{path}, which can be considered, for example, to be the control group of the cubic Hermite curve. Similarly, the un-deformed profile is the straight line; i.e., with internal group $\mathbb{R}$. This is deformed by Ext_{path}, which again can be considered, for example, to be the control group of the cubic Hermite curve. The successive section planes along the sweeping correspond to the successive fibers of the group G_{path}, and one can think of each plane as containing its own copy of $Ext_{profile}$. In each fiber, some member of $Ext_{profile}$ creates the deformed version of the profile appearing in that fiber.

Note that the above analysis of swept structures exemplifies our method of describing asymmetries in terms of higher order symmetry spaces. For example, deformations of the profile are described in terms of the group $Ext_{profile}$ which is a higher order symmetry group acting with respect to

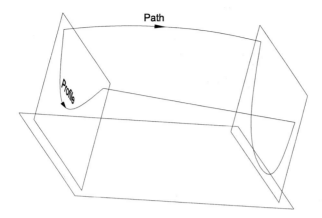

Fig. 18.4. Deformation of the profile along the path.

the lower order symmetry described by the group $G_{profile}$. What allows this characterization purely in terms of symmetry groups is the fact that "higher" and "lower" are defined with respect to a wreath hierarchy, and therefore the group action of the higher group is on a fiber-group product of the lower group, where each fiber copy is one of the deformed versions. For example, the fiber-group product can be the space of cubic Hermite curves, and the higher symmetry group be $\mathcal{H}$. What is being exemplified here is the fact that, in a wreath product, the higher group does not act as an automorphism group on the fiber itself, but on the fiber-group product. This means that one gets the following double effect: (1) The asymmetrizing deformation is actually a movement of the fiber copy off itself onto another fiber copy; i.e., a failure in symmetry of the actual fiber. However, (2) this effect is described purely as an automorphic action, i.e., symmetry of the space of fiber copies.

18.2   Aesthetics and Sweep Representations

Choi & Lee [22] say that one of the important motivations for using a sweep representation is that "the resulting surface is aesthetically pleasing". This greatly reinforces our theory of aesthetics as the maximization of transfer. When one looks at Fig. 18.4, one finds it aesthetically pleasing exactly because its structure of transfer is very visible: i.e., one sees the surface to be the result of transferring the profile curve P_0P_1 along the sweep direction. Thus the surface is *defined* as a structure of transfer - and according to our rigorous definition of aesthetics, this is an *aesthetical* structuring of the surface.

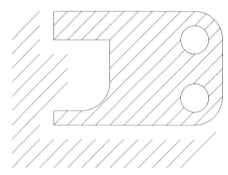

Fig. 18.5. A ray-representation of an object, based on Hartquist, et al. [52].

18.3 Ray Representations

Ray representations have been commonly used for rendering in computer graphics, and have been advocated for a number of demanding CAD applications, such as NC verification, by Hartquist, Menon, Suresh, Voelcker, & Zagajac, [52] (see also Ellis, Kedem, Lyerly, Thielman, Marisa, Menon & Voelcker [30]; Hartquist [51], Menon, Marisa & Zagajac [107]).

Consider Fig. 18.5 which illustrates the scheme of Hartquist, et al, as follows: The three-dimensional ambient space is divided into a regular grid G of parallel lines. These lines are clipped by the boundary of an object within that space. Set-theoretically, a ray representation is specified like this:

$$RR(A;G) = \bigcup_{m,n} (L_{m,n} \cap A) \qquad (18.6)$$

where $L_{m,n}$ is a grid-line indexed by m and n, and A specifies the object. Hartquist, et al. ([52] p181), argue that "the very simple and regular structure of ray-reps makes them easy to store, retrieve, and address." Furthermore, these representations satisfy the valuable property of Boolean simplification illustrated in Fig. 18.6: Part (a) of the figure shows a Boolean formula which some chosen shapes must satisfy. Part (b) shows the way in which the shapes themselves satisfy the formula. Part (c) then illustrates the crucial property of Boolean simplification: When one performs the Boolean composition of two objects, the ray representation of the composite is obtained by composing the ray representation of the two objects. In this, the 3D problem will be reduced to a series of independent 1D problems. Hartquist, et al, also argue that the highly regular structure of ray representations makes them appropriate to *parallel processing* architectures, which these researchers have exploited (see also Voelcker & Riquicha [152]).

The scheme of Hartquist, et al, is expressed by them in a *set-theoretic* form, as exemplified by equation (18.6) above. We argue that, behind expressions of this form is actually a *symmetry structure* organized in accord

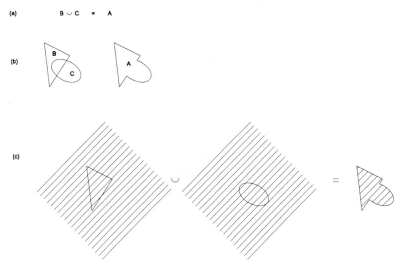

Fig. 18.6. Boolean simplification in ray-representations, based on Hartquist, et al. [52].

with the generative theory of this book, and that this symmetry structure should be used as the explicit representation of the design. After all, by our Symmetry-to-Trace Conversion Principle (Sect. 2.15), the symmetry structure will be what is exploited in the planning and action that is to be based on the design.

Let us therefore give a wreath-symmetry representation of the scheme of Hartquist, et al. In particular, we convert equation (18.6) into what we argue, is its wreath-symmetry structure. First the 2-parameter (m, n) grid structure is really the wreath product hierarchy:

$$\mathbb{R} \circledw \mathbb{R}^2$$

where the fiber group $\mathbb{R}$ represents a grid line, and the control group $\mathbb{R}^2$ represents the 2-parameter structure (m, n) which parameterizes these lines through 3D space. However, Hartquist, et al, also parameterize the individual grid line, and therefore we extend the above structure like this:

$$\mathbb{R}^0 \circledw \mathbb{R} \circledw \mathbb{R}^2$$

where $\mathbb{R}^0$ is the starting point. This is almost completely a generative structure, i.e., a wreath c-polycyclic group, except the final $\mathbb{R}^2$ control group. To make the entire structure generative, we wreath re-constitute the final group. The total structure now becomes

$$\mathbb{R}^0 \circledw \mathbb{R} \circledw \mathbb{R} \circledw \mathbb{R}$$

which is, of course, our standard wreath c-polycyclic group for the infinite 3-solid.

Now, having established the grid structure, we can then add the object A which will clip the infinite grid lines. Observe that this clipping is the concept explicitly used by Hartquist, et al, and it exactly accords with our theory: One starts with the infinite symmetry structure of the grid and breaks its symmetry by bringing in the boundary object A.

Our theory of solid modeling provides a detailed analysis of how this is done: In order to maximize transfer and recoverability, one must use *unfolding groups*. This means that the grid structure just given is part of an alignment kernel, and we will unfold this kernel by various control groups. In Sect. 16.16, we argued that the complete alignment kernel of a solid model is a 4-fold wreath product:

$$\mathbb{R}^0$$
$$\textcircled{w}\ [[SO(2)]_{P_2} \times [\mathbb{R}]_{P_1}]_{\mathcal{U}}$$
$$\textcircled{w}\ [\mathbb{R} \ \textcircled{w} \ \mathbb{R}]^{P_1}$$
$$\textcircled{w}\ [G_{K_n} \times \ldots \times G_{K_2} \times G_{K_1}]. \tag{18.7}$$

The successive four levels are given on the successive four lines, and these are (1) the starting point $\mathbb{R}^0$, (2) the aligned continuous roots $SO(2)$ and $\mathbb{R}$, (3) the remainder $\mathbb{R} \ \textcircled{w} \ \mathbb{R}$ of the infinite solid structure $\mathbb{R}^0 \ \textcircled{w} \ \mathbb{R} \ \textcircled{w} \ \mathbb{R} \ \textcircled{w} \ \mathbb{R}$, and (4) the direct product of the surface kernels G_{K_i} of any of the solid primitives to be used in generating the solid. Notice that the wreath c-polycyclic translation group $\mathbb{R}^0 \ \textcircled{w} \ \mathbb{R} \ \textcircled{w} \ \mathbb{R} \ \textcircled{w} \ \mathbb{R}$ given for the grid structure is now distributed over the first three lines of this expression.

Recall, according to our theory of solid modeling, that in order to unfold the solid primitives, one first has to *compile* them from the alignment kernel - this is fully explained in Sect. 16.16. One then wreath super-appends the control groups to unfold the compiled primitives, and obtain the required boundary object A which will break the symmetry of the infinite solid. Thus for example, in Fig. 18.6, which is similar to the illustrating example given by Hartquist, et al., we would have this: (1) The entire object would be unfolded as a rectangular block from the primitive block compiled from the alignment kernel. (2) The two circular holes would be unfolded as cylinders from the primitive cylinder compiled from the kernel. (3) Similarly, the two fillets (rounded corners), on the right, would be unfolded as cylinders from the primitive cylinder compiled from the kernel. (4) The removed section on the left, would be an unfolded compiled block and unfolded compiled cylinder as fillet.

Notice that the requirement by Hartquist, et al, that the scheme has Boolean simplification, is easily satisfied in our system. The initial ray-structure is an infinite wreath c-polycyclic group and unfolding will cut out from the infinite structure as many bounded objects as are needed.

Let us now turn to the fact that Hartquist, et al. have argued that the highly regular structure of ray representations makes them appropriate to *parallel processing* architectures. At this stage the reader should recall that, in Sect. 3.6, we argued that the independence of fiber copies in the fiber-group product, together with the symmetry which the control group has over the fiber-group product, means that the fiber level is best handled by a parallel architecture. In the case being discussed here, the grid structure is a wreath c-polycylic group, and the wreath aspect of the structure makes any level a direct product, for example:

$$\mathbb{R} \times \mathbb{R} \times \ldots \times \mathbb{R}. \qquad (18.8)$$

Recall also that Hartquist et al. exploit the independence of the fibers, for example, when they say (in [52] p178) "This property reduces an n-dimensional (usually 3D) problem into a series of independent 1D problems". In our approach, this independence is captured in the group-theoretic direct-product symbol $\times$ in the expression (18.8). Notice that the fact that the individual units within the parallel structure are identical comes from the symmetry expressed in the control group. That is, the very uniformity inherent in the parallelism is captured by the wreath product in its semi-direct expression, for example:

$$[\mathbb{R} \times \mathbb{R} \times \ldots \times \mathbb{R}] \circledS \mathbb{R} \ldots$$

18.4 Multiple Sweeping

Abdel-Malik & Othman [1] define *multiple-sweeping* as hierarchical successive sweeping. To quote one of their illustrations: "A circle, for example, is extruded along an axis to produce a cylindrical surface characterized by two parameters. This surface is revolved about another axis to yield a volume characterized by three parameters. Again, the volume is now extruded to yield a more complex solid in four parameters" (p567, [1]).

Clearly such a structure is what our generative theory calls a hierarchy of nested control, and can therefore be described conveniently by wreath products. Furthermore, one should note this: Abdel-Malik & Othman's paper is concerned with hierarchical sweeps where each successive sweep is either an extrusion or revolution. This is a frequently occuring situation in computer-aided design. Abdel-Malik & Othman develop a valuable formulation of this using the Denavit & Hartenberg [26] representation commonly applied in robot kinematics. They observe that an extrusion can be modeled as prismatic joint, and a revolution can be modeled by a revolute joint. Then using the standard assignment of Cartesian frames for the Denavit-Hartenberg approach, they develop a representation of the swept structure using the standard serial-link kinematic matrix composition:

$$_E T^B \;=\; _E T^1 \,*\, _1 T^2 \,*\, \ldots \,*\, _{n-2} T^{n-1} \,*\, _{n-1} T^B \qquad (18.9)$$

where the successive indexes give the successive Cartesian frames. In Sect. 6.4, we showed that this structure is conveniently expressed by the wreath product:

$$SE(3) \textcircled{w} \ldots \textcircled{w} SE(3) \textcircled{w} SE(3)$$

exploiting our claim that the algebraic structure of parent-child relationships is best given by wreath products.

19. Process Grammar

19.1 Introduction

The purpose of this chapter is to describe a grammar that we elaborated in Leyton [94], [96], to describe curve evolution with respect to a set of landmarks in that evolution. These landmarks are given by the curvature extrema. The grammar therefore relates the curvature extrema along the curve evolution. It accords with the theory of this book, in that the grammar is a **symmetry-breaking grammar**; i.e., it proceeds by a succession of symmetry-breaking phase-transitions. Since the grammar has been substantially discussed in Chapters 1 and 2 of our previous book (Leyton [96]), it will be described only briefly in the present chapter.

The shapes analyzed by the grammar are smooth closed curves in the 2D plane.

19.2 Inference from a Single Shape

According to Sect. 2.9, there are exactly two types of inference problems with respect to recovering a generative history: (External Inference) the data set contains what is assumed to be a record of a single state, and (Internal Inference) the data set contains what is assumed to be records of multiple states. This section considers the first of these two problems, and Sect. 19.3 considers the second.

The concept of symmetry is basic to all aspects of our generative theory. It is therefore necessary to establish the way in which symmetry is to be defined for smooth closed curves. There are three types of analysis that have been invented - all based on pushing a circle through the shape, while ensuring the circle is always tangential to the shape at two points, A and B, simultaneously, as shown in Fig. 19.1. The first analysis, the *Symmetric Axis Transform (SAT)*, due to Blum [12], defines a symmetry axis to be the locus of circle centers. The second analysis, the *Smooth Local Symmetry (SLS)*, due to Brady [14], defines a symmetry axis to be the locus of the midpoint of the chord AB. The third analysis, the *Process-Inferring Symmetry Analysis (PISA)*, due to Leyton [94], is actually shown in Fig. 19.1, and defines a symmetry axis to be the locus of the point Q, which is the midpoint of the arc AB. Extensive discussion of these three different symmetry axes is given in Leyton [96].

We now present the three inference rules that accomplish the task of inferring history from an individual smooth shape:

Rule 1.

A theorem proved by us in Leyton [92], constitutes the first rule in this inference system.

SYMMETRY-CURVATURE DUALITY THEOREM (Leyton [92]). *Any smooth section of curve, that has only one curvature extremum, has only one symmetry axis. This is forced to terminate at the extremum itself.*

Fig. 19.2 illustrates the theorem: The section of curve between the two letters m, has only one curvature extremum - that indicated by the letter M. The theorem says that this section of curve can have only one symmetry axis, and that the axis is forced to terminate at the extremum, as shown.

The Symmetry-Curvature Duality Theorem is the first of our inference rules, and says that each curvature extremum can be assigned a unique symmetry axis leading to the extremum.

Rule 2.

The second rule in the inference system is a particular consequence of the Symmetry Principle (p. 43), which says that, given a data set $\mathcal{D}$, a program for generating $\mathcal{D}$ is *recoverable from* $\mathcal{D}$ only if each symmetry in $\mathcal{D}$ is preserved backwards through the generative history. The particular consequence is that symmetry axes must be preserved backwards in time. In fact, this means that, in running time backwards, processes "withdraw" along the symmetry axes. For example, in running time backwards in the growth of the human hand, the growth processes "withdraw" along the symmetry axes. Thus, in the forward-time direction, the growth processes must have gone along the symmetry axes - which indeed is exactly what happened in the growth of the fingers. As an inference rule, this consequence is stated as follows:

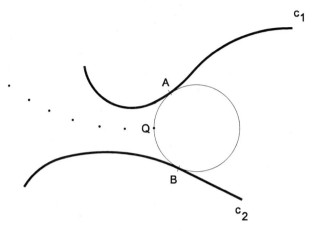

Fig. 19.1. The PISA symmetry analysis.

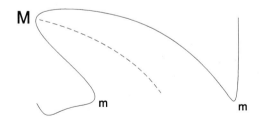

Fig. 19.2. Illustration of the Symmetry-Curvature Duality Theorem.

INTERACTION PRINCIPLE. *Symmetry axes are the directions along which processes are hypothesized as most likely to have acted.*

Rule 3.

The third rule to be used in the inference system is the Asymmetry Principle (p. 42), which says that, given a data set $\mathcal{D}$, a program for generating $\mathcal{D}$ is *recoverable from* $\mathcal{D}$ only if the program is symmetry-breaking on the successively generated states. The particular kind of symmetry we are going to consider is the rotational symmetry around a curve, and reflectional symmetry of each of its normals. These symmetries are maximized on a circle, which has constant curvature, and are broken by introducing curvature variation. Furthermore, the symmetry-breaking increases with greater curvature variation.

Putting this rule together with the previous two rules, the consequence is that, on an arbitrary smooth shape (e.g., an embryo), the processes are assumed as having gone along symmetry axes in the direction of increasing curvature variation: in fact, they created the curvature extrema at the ends of the symmetry axes. For example, in the human hand, the growth processes went along the symmetry axes leading to the curvature extrema at the tips of the fingers, and were responsible for creating those curvature extrema. Notice that the starting state was the human egg in which there were no curvature extrema, i.e., curvature was constant. Generally, according to Leyton [96], biological growth concerns the successive creating of curvature extrema, each one of which corresponds to a limb.

Application of the three rules.

To see that the three rules consistently yield appropriate process-histories, let us obtain the processes that these rules give for a large set of shapes. Figures 19.3-19.5 take all smooth shapes with up to and including, eight curvature extrema, and apply the inference rules to each of these shapes. The processes inferred by the rules are given by the arrows shown.

Note that the total set of shapes fall into three levels: shapes with four extrema, shapes with six extrema, and shapes with eight extrema. The reason is that there cannot be shapes with an odd number of extrema - because maxima have to alternate with minima of curvature - and there cannot be shapes with less than four extrema by the four-vertex theorem in differential geometry. There are a total of 21 shapes, of successively increasing complexity.

Observe that each extremum is marked by one of four symbols: M^+, M^-, m^+, m^-, meaning respectively positive maximum, negative maximum, positive minimum, and negative minimum.

Surveying the shapes, one finds that there is the following simple rule that relates the type of extremum to an English word for a process:

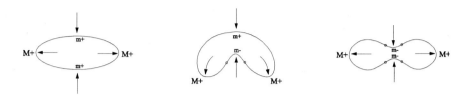

Fig. 19.3. The inferred histories on the shapes with 4 extrema.

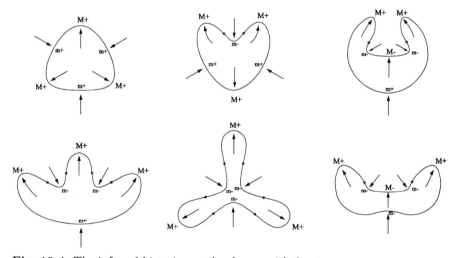

Fig. 19.4. The inferred histories on the shapes with 6 extrema.

SEMANTIC INTERPRETATION RULE.

$$
\begin{aligned}
M^+ &\longleftrightarrow \textit{protrusion} \\
m^- &\longleftrightarrow \textit{indentation} \\
m^+ &\longleftrightarrow \textit{squashing} \\
M^- &\longleftrightarrow \textit{internal resistance.}
\end{aligned}
$$

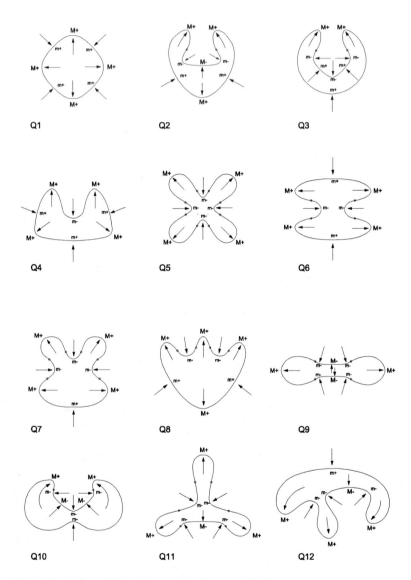

Fig. 19.5. The inferred histories on the shapes with 8 extrema.

19.3   Intervening History

Let us now turn to a type of problem occurring, for example, in a doctors office: The doctor is examining two X-rays of a tumor - one taken a month ago, and the other taken today. The doctor is trying to infer what happened in the intervening time.

Observe now that, since the later shape is assumed to emerge from the earlier shape, the doctor will try to explain it, as much as possible, as the outcome of processes seen in the earlier shape. In other words, he or she will try to explain the later shape, as much as possible, as the *extrapolation* of what can be seen in the earlier one.

As a simple first cut, let us divide all extrapolations of processes into two types:

(1) Continuations.
(2) Bifurcations (i.e. branchings).

What we will do now is elaborate the only forms that these two alternatives can take. There are four types of extrema, M^+, m^-, m^+, M^-. We will look at continuations at each of these four, and then at bifurcations at each of these four. This gives a total of eight cases to consider.

However, the first two are structurally trivial with respect to alteration of curvature extrema, as follows: Consider any one of the M^+ extrema in Fig. 19.6. It is the tip of a protrusion, as predicted by the Semantic Interpretation Rule, above. Observe that, if one continued the process creating that protrusion, i.e., continued pushing out the boundary in the direction shown, the protrusion would remain a protrusion. That is, the M^+ extremum would remain a M^+ extremum. This means that continuation at a M^+ extremum does not structurally alter the boundary, and can therefore be ignored. Exactly the same argument applies to continuation at a m^- extremum (indentation), and therefore continuation at this extremum can also be ignored.

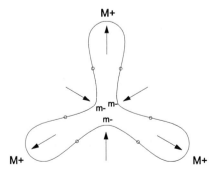

Fig. 19.6. Continuation at M^+ and m^- do not change extremum-type.

Therefore, from the eight cases, we are left with a total of six, each of which is structurally non-trivial with respect to alteration of curvature extrema, as follows: We derived the six cases by analyzing the singularities in the curvature function. This derivation is omitted here, and only the six results are stated:

Operation Cm^+. This is illustrated in Fig. 19.7. Here, the downward process, at the top of the left shape, continues pushing on the boundary, changing the extremum from the m^+ (left shape) to the m^- (right shape). Necessarily, two zeros of curvature (indicated by dots) are introduced on either side of the extremum (right shape). Therefore, the transition is given by this operation:

$$Cm^+ \; : \; m^+ \longrightarrow 0m^-0.$$

Operation CM^-. This is illustrated in Fig. 19.8. Here, the upward process, in the center of the left shape, continues pushing on the boundary, changing the extremum from the M^- (left shape) to the M^+ (right shape). Necessarily, two zeros of curvature (indicated by dots) are introduced on either side of the extremum (right shape). Therefore, the transition is given by this operation:

$$CM^- \; : \; M^- \longrightarrow 0M^+0.$$

Operation BM^+. This is illustrated in Fig. 19.9. Here, the upward process, at the top of the left shape, branches, sending the M^+ (left shape) to two copies of itself (right shape). Necessarily, a new extremum m^+ is introduced at the top, between the two M^+ copies (right shape). Therefore, the transition is given by this operation:

$$BM^+ \; : \; M^+ \longrightarrow M^+m^+M^+.$$

Operation Bm^-. This is illustrated in Fig. 19.10. Here, the downward process, in the center of the left shape, branches, sending the m^- (left shape) to two copies of itself (right shape). Necessarily, a new extremum M^- is introduced between the copies (right shape). Therefore, the transition is given by this operation:

$$Bm^- \; : \; m^- \longrightarrow m^-M^-m^-.$$

Operation Bm^+. The above two bifurcation operations indicate that a bifurcation has the following form: Send the extremum to two copies of itself with a new extremum in between of the same sign but the opposite type (Max vs. min). Thus one can immediately write bifurcation at m^+ as:

$$Bm^+ \; : \; m^+ \longrightarrow m^+ M^+ m^+.$$

Operation BM^-. Similarly, bifurcation at M^- can be immediately given as:

$$BM^- \; : \; M^- \longrightarrow M^- m^- M^-.$$

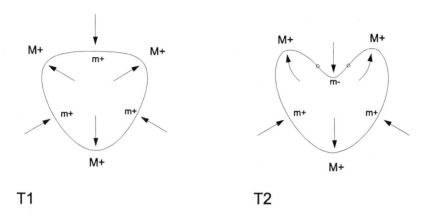

T1   T2

Fig. 19.7. Continuation at m^+.

19.4   Process Grammar

The six operations given above form a grammar that we call, the process grammar:

$$
\begin{aligned}
Cm^+ &: m^+ \longrightarrow 0m^-0 \\
CM^- &: M^- \longrightarrow 0M^+0 \\
BM^+ &: M^+ \longrightarrow M^+ m^+ M^+ \\
Bm^- &: m^- \longrightarrow m^- M^- m^- \\
Bm^+ &: m^+ \longrightarrow m^+ M^+ m^+ \\
BM^- &: M^- \longrightarrow M^- m^- M^-.
\end{aligned}
$$

Fig. 19.11 gives an example of shape evolution as inferred by the grammar. The first and last figures represent two shapes of a tumor that a doctor might

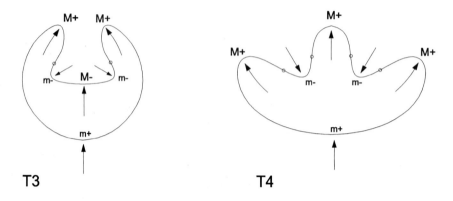

Fig. 19.8. Continuation at M^-.

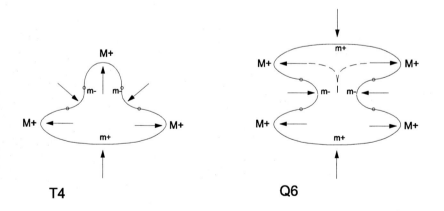

Fig. 19.9. Bifurcation at M^+.

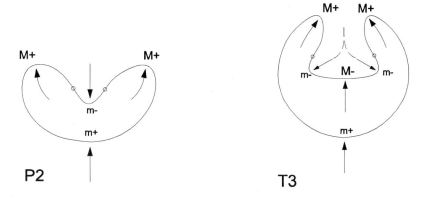

Fig. 19.10. Bifurcation at m^-.

be examining - taken a month apart. The grammar infers the intervening history. It says that the history (from the first to the last shape) is generated by three successive operations: CM^-, BM^+, and Cm^+. The inferred evolution is this:

(1) The entire history is dependent on a single crucial process: the *internal resistance* represented by the bold upward arrow in the first shape of Fig. 19.11.

(2) This process continues upward until it creates the protrusion shown in the second shape.

(3) This process then bifurcates, creating the upper lobe shown in the third shape. The bifurcation is due to the downward squashing process at the top of this shape.

(4) Finally, that downward squashing process continues till it causes the indentation in the top of the fourth shape.

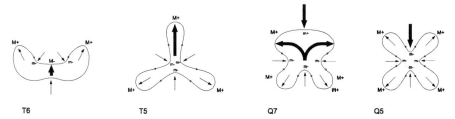

Fig. 19.11. A sequence of grammatical operations.

19.5   Other Literature

Smooth curves are used not only to model biological forms, but also in the design of highway and railway routes, Baas [9], Hartman [50]; as well as robot trajectories, Schmitt [135]; and general planar curve design in CAD, Farin [34], Hoschek & Lasser [61], Walton & Meek [153], [154].

Important work on smooth symmetry axes has been done by Leymarie & Levine [85], Leymarie & Kimia [84], Giblin & Sapiro [43]. Related to their work is the shock-based approach to shape representation: Kimia, Tannenbaum, Zucker [76], Siddiqi, Shokoufandeh, Dickinson & Zucker [144], Giblin & Kimia [41], [42]. A crease-based approach has been substantially investigated by Brassard [15]. Also, because of the importance of curvature extrema in biological shape, their significance can be seen in computer-aided design of garment's, Au & Yuen [7], [8]. Finally, Hayes & Leyton [55] extended the grammar described in this chapter by a further rule that creates smoothness-breaking. In addition, the reader should see the paper by Hayes [54] which presents a view of smooth liquid boundaries as historical entities.

20. Conservation Laws of Physics

20.1 Wreath Products and Commutators

This chapter considers very briefly a topic to be examined in much greater depth in Volume II: the conservation laws of physics. To keep the discussion brief, we will try to avoid the full technical details.

Let V and W be two smooth vector fields on a differentiable manifold. Let G_V and G_W be the 1-parameter groups that they generate, respectively. These groups are the flows, as illustrated in Fig. 20.1. Suppose now that the vector fields commute; that is, $[V, W] = 0$. This implies, in particular, that V pushes the flow-lines of W onto each other. What we propose doing is this:

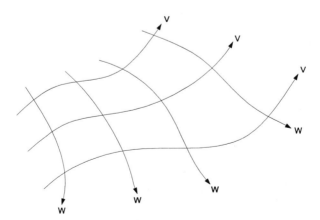

Fig. 20.1. Two flows V and W.

Create a wreath product in which G_V is the control group and G_W is the fiber group. That is:

COMMUTATOR WREATH-PRODUCT CORRESPONDENCE.
Set up a correspondence between commuting vector fields and the wreath product of their 1-parameter groups:

$$[V, W] = 0 \qquad \longleftrightarrow \qquad G_W \textcircled{w} G_V. \qquad (20.1)$$

Notice that the arrow links two different kinds of organizations of the manifold. For example, with respect to the structure $[V, W] = 0$, the group G_W moves simultaneously along all its flow lines. However, in the wreath product $G_W \textcircled{w} G_V$, there is an independent copy of G_W on each individual flow-line. That is, a copy of G_W can slide along its own flow-line, without affecting the copy of G_W on any other flow line. This can be thought of as representing experiments that were independently done before a process of induction discovered a relation between these experiments. Then after induction had established the control group G_V, experiments could be coordinated and one could, for example, establish a single "wave front" of points moving along the flow. This in fact, corresponds to the diagonal of the wreath product. Thus all the stages of scientific discovery are contained in the wreath structure, as opposed to the commutator. We will consider Chapter 3 as describing fully how wreath products capture the process of scientific discovery. This is because Chapter 3 described wreath products as hierarchies of detection.

Therefore, in the present chapter, one should understand that, whenever a commuting pair of operators is converted into a wreath product, we are importing the entire description of scientific discovery that a wreath product gives; i.e., all of Chapter 3. This will not be pointed out again; but the reader should be strongly aware of it.[1]

With respect to the topic of detection, note that Ishida [62] has developed a computational system for specifying, deriving, and generalizing mathematical equations in physics using symmetry-based production rules.

---

[1] The correspondence given above, between commutation and wreath products, has been stated without specifying the localness or globalness of the action. For example, on certain domains, the correspondence might hold only in some local version, in which case one can define a local wreath product, etc. Because of brevity, the local vs. global issue will not be mentioned again, in the remainder of this chapter. In fact, this is quite typical of books on mathematical physics. The issue of local boundedness of operators is so complicated in mathematical physics that to mention the vast labyrinth of results - needed to make even a quite simple statement - would be an enormous obstactle to following the main line of the argument.

20.2   Transfer in Quantum Mechanics

The remainder of this chapter illustrates the above concepts in the area of Quantum Mechanics.

The role of transfer is a powerful one in classical mechanics, for example, because of the fundamental importance of the three conservation principles: conservation of energy, linear momentum, and angular momentum. These are all derivable from Newton's 3rd Law, which states that action and reaction are equal and opposite. However, remarkably, conservation laws are even more important in quantum mechanics than in classical mechanics. The need for measurement operators that commute with the Hamiltonian underlies virtually every line of quantum mechanics since without commuting observables, no proper measurements can take place. Because commuting operators correspond to conservation principles, every step of the way in quantum mechanics uses a conservation principle. Furthermore, since our theory says that the commutation of operators corresponds to *transfer*, every step in quantum mechanics is based on transfer. Quantum mechanics manifests our *principle of the maximization of transfer* in the strongest possible way. It is not surprising therefore that quantum mechanics is regarded as the most aesthetic of all sciences - for our formal theory of aesthetics states that aesthetics is the maximization of transfer.

20.3   Symmetries of the Schrödinger Equation

Schrödinger's equation merely rotates Hilbert space; i.e., its action is unitary. Therefore it generates a flow in which each flow-line is simply the continuous rotation of a state through Hilbert space. We want to consider the *transfer* of flow-lines onto each other. This is given by a symmetry group of the Schrödinger equation. The symmetry group is also unitary. That is, it simply *rotates* the Schrödinger flow-lines onto each other.

Now, given a physical system, its Schrödinger equation

$$i\hbar\frac{\partial}{\partial t}|\psi(t)\rangle \;=\; H|\psi(t)\rangle \tag{20.2}$$

is particular to that system in the sense that the Hamiltonian H, in this equation, is specific to the system. That is, the Hamiltonian contains the potential energy function of that system, and this function is different for each different system. Therefore, since the Hamiltonian H is determined by the potential energy function, and the Schrödinger equation is determined by the Hamiltonian, the Schrödinger flow on Hilbert space will have a different shape for different systems.

We are considering symmetries of the particular flow induced by the particular Schrödinger equation for a particular system. A symmetry is provided by a symmetry operator U which rotates flow-lines onto flow-lines for that particular system. This means that, given any one of the flow-lines $|\psi(t)\rangle$, its rotated version under U, that is, $U|\psi(t)\rangle$, will also be a flow-line in the flow; i.e., it will satisfy the same Schrödinger equation, thus:

$$i\hbar\frac{\partial}{\partial t}U|\psi(t)\rangle \; = \; HU|\psi(t)\rangle. \tag{20.3}$$

In order to see what this condition requires, multiply both sides of equation (20.2) from the left by U, and insert the trivial identity operator $U^{-1}U$ as shown thus:

$$i\hbar\frac{\partial}{\partial t}U|\psi(t)\rangle \; = \; UHU^{-1}U|\psi(t)\rangle. \tag{20.4}$$

Now, equating (20.4) with the symmetry condition (20.3), one gets

$$H \; = \; UHU^{-1} \tag{20.5}$$

which means that the following commutation relation holds:

$$[U, H] \; = \; 0. \tag{20.6}$$

The reverse argument is also true. Thus, U is a symmetry of the Schrödinger flow, if and only if it commutes with the Hamiltonian. This is our condition for *transfer*: U transfers flow-lines onto flow-lines, if and only if $[U, H] \; = \; 0$.

As a consequence, it is now possible to use our rule that commuting actions can be converted into wreath products, thus: If G_{sym} is a symmetry group of the Schrödinger equation, the following wreath product can be formed:

$$G_H \;\textcircled{w}\; G_{sym} \tag{20.7}$$

where G_H is the 1-parameter group generated by the Hamiltonian.

20.4 Space-Time Transfer in Quantum Mechanics

The remainder of the chapter will look closely at various examples of the wreath product in (20.7) above. Let us first consider the space-time symmetries of (non-relativistic) quantum mechanics. Here, the group of continuous symmetries is the Galilean group, which is a 10-parameter group: The ten parameters are: (1-3) translations along the three spatial axes; (4-6) rotations around the three spatial axes; (7-9) boosts in velocity along the three spatial axes; (10) translations along the time axis. The situation of a free-particle (i.e., without external fields) has all these symmetries. The addition of external fields successively breaks these symmetries. Nevertheless, some of the

Table 20.1. Going from the Galilean group to Quantum-mechanical observables.

Space-time representation	Unitary representation	Observables
Spatial translations along axes $\mu = 1, 2, 3$ $x_\mu \to x_\mu + a_\mu$	$e^{-ia_\mu P_\mu}$	P_μ
Spatial rotations around axes $\mu = 1, 2, 3$ $v \to r_\mu(\theta_\mu)v$	$e^{-i\theta_\mu J_\mu}$	J_μ
Velocity boosts along axes $\mu = 1, 2, 3$ $x_\mu \to x_\mu + v_\mu t$	$e^{iv_\mu G_\mu}$	G_μ
Temporal translations $t \to t + \tau$	$e^{i\tau H}$	H

symmetries might remain. For example, in a spherically symmetric field (e.g., the basic hydrogen atom), the rotational symmetry has remained.

By Noether's theorem, any continuous symmetry in a system corresponds to a conservation law. This means that the components of the Galilean group provide significant observables which one would wish to measure in a system. One goes from the parameters of the Galilean group to their corresponding observables in the following way: Any parameter of the Galilean group is a 1-parameter group acting on space-time. This corresponds to a 1-parameter unitary group acting on Hilbert space. The generator of this group is a vector in the (1-dimensional) Lie algebra of that unitary group. This generator *is* the quantum-mechanical observable. The relation between the observable and its 1-parameter unitary group is given simply by exponentiation (which is the standard means of going from a Lie algebra to a Lie group). Thus, Table 20.1 shows the 10 parameters of the Galilean group, their corresponding unitary operators, and their corresponding observables. The observables are chosen to be Hermitian. The Lie algebra generated by the full set of 10 observables will be called the Galilean Lie algebra, $\mathfrak{G}$.

Table 20.2. The structure of the Galilean Lie algebra $\mathfrak{G}$.

$[P_\mu, P_\nu] = 0$	$[P_\mu, H] = 0$	$[J_\mu, H] = 0$
$[J_\mu, J_\nu] = i\epsilon_{\mu\nu\gamma} J_\gamma$	$[J_\mu, P_\nu] = i\epsilon_{\mu\nu\gamma} P_\gamma$	$[J_\mu, G_\nu] = i\epsilon_{\mu\nu\gamma} G_\gamma$
$[G_\mu, G_\nu] = 0$	$[G_\mu, H] = iP_\mu$	$[G_\mu, P_\nu] = i\delta_{\mu\nu} MI$

Now, we are interested in the capacity to *transfer* experiments. Transfer is given by symmetries, in fact, by the commuting of operators. Therefore, it is necessary to give the commutation relationships between the 10 generators of the Galilean Lie algebra $\mathfrak{G}$. These relationships are given in Table 20.2.

Notice the entries of the table where a commutator equals 0. In each of these cases there is transfer, and hence a wreath product, as described in the correspondence on p. 468. Most significantly, observe from the top line in Table 20.2, that the three generators P_μ of spatial translations, and the three generators J_μ of spatial rotations, each commute with the generator H of temporal translations. Since H corresponds to the Schrodinger equation, this means that these six generators define symmetries of that equation; i.e., map flow-lines to flow-lines. These therefore correspond to wreath products that represent conservation laws.

Table 20.2 defines the Lie algebra of the Galilean group. As noted above, the dynamics of a *free* particle are invariant under this *entire* group.[2] Now let us introduce a field, making the particle non-free. The fundamental consequence of this is that the form of the temporal generator H changes. However, the six generators P_μ and J_μ do not change. This means that any of the commutation relations that H had with P_μ and J_μ might now be altered. Since each of these commutation relations were zero, an altered commutation relation means that there is a loss of symmetry. That is, there is no longer the corresponding *transfer* of flow-lines onto flow-lines in the Schrödinger equation. This means that the associated wreath product has been lost. Volume II will consider means of expressing this not as the loss of the wreath product (i.e., symmetry-breaking) but as the transfer of the wreath product under a still higher wreath level (i.e., asymmetry-building). However, the present discussion will talk of symmetry-breaking and the loss of the wreath product.

---

[2] In fact, they are invariant under the larger Lie group associated with the Schrödinger algebra. This algebra consists of the Galilean algebra together with two more generators, that which generates dilatations, and that which generates conformal transformations. Each acts on space and time independently.

20.5   Non-solvability of the Galilean Lie Algebra

It is necessary to observe the following significant complexity in the structure of the Galilean Lie algebra $\mathfrak{G}$. The complexity is given by the entire second line in Table 20.2. This line shows that, under the commutation process, the generators J_μ keep on re-generating themselves as well as the P_μ and G_μ.

The consequences of this should be understood, as follows: Recall that an ideal $\mathfrak{h}$ of a Lie algebra $\mathfrak{g}$ is a subalgebra which "collapses" into itself under commutation with the entire algebra $\mathfrak{g}$; that is:

$$[\mathfrak{g}, \mathfrak{h}] \subseteq \mathfrak{h}.$$

An ideal corresponds to the concept of normal subgroup in group theory. Recall also the definition of the solvability of a Lie algebra: That is, define $\mathfrak{g}^{(1)} = [\mathfrak{g}, \mathfrak{g}]$, which is an ideal of $\mathfrak{g}$; and then define $\mathfrak{g}^{(2)} = [\mathfrak{g}^{(1)}, \mathfrak{g}^{(1)}]$, which is an ideal of $\mathfrak{g}^{(1)}$; and so on. Then $\mathfrak{g}$ is said to be *solvable* if the sequence terminates at 0. The *radical* of a Lie algebra is the maximal solvable ideal. It is unique and contains all other solvable ideals.

What can be seen therefore is that the second line of Table 20.2 shows that the angular momentum operators J_μ prevent the Galilean Lie algebra from being solvable. Thus consider the subalgebra $\mathfrak{r}$ generated by the remaining generators. It is the radical of the Lie algebra; i.e., the maximal solvable ideal. Therefore, the factor algebra $\mathfrak{G}/\mathfrak{r}$ is isomorphic to the subalgebra generated by the three angular momentum observables J_μ. This is isomorphic to $\mathfrak{so}_3$, the Lie algebra of the rotation group $SO(3)$. This means that $\mathfrak{G}$ can be decomposed as the semi-direct sum of the radical $\mathfrak{r}$ and the angular momentum algebra $\mathfrak{so}_3$, thus:

$$\mathfrak{G} \;=\; \mathfrak{r} \;\oplus_s\; \mathfrak{so}(3).$$

This is an example of what is called the Levi decomposition of a Lie algebra: Every Lie algebra can be decomposed as the semi-direct sum of its radical and a semisimple subalgebra. A semisimple Lie algebra is a Lie algebra that does not have an *abelian* ideal except $\{0\}$. The reader will find Fig. 20.2 useful for the different kinds of Lie algebras that are being discussed.

It is easy to prove that $\mathfrak{so}(3)$ is semisimple. In fact, the first entry on the second line of Table 20.2 shows that it is simple (i.e., there are no non-trivial ideals). That is, if one tried to construct an ideal of only two of the generators it would produce the third, which again constructs all of $\mathfrak{so}(3)$. Thus, since $\mathfrak{so}(3)$ is simple, it must be semisimple: That is, since it does not have *any* ideals, it does not, in particular, have any *abelian* ideals.

Most of the structure of the physical universe comes from the structure of semisimple Lie algebras. These algebras have extensive symmetry properties with respect to the Schrödinger equation, and this allows them to yield crucial conservation laws. With respect to our considerations, this means that they have significant *transfer* structures, as will now be seen.

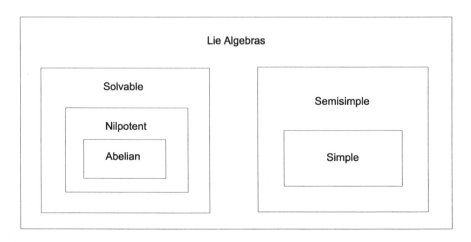

Fig. 20.2. Containment relations between the different types of Lie algebras.

20.6 Semisimple Lie Algebras in Quantum Mechanics

We are interested in *transfer*, which means that a symmetry group of the Hamiltonian should be considered. Let G be such a group. Consider its action on the chosen Hilbert space. It is necessary to look first at those subspaces of the Hilbert space that remain invariant under G. In particular, the fundamental concern is with the *irreducible* invariant subspaces; i.e., those invariant subspaces that do not contain proper subspaces that are also invariant. An irreducible invariant subspace is called a *multiplet* of G.

In particular, let us suppose that G is the most important type of symmetry group encountered in quantum mechanics:

G *is a semisimple Lie group with n generators and rank k.*

The rank k is the maximal number of mutually commuting generators; i.e., they generate the Cartan subalgebra.

Now for a fundamentally important theorem by Racah: For any such group G, there exist k operators C_i, which are functions of the n generators, and which commute with every operator of the group. The eigenvalues of the operators C_i uniquely characterize the multiplets of the group. The operators C_i are called *invariant* operators or *Casimir* operators. Note that the Casimir operators are not members of the group, nor usually members of the Lie algebra, but belong to the center of the universal enveloping algebra of the Lie algebra.

To illustrate the power of Racah's theorem, consider the most frequently encountered example of a Casimir operator: the square J^2 of the angular momentum operator. To understand its relation to the theorem, let us assume (for ease of exposition) that J represents only *orbital* angular momentum. The symmetry group being considered is $SO(3)$. It is semisimple, as shown in the previous section. Clearly its Lie algebra has rank 1, since any generator of $SO(3)$ commutes with only itself. Racah's theorem then states that a set of Casimir operators for this group consists of only one element. In fact, it is J^2. Observe that J^2 is a function of the three generators, thus: $J^2 = J_1^2 + J_2^2 + J_3^2$. Observe still further that its irreducible invariant subspaces are the subspaces generated (as vector spaces) by the spherical harmonics $Y_{lm}(\theta, \pi)$, for fixed l. The fixed l is associated with the fixed eigenvalue of J^2 on this subspace. This value *uniquely* selects this eigenspace, which is the associated multiplet. We know that the multiplet is $(2l + 1)$-fold degenerate. Notice that, being a Casimir operator, J^2 commutes with the entire set of operators. Therefore, it is its status as a Casimir operator that allows us to choose it together with J_3 to produce the good quantum numbers (l, m) for the angular momentum eigenstates.

Let us now turn fully to the issue of *transfer*. In the above example, $SO(3)$ is being assumed to be a symmetry group. For example, the system could be the spherically symmetric Hamiltonian of a spinless particle in a central field. The symmetry effect of $SO(3)$ on the Hamiltonian means that the *entire* group (in its unitary form) will commute with H. In particular, the generators J_1, J_2, and J_3, will commute with H. Therefore, to each of these operators, there will be a *conservation law* (since the operator is a conserved quantity). This means that the 1-parameter group generated by any of these conserved generators can be the control group $G_{conserved}$ in the following wreath product

$$G_H \ \circledW \ G_{conserved} \tag{20.8}$$

where the fiber G_H represents the symmetry action along a flow-line of the Schrödinger equation.

The important thing however to observe is that these three conservation laws cannot hold simultaneously. This is because the rank of the algebra is 1. In other words, when setting up the wreath product (20.8) representing one of the conservation laws, the possibility has been excluded of setting up any of the other wreath products representing a conservation law. In contrast, in a Lie algebra of rank k, we can set up k such wreath products simultaneously using each of the k commuting generators. Notice why this works: The whole group is a symmetry group of the Hamiltonian, which means that any member of the group commutes with H. However, the k generators also commute *with each other*! Therefore, they will correspond to k simultaneous conservation laws. Thus they will correspond to k simultaneous wreath products of the form (20.8).

Let us go back to $SO(3)$. Even though the rank of $SO(3)$ is 1, which means that only one of the generators can be used to set up a conservation wreath product at any one time, we can actually set up one other conservation wreath product: that corresponding to the Casimir operator.

More generally therefore, the situation is as follows: Let the symmetry group of the Hamiltonian be a semisimple unitary Lie group on n generators with rank k. Then, $2k$ wreath products can be set up, of the form (20.8), simultaneously. The control groups of k of these wreath products will come from the k commuting generators. The control groups of the remaining k of these wreath products will come from the k Casimir operators. Each of these $2k$ operators pushes flow-lines onto flow-lines in the Schrödinger equation; i.e., creates a *transfer* structure in the equation. In Volume II, we will describe a means of stacking the 2n operators hierarchically within a single $(2n+1)$-fold wreath product.

21. Music

21.1 Introduction

Section 2.22 gave the fundamental proposal of our formal theory of aesthetics:

Aesthetics is the maximization of transfer and recoverability.

In addition, we proposed the following concerning art-works:

Art-works are maximal memory stores.
The rules of aesthetics are therefore the rules of memory storage.

Chapters 2 and 20 showed how these proposals explain the role of aesthetics in science. That argument will be continued in Volume II. The present chapter will show how these proposals explain the structure of music. Our corresponding theory of the structure of paintings was given in Chapter 8 of the previous book, Leyton [96].

21.2 Motival Material

A great work of art is based on only a minimum of motival material, which is used and re-used throughout the work in different forms and positions. This chapter will show that the use and re-use conforms to our rigorous mathematical theory of transfer.

Thus the work is constructed out of transfer. Schönberg, in his enormously insightful study of classical music, stated:

"The *motive* generally appears in a characteristic and impressive manner at the beginning of the piece. ... And since it is included in every subsequent figure, it could be considered the 'greatest common factor' "(p8, [137]).

Anyone who has written a symphony knows that the first few notes of a movement are taken by the ear to be motival and, if the remainder of the movement is not the continual use of those first few notes, the piece will simply sound bad. Indeed, in the process of composition, if one happens to add new material that does *not* come from existing material, the effect is catastrophic. The piece suddenly fragments and one looses one's way. No matter how beautiful the new material might intrinsically be, a great composer like Beethoven learns that it has to be removed if it is not heard as the transfer of the motival material presented at the outset of the piece.

This chapter gives a mathematical theory of how this works.

21.3 Modulation as a Wreath Product

One of the basic principles of Western music is that, given a scale, one can transpose the scale to any position in that scale. For example, given the diatonic scale, one can transpose it from the tonic to the dominant, so that the dominant now becomes the tonic. The same applies to the chromatic scale, the harmonic series, etc. The transposition of scales is the basis of the compositional procedure called *modulation*. In fact, for ease of exposition, we shall refer to scale-transposition simply as modulation, even though the former is more general.

Modulation is clearly a structure of *transfer*. We argue that it is best described by a wreath product. Let $\mathbb{S}$ be the group of movements in a scale. Then the ability to move the scale to any position within the scale, i.e., modulation, can be described as the following wreath product:

$$\mathbb{S} \textcircled{w} \mathbb{S}.$$

The control group represents the home scale, often called the home key. The fiber group represents the key into which modulation occurs. There is one such fiber for each member of the home key (control group).

Notice, of course, that when one is in the modulated key, one can then modulate to any position within that key, and so on. Thus, the structure of modulation is:

$$\ldots \textcircled{w} \mathbb{S} \textcircled{w} \mathbb{S} \textcircled{w} \mathbb{S} \textcircled{w} \mathbb{S}.$$

Notice that this is very similar to the type of group we gave for the serial-link manipulator in Chapter 6:

$$\ldots \textcircled{w} SE(3) \textcircled{w} SE(3) \textcircled{w} SE(3) \textcircled{w} SE(3).$$

The reason is simple but profound: In both cases, the hierarchy represents a hierarchy of *workspaces*. Furthermore, in both cases, the workspace on any level has the same structure as the workspace on any other level. This means that a workspace moves an identical workspace about itself. We have called the resulting type of group hierarchy a **wreath poly-X group** (Definition 4.11, p. 132), where X is the group that is repeated down the hierarchy.

ALGEBRAIC STRUCTURE OF MODULATION. *Modulation is structured by a wreath poly-$\mathbb{S}$ group.*

Notice that the above algebraic theory of modulation conforms to our theory of object-oriented inheritance given in Chapter 7: The fiber group is the child and the control group is the parent. Notice also that this algebraic theory of modulation exactly corresponds to the theory of relative motion developed in Chapter 9.

21.4 Psychological Studies of Sequential Structure

Most of this chapter will be concerned with developing a group theory of sequential structure in music. For this, it is first necessary to recapitulate some previous research:

Significant progress has been made in understanding sequential structure by psychologists working on the generation of serial patterns. Herbert Simon, himself an outstanding musician, together with collegues, was the first to consider rule-systems for psychological sequence generation, Simon & Kotovsky [145], Kotovsky & Simon [78]. A further advance was made by Restle [125], who used hierarchies of rules. Fig. 21.1 shows a example typical of one of Restle's hierarchies. Three generative rules are used in this hierarchy: $T = $ transpose by one unit upwards in the scale; $R = $ repeat; and $M = $ mirror about the scale center. Here the scale is assumed to consist of 12 notes. Each of the operators takes the entire subsequence that it dominates via its left node and maps it to the entire subsequence dominated by its right node. Notice that the tree is *binary*, and that it is *strictly nested*, the term used by Greeno & Simon [47], meaning that all the operators within a level are exactly the same. The condition of strict nesting is equivalent to the fact that the tree can be represented by a recursive formula. In the example shown, the formula is:

$$M(T(R(T(1)))). \tag{21.1}$$

The symbol 1 in this formula, is the left most 1 in Fig. 21.1; and the formula generates the remainder of the sequence.

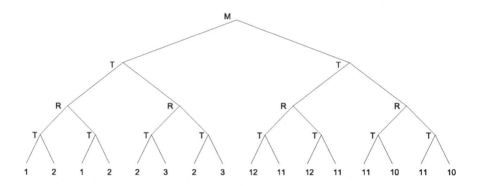

Fig. 21.1. An example of one of Restle's rule hierarchies.

An additional advance came when a number of researchers independently started to use groups to structure the rules; Babbit [10], Leyton [86], Greeno & Simon [47], Jones [66]. The major school for the use of group theory in music has become that of Guerino Mazzola in Switzerland: Mazzola [104], [105], [106]. See also the work of Thomas Noll [113], [114].

21.5   Transfer in Musical Sequence Structure

Our theory of aesthetics says that aesthetics is the maximization of transfer and recoverability. This section will concentrate on transfer.

We argue that, when one examines the hierarchical theory of Restle, one must conclude that the human mind is maximizing transfer. That is, the process of sequence comprehension or generation is a process of transferring previous structure onto future structure. The fact that the mind tries to maximize this can be seen by the psychological studies carried out by Restle to support his hierarchical rules - e.g., profiles of *anticipation errors* showed that subjects were mapping previous structure onto the anticipated structure, Restle & Brown [126].

This chapter will develop a group-theoretic approach to such sequence structure. Our claim is that, because one seeks to maximize transfer, such structure should be modeled by wreath products, as follows:

Let us call a group generated by a set of compositional operators, a **rule group** $\Re_i$. Given a hierarchy of the type shown in Fig. 21.1, the levels will be numbered upward from 1 to n. Now assign a rule group to each node.

Within any level i, the nodes should each receive the *same* rule group $\Re_i$. We argue that the rule-structure of the hierarchy is given by taking the wreath product of the groups $\Re_i$, thus:

$$\Re_1 \ \textcircled{w} \ \Re_2 \ \textcircled{w} \ \ldots \textcircled{w} \ \Re_n.$$

Notice that, in this formulation, there is no reason why the tree should be binary. The requirement is merely that, given a node x on Level i, the number of nodes N_{i-1} immediately dominated by x is the same for all nodes on Level i.

In the remainder of this chapter, we will create an important class of such groups, those dealing with meter. Jones [67] has given an extensive psychological analysis of meter structure using rule hierarchies similar to those of Restle. Our purpose will be to lay out the group theory of such structures.

21.6 Meter

Starting with the next section, we will develop an algebraic theory of meter. The present section will review the basic facts of meter, and give the terminology to be used. There are considerable differences among various researchers, in the terminology by which meter is described. Thus, clarifying the terminology, to be used here, is necessary.

In the West, a musical work moves forward with a regular beat, called the *beat stream*. A very obvious example is given by the first movement of Mahler's 6th symphony shown in Fig. 21.2. Here the beats are successive quarter note steps.

Fig. 21.2. Mahler: Symphony No. 6, first movement.

Let us now review the standard *groupings* in the beat stream in Western music. As with all perception, the structure of grouping is equivalent to the structure of division; i.e., the percept is divided into its groupings.

Primary accent grouping.

Within the beat stream, there are particular beats that are perceived as accented, and are marked as the beginning of a grouping within the stream. These are the groupings that are standardly called bars. In the Mahler example, they have four beats. Generally, the bar-grouping of the beat stream is called *meter*. Standardly, a bar consists of two, three or four beats - referred to respectively as duple, triple and quadruple meters.

Meter	*Simple*			*Compound*		
two beats (duple meter)	2/8	2/4	2/2	6/16	6/8	6/4.
three beats (triple meter)	3/8	3/4	3/2	9/16	9/8	9/4.
four beats (quadruple meter)	4/8	4/4	4/2	12/16	12/8	12/4.

Table 21.1. Time signatures classified with respect to meters.

Table 21.1 shows how the standard time signatures are classified with respect to meters. The terms "simple" and "compound", along the top of the table, will be explained later. For the moment, only the vertical classification of duple, triple, and quadruple, will be needed. Notice, for example, that 6/4 is considered to be a duple meter, whereas 12/8 is considered to be a quadruple meter; even though their bar-length is the same.

An example of 6/4 is the first movement of Brahms Piano Concerto No.1, illustrated in Fig. 21.3. There are only two beats in each bar. Each beat is a dotted half note. In contrast, the Bach Two-Part Invention No. 12, shown in Fig. 21.4, is in 12/8, which means that the bar has four beats. Each beat is a dotted quarter note.

Fig. 21.3. Brahms: Piano Concerto No. 1, first movement.

Secondary accent grouping.

Meters that have more than three beats in a bar, tend to be perceived as having secondary accents that organize the bar into subgroupings of two or three beats. For instance, bars with a 4/4 time signature are perceived as divided into two successive subgroupings each of two beats. An example of this is Fig. 21.5 which shows a passage from the first movement of Bruckner's

Fig. 21.4. Bach: Two-Part Inventions, No. 12.

9th symphony. The subgrouping here is supported by a number of devises: First there is the two-note descending octave motive that begins the first, third and fifth measures shown. This is on the tonic (the work is in D minor). The octave interval of this descent is then vertically contracted to produce the two-note descent that begins the second and fourth bars. That two-note descent is then followed by its inversion, the two-note ascent. The base line further supports the grouping.

Fig. 21.5. Bruckner: Symphony No. 9, first movement.

Now let us turn from the 4/4 to the 12/8 time signature. The reader will recall that a 12/8 signature means a meter of four beats to the bar. The secondary accent in 12/8 partitions the bar into two. The previous example from Bach (Fig. 21.4) illustrates this. Notice the end of the base line at the middle of the bar. This emphasizes the division. The reader can find many other examples of this division of 12/8, for instance in the woodwind theme of the first movement of Sibelius's Symphony No. 5 (bars 20-21); also the third movement of Prokofieff's Piano Sonata No. 8, and so on.

Now let us turn to the 6/4 time signature. Generally, this is considered to be duple meter, as listed in Table 21.1. However, when used in a slow tempo, it can be perceived as a sextuple meter; i.e., as six beats to a bar, each beat being a quarter note. The distinction is illustrated with the first and second movements of the Brahms Piano Concerto No. 1. Both movements are in 6/4, but the first is in duple meter, and the second is in sextuple meter. The first was illustrated in Fig. 21.3, where the beat is half the bar. The second is illustrated in Fig. 21.6. Here the successive quarter-note beat is carried, for example, by the chords in the right hand in the second bar.

Fig. 21.6. Brahms: Piano Concerto No. 1, second movement.

The reason for choosing this example here is that it illustrates the fact that, when 6/4 is used as a sextuple meter, it has a subgrouping structure: The bar is divided in two, with three beats to each subgrouping.

Beat division.
Not only can a beat be grouped with others in the above way, but it can be divided; i.e., the single beat forms a grouping in its own right. There are two standard divisions of a beat: (1) *simple meter*, which is division into two; and (2) *compound meter*, which is division into three.

The time signature 12/8 is an example of compound meter, as illustrated by the previous Bach example Fig. 21.4. The beat, which is the dotted quarter note, is divided by three, as shown in the base. Similarly, the time signature 6/4 is an example of compound meter in the case where it is used as duple meter, as illustrated with the first movement of the Brahms Piano Concerto No.1, in Fig. 21.3. The beat, which is half the bar, is divided into three quarter notes.

Now let us consider again the second movement of this concerto, illustrated in Fig. 21.6. As noted earlier, this is in sextuple meter, i.e., each beat is a quarter note. What can be observed here is that there is a *simultaneous* use of simple and compound meter: The left hand has a division of the beat into two notes (simple meter), and the right hand has a division of the beat into three notes (compound meter).

This phenomenon of a simultaneous use of simple and compound meter is illustrated very strongly in Fig. 21.7 from the second movement of Bartok's String Quartet No. 4. The second violin and viola are in 2/4; and the first violin and cello are in 6/8. Both 2/4 and 6/8 are duple meters (recall Table 21.1). Thus the beats coincide in all four instruments. Nevertheless, the divisions do not. In the second violin and viola, the division is into two; whereas in the first violin and cello, the division is into three.

Beat subdivision.
The beat can be further divided into subdivisions. The previous Bach example Fig. 21.4 illustrates subdivision: Because the time-signature is 12/8, there are four beats to the bar; i.e., the beat is the dotted quarter note. The 1/8th note

Fig. 21.7. Bartok: String Quartet No. 4, second movement.

division of the beat is carried in the base. The 1/16 note *subdivsion* is carried in the treble.

21.7   Algebraic Theory of Meter

Section 21.5 said that, because the mind tries to maximize transfer in a musical sequence - indeed in any sequence - the sequence is best described as a wreath product

$$\Re_1 \ \textcircled{w} \ \Re_2 \ \textcircled{w} \ \dots \textcircled{w} \ \Re_n$$

where $\Re_i$ are rule groups used to generate levels of transfer in the sequence. An important class of such wreath products will now be developed to describe musical meter. The main principle of our theory of meter is as follows:

ALGEBRAIC THEORY OF METRICAL STRUCTURE. *Given a metrical unit (e.g., a bar, a subgrouping, a beat), its occurrence within the next higher unit is given by a cyclic group $\mathbb{Z}_i$, and its subdivision is given by a cyclic group $\mathbb{Z}_j$. The upper group $\mathbb{Z}_i$ transfers copies of the lower group $\mathbb{Z}_j$ as fiber, along the musical work. Therefore, the relation between the upper and lower group is that of a regular wreath product:*

$$\mathbb{Z}_j \ \textcircled{w} \ \mathbb{Z}_i.$$

The full metrical structure (encompassing all levels) is therefore given by an n-fold wreath product $\mathbb{Z}_1 \textcircled{w} \mathbb{Z}_2 \textcircled{w} \dots \textcircled{w} \mathbb{Z}_n$. If one defines the standard invariant metric on time, then this wreath product is an iso-regular group (wreath c-polycyclic, wreath-isometric).

A particular aspect of this statement can be given as follows:

THEORY OF DIVISION.   *Division by j is wreath sub-appendment by $\mathbb{Z}_j$.*

This, in fact, is simply our theory of refinement given in Sect. 15.22. To illustrate the above theory of meter, let us first concentrate on a *single* bar, and look at its internal grouping. The analysis will go down successive divisions of the bar.

Primary and secondary accent grouping.
The first subdivision of the bar is given by the secondary accent structure. Examples of standard time signatures that have such subgroupings are 4/4, 12/8, and 6/4 (in the sextuple version) - all of which divide the bar into two. According to our theory of division, the division of the single bar into S number of subgroupings, is wreath sub-appendment by the cyclic group $\mathbb{Z}_S$. Since, for the moment, only a single bar is being considered, the wreath product is

$$\mathbb{Z}_S \;\textcircled{w}\; \{e\}. \tag{21.2}$$

When one considers a sequence of bars, e.g., represented by the 1-dimensional lattice $\mathbb{Z}$, then the group of the sequence will be substituted for the control group $\{e\}$; for instance, yielding the group $\mathbb{Z}_S \;\textcircled{w}\; \mathbb{Z}$, which is an example of what we call a generative crystallographic group (Definition 5.1, p. 140). However, for the moment, only a single bar is being considered; i.e., the control group is $\{e\}$. To save space, $\{e\}$ will be omitted from the notation.

Now, the action of $\mathbb{Z}_S$ is to cycle *between* the subgroupings in a bar. Thus, in the standard cases of 4/4, 12/8, and 6/4 (sextuple version), the group $\mathbb{Z}_S$ is actually $\mathbb{Z}_2$.

Next, going down one level, each subgrouping is a grouping of regular beats. Thus, within a subgrouping, the beat structure will be given by a cyclic group $\mathbb{Z}_B$, where B is the number of beats in a subgrouping. Therefore, in accord with our theory of division, the beat structure of a bar is given by the following wreath product:

$$\mathbb{Z}_B \;\textcircled{w}\; \mathbb{Z}_S. \tag{21.3}$$

Most crucially, notice that this means that $\mathbb{Z}_S$, which cycles between subgroupings, acts as a control group, transferring the beat structure of one subgrouping onto the beat structure of another subgrouping. This, for example, is exactly what can be seen in the Bruckner example Fig. 21.5. Here, the subgrouping is the two-note motive; and it is mapped from one subgrouping to another. Therefore, for this example, the wreath product, $\mathbb{Z}_B \textcircled{w}\mathbb{Z}_S$, becomes:

$$\mathbb{Z}_2 \;\textcircled{w}\; \mathbb{Z}_2$$

which is the hyperoctahedral wreath group of degree 2. This then is the primary-secondary structure of the 4/4 time signature. Exactly the same group $\mathbb{Z}_2 \, \textcircled{w} \, \mathbb{Z}_2$ applies to the time signature 12/8, because this has four beats divided into two halves.

When one regards 6/4 as a sextuple meter, the bar is divided into two subgroupings, each of which has three beats. In this case, the group $\mathbb{Z}_B \textcircled{w} \mathbb{Z}_S$ becomes:

$$\mathbb{Z}_3 \, \textcircled{w} \, \mathbb{Z}_2.$$

In time signatures where there is no subgrouping structure between the level of the bar and the level of the beat, we will simply omit $\mathbb{Z}_S$, and include only $\mathbb{Z}_B$. Section 5.10 discussed exactly this type of situation (inclusion and omission of levels) in the visual domain. Notice, for example, how Fig. 5.4 (p. 153) is an example of the same kind of phenomenon.

Beat division.

The fact that the beat itself can be divided will now be modeled, in accord with our theory of division, by wreath sub-appending the cyclic group $\mathbb{Z}_D$ to the previous sequence $\mathbb{Z}_B \textcircled{w} \mathbb{Z}_S$, where D is the number into which the beat is divided. That is, the following is obtained:

$$\mathbb{Z}_D \, \textcircled{w} \, \mathbb{Z}_B \, \textcircled{w} \, \mathbb{Z}_S.$$

Thus, there have now been three successive divisions, each according with our theory of division.

As an example, recall that, in the time signature 12/8, the subgrouping structure $\mathbb{Z}_B \textcircled{w} \mathbb{Z}_S$ is the hypeoctahedral group $\mathbb{Z}_2 \textcircled{w} \mathbb{Z}_2$. The 12/8 signature requires that the beat is divided by three, and therefore one must wreath sub-append the group $\mathbb{Z}_3$ to the hyperoctahedral group, as follows:

$$\mathbb{Z}_3 \, \textcircled{w} \, \mathbb{Z}_2 \, \textcircled{w} \, \mathbb{Z}_2.$$

Now let us turn to cases where there is simultaneous use of simple and compound meter, for instance, the Brahms and Bartok examples, Fig. 21.6 and Fig. 21.7 respectively. Let us deal with the general case first:

SIMULTANEOUS DIVISION. *Simultaneous division of an interval by different numbers D_1, D_2, $\ldots D_n$, will be given by wreath sub-appendment by the direct product $\mathbb{Z}_{D_1} \times \mathbb{Z}_{D_2} \times \ldots \times \mathbb{Z}_{D_n}$.*

Therefore, the simultaneous use of simple and compound meter is given by wreath sub-appending the group $\mathbb{Z}_2 \times \mathbb{Z}_3$ to the beat-division group $\mathbb{Z}_B \textcircled{w} \mathbb{Z}_S$. This allows us to give the full metrical groups for the Brahms and Bartok examples, as follows:

Brahms Piano Concerto No. 1, second movement (Fig. 21.6). The time signature is 6/4, which is interpreted here as a sextuple meter due to the slowness of the tempo. Such 6/4 meter decomposes the bar into two subgroupings, each of three quarter notes, thus yielding $\mathbb{Z}_B \textcircled{w} \mathbb{Z}_S = \mathbb{Z}_3 \textcircled{w} \mathbb{Z}_2$. Then the simultaneous use of simple and compound meter shown in Fig. 21.6, subdivides the beat into two (in the right hand) and into three (in the left hand), which results in wreath sub-appending $\mathbb{Z}_2 \times \mathbb{Z}_3$. That is, the full group of the metrical structure is:

$$[\mathbb{Z}_2 \times \mathbb{Z}_3] \textcircled{w} \mathbb{Z}_3 \textcircled{w} \mathbb{Z}_2.$$

Bartok String Quartet No. 4, second movement (Fig. 21.7). The time signature 2/4 does not have a subgrouping structure intervening between the level of the bar and the level of beats. Therefore $\mathbb{Z}_B \textcircled{w} \mathbb{Z}_S$ is simply $\mathbb{Z}_B = \mathbb{Z}_2$. Then the simultaneous use of simple and compound meter shown in Fig. 21.7, subdivides the beat into two (in the second violin and viola) and into three (in the first violin and cello), which results in wreath sub-appending $\mathbb{Z}_2 \times \mathbb{Z}_3$ to the beat group. That is, the full group of the metrical structure is:

$$[\mathbb{Z}_2 \times \mathbb{Z}_3] \textcircled{w} \mathbb{Z}_2.$$

The reader should note that the groups given above for these Brahms and Bartok cases are examples of what we call *semi-rigid groups* (Definition 6.1, p. 172). Recall that we first introduced such groups in the context of robotics.

Beat subdivision.
Clearly, beat subdivision takes place by wreath sub-appending a further cyclic group, in accord with our theory of division. An example is the following

Bach Two-Part Invention No. 12 (Fig. 21.4). The time signature 12/8 has a subgrouping structure that divides the bar into two halves each of which has two beats; thus yielding $\mathbb{Z}_B \textcircled{w} \mathbb{Z}_S = \mathbb{Z}_2 \textcircled{w} \mathbb{Z}_2$. The right hand in the Bach example further divides the beat by three, and the left hand creates an additional division by two. This gives the following as the full metrical group:

$$\mathbb{Z}_2 \textcircled{w} \mathbb{Z}_3 \textcircled{w} \mathbb{Z}_2 \textcircled{w} \mathbb{Z}_2.$$

Musical forms. This section has considered the successive division of a bar. To get higher-order structures, one simply wreath super-appends higher order control groups onto the wreath products considered. Examples are the standard musical forms - which can obviously be given by wreath products

THEORY OF METRICAL MOVEMENT (PULSE)

(1) Metrical movement is given by *transfer* along the metrical sequence.

(2) According to our generative theory of shape, transfer is given by control-nested τ-automorphisms. For example, the structure of arrows in Fig. 21.8 (p. 490) illustrates a control-nested τ-automorphism acting on a metrical hierarchy.

(3) When such an automorphism is applied to the left-most element, the selective effect in wreath products results in the selection of the particular arrows shown in Fig. 21.9 (p. 491).

(4) This moves the left-most element in the following way: first within the smallest box shown, then out of the smallest box, then out of the next larger box, then out of the next larger box, . . . , and so on. This results in the movement of the element across the sequence.

(5) The successive arrows upwards (i.e., components of the automorphism) yield the accent structure of the movement; i.e., the pulse.

(6) Notice that the movement comes from the symmetry structure: The symmetries of the meter are its control-nested τ-automorphisms. The Symmetry-to-Trace Conversion Principle (p. 63) states that traces, i.e., temporal asymmetries, come from the symmetries of an organization.

(representing the various levels - phrases, periods, etc.) that can be wreath super-appended, extending the hierarchy upward.

21.8   Theory of Metrical Movement (Pulse)

The above argument allows us to give a theory of metrical movement, i.e., pulse. The main points of the theory are presented in the table on p. 489. The reader should first read that table and then return here.

The remainder of this section will present certain additional details of this theory, as follows: The bottom row of dots in Fig. 21.8 represents a sequence of equivalent metrical units in a musical work. Now these units can themselves be divided, so that the tree can be extended downward. Thus let us assume that the entire hierarchy is of height n, and that the diagram represents Level i up to n. According to our theory in Sect. 21.7, the entire symmetry group of the metrical structure is given by an iso-regular group

$$\mathbb{Z}_{m_1} \textcircled{w} \mathbb{Z}_{m_2} \textcircled{w} \ldots \textcircled{w} \mathbb{Z}_{m_n}$$

where each $\mathbb{Z}_{m_j}$ is a cyclic group of order m_j.

Now, since each node at the bottom of Fig. 21.8 is itself divided downwards to Level 1, each of the bottom nodes shown in the figure represents a copy of the fiber group $\mathbb{Z}_{m_1}\textcircled{w}\ldots\textcircled{w}\mathbb{Z}_{m_i}$ from Level i down to 1. On the level above this fiber, each node represents a copy of the group $\mathbb{Z}_{m_{i+1}}$; on the level above this, each node represents a copy of the group $\mathbb{Z}_{m_{i+2}}$; and so on ... till the single node at the top level which represents the final control group $\mathbb{Z}_{m_n}$.

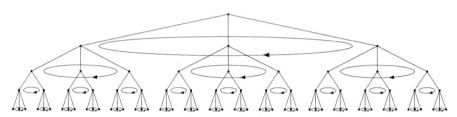

Fig. 21.8. A control-nested τ-automorphism used to structure meter.

Now, as in any wreath product, each level of $\mathbb{Z}_{m_1}\textcircled{w}\mathbb{Z}_{m_2}\textcircled{w}\ldots\textcircled{w}\mathbb{Z}_{m_n}$, that is, each control group $\mathbb{Z}_{m_j}$, acts on a corresponding control set C_j. In the meter situtation, the control set represents a set of grouped metrical units; e.g., the set could be the four beats of a bar; or the three notes in a triplet. Such a set will be called a **metrical set**. Furthermore, as in any wreath product, there is a group action of the control group on the control set (metrical set). Notice that the size of the control group and its metrical set are the same.

Now let us look at the structure of the bottom set of nodes in Fig. 21.8. These represent a sequence of equivalent metrical groupings, e.g., bars. Each node (e.g., each bar) is structured by the fiber group $\mathbb{Z}_{m_1}\textcircled{w}\ldots\textcircled{w}\mathbb{Z}_{m_i}$ up to Level i. For example, the node can be a bar divided into subgroupings, which are divided into beats, which are divided into triples. Now, since the structure of the bottom level shown is *parallel*, the group of the entire bottom level is the direct product of all these copies of the fiber group $\mathbb{Z}_{m_1}\textcircled{w}\ldots\textcircled{w}\mathbb{Z}_{m_i}$. In other words, the group of the bottom level shown (e.g., the bar level) is $Det_i = [\mathbb{Z}_{m_1}\textcircled{w}\ldots\textcircled{w}\mathbb{Z}_{m_i}]^{m_{i+1}\times m_{i+2}\times\ldots\times m_n}$.

Claim. The **metrical structure** *up to Level i is isomorphic to Det_i for some wreath product $\mathbb{Z}_{m_1}\textcircled{w}\ldots\textcircled{w}\mathbb{Z}_{m_n}$. In the context of musical discussion, the group Det_i will be re-labelled Met_i (for metrical structure). That is:*

$$Met_i = [\mathbb{Z}_{m_1} \textcircled{w} \ldots \textcircled{w}\mathbb{Z}_{m_i}]^{m_{i+1}\times m_{i+2}\times\ldots\times m_n}.$$

The hierarchy of metrical structures in a meter forms a group-theoretic subnormal series:

$$Met_1 \vartriangleleft Met_2 \vartriangleleft Met_3 \vartriangleleft \ldots \vartriangleleft Met_n.$$

Notice, if one considers only an individual level j, then its group-theoretic structure is

$$\mathbb{Z}_{m_j}^{m_{j+1} \times \dots \times m_n} \;=\; Met_j/Met_{j-1}.$$

The direct product structure here also indicates that the level has a parallel organization.

Now return to Fig. 21.8. Let each circular arrow in the hierarchy in Fig. 21.8 be some chosen *element* from the control group $\mathbb{Z}_{m_j}$ above it. Then the entire collection of circular arrows, shown in Fig. 21.8 (one element selected from each node), represents a control-nested τ-automorphism.

The automorphic action expresses a symmetry of the metrical structure Met_i. This will be called a **transfer symmetry**, to distinguish it from the internal symmetry within any node, i.e., within any fiber-group copy of $\mathbb{Z}_{m_1} \textcircled{w} \dots \textcircled{w} \mathbb{Z}_{m_i}$. Such an internal symmetry can be thought of as analogous to a guage symmetry in quantum field theory. In contrast, the transfer symmetry is a re-arrangement of the bottom level set of nodes.

It is the control-nested τ-automorphic action (transfer symmetry) that will allow the *movement* of metrical units along the bottom of the diagram. The musical sequence begins with the left-most node. When one applies the control-nested τ-automorphism shown in Fig. 21.8, to this node, the node selects upwards only those circular arrows which dominate it. The result is Fig. 21.9, and one obtains movement along the time dimension. The selective effect involved was fully formalized in Sect. 4.5.

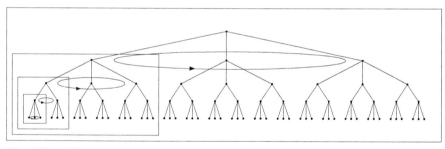

Fig. 21.9. How a control-nested τ-automorphism achieves movement.

21.9   Algebraic Structure of Grouping

Chapter 5 developed an algebraic theory of grouping. It will now be shown that the structure of meter conforms exactly to that theory. The basic principles of the theory were given by the Law of Grouping on p. 138. The reader should read that page and then return here.

The grouping law will now be applied to musical meter in order to understand the structure of the latter. First let us take item (2) in the law (p. 138); i.e., the statement that the groupings in an organization are enumerated as the left-subsequences $G_1 \textcircled{w} \ldots \textcircled{w} G_i$ of an n-fold wreath product. In order to understand this with respect to meter, return to Fig. 21.8 (p. 490). A left-subsequence corresponds to the tree hanging down from a single node in that diagram. The node is a metrical unit and the tree hanging down from it are its successive divisions; e.g., a node could be a bar which is divided downwards successively into subgroupings then beats and then triplets. It is clear therefore that, in accord with the law, a grouping does indeed correspond to a left-subsequence $G_1 \textcircled{w} \ldots \textcircled{w} G_i$.

Now let us consider item (3) in the law (p. 138). This says that the cohesive structure of a grouping $G_1 \textcircled{w} \ldots \textcircled{w} G_i$, is given by its wreath product $(G_1 \textcircled{w} \ldots \textcircled{w} G_{i-1}) \textcircled{w} G_i$, where the relevant wreath product symbol is that after the parentheses. The various factors in the cohesive structure can then be read directly from this wreath product using items (4)-(7) in the law, as follows:

First, item (4), which says that a perceptual *element* of the grouping $G_1 \textcircled{w} \ldots \textcircled{w} G_i$ is the fiber $G_1 \textcircled{w} \ldots \textcircled{w} G_{i-1}$. Thus, having chosen a node G_i in Fig. 21.8, consider any node beneath it. It corresponds to the fiber $G_1 \textcircled{w} \ldots \textcircled{w} G_{i-1}$. It is clear that this represents a perceptual element of the grouping.

Next turn to item (5), which says that the *set of elements grouped* is the set of fiber copies $(G_1 \textcircled{w} \ldots \textcircled{w} G_{i-1})_{g_1}, \ldots, (G_1 \textcircled{w} \ldots \textcircled{w} G_{i-1})_{g_n}$. Clearly this corresponds to the set of nodes dominated by the node G_i. And again, this is the set of elements grouped.

Next turn to item (6) which says that the *grouping factor* is transfer, i.e., the control group G_i. Clearly, the group action at node G_i has the effect of grouping the fibers it dominates by mapping them onto each other, e.g., beats are mapped onto each other.

Next turn to item (7) which says that the *grouping action* is the τ-automorphic action of G_i on its fiber-group product. Notice that the algebraic structure of the set of nodes dominated by G_i is the fiber-group product consisting of those nodes. Clearly, the G_i acts as a τ-automorphism group on this fiber-group product; and it is this that gives the transfer action on the grouping.

Finally, apply the Extension of the Law of Grouping, page 148. This says that all occurrences of a grouping are its control-nested τ-conjugates. This can again be seen in the case of meter. All occurrences of the grouping defined by G_i (e.g., a bar) are its control-nested τ-conjugates; i.e., the other nodes on that level. These are reached by control-nested τ-conjugation, e.g., in the manner shown in Fig. 21.9.

In relation to this, it is worth considering the visual example in Fig. 5.2 (p. 147). This concerns calculating the Gestalt relationship between two elements, the two sides highlighted in different regions of the entire configuration. Section 5.7 (p. 146) shows how to perform the calculation, and it is worth reviewing this section in relation to the issue of meter.

22. Against the Erlanger Program

22.1 Introduction

Klein's Erlanger program has constituted much of the basis of 20th century geometry and physics. According to that program, geometry is the study of invariants under some assumed group of transformations.

This chapter will show that our generative theory - which we will call **generative geometry** - takes a fundamentally opposite view of the nature of geometry. The chapter will define the main differences between the two approaches. The chief difference centers around the issue of *recoverability*, as follows:

Klein defines geometry as the study of invariants under transformation groups. The phenomenon of invariance is really one of *memorylessness*, as follows: When one applies a transformation to an invariant, one will not be able to recover the transformation from the invariant. In other words, an invariant is not a memory store for the transformation. Klein's geometry therefore is a study of those properties that cannot be used as memory stores for actions. A geometric object, in Klein's theory, is a memoryless object.

Our generative theory is the extreme opposite. In this theory, a geometric object is that from which one can maximally recover applied actions. This is because the object is defined by the sequence of generative actions that produced the object, and these actions are set up to be recoverable. In this way, a geometric object is a memory store for past action. Thus the fundamental difference between Klein's theory and ours is this:

KLEIN'S GEOMETRY: A geometric object is one from which the transformations are non-recoverable; i.e., a geometric object is memoryless.

GENERATIVE GEOMETRY: A geometric object is one from which the transformations are recoverable; i.e., a geometric object is a memory store. In fact, we claim the following equivalence:

$$\text{Geometry} \quad \equiv \quad \textbf{Memory Storage.}$$

We propose that Klein's theory of geometry is an inadequate and incorrect one for the scientific, computational, and design disciplines. We argue that these disciplines require the very different approach elaborated in this book. The reason is that all these disciplines fundamentally rely on the recovery of causal or generative actions from shape. Thus, with respect to the *scientific disciplines*, the entire program of a science is aimed at the recovery of the sequence of environmental events that lead up to the state appearing on the measuring instruments. Again, in the *computational disciplines*, computer vision is set up to recover the environmental structure from the image. Furthermore, in machine learning, Carbonnell [17] has argued that any advanced computational system needs to be able to recover its own computational history, in order to be able to modify that history, if the current state is not the desired one. Again, in the *design disciplines*, Hoffmann [58], [20], has argued that keeping a record of the design history is essential because it allows editability of the design decisions. Indeed, all advanced design programs now offer the user access to the design history of an object - exactly for the purpose stated by Hoffmann. In fact, the recovery of design history is one of the most frequently used operations by any computer-based designer. Also observe that computer-aided manufacturing is really the inference of construction and milling operations from the design geometry. In our chapters on computer-aided manufacturing, we showed that this requires geometry to be defined in a generative-recoverable way.

Thus, in complete contrast to the Klein approach, which is non-recoverable, we argue that each of the above disciplines requires geometry to be defined in a recoverable way.

Other important differences between Klein's theory and the generative theory follow from this basic contrast between non-recoverability and recoverability. For example, this chapter shows that the two theories of geometry oppose each other fundamentally on the following issues: (1) symmetry, (2) geometric levels, (3) transformational exhaustiveness, (4) transitivity, (5) observer independence, (6) coordinate freedom, and (7) uniqueness of description.

In order to fully understand the difference between the two theories, the chapter will carefully study how they differ on these important issues.

22.2   Orientation-and-Form

It will be possible to illustrate most of the differences between Klein's theory and the generative theory by examining a particular example of the orientation-and-form phenomenon in perceptual psychology: the example shown in Fig. 22.1 due to Goldmeier [45]. The reader might need to review Sect. 8.8 on orientation-and-form.

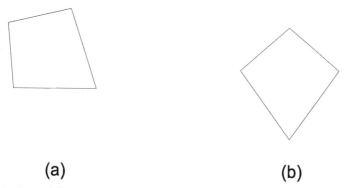

(a)   (b)

Fig. 22.1. One of Goldmeier's orientation-and-form examples.

We are going to consider the relation between the orientation-and-form phenomenon and Klein's Erlanger program. According to the Erlanger program, the two figures shown in Fig. 22.1 are the same because they are invariant under rotations. One cannot simply argue that human subjects do not see the rotation group - they do, endlessly. The rotation group is hard-wired into the human perceptual system. Thus, according to the Erlanger program, a "figure" should not be orientation-dependent.

What this chapter will show is that the human visual system is not based on the Klein theory of geometry, but on the very opposite theory proposed in this book.

22.3   The Generative Structure of Quadrilaterals

We are now going to explain the Goldmeier effect. It will be argued that the first figure is seen as a deformation of a square, and the second figure is seen as the deformation of a diamond. This argument will powerfully undermine Klein's program for geometry - because the two figures, being simply rotations of each other, ought to be seen as a deformation of the same object.

Let us begin by concentrating on the first object. It is clear that this is not an affine transformation of a square - since the parallel lines of a square

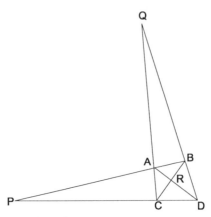

Fig. 22.2. A complete quadrangle.

are destroyed in the deformation - and affine transformations preserve parallelism. Thus it is necessary to go to a less restrictive deformation. The minimal deformation which will account for the figure is a projective transformation. Therefore assume that the figure is a subset of the projective plane $\mathbb{P}^2$.

$$\mathbb{P}^2 \;=\; \mathbb{R}_1^2 \;\cup\; \mathbb{P}^1. \tag{22.1}$$

In fact, more precisely, the sides will be described as parts of infinite lines that are independent subsets of the plane. The notation means this: Generally, for $\mathbb{R}^{n+1}$, the final coordinate will be labelled w. Define $\mathbb{R}_0^n$ to be the hyperplane given by $w = 0$, and $\mathbb{R}_1^n$ to be the coset at $w = 1$, parallel to that hyperplane. Throughout the chapter, this coset will also be referred to as a hyperplane.

Now, rather than evoking "extrinsic" perspective transformations that map between different projective planes slanted at different angles in projective 3-space, the "intrinsic" view will be taken; i.e., that the perspective transformations are sending the projective plane to itself. In other words, assume that the group which is acting is $PGL(3, \mathbb{R})$ as an automorphism group of $\mathbb{P}^2$.

Now the Goldmeier Fig. 22.1a can be understood as a quadrangle of points A, B, C, D, in the projective plane. Fig. 22.2 shows what, in projective geometry, is called the *complete quadrangle ABCD*. This is constructed by taking all *six lines* determined by the four points A, B, C, D, and finding the three extra intersection points, which are

$$P = AB \cap CD, \quad Q = AC \cap BD, \quad R = AD \cap BC. \tag{22.2}$$

These three points are usually called the *diagonal* points. The reader should note that our analysis can use equally the complete quadrangle, or complete quadrilateral, since these figures are dual to each other (points dual to lines). However, as illustration, the complete quadrangle will be used.

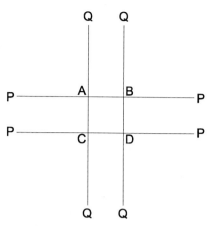

Fig. 22.3. The square on the projective plane.

Now, the fundamental theorem of planar projective geometry says that $PGL(3, \mathbb{R})$ is transitive on ordered complete quadrangles. This means that there is some projective transformation that takes the quadrangle $ABCD$ shown in Fig. 22.2 to the square shown in Fig. 22.3. For ease of viewing, the central point R will be left out of the diagram; but will be brought back later in the discussion.

Notice, most crucially, that two of the diagonal points P and Q have gone to infinity. However, observe that the intersection equations (22.2) above, nevertheless hold. Most particularly, consider P. In Fig. 22.3, this is still the intersection of the two lines AB and CD. However, because the two lines are parallel, their intersection P is at infinity. Furthermore, this intersection point lies in both directions. Therefore, the single point P is at the end of the two parallel lines in both directions, as is shown by having the letter P label *four* ends of lines in Fig. 22.3. Exactly the same argument applies to the intersection Q of lines AC and BD in Fig. 22.3. Therefore, the single point Q is shown as labeling *four* ends of lines in that figure.

We now want to define the *generative* structure of *quadrilaterals*. According to our theory, the generative structure must start with visually the most symmetrical state, which is the square configuration Fig. 22.3. If the four infinite lines shown are considered to be Euclidean lines, $\mathbb{R}^1$, then, according to our theory, the structure of this configuration is the hyperoctahedral wreath hyperplane group,

$$HWH(2) = \mathbb{R}^1 \textcircled{w} \mathbb{Z}_2 \textcircled{w} \Sigma_2.$$

Now let us extend each of these Euclidean lines $\mathbb{R}^1$ by their ideal point $\mathbb{P}^0$, thus obtaining the projective line:

$$\mathbb{P}^1 = \mathbb{R}^1_1 \cup \mathbb{P}^0. \tag{22.3}$$

Using the projective line as the fiber in the $HWH(2)$ group, the following "group" is obtained:

$$\mathbb{P}^1 \ \textcircled{w} \ \mathbb{Z}_2 \ \textcircled{w} \ \Sigma_2.$$

However, there is a problem in this expression. The fiber $\mathbb{P}^1$ is not actually a group; it is the set shown in expression (22.3). Thus it must be given a group structure; and this will be done by giving the set a group action that corresponds to its partition $\mathbb{P}^1 = \mathbb{R}_1^1 \ \cup \ \mathbb{P}^0$. Thus we do this:

Definition 22.1. *The group* $\mathbb{GP}^1$ *is defined by the direct product*

$$\mathbb{GP}^1 = \mathbb{R}^1 \ \times \ \{e\}$$

together with the group action corresponding to the partition, $\mathbb{P}^1 = \mathbb{R}_1^1 \cup \mathbb{P}^0$, *of the projective line; that is, the group action*

$$\mathbb{GP}^1 \times \mathbb{P}^1 \longrightarrow \mathbb{P}^1$$

is defined as the composition of the two natural group actions:

$$\mathbb{R}^1 \times \mathbb{R}_1^1 \longrightarrow \mathbb{R}_1^1$$

and

$$\{e\} \times \mathbb{P}^0 \longrightarrow \mathbb{P}^0.$$

Now, using this as a recursive basis, the n-dimensional case can be defined as follows:

Definition 22.2. *The group* $\mathbb{GP}^n$ *is defined recursively by the direct product*

$$\mathbb{GP}^n = \mathbb{R}^n \ \times \ \mathbb{GP}^{n-1}$$

together with the group action corresponding to the partition, $\mathbb{P}^n = \mathbb{R}_1^n \cup \mathbb{P}^{n-1}$, *of projective n-space; that is, the group action*

$$\mathbb{GP}^n \times \mathbb{P}^n \longrightarrow \mathbb{P}^n$$

is defined as the composition of the two natural group actions:

$$\mathbb{R}^n \times \mathbb{R}_1^n \longrightarrow \mathbb{R}_1^n$$

and

$$\mathbb{GP}^{n-1} \times \mathbb{P}^{n-1} \longrightarrow \mathbb{P}^{n-1}.$$

With this in mind, return now to the generative structure of the square (Fig. 22.3) which we propose is the hyperoctahedral wreath hyperplane group, but where the hyperplane is now given by group $\mathbb{GP}^1$. Thus, the transfer structure of the square is now

$$\mathbb{GP}^1 \textcircled{w} \mathbb{Z}_2 \textcircled{w} \Sigma_2. \tag{22.4}$$

This gives the most *symmetrical* quadrilateral, the square. Now the goal is to generate, from this, any arbitrary quadrilateral. According to our theory, this is done by wreath appending an asymmetrizing group as a control group. This will transfer the structure of the square configuration onto each non-square one.

To achieve this, we wreath super-append the projective group $PGL(3, \mathbb{R})$ to the group in expression (22.4), thus:

$$\textit{Proj-HWH(2)} \ = \mathbb{GP}^1 \textcircled{w} \mathbb{Z}_2 \textcircled{w} \Sigma_2 \textcircled{w} PGL(3, \mathbb{R}). \tag{22.5}$$

This will be called the *projective hyperoctahedral wreath hyperplane group of degree 2*. The degree corresponds to the degree of the hyperoctahedral component, which is the same as the projective dimension of the highest projective space involved, which is the plane; i.e., not the projective lines within the plane. That is, our *intrinsic* structure is that of the plane rather than the lines. This fact is important for the wreath structure; that is, the action of the projective control group will move the lines within the projective plane.

Before continuing, let us generalize the above structure to the n-dimensional case, thus:

Definition 22.3. *The* **projective hyperoctahedral wreath hyperplane group of degree** n, *will be defined as*

$$\textit{Proj-HWH(n)} \ = \mathbb{GP}^{n-1} \textcircled{w} \mathbb{Z}_2 \textcircled{w} \Sigma_n \textcircled{w} PGL(n+1, \mathbb{R}).$$

The reader should notice carefully the dimensionalities involved. The underlying real space is of dimension $n+1$. We are dealing with a projective n-cube. Its faces are projective $n-1$ spaces, $\mathbb{P}^{n-1}$, as can be seen from this expression. The permutational action sending reflectional pairs of $\mathbb{P}^{n-1}$, is the symmetric group of degree n.

The present section will be dealing with the degree-2 case, given in expression (22.5), although the dimension will be increased in later sections.

It is now necessary to understand the group action in the degree-2 case, as follows: Observe first that, according to the generative theory, the square configuration shown in Fig. 22.3 is associated with the copy of fiber group $\mathbb{GP}^1 \textcircled{w} \mathbb{Z}_2 \textcircled{w} \Sigma_2$ that corresponds to the identity element in the projective control group $PGL(3, \mathbb{R})$ in expression (22.5):

$$\text{square} \ \longleftrightarrow \ [\mathbb{GP}^1 \textcircled{w} \mathbb{Z}_2 \textcircled{w} \Sigma_2]_e \ .$$

Then apply a projective transformation obtaining the Goldmeier figure shown in Fig. 22.2 (point R is still being ignored but will be considered later). This means that a particular element g is being applied from the $PGL(3, \mathbb{R})$ control group in expression (22.5) above. Thus the non-square configuration is associated with the copy of fiber group $\mathbb{GP}^1 \textcircled{w} \mathbb{Z}_2 \textcircled{w} \Sigma_2$ that corresponds to the element g in the projective control group $PGL(3,\mathbb{R})$. thus:

$$\text{non-square} \longleftrightarrow [\mathbb{GP}^1 \textcircled{w} \mathbb{Z}_2 \textcircled{w} \Sigma_2]_g \quad .$$

With this in mind, it is necessary to look at the group action of $\mathbb{GP}^1$ on any one of the lines, for example, the line along PCD. The line is a projective line $\mathbb{P}^1$, and the group action of $\mathbb{GP}^1$ on $\mathbb{P}^1$ was defined to correspond to the partition of the projective line $\mathbb{P}^1 = \mathbb{R}_1^1 \cup \mathbb{P}^0$. This action was given in Definition 22.1. In the original $square$ configuration, the group component $\mathbb{R}^1$ of $\mathbb{GP}^1$ in Definition 22.1 acted on the Euclidean line through CD in Fig. 22.3. Furthermore the group component $\{e\}$ of $\mathbb{GP}^1$ in Defintion 22.1 acted on the point at infinity P. To define the action of $\mathbb{GP}^1$ on the $projected$ version of $\mathbb{P}^1$ in the non-square configuration Fig. 22.2, simply use the $wreath$ $product$ structure. In other words, to find out the effect of a group member $x \in \mathbb{GP}^1$ on the projected line $\mathbb{P}^1$, simply look at the action of $x \in \mathbb{GP}^1$ in the original $square$ configuration, and the apply the projective transformation $g \in PGL(3,\mathbb{R})$. That is, simply use the $conjugacy$ by which the control group (in any wreath project) sends fiber group copies onto each other.

In particular, the direct-product component $\{e\}$ of $\mathbb{GP}^1$, which acted on the point at infinity P in the square configuration, now acts on the visible point P in the non-square configuration. And the direct-product component $\mathbb{R}^1$ (of $\mathbb{GP}^1$) which acted on the $Euclidean$ line through CD in the square configuration, now acts on the non-Euclidean line $\mathbb{P}^1 \backslash P$ in the non-square configuration. So it acts in a deformed way on this line. However, the action is well defined by simply using conjugacy between the square and non-square fiber-group copies.

Let us now examine the asymmetrization process. To do this, it is necessary to point out some deeper issues in the symmetric configuration, the square. Thus return to Fig. 22.3. First observe that the mirror reflecting the two infinite vertical lines is half-way between them, and parallel to them. Therefore the mirror actually meets these two lines at their ideal point Q, both in the up direction and down direction. Similarly, the two infinite horizontal lines and the mirror between them all intserest at the ideal point P, both in the left direction and the right direction.

Then apply a projective transformation to obtain the Goldmeier figure shown in Fig. 22.2. The two ideal points P and Q have now moved in from infinity and have become vanishing points.

The crucial thing to understand is that this movement of the ideal points has created a visual $asymmetrization$. To see this, observe first that, in the square configuration (Fig. 22.3), the ideal point Q was symmetric about the horizontal mirror, because it was at infinity both at the top and bottom of the figure. However, after the projective transformation producing Fig. 22.2, this visual symmetry with respect to the horizontal mirror has been broken - i.e., point Q is now only above the horizontal mirror. Similarly, the change of P from being a point at infinity in the square configuration - both left and right - to being a vanishing point on the left has destroyed the left-right symmetry with respect to the vertical mirror plane.

We conclude therefore that the projective transformation has caused two asymmetrizations - and that this is embodied in the two vanishing points. That is, there is this correspondence:

Asymmetrizations $\longleftrightarrow$ Vanishing points.

Notice that this view only makes sense because of three propositions which are basic to our theory:

(1) Recoverability requires that the starting states of generativity are symmetries.
(2) The gravitational frame is a symmetry.
(3) The observer imposes the gravitational symmetry frame.

These three propositions allow us to define the Goldmeier figure Fig. 22.2 generatively; and furthermore to make that generativity an asymmetrization, which means that it is recoverable.

Now let us contrast this with Klein's view of geometry. As far as the Klein view is concerned, each configuration created by the projective group is "equal", because the projective space on which it acts is symmetric in the following important sense: Return, for example, to the projective line PCD which was discussed earlier. Topologically a projective line is a circle with no distinguished point. This means that with respect to conventional projective geometry, a point at infinity has the same status as a point not at infinity. In contrast, in our generative theory of geometry, the observer's imposition of the gravitational symmetry frame, results in a great difference between a point being an ideal point versus a vanishing point. Most crucially, our view requires all operations to be recoverable.

To emphasize: *Klein's geometry attempts to achieve non-recoverability; whereas the generative geometry attempts to achieve recoverability.*

22.4   Non-coordinate-freedom

One of the fundamental doctrines of 20th century geometry is that geometry is *coordinate free*. This concept is not only associated with the Kleinian view of geometry, but it becomes the criterion for defining physics - for example, Einstein defined the proper objects of physics to be those objects that are frame independent (e.g., the electromagnetic field tensor).

However, our generative geometry takes a very different view: Because an object, in that view, is a structure generated from a symmetric ground state, and because a coordinate frame is really a symmetric ground state, one cannot remove the coordinate frame without removing the object's structure (Sect. 8.4). In fact, in more detail, the object is the *transferred* structure of the symmetry ground state, since asymmetrization takes place by a wreath control group. Thus without the coordinate system, the object looses the structure of which it is a transferred image.

This section will begin to look at the non coordinate-free nature of our generative theory of projective geometry. It is necessary to understand the relation between the asymmetrization defined in Sect. 22.3 and the notion of coordinate frame.

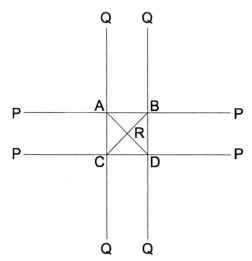

Fig. 22.4. The complete quadrangle square on the projective plane.

To explore this, let us return to the square configuration, which is shown again in Fig. 22.4, but with the point R included, making it a complete quadrangle. Let us define this configuration with respect to the standard projective basis $[1, 0, 0]$, $[0, 1, 0]$, $[0, 0, 1]$, $[1, 1, 1]$, where the first three points are the standard projective *simplex*, i.e., the ordinary vector basis of $\mathbb{R}^3$, and the final point is the *unit* point *determined* by that simplex. In terms of Fig. 22.4, the projective basis is naturally described as:

$$[1, 0, 0] = P$$
$$[0, 1, 0] = Q$$
$$[0, 0, 1] = R$$
$$[1, 1, 1] = B \qquad (22.6)$$

To be clear about this: The underlying space is $\mathbb{R}^3 = \{(x, y, z)\}$, with the z-axis perpendicular to the plane of the paper at point R. The plane of the paper is the plane $\mathbb{R}^2_1$ at position $z = 1$ in $\mathbb{R}^3$. The first three of the members of the above projective basis correspond to the ordinary vector basis of the underlying 3-space, i.e., the basis $[1,0,0]$, $[0,1,0]$, $[0,0,1]$. Notice therefore, the first basis member $[1,0,0]$, is horizontal and parallel to the plane of the paper, and one unit away from it. Similarly, the second basis member $[0,1,0]$ is vertical and parallel to the plane of the paper, and one unit away from it. Finally, the third basic member $[0,0,1]$, corresponds to the origin R of the plane $\mathbb{R}^2_1$ of the paper.

Let us consider symmetry issues. First consider the basis member $P = [1,0,0]$. In terms of the projective plane represented by the plane of the paper, $P = [1,0,0]$ is the point at infinity directly leftward and rightward. In Sect. 22.3, it was seen that this point is symmetric with respect the vertical reflection axis. However, it is necessary to be a little more careful about this: The point is actually symmetric about *any* of the vertical lines, since each is infinitely far from the point, in both the left and right directions. The role of the origin basis member $R = [0,0,1]$ can now be seen. By creating a distinguishable point, i.e., the origin $R = [0,0,1]$, one destroys all these possible reflectional symmetries, except the one through the origin itself.

Exactly the corresponding argument holds for the other ideal basis member $Q = [0,1,0]$. Together with the origin, it creates reflectional symmetry about the x-axis.

Thus, the three ordinary vector basis members in $\mathbb{R}^3$ correspond to two reflection axes in the projective plane. Note that these two reflection axes correspond to the two $\mathbb{Z}_2$ fibers of the hyperoctahedral wreath group. In fact, this is easily generalized to projective n-space, as follows:

The standard vector basis of $n+1$ real space, $\mathbb{R}^{n+1}$, expresses the hyperoctahedral wreath symmetry of projective n-space $\mathbb{P}^n$.

Notice that this is because n of the vector basis members are points at infinity in projective n-space, and the remaining member is the origin.

The above statement accounts for the $n + 1$ members of the projective basis. There is only one other member of the basis. It is the unit element $[1,1,\ldots,1]$.

In order to understand its role, return to the example of projective 2-space, which has the four basis members listed at (22.6). Notice that the first basis member $P = [1,0,0]$ is the point at which *all* horizontal lines meet. Similarly, the second basis member $Q = [0,1,0]$ is the point at which *all* vertical lines meet. Thus the first two basis members $[1,0,0]$ and $[0,1,0]$ together define the infinite grid - i.e., two crossing parallel systems of lines. This grid has *infinite* vertical/horizontal reflectional symmetry - i.e., vertical/horizontal reflectional symmetry at each of its points. Therefore, one way of understanding the third basis member, the origin $[0,0,1]$, is that it establishes which reflectional vertical/horizontal symmetry will be chosen for this grid.

Now, with the grid structure in mind, it is possible to understand the role of the fourth member of the projective basis, the unit element $[1, 1, 1]$. What this point does is to select two of the lines AB and BD (see Fig. 22.4) from this infinite grid; i.e., because $[1, 1, 1]$ is at the intersection of these two lines. Furthermore, by the symmetry operations, point $[1, 1, 1]$ implies the other corners of the square. Thus the point $[1, 1, 1]$ selects one square from the full grid. However, each of the lines of the square is a hyperplane of the plane containing the square. Therefore, one can wreath sub-append the hyperplane to the hyperoctahedral wreath group. In conclusion: The standard projective basis corresponds to the hyperoctahedral wreath *hyperplane* group. Of course, what is actually meant by hyperplane, in this statement, is the group $\mathbb{GP}^{n-1}$

It can be seen therefore that the purpose of the standard projective basis is to set up a particular symmetry structure - in fact, the hyperoctahedral wreath hyperplane group centered at the origin - and define a particular square on which it should act. This accords with our theory in Chapter 8 that reference frames actually arise out of symmetry structures: They express the maximally symmetric state with respect to which actions are going to be applied and be understood as destroying that symmetry.

22.5   Theorem-Proving in Geometry

While the study of affine, Euclidean, and projective geometry claims to be founded on Klein's Erlanger program, close examination reveals something much deeper that accords more with our generative theory of shape.

To illustrate, let us begin by looking at affine geometry. Consider the theorem which states that the medians of a triangle intersect two-thirds the way down those medians. The standard proof is simply this: Begin by proving the theorem for the easy case: for equilateral triangles, i.e., the symmetric case. Then use the fact that an invariant of the affine group is ratios of lengths along a line; and use the fact that the affine group can map any triangle onto any other triangle. This extends the result from the symmetric case to the asymmetric case.

This style of proof is used also with enormous frequency in projective geometry - where one usually proves a theorem in the standard projective basis - i.e., the most symmetric case - and then extends to asymmetric projective bases, using the projective group.

Clearly, this frequent style of proof *does* use Klein's program. However, when we examine how Klein's program is being used here, we find that it is being used in the service of our generative theory of geometry, as follows:

Observe that these proofs *generate* the arbitrary case by first starting with the most symmetric case, and then applying the geometry group that extends the result to the asymmetric case. Most crucially, the style of proof follows the generative order prescribed by our generative theory; i.e., symmetry $\longrightarrow$ asymmetry.

Furthermore, as prescribed by the theory, the geometry group is used as the *symmetry group of an asymmetrizing action.*

Notice also that, while the Erlanger program does not recognize distinguished figures within the orbit of the geometry group - i.e., a symmetric shape has no privileged status in the orbit - the *proofs* nevertheless identify and exploit distinguished starting states. This recognition of distinguished starting states is again a feature of our generative theory; i.e., the geometry group is actually used asymmetrically rather than symmetrically. Notice that this is emphasized in our wreath formulation of geometry, where a symmetric state is described by the fiber copy corresponding to the identity element of the control group.

What role do we give to the *invariants* of the geometry group? In fact, they correspond to the use of our Symmetry Principle on internal structure, as follows: The Symmetry Principle (p. 43) says that, for *recoverability*, indistinguishabilities must be preserved backwards in time. In the case of internal structure - i.e., trace structure - these indistinguishabilities must be inter-state indistinguishabilities (Sect. 2.9). But inter-state indistinguishabilities are invariants. Thus the invariants of Klein's geometry are actually the inter-state indistinguishabilies of our inferential theory of shape.

Klein's geometry can therefore be understood as a part of our generative theory. However, our generative theory does considerably more than Klein's; for example, it accounts for the structure of proofs in geometry - whereas Klein's does not.

22.6 The Geometry Hierarchy

Standardly in geometry, one has this hierarchy:

1. Euclidean Geometry
Transformations: Isometries
2. Affine Geometry
Transformations: Affinities (products of stretches, shears, and isometries)
3. Projective Geometry
Transformations: Collineations.

By the Kleinian view, this hierarchy is one of successively less restrictiveness. In contrast, from our point of view, because we are interested in *recoverable generativity*, and because this depends on the Asymmetry Principle, the hierarchy represents successively greater *asymmetrization*. The difference between the Klein view and our view of this hierarchy is profound, as will be seen:

The consequence of our asymmetrization view is that the hierarchy represents the order in which operations are applied in the nested hierarchy. Thus the generative order is:

1. Euclidean. According to our theory, the underlying structure of any shape is a control-nested hierarchy of Euclidean operations; i.e., an iso-regular group. That is, according to the Externalization Principle, when one removes all external operations, one arrives back at Euclidean operations.

2. Affine. After generating the underlying structure using Euclidean operations, one then applies affine transformations. These do not preserve Euclidean metric but nevertheless preserve parallelism.

3. Projective. Finally one applies the projective operations. These do not even preserve parallelism. Thus one obtains distortions such as the Goldmeier figure.

A major difference between the generative theory and the Kleinian theory can now be seen: In the generative theory, each level provides a symmetric structure, relative to which the next level is an asymmetrization. However, in the Kleinian view, the next level does not recognize the difference between symmetric and asymmetric structures, and therefore it is irrelevant which structures on that level were inherited from the previous level.

Most crucially, in the generative theory, figures created on any level are *transferred* upward onto figures created on any higher level. For example, the structure of a square is transferred onto the first Goldmeier figure. Thus the geometry hierarchy represents a hierarchy of *nested control*; i.e., a wreath hierarchy.

It is now possible to understand the way in which Klein's theory of geometry fits as a component in the generative theory. As was said, each successive level in this control-nested hierarchy, acts as an asymmetrization of any structure defined at a lower level. However, in the generative system, an asymmetrization is carried out by what we have called the *symmetry group of the asymmetrizing process*. As a symmetry group, the control level acts in the way intended by Klein - that is, the level defines what he would call a *geometry*. However, because, in the generative system, the levels are glued together into a wreath hierarchy, this means that, because any lower structure is symmetric relative to anything produced by the higher control, the lower structure must correspond to the identity element in the higher control.

Thus Klein's "symmetry group" is forced to have an asymmetrizing action. Of course, what gives the hierarchy this order is our fundamental constraint of maximizing recoverability. *Klein's geometry-group becomes, in the generative theory, the symmetry group of an asymmetrizing action. In this capacity, it becomes a control group in a wreath hierarchy, and thus maximizes transfer and recoverability of generative structure.*

22.7   Projective Asymmetrization: Extrinsic View

It will now be argued that projective structure destroys the observer's symmetry. First, we shall see that the image plane is an unfolded version of the hyperoctahedral symmetry of the observer - destroying the observer's symmetry by misalignment (which is part of the basis of unfolding) - and then projective asymmetrization further destroys the observer's symmetry by additional misalignment and perspective transformation.[1]

To demonstrate this, it is now necessary to go to an *extrinsic* description of the image plane. So far, asymmetrization has been discussed from an intrinsic point of view; i.e., the mapping of the projective plane to itself. An extrinsic approach involves the mapping of one projective plane to another, each embedded in projective 3-space, each at a different orientation with respect to an observer who is in that projective 3-space. To keep the figures simple, the situation will be represented using diagrams of one dimension lower - i.e., projective lines in the projective 2-space. However, it must be understood that we are always representing a situation of one dimension higher.

In Fig. 22.5, the plane of the paper represents projective 3-space with coordinates $[x, y, z, w]$ in $\mathbb{R}^4$. The computer graphics convention will be used of making the last coordinate w the "dummy" coordinate. In other words, the hyperplane $\mathbb{R}_0^3$ that yields the points at infinity will be considered to be positioned at $w = 0$. Thus the "Euclidean" part of $\mathbb{P}^3$ will be the parallel "hyperplane" $\mathbb{R}_1^3$ situated at $w = 1$. Projective 3-space therefore decomposes like this:

$$\mathbb{P}^3 \quad = \quad \mathbb{R}_1^3 \quad \cup \quad \mathbb{P}^2. \tag{22.7}$$

Thus, in Fig. 22.5, the plane of the paper can be considered to be in $\mathbb{R}_1^3$, and the y coordinate perpendicular to the plane of the paper will also be in that hyperplane. Finally the coordinate w, also perpendicular to the plane of the paper, is the axis on which the two hyperplanes $\mathbb{R}_0^3$ and $\mathbb{R}_1^3$ are positioned.

---

[1] Strictly, unfolding involves misalignment and selection. We will concentrate, in this chapter, on the misalignment aspect. The selection here can be considered to be trivial, in the sense that the whole structure is selected. Note that we will not commit here to a particular kind of unfolding group.

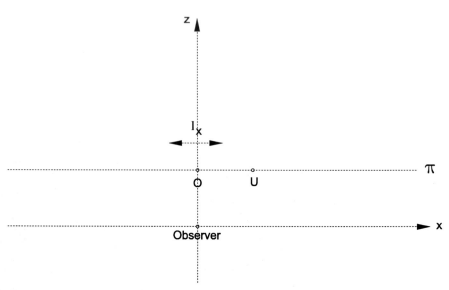

Fig. 22.5. Image plane preserves the observer's symmetries as much as possible.

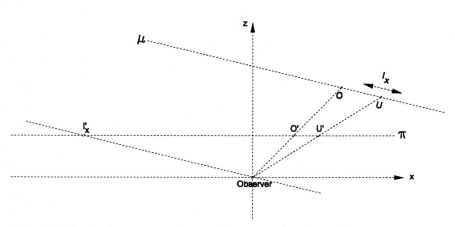

Fig. 22.6. Projective asymmetrization destroying the observer's symmetries.

The observer will be placed at the origin of $\mathbb{P}^3$, as shown. For ease of exposition, the observer will be referred to as "he". Most crucially, the observer will view the entire $\mathbb{P}^3$ via the image plane π shown in the diagram. This will be the standard embedded projective plane $\pi[0, 0, 1, -1]$.

Now we come to the fundamental issue of symmetry. The observer has the symmetry structure given by the hyperoctahedral wreath group, which Chapter 8 argued is the main structure of the Cartesian frame in $\mathbb{R}_1^3$. The observer's three symmetry planes can be seen in Fig. 22.5 as follows: (1) the plane of the page corresponds to one symmetry plane; (2) the x-axis is aligned with another symmetry plane perpendicular to the plane of the page; and (3) the z-axis is aligned with another symmetry plane perpendicular to the plane of the page. These planes will be referred to respectively as the y-, x-, and z-symmetry planes.

Now, the observer structures the image plane π with the same symmetry group - the hyperoctahedral wreath group. What this actually means is that the observer *unfolds*, from his own position, a copy of the hyperoctahedral wreath group to structure the image plane. Unfolding is a process of misalignment. Thus what the observer is doing is misaligning two copies of the hyperoctahedral wreath group.[2] This is the beginning of the symmetry-breaking structure.

Now, by the History Minimization Principle, the observer creates the minimal asymmetrization that is compatible with the data, and therefore the unfolding tries to keep the misaligned copy of the hyperoctahedral wreath group as similar to the original copy - in fact, it will keep two of the same mirror planes as those at the origin, and the third will be kept parallel to that at the origin.

Now let us assign coordinates to the embedded projective plane π. In computer graphics, one says that the natural coordinates are those determined by the standard projective simplex. Why? From the previous sections, we see that this is because the projective simplex is the most symmetrical in projective space. In fact, we see that it is the one that most preserves the symmetry of the observer - since the plane π is unfolded from the observer by minimal misalignment. The projective basis for π therefore consists of the four points Fig. 22.5,

$$O = [0, 0, 1, 1]$$
$$U = [1, 1, 1, 1]$$
$$I_x = [1, 0, 0, 0]$$
$$I_y = [0, 1, 0, 0] \tag{22.8}$$

The first three are shown in Fig. 22.5. The fourth is in the y-direction perpendicular to the plane of the page.

---

[2] Notice, in this case, we do not need an alignment kernel (for example, as in canonical unfoldings). The misalignment can be considered to occur here between two copies within one fiber-group product.

Fig. 22.5 tries to indicate some of the important symmetry structure, as follows: Notice that the ideal point I_x, shown by the double arrow, is indicated as being reflectionally symmetric about the z-axis; and is therefore reflectionally symmetric about the z-symmetry plane. The same is true of the other ideal point I_y, coming vertically out of the plane of the paper. And again, the same is true of the origin O. That is, the projective simplex, O, I_x, I_y, is reflectionally symmetric about the z-symmetry plane. In fact, further consideration reveals that the projective simplex is reflectionally symmetric about each of the symmetry planes of plane π.

One member of the basis exists outside the simplex, and this is the unit U. This, as seen before, selects the square in π from infinite grid in π.

Now let us look at how one gets projective asymmetrization extrinsically. Unfold from the plane π another plane μ as shown in Fig. 22.6. Notice that it takes along with it, a copy of the projective basis O, I_x, I_y, U, and therefore has internally the same symmetry structure as the plane π.

Now, the observer is viewing this plane via the image plane π. The crucial thing to notice is that the basis O, I_x, I_y, U on plane μ is projected to a new basis O', I_x', I_y', U' on the image plane π. This new basis destroys the symmetry structure of the plane π. Notice for example, that point I_x at infinity, now comes in from infinity to become the point I_x' which is therefore a vanishing point. We saw this in the Goldmeier effect, where this was point P in Fig. 22.2 (p. 498). Notice that ultimately, one has further destroyed the symmetry structure of the observer.

Thus one sees here again how the Klein view of geometry is not natural. The research literature talks about projective distortion, which we called projective asymmetrization.

PROJECTIVE ASYMMETRIZATION. *Projective asymmetrization occurs because one is assuming that the observer is structured by the hyperoctahedral wreath group which is successively broken as follows: (1) The image plane is an unfolded version of the hyperoctahedral symmetry of the observer - destroying the observer's symmetry by misalignment (which is the basis of unfolding) - and then (2) the observer's symmetry is further destroyed by additional unfolding of the plane and perspective transformation.*

Notice that the Klein view, in which all projective bases are equally valid, does not allow for recoverability. That is, in accord with our theory, recoverability of an action is possible only if the action breaks symmetry. This means that not all projective bases are equal. Notice furthermore, that because recoverability now becomes possible, we are building the observer up as a memory object - i.e., after the two successive symmetry-breaking actions, the observer's image now has asymmetry, and thus the observer has a layered structure of asymmetry. This conforms generally with our view of science as adding structure to the observer as a memory object - and in particular as asymmetrizing the measuring instruments so that they can be causal memory of the environmental structure.

22.8   Deriving Projective Coordinate Systems

It will now be argued that the Goldmeier effect shows that the derivation of projective coordinate systems is more complicated than has previously been suspected. This again will be based on our proposal that the projective plane is structured, as near as possible to the hyperoctahedral wreath structure of the observer.

First observe this: Our research program, Leyton [87], [88], [89], [90], [91], [96],, has demonstrated the enormous psychological salience of the quadrilateral and its reference to a square. One might ask why the quadrilateral is so salient to the human perceptual system. We conjecture that the answer is this: The quadrilateral is associated with a projective basis, and this defines a projective coordinate system on the entire plane.

This however, is only the first stage in the argument. The second stage is this: We argue that the derivation of a projective coordinate system from a quadrilateral is different depending on the orientation of the quadrilateral. This is basic to understanding the human visual system.

Let us however pretend that the second stage does not exist, and that there is a single way to derive a projective coordinate system from a quadrilateral. The most direct way possible will be developed. The reader should note the following: The *intrinsic* projective plane will be used, and this turns out to be critical in later sections. Furthermore, it will not matter whether one uses the quadrilateral or quadrangle - and thus no distinction between the two will really be made. Also the computer graphics convention will be used of choosing the last coordinate to the "dummy" variable.

To begin: Return to the first Goldmeier figure, and look again at the construction we gave it as a complete quadrangle in Fig. 22.2 (p. 498). Using this structure, let us now construct a projective coordinate system. This will be done by assigning a projective basis to the figure. This basis, which is the same as that used before, is the natural one arising from the symmetry structure of the observer. The four points of the projective basis are therefore as follows: The point R is chosen to be the origin $[0, 0, 1]$; the points P and Q are chosen to be the two vanishing points $[1, 0, 0]$ and $[0, 1, 0]$ respectively; and the point B is chosen to be the unit point $[1, 1, 1]$. The x- and y-axes are thereby completely determined, as follows: The x-axis must go through P and R; and the y-axis must go through Q and R. This is shown in Fig. 22.7. The symmetry structure now becomes very apparent: That is, the x- and y-axes now obviously represent the symmetry axes. That is, if the figure is

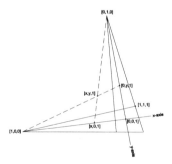

Fig. 22.7. A projective coordinate system from the first Goldmeier figure.

un-distorted, then they become the symmetry axes of the figure and the observer's coordinate system.

From this diagram, one can quite easily derive the projective coordinates for any point in the plane. First establish the coordinates of points along axes, as follows: Consider an arbitrary point on the x-axis; for example, the point indicated as $[x, 0, 1]$. The value x is simply calculated as the cross-ratio of the following four points on the x-axis: The vanishing point, the origin, the unit point to the right of the origin (i.e., the point where the x-axis first crosses the quadrilateral from the right); and the point $[x, 0, 1]$ itself. The method works because one can regard the cross ratio as the projective scaling along the axis. Similarly, use this method to calculate the value y of an arbitrary point $[0, y, 1]$ on the y-axis.

Thus, having calculated the values of any points on the two axes, the value can be calculated of an arbitrary point shown as $[x, y, 1]$. This point is determined as the intersection of the two dashed lines shown. Therefore one merely reads off the values where the dashed lines hit the coordinate axes.

Therefore, what has been seen above is how to construct, from a quadrilateral, a projective coordinate system for the entire plane. Notice again that it exploits the hyperoctahedral symmetry structure of the observer.

Now, if Klein's view of geometry were correct, then this method would be the appropriate one for *all* quadrilaterals. In fact, according to our theory, it is appropriate for only half of them. A different method must be chosen for the other half. Let us proceed.

Turn now to the second Goldmeier figure; i.e., that shown in Fig. 22.1b (p. 497). If one applied exactly the same method to this figure, one would obtain the coordinate system shown in Fig. 22.8. In other words, this figure is merely a rotated version of the previous one (Fig. 22.7).

The problem is that this does not correspond to the symmetry structure of the observer. In fact, it is the furthest possible rotational distance from the observer's symmetry structure, since the x- and y-axes are now maximally diagonal. In contrast, in this quadrilateral, the nearest axes that would correspond to the observer's symmetry structure would be horizontal and vertical

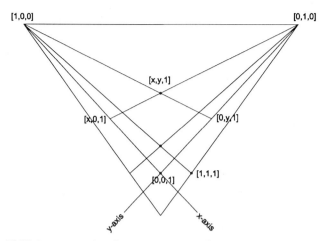

Fig. 22.8. If Klein were right: A projective coordinate system.

- i.e., through the vertices of the figure. Thus the x- and y-axes should be as shown in the next diagram, Fig. 22.9.

We now want to construct a coordinate system based on these two axes. Our argument is as follows: First construct the vertical coordinate lines - i.e., those parallel to the y-axis. Observe that, whatever the coordinate system, the sides of the figure produce the same two vanishing points, i.e., the top-left and top-right converging points in Fig. 22.9. The line shown as connecting these two points along the top is called the horizon line. As shown, take the y-axis (vertical) as crossing this line at the point $[0, 1, 0]$. Now all coordinate lines parallel to the y-axis will converge at this point. In particular, the two vertical parallel lines through the left and right vertices of the quadrilateral will converge at that point.

Next construct the coordinate lines parallel to the x-axis. Symmetry is not broken about the y-axis. Therefore the coordinate lines parallel to the x-axis converge at infinity, in fact at the point $[1, 0, 0]$. This is shown both on the left and right of the diagram as the double-headed arrows labeled $[1, 0, 0]$. The parallel horizontal lines represent the lines parallel to the x-axis, and this single ideal point is at the end of all of them (at both ends).

The figure therefore shows the three points $[1, 0, 0]$, $[0, 1, 0]$, $[0, 0, 1]$, of the projective simplex that arises by corresponding the symmetry structure of the figure maximally with that of the viewer. The point outside the simplex, i.e., the unit point $[1, 1, 1]$, which sets the scale, can be chosen to be the intersection of the positive x- and y-coordinate lines that go through the vertex. This point is less important than the others, because the others determine the parallel grid structure of the plane. However, this point does let us assign coordinate numbers to those lines. Thus, a completely specified coordinate system has now been obtained, created by a projective basis $[1, 0, 0]$, $[0, 1, 0]$, $[0, 0, 1]$, $[1, 1, 1]$.

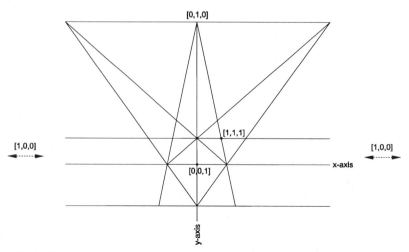

Fig. 22.9. The appropriate projective coordinate system for the second Goldmeier figure.

A remarkable fact now emerges from what has been done. To clarify this, let us extract from the figure, the coordinate system just created. This coordinate system is shown separately in the next diagram Fig. 22.10. To emphasize however: This is the coordinate system inferred from the *second* Goldmeier figure. The remarkable thing that can be seen is that this coordinate system is much more symmetric than the coordinate system derived from the first Goldmeier figure Fig. 22.7.

This symmetry can be precisely understood as follows: In the coordinate system for the first Goldmeier figure, the coordinate lines in the x-direction converge at a vanishing point; and the coordinate lines in the y-direction converge at a vanishing point. However, in the coordinate system for the second Goldmeier figure, the coordinate lines in the x-direction do not converge at a vanishing point; i.e., they remain parallel. Only the coordinate lines in the y-direction converge at a vanishing point.

The issue of vanishing points relates to the issue of symmetry. The more vanishing points in the coordinate lines, the greater the asymmetry - because the coordinate lines correspond to the symmetry structure. That is, in the first Goldmeier figure, the existence of a vanishing point for the coordinate lines in the x-direction means that symmetry about the y-axis is broken. Similarly, the existence of a vanishing point for the coordinate lines in the y-direction means that symmetry about the x-axis is broken. Now, in the second Goldmeier figure the non-existence of a vanishing point for the coordinate lines in the x-direction means that symmetry *is not broken* about the y-axis, as can be seen by directly inspecting Fig. 22.10. Similarly, the existence of a vanishing point *only* for the coordinate lines in the y-direction means that symmetry is broken *only* about the x-axis.

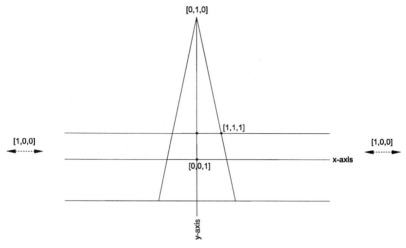

Fig. 22.10. The projective coordinate system for the second Goldmeier figure.

Here, therefore, is another crucial violation of Klein's view of geometry: The projective coordinate systems derived from the two Goldmeier figures imply that the first figure is more asymmetric than the second. This conclusion violates Klein's view - in which the figures would be equally symmetric.

22.9   Non-transitivity of the Geometry Group

The above concepts in mind, let us take a moment to consider the relation of our approach to that of Klein's. First observe that our approach is made possible by *transfer* - i.e., the transfer of symmetric structures onto asymmetric structures. This is realized as a wreath product in which the fiber is the symmetry group of the symmetrical structure, and the symmetrical structure is then identified with the identity element of the control group. The control group is Klein's geometry group but realized, via the wreath action, as an asymmetrizing group.

Notice that this can result in the *non-transitivity* of the Klein geometry group. For example, because of the symmetry assignment in the two Goldmeier figures, they are no longer in the same orbit of the Klein group. In contrast, transitivity is basic to the Klein view. For example, the fundamental theorem of Kleinian projective geometry states that the projective group is transitive on projective bases - e.g., on complete quadrilaterals. This cannot be allowed in our system.

The reasons for the non-transitivity of our generative geometry is extremely profound, and are as follows:

The generative structure is inferred from a data set of points, using a system of interpretation rules (i.e., our recovery rules). The application of these rules implies the existence of an observer. Therefore application depends on the relation of the structure of the observer to the data set. Furthermore, generative geometry, being exhaustive does not allow any ambiguity in the inferred generative sequence assigned to the data set - and therefore all aspects of the generative sequence are dependent on the relation of the structure of observer to the data set.

In contrast, the structure of two data sets in Klein geometry can be equivalent simply because such geometry is not generatively exhaustive with respect to rules applied by an observer and dependent on the observers relation to the data set. Equivalence of this trivial nature cannot occur in our generative geometry.

22.10   Regular Translation Structure

Let us concentrate on the grid-line structure created in the previous section. Grid lines correspond to translational symmetry. The action of the projective group on this translation structure will now be considered. Since one of the goals of our theory is generative exhaustiveness, the structure of the grid will be described as the wreath c-polycyclic group $\mathbb{R} \textcircled{w} \mathbb{R}$, and therefore, together with the projective group as the control group, the following is obtained:

$$\mathbb{R} \textcircled{w} \mathbb{R} \textcircled{w} PGL(3, \mathbb{R}).$$

By our anti-Klein view, the fiber copy of $\mathbb{R} \textcircled{w} \mathbb{R}$ corresponding to the identity element of the projective control group is that which corresponds to the Euclidean translation action on the embedded Euclidean plane.

Notice that the above sequence can be wreath sub-appended by an iso-regular group G, and the result corresponds what we called a *projective generative crystallographic group* (p. 159), when $\mathbb{R} \textcircled{w} \mathbb{R}$ is discretized to $\mathbb{Z} \textcircled{w} \mathbb{Z}$. Thus, the total structure is this:

$$G \textcircled{w} \mathbb{Z} \textcircled{w} \mathbb{Z} \textcircled{w} PGL(3, \mathbb{R}).$$

Again, the fiber copy of $G \textcircled{w} \mathbb{Z} \textcircled{w} \mathbb{Z}$ corresponding to the identity element in the projective group, would correspond to the configuration which allows the maximal action of the Euclidean group.

The issue of crystallographic structure is mentioned here because it is our argument that measuring instruments and other recording devises are necessarily structured as regular hierarchies, that is, wreath c-polycylic wreath-isometric groups, because such groups allow the maximal recovery of any

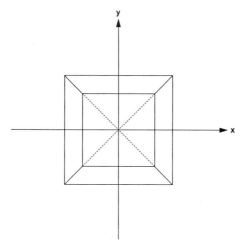

Fig. 22.11. Projective asymmetrization of centered cube.

applied action. In accord with our Externalization Principle, an external inference of past action goes back to a wreath c-polycylic wreath-isometric group because such a group is maximally "clean" - and therefore maximally receptive to recoverability, and hence data registration and memory storage.

22.11   3D Projective Asymmetrization

Let us now move from 2D to 3D perspective. All textbooks explain 3D perspective using cubes. Why? The answer we propose follows from our generative theory and its anti-Erlangen property: Despite the fact that the projective group is transitive on projective bases, one uses the most symmetric projective basis as the starting point for generating any arbitrary basis, and applies the projective group as an asymmetrizing action.

This, we argue, is behind all textbook discussions of how it is possible to make images more *informative*. Standardly, the discussion begins with a cube centered at the origin and facing the viewer, as shown in Fig. 22.11. The standard textbook points out that this is the least informative view.

To gain more information, the textbook then argues that one must apply a translation from the origin, e.g., along the x-axis, as shown in Fig. 22.12. Notice however that if the cube were solid, then all that would be revealed here would be the front side, and the left side.

Thus, to gain more information, the textbook argues, one must apply a translation from the origin, that is not along an axis, e.g., as shown in Fig. 22.13. Now, three sides are visible: the front, left, and bottom sides. The

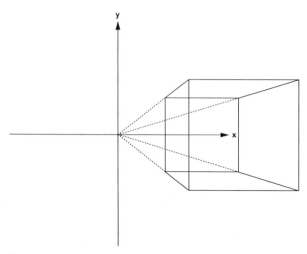

Fig. 22.12. Asymmetrization caused by translation of cube along one symmetry axis.

figure has kept the back side partially visible to reveal that is still parallel to the viewplane, as is the front side.

Now the fact that the front and back planes are still parallel to the viewplane means that less depth-cue information is available than one wants for a compelling image of the cube. Thus the standard textbook then argues that, to overcome this, one must rotate the cube. Fig. 22.14 shows the cube rotated about the vertical axis. The front face itself can be seen to contain a compelling sense of depth. Similarly, Fig. 22.15 shows the cube rotated about the horizontal axis. Again, the front plane has a compelling sense of depth, but in a different direction.

The textbook then goes onto to explain that, in each of the rotation figures just considered (p. 522), only two sides can be seen. Thus to reveal a third side, a combination of these two rotations must be applied. This is shown in Fig. 22.16. The textbook points out that, clearly, this view is the most *informative*.

This completes the standard argument given in textbooks. The question we now wish to ask is this: What is actually going on in the standard argument? What do the writers actually mean by a diagram being "informative" and by the successive "increase of information" over the series of diagrams?

We argue that the term information actually means *asymmetry*. The value of expressing it in this way is that the information can be rigorously defined in terms of group theory. In addition, we argue that the informative nature of the asymmetry is that it is used by the Asymmetry Principle to infer the generative structure that encodes the configuration. Our argument proceeds as follows:

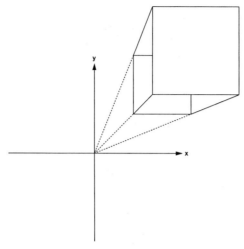

Fig. 22.13. Additional asymmetrization caused by translation of cube not along a symmetry axis.

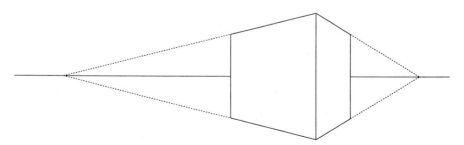

Fig. 22.14. Asymmetrization caused by rotation of cube about vertical axis.

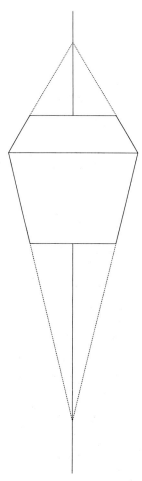

Fig. 22.15. Asymmetrization caused by rotation of cube about horizontal axis.

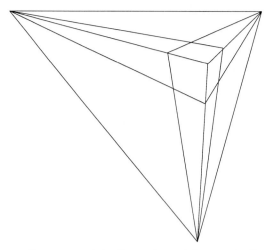

Fig. 22.16. Perspective asymmetrization using arbitrary rotation.

First, one can use the viewing configuration defined in Sect. 22.7, and illustrated in Figs 22.5 and 22.6 on p. 510. That section was examining the projective asymmetrization in the Goldmeier effect, which we argued is in the projective plane. However, this was being done from an extrinsic point of view, which meant that the projective plane was embedded in projective 3-space. The current section concerns intrinsic 3D projective action, which means that we are actually using, once again, projective 3-space. Thus the structure illustrated in Figs 22.5 and 22.6 on p. 510 can be used. If the reader has forgotton the argument in that section, then he or she should review it - i.e., starting p. 509.

The only difference between that previous section and the present one is this: The previous section considered only a 2D projective basis O, I_x, I_y, U. Now it is necessary to consider a 3D projective basis, and we do so simply by adding the ideal point I_z to the 2D basis. This will be a "double-arrow" in the z direction in Fig. 22.5. Most importantly, observe that the projective simplex O, I_x, I_y, I_z, of this basis, is the most symmetrical simplex possible relative to the observer's structure, which is the hyperoctahedral wreath group of degree 3. In fact, the degree 3 version was already being used in that previous section because we wanted to consider symmetry with respect to the 3D extrinsic structure.

Now the cube starting-state maximizes the observer's symmetry, in the manner described in Sect. 22.7. The cube is obtained from the hyperoctahedral wreath group by wreath sub-appending the hyperplane representing the cube face as a fiber. That is, the cube is given by the hyperoctahedral wreath hyperplane group $HWH(3)$. The hyperoctahedral structure within this group is a unfolded version of the hyperoctahedral structure of the observer - again in the manner described in Sect. 22.7.

With this in mind, now turn to the sequence of figures in the present section. To fully understand them, it will be sometimes valuable to consider them to be 3D structures obtained by a *perspective* transformation - i.e., prior to usual *parallel projection* onto the viewing plane.

According to our analysis, each of the figures breaks one or more of the three fiber reflectional symmetries of the hyperoctahedral wreath group. This group should be considered fixed to the starting state of the cube. Now let us interpret each of the figures in terms of symmetry-breaking.

Fig. 22.11: Here the reflectional symmetry between the front and back planes is broken - as represented by the fact that they are of different sizes in the image.

Fig. 22.12: Here, in addition, the left-right reflectional symmetry is broken.

Fig. 22.13: Here, in addition, the top-bottom reflectional symmetry is also broken.

Thus the above three figures successively break the three reflectional fiber symmetries of the hyperoctahedral wreath group.

Before turning to the remaining figures - the rotations - a new level of understanding needs to be introduced: The issue that needs to be considered is this: Although the symmetries of the cube are being broken, it is not necessarily the case that the symmetries of the 2D fibers, the faces, are being broken, or the symmetries of the lines that are the edges. To understand what is going on here, we need to use the recursive structure of the *recursive solid hyperoctahedral wreath hyperplane group, RS-HWH(n)*, which we proposed as being the symmetry group of the solid n-cube (Sect. 16.6). Recall that a basic aspect of this group is that each of the hyperplane faces are themselves solid $n-1$ cubes, which have solid $n-2$ cubes as faces; and so on downwards. Thus, in the case of the solid 3-cube, begin considered here, each face is a solid 2-cube, and face edge is a solid 1-cube. They are recursively nested within each other, in the manner described in Sect. 16.6. Note, generally, the reader should not confuse the recursive nature of the *RS-HWH(n)* group, with the recursive nature of the $\mathbb{GP}^n$. The former is recursive on the hyperoctahedral structure - recognizing that each face, sub-face, etc., is structured by a hyperoctahedral group. The latter is recursive on the translation structure of a projective sub-space. It is essentially used as a fiber group, substituting for each $\mathbb{R}^j$ (of any level) in the former recursive group, an extended group $\mathbb{GP}^j$.

The overall action of the projective group can now be considered as affecting each of the recursive levels. For example, in Fig. 22.13, the most asymmetrical of the configurations just described, even though the cube has been asymmetrized, the front face has not; i.e., it does not have a vanishing point. Thus the projective action has not affected this fiber copy; i.e., it has not broken the *RS-HWH(2)* structure of this fiber copy.

With this in mind, one can now understand what is going on in the rotation figures, as follows:

Fig. 22.14: In this figure, let us go successively down the recursive hierarchy. At the top level, the cube has retained its reflectional symmetry about the horizontal plane, but lost its symmetries about the other Cartesian planes. Now go down one recursive level. Here the *RS-HWH(2)* group of each fiber face (2-cube) has been broken, because each fiber face is a plane that contains a vanishing point. Observe furthermore, that the top and bottom faces each contain two vanishing points, whereas the side faces contain only one vanishing point. Therefore there has been greater symmetry-breaking in the top and bottom faces than the side-faces. We emphasize that the symmetries being broken here are the 2D reflectional symmetries associated with the square - i.e., that map opposite edges of the square onto each other. Now go down one more recursive level. Observe that, while the *RS-HWH(1)* group on each "horizontal" fiber line (1-cube) has been broken - by the presence of a vanishing point - the *RS-HWH(1)* group on each vertical fiber line has not. We emphasize that the symmetries being broken here are the 1D reflectional symmetries associated with the internal structure of lines; i.e., not *between* lines as in the level above. That is, the symmetry-breaking of a line; is due to the ideal point being pulled along the line, inwards from infinity.

Fig. 22.15: Here, there is exactly the same structure as that just defined, except that horizontal is exchanged for vertical.

Fig. 22.16. Here all hyperoctahedral levels in the recursive hierarchy have been fully broken. That is, the cube *RS-HWH(3)* has been broken in all its three reflection planes; each face *RS-HWH(2)* has been broken in both its reflection axes; each edge *RS-HWH(1)* has been broken in its reflection point.

INFORMATIVENESS OF A PROJECTIVE IMAGE. *The informativeness of a projective image can be fully characterized by its particular symmetry-breaking of the RS-HWH(n) group, including all recursive fibers within that group.*

Notice also, on each of the three recursive levels, the visual salience of the symmetry-breaking of that level, means that the visual system is anti-Kleinian on all levels.

Notice finally that, by using the Asymmetry Principle, one can thereby recover the unfolding actions - translations and/or rotations - that produced the configuration by movement from its original location and orientation at the origin of the world coordinate system (identified with the observer). This is exactly how a human observer defines the configuration.

22.12   Against the Erlanger Program: Summary

NON-MEMORY VS. MEMORY

Klein geometry: This geometry concerns *memorylessness*. A geometrical object is defined as an invariant under transformations. An invariant is that part of a configuration that is memoryless with respect to any applied transformation.

Generative geometry: This geometry concerns the *creation of memory*. A geometric object is defined as a structure from which one can recover the sequence of actions that created its current state. In other words, in our view, there is an *equivalence between the concept of geometry and the concept of memory-storage*.

NON-RECOVERABILITY VS. RECOVERABILITY

Klein geometry: Concerns non-recoverability.

Generative geometry: Concerns recoverability.

SYMMETRIZATION VS. ASYMMETRIZATION

Klein geometry: The group of a geometry acts symmetrically on the chosen space.

Generative geometry: The group of a geometry acts asymmetrically on the chosen space, converting symmetric figures into asymmetric ones; not the reverse.

EQUALITY VS. INEQUALITY OF FIGURES

Klein geometry: Figures linked by the transformation group are equal.

Generative geometry: Figures linked by the transformation group can be unequal, in the sense that one comes from the level below (i.e., is more symmetric)

EUCLIDEAN $\longrightarrow$ AFFINE $\longrightarrow$ PROJECTIVE HIERARCHY

Klein geometry: A level in this hierarchy does not recognize the difference between structures generated on the previous level and structures not generated on the previous level, i.e., symmetric vs. asymmetric structures.

Generative geometry: Each level provides the symmetric structures, relative to which the next level is an asymmetrization. Furthermore, each asymmetric structure, in a level, is a transferred version of a symmetric structure on the previous level.

NON-EXHAUSTIVENESS VS. EXHAUSTIVENESS OF TRANSFORMATIONAL DESCRIPTION

Klein geometry: A geometry uses a single group of transformations; and on only one level.

Generative geometry: Geometry must be exhausively generative. Therefore it must use a group of transformations on every level. Furthermore, it can use different groups on different levels.

THE GEOMETRY GROUP ACTS ON A SET VS. A GROUP

Klein geometry: The geometry group acts on an ordinary set. This is because structure is not generative all the way down.

Generative geometry: Structure is defined as a wreath product in which the fiber is the symmetry group of a symmetrical structure, and the symmetrical structure is then equated with the identity element of the control group. Thus the action of the geometry group on figures, is the action of a group on a *group*, rather than a set, and this action is algebraically complex.

INVARIANCE VS. TRANSFER

Klein geometry: What is shared between a figure A and its transformed version B is an invariant.

Generative geometry: What is shared between a figure A and its transformed version B is actually figure A. This is because figure B is a transferred version of figure A.

TRANSITIVITY VS. NON-TRANSITIVITY

Klein geometry: The group has wide transitivity in its action, since figures are merely point sets. For example, the two Goldmeier figures can be mapped onto each other.

Generative geometry: The group has limited transitivity in its action, since figures are themselves generative structures. For example, the two Goldmeier figures cannot be mapped onto each other, and indeed one is more symmetric than the other, despite having the same point sets.

OBSERVER-INDEPENDENT VS. OBSERVER-RELATIVE

Klein geometry: The observer structure is irrelevant.

Generative geometry: All structure is ultimately a transferred version of the observer structure.

COORDINATE-FREE VS. COORDINATE-FREE

Klein geometry: A geometric object is coordinate free.

Generative geometry: A geometric object is non-coordinate free. Because an object is a structure generated from a symmetric ground state, and because a coordinate frame is really a symmetric ground state, one cannot remove the coordinate frame without removing the object's structure.

EQUIVALENCE VS. NON-EQUIVALENCE BETWEEN IDEAL AND VANISHING POINTS

Klein geometry: Points at infinity have the same structural status as the vanishing points into which they are transformed after projective action.

Table 22.1. Comparison between Klein's theory of geometry and the generative theory.

Klein's	*Generative*
Memoryless	Memory storage.
Non-recoverability	Recoverability.
Symmetric action	Asymmetric action.
Figure equality under group action	Figure inequality under group action.
Levels unlinked	Levels linked.
Non-exhaustive	Exhaustive
Group acts on set	Group acts on group
Source and target share invariant	Source and target share source
Wide transitivity	Narrow transitivity
Observer independent	Observer dependent
Coordinate free	Non coordinate free
Unique description	Multiple descriptions

Generative geometry: The points at infinity correspond to observer symmetries, and the vanishing points into which they are transformed are observer asymmetries. Therefore the former are generative ground states, and the latter are derived states.

ONE DESCRIPTION PER SET VS. SEVERAL DESCRIPTIONS PER SET

Klein geometry: A particular set always has the same description.

Generative geometry: A particular set can have a number of non-equivalent descriptions. This is because the description given to the data set is a generative version of the observer's structure, and thus, if the data set has a different relationship (e.g., orientation) to the observer, then the data set will receive a different structural description.

These points are listed in Table 22.1.

A. Semi-direct Products

A.1 Normal Subgroups

A group extension is a means of extending one group by another. A semi-direct product is a particular type of group extension, called a splitting extension. Standardly, a group extension is built up by starting with a type of group called a normal subgroup, and extending the latter. Therefore we need to understand what a normal subgroup is.

Any subgroup S of a group G is a subset that is also a group. One writes

$$S < G$$

where $<$ means "subgroup of". Amongst the subgroups of a group, a particular type, called *normal subgroups*, are the most significant. The reason is that they provide maps from the group, as follows:

A homomorphism $\pi : G \longrightarrow H$ from a group G onto a group H is a map which preserves the group operation; that is:

$$\pi(g_1 g_2) \quad = \quad \pi(g_1)\pi(g_2).$$

Now consider the subset N of elements of G that are sent to the identity element e in H, as illustrated in Fig. A.1. The subset N forms a subgroup of G, called the *kernel* of the homomorphism. A normal subgroup is a subgroup that can be the kernel of some homomorphism. One writes:

$$N \triangleleft G$$

where $\triangleleft$ means "normal subgroup of".

A direct way of defining a normal subgroup is this: A normal subroup N of G is a subgroup such that, for any element g in G,

$$gN \quad = \quad Ng \qquad\qquad (A.1)$$

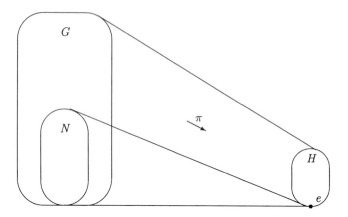

Fig. A.1. The kernel N of a homomorphism.

By this one means: Suppose the subgroup N can be written like this:

$$N \;=\; \{\, n_1,\, n_2,\, n_3, \ldots \,\}.$$

Then the set gN can be written like this:

$$gN \;=\; \{\, gn_1,\, gn_2,\, gn_3, \ldots \,\},$$

and the set Ng can be written like this:

$$Ng \;=\; \{\, n_1 g,\, n_2 g,\, n_3 g, \ldots \,\}.$$

The condition $gN = Ng$ states that gN and Ng are equal as sets. The two sets are called left and right *cosets* of N. They can be considered to be translates of the subgroup N under the action of g. The element g is called the coset leader.

The normal subgroup condition (A.1) can obviously be re-written like this:

$$gNg^{-1} \;=\; N. \tag{A.2}$$

This is an extremely important expression. One calls the bracketing $g - g^{-1}$ around N a *conjugation* operation. Expression (A.2) states that a normal subgroup N is invariant under conjugation by any member g of G. By this one really means that $g - g^{-1}$ sends N isomorphically onto itself. An isomorphism is a homomorphism that is 1-1 and onto. An isomorphism of a group onto itself is called an *automorphism* of the group. Thus, expression (A.2) states that *a subgroup N is normal if conjugation of N by any member $g \in G$ acts as an automorphism of N.*

A.2 Semi-direct Products

A group extension is a means of extending one group by another. Standardly, a group extension is built up by starting with a normal subgroup, and adding another group. A semi-direct product is a particular type of group extension, called a splitting extension.

Semi-direct products are by far the most frequent means of combining groups in group theory. This section will give an easy example of a semi-direct product. It is very important, for the remainder of the book, that the reader fully grasp this example.

Consider the group D_4, which can be regarded as the symmetry group of the square, i.e., the group of transformations that send the square back onto itself. D_4 has eight elements which will be written in this way:

$$D_4 = \{ \ e, \quad r_{90}, \quad r_{180}, \quad r_{270}, \quad (A.3)$$
$$m_V, \quad m_D, \quad m_H, \quad m_d \ \}.$$

The top row are the *four rotations* r_θ that send the square to itself, and the bottom row are the *four reflections* that send the square to itself, where

m_V = reflection about the vertical axis,
m_H = reflection about the horizontal axis,
m_D = reflection about one of the diagonal axes D,
m_d = reflection about the other diagonal axis d.

Two particular subgroups of D_4 will now be considered: The first is the subgroup $\mathbb{Z}_4$ of four rotations:

$$\mathbb{Z}_4 = \{ \ e, \quad r_{90}, \quad r_{180}, \quad r_{270} \ \}. \quad (A.4)$$

The successive group elements here are obviously rotations by successive 90^0 increments. The other subgroup is the group generated by reflection m_V about the *vertical* axis:

$$\mathbb{Z}_2 = \{ \ e, \quad m_V \ \} \quad (A.5)$$

which is obviously a cyclic group of order 2.

Now the subgroup $\mathbb{Z}_2$, just defined, is not a normal subgroup of D_4. However, the subgroup $\mathbb{Z}_4$ is a normal subgroup. This means that, in D_4 the subgroup $\mathbb{Z}_4$ can be the kernel of a homomorphism. In fact, let us construct a homomorphism

$$\pi \ : \ D_4 \ \longrightarrow \ \mathbb{Z}_2.$$

as shown in Fig. A.2. Notice that the kernel is $\mathbb{Z}_4$. Notice also that, in the particular case being considered, the image $\mathbb{Z}_2 = \{e, m_V\}$ of the homomorphism happens to be a subgroup of D_4.

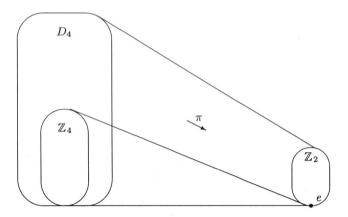

Fig. A.2. A homomorphism from D_4 onto $\mathbb{Z}_2$ with kernel $\mathbb{Z}_4$.

We now have the ingredients for constructing a semi-direct product. The group D_4 will be constructed by starting with the normal subgroup $\mathbb{Z}_4$ and adding the subgroup $\mathbb{Z}_2$. This is written as follows:

$$D_4 \;=\; \mathbb{Z}_4 \, \circledS \, \mathbb{Z}_2$$

where the symbol $\circledS$ means *semi-direct product*. Notice that the normal subgroup $\mathbb{Z}_4$ is written on the left of $\circledS$ and the non-normal subgroup $\mathbb{Z}_2$ is written on the right of $\circledS$.

Other features should be noticed about this construction. The first is that the two component groups $\mathbb{Z}_4$ and $\mathbb{Z}_2$ intersect only at the identity element. This can be seen instantly by recalling that $\mathbb{Z}_4 = \{\, e,\; r_{90},\; r_{180},\; r_{270}\,\}$ and $\mathbb{Z}_2 = \{\, e,\; m_V\,\}$.

The second feature is that, if these two subgroups are multiplied element-wise, we get the entire 8 elements of D_4. That is $\mathbb{Z}_4\mathbb{Z}_2 = D_4$. Generally therefore, the following gives the definition of a semi-direct product:

Definition A.1.   *Consider two subgroups N and H of G. Then G is a called a* **semi-direct product** *of N and H if*

(1) $N \triangleleft G$
(2) $N \cap H = \{e\}$
(3) $NH = G$.

G is also called a **splitting extension** *of N by H. Any subgroup H, fulfilling conditions (2) and (3), is called a* **complement** *of N in G.*

Splitting extensions therefore have a particularly "clean" property: One can think of them as allowing G to be split into N and a complement H

which is also part of G. There also exist non-splitting extensions. In such cases, N does not have a complement H in G. For details on such extensions, the reader should consult our book on group extensions, Leyton [98].

A.3 The Extending Group H as an Automorphism Group

The reader will recall from Sect. A.1 that a normal subgroup N of G has the property that, for any element g of G,

$$gNg^{-1} \;=\; N. \tag{A.6}$$

That is, each element g acts on N as an automorphism, when g is used as the conjugation $g - g^{-1}$.

In particular, it is clear that, when G is constructed as a splitting extension of N by H, the elements of H will act on N as automorphisms of N. Now the set of automorphisms of N form a group, simply by composition. Thus if $Aut[N]$ denotes the group of automorphisms of N, then the fact that each member of H is understood as an automorphism of N is equivalent to saying that there is a map

$$\sigma : H \longrightarrow Aut[N].$$

This map is in fact a homomorphism from H into $Aut[N]$.

Let us illustrate how this works with our example of D_4. Here we are creating the splitting extension:

$$D_4 \;=\; \mathbb{Z}_4 \;ⓢ\; \mathbb{Z}_2$$

which means that $\mathbb{Z}_2$ acts as an automorphism group of $\mathbb{Z}_4$. In other words, there is a map

$$\sigma : \mathbb{Z}_2 \longrightarrow Aut[\mathbb{Z}_4].$$

This map is easy to understand. The identity element of $\mathbb{Z}_2$ is sent to the identity element of $Aut[\mathbb{Z}_4]$, that is, the element that has no effect on $\mathbb{Z}_4$. The other element of $\mathbb{Z}_2$ is sent to the automorphism of $\mathbb{Z}_4$ which makes a vertically reflected copy of $\mathbb{Z}_4$. That is, $\mathbb{Z}_4$ can be defined as the group of *clockwise* rotations. The automorphism creates the reflected version of this; i.e., it sends the clockwise version of $\mathbb{Z}_4$ onto the counter-clockwise version. This automorphism is shown in Fig. A.3. In this figure, the four elements down the left side comprise $\mathbb{Z}_4$. Similarly, the four elements down the right side comprise $\mathbb{Z}_4$. The arrows therefore map $\mathbb{Z}_4$ to itself. This map sends every clockwise rotation on the left to its reflected version on the right. For example, r_{90} is sent to r_{270}, which is r_{90} in the reverse direction.

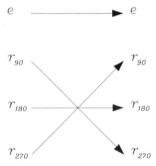

Fig. A.3. Reflection acting as an automorphism on $\mathbb{Z}_4$.

Now let us return to the general case of a splitting extension of N by H. It was said that, associated with the splitting extension, there is a homomorphism of the form

$$\sigma : H \longrightarrow Aut[N].$$

This *represents* H within $Aut[N]$. There are of course several alternative *representations* that H can have within $Aut[N]$, that is, several alternative maps σ. Each one of these different representations of H can be chosen to produce a different splitting extension of N. That is, the extension, which is being created, depends on the particular representation σ of H that is chosen. Thus, it is necessary to indicate which representation σ was chosen in the extension. This is shown thus:

$$G = N \circledS_\sigma H.$$

A.4   Multiplication in a Semi-direct Product

It is now necessary to see how group multiplication is defined in a group G that has been created as a semi-direct product $N \circledS_\sigma H$.

Let any element from the group $N \circledS_\sigma H$ be written like this:

$$\langle\, n \mid h \,\rangle$$

where n is from the group N, and h is from the group H. It is necessary to specify what happens when *two* elements $\langle\, n_1 \mid h_1 \,\rangle$ and $\langle\, n_2 \mid h_2 \,\rangle$ are multiplied together, thus:

$$\langle\, n_1 \mid h_1 \,\rangle \circ \langle\, n_2 \mid h_2 \,\rangle.$$

Obviously, the simplest method would be to do componentwise multiplication thus:

$$\langle\, n_1 n_2 \mid h_1 h_2 \,\rangle. \tag{A.7}$$

However, in a semi-direct product, an ammendment has to be made to the first element n_1 in this expression. Instead of using n_1, one uses the automorphic version of n_1 supplied by h_2, as follows: Recall that h_2 acts as an automorphism of N. Its automorphism is called $\sigma(h_2)$, using the map $\sigma : H \longrightarrow Aut[N]$. Thus one supplies $\sigma(h_2)[n_1]$ instead of n_1 to the first position in the product. That is, instead of (A.7), one gets

$$\langle\, \sigma(h_2)[n_1]\, n_2 \mid h_1 h_2 \,\rangle.$$

The reader can see therefore that multiplication of group elements in a semi-direct product $G = N \circledS_\sigma H$ uses the particular way in which H is represented as an automorphism group of N.

A.5   Direct Products

As said above, each splitting extension is determined by a choice of representation $\sigma : H \longrightarrow Aut[N]$. Now let us consider the simplest example of a representation. Let all of H be mapped to the identity element $\mathbb{I}_d$ of $Aut[N]$. Clearly $\mathbb{I}_d$ is the automorphism that does nothing to N.

This particular choice of σ produces the simplest possible multiplication structure within $G = N \circledS_\sigma H$. Recall that generally, group multiplication $\langle\, n_1 \mid h_1 \,\rangle \circ \langle\, n_2 \mid h_2 \,\rangle$ is defined thus:

$$\langle\, \sigma(h_2)[n_1]\, n_2 \mid h_1 h_2 \,\rangle.$$

However, in the particular choice we have now made for σ, the automorphism $\sigma(h_2)$ is simply the identity map $\mathbb{I}_d$. This means that multiplication becomes this:

$$\langle\, n_1 n_2 \mid h_1 h_2 \,\rangle.$$

Thus, the multiplication $\langle\, n_1 \mid h_1 \,\rangle \circ \langle\, n_2 \mid h_2 \,\rangle$ has been carried out *componentwise*. This gives the simplest of all types of splitting extension, called a *direct product*. It is written thus:

$$G \quad = \quad N \times H.$$

In a direct product, *both* the subgroups N and H are normal subgroups of G.

As an example, return to the case of D_4 as a semi-direct product

$$D_4 \quad = \quad \mathbb{Z}_4 \circledS_\sigma \mathbb{Z}_2.$$

We saw that, here, the map $\sigma : \mathbb{Z}_2 \longrightarrow Aut[\mathbb{Z}_4]$ was not trivial. In particular, one of the members of $\mathbb{Z}_2$ is not sent to $\mathbb{I}_d$, but to the reflection automorphism in $Aut[\mathbb{Z}_4]$. However, another splitting extension could be created in which $\sigma : \mathbb{Z}_2 \longrightarrow Aut[\mathbb{Z}_4]$ sends all of $\mathbb{Z}_2$ to the identity element $\mathbb{I}_d$ of $Aut[\mathbb{Z}_4]$. In this case, one would get the direct product

$$\mathbb{Z}_4 \times \mathbb{Z}_2.$$

This is *not* the group D_4. Thus we have constructed two different groups, D_4 and $\mathbb{Z}_4 \times \mathbb{Z}_2$. Both were created as splitting extensions of $\mathbb{Z}_4$ and $\mathbb{Z}_2$. But they used different maps $\sigma : \mathbb{Z}_2 \longrightarrow Aut[\mathbb{Z}_4]$ and therefore lead to different resulting groups.

B. Symbols

$\triangleleft$ normal subgroup, p. 531

C control set p. 82

D_4 the dihedral symmetry group a square, p. 533

D_n the dihedral symmetry group of a regular n-sided polygon.

Det_i the detected symmetry group, p. 121

F fiber set p. 81

$G(C)$ control group p. 82

$G(F)$ fiber group p. 81

$HWH(n)$, hyperoctahedral wreath hyperplane group degree n p. 398-409

$\mathfrak{R}$, a group generated by a chosen set of compositional operators, p. 480

$RS\text{-}HWH(n)$, recursive solid wreath hyperoctahedral wreath hyperplane group, p. 408

$\circledS$ semi-direct product, p. 531-538

$\circledS_\tau$ semi-direct product via the τ-representation p. 91

$SE(3)$, the special Euclidean group on 3-space

$S\text{-}HWH(n)$, solid hyperoctahedral wreath hyperplane group, p. 405

$SO(2)$, the continuous group of rotations of the plane

$SO(3)$, the continuous group of rotations of the 3-space

$\tau = $ transfer p. 89

$\circledW$ wreath product p. 4, p. 91

$\mathbb{Z}_4$ the cyclic group of order 4 represented as the group of 90^0 rotations, p. 7

$\mathbb{Z}_n$ the cyclic group of order n

References

1. Abdel-Malek, K. & Othman, S. (1999). Multiple sweeping using the Denavit-Hartenberg representation method. *Computer-Aided Design*, **31(9)**, 567-583.
2. Arbib, M. (1968). (Editor). *Algebraic Theory of Machines, Languages, and Semigroups*. New York: Academic Press.
3. Ascher, E. & Janner, A. (1965). Algebraic aspects of crystallography. *Helvetica Physica Acta*, **38(6)**, 551-572.
4. Ascher, E. & Janner, A. (1968). Algebraic aspects of crystallography, II. Non-primitive translations in space groups. *Commun. math. Phys.*, **11**, 138-168.
5. Attneave, F. (1968). Triangles as ambiguous figures. *American Journal of Psychology*, **18**, 447-453.
6. Attneave, F. (1971). Multistability in perception. *Scientific American*, **225**, 62-71.
7. Au, C. K.& Yuen, M. M. F. (1999). Feature-based reverse engineering of mannequin for garment design. *Computer-Aided Design*, **31(12)**, 751-759.
8. Au, C. K.& Yuen, M. M. F. (2000). A semantic feature language for sculptured object modelling. *Computer-Aided Design*, **32(1)**, 63-74.
9. Baass, K. G. (1984). The use of chlothoid techniques in highway design. *Transportation Forum*, **30(3)**, 47-52.
10. Babbit, M. (1961). Set structure as a compositional determinant. *Journal of Music Theory*, **5**, 72-94.
11. Blaschek, G. (1994). *Object-Oriented Programming with Prototypes*. Berlin: Springer-Verlag.
12. Blum, H. (1973). Biological shape and visual science. *Journal of Theoretical Biology*, **38**, 205-287.
13. Bouma, W., Chen, X., Fudos, I., Hoffmann, C., & Vermeer, P. J. (1999). *An Electronic Primer on Geometric Constraint Solving*, http://www.cs.purdue.edu/homes/cmh/electrobook/intro/html.
14. Brady, J. M. (1983). Criteria for representations of shape. In: A.Rosenfeld & J. Beck, editors. *Human and Machine Vision: Vol 1*. Hillsdale, NJ: Erlbaum.
15. Brassard, L. (1998). *The Perception of the Image World*. PhD Thesis, Simon Fraser University.
16. Brockett, R. (1990). Some mathematical aspects of robotics. In *Robotics*. AMS Short Course Lecture Notes, Vol 41. Providence: American Mathematical Society.
17. Carbonell, J. G. (1986). Derivational analogy: A theory of reconstructive problem solving and expertise acquisition. In R. S.Michalski, J. G.Carbonell, & T. M.Mitchell (Ed.), *Machine Learning: An Artificial Intelligence Approach. Vol 2*. Los Altos, CA: Morgan-Kaufman.
18. Carlton, E. & Shepard, R. N. (1990a). Psychologically simple motions as geodesic paths. *Journal of Mathematical Psychology*, **34**, 127-188.

19. Carlton, E. & Shepard, R. N. (1990b). Psychologically simple motions as geodesic paths. *Journal of Mathematical Psychology*, **34**, 189-128.

20. Chen, X., & Hoffmann, C. M. (1995). On editability of feature-based design. *Computer-Aided Design*, **27(12)**, 905-914.

21. Chevallier, D. P. (1991). Lie algebras, modules, dual quaternions and algebraic methods in kinematics. *Mech. Mach. Theory*, **26(6)**, 613-627.

22. Choi, B. K., & Lee, C. S. (1990). Sweep surfaces modelling via coordinate transformations and blending. *Computer-Aided Design*, **22**, 87-96.

23. Choi, B. K., Barash, M., & Anderson, D. (1984). Automatic recognition of machined surfaces from a 3-D solid model. *Computer-Aided Design*, **16(2)**, 81-86.

24. Collins, J. J.& Stewart, I. N. (1993). Coupled nonlinear oscillators and the symmetries of animal gaits. *Journal of Nonlinear Science*, **3**, 349-392.

25. Cunningham, J. J., & Dixon, J. R. (1988). Designing with features: the origin of features. *ASME computers in engineering,*San Fransisco, USA, July/August, p237-43.

26. Denavit, J., & Hartenberg, R. S. (1955). A kinematic notation for lower-pair mechanisms based on matrices. *Journal of Applied Mechanics*, ASME, **22**, 215-221.

27. Dias, A. P. (1998). Hopf bifurcation for wreath products. *Nonlinearity*, **11**, 247-264.

28. Dionne, B., Golubitsky, M. & Stewart, I. (1996). Coupled cells with internal symmetry. Part I: wreath products. *Nonlinearity*, **9**, 5559-574.

29. Duncker, K. (1929). Uber induzierte Bewegung. *Psychologische Forschung*, **12**, 180-259. Translated and condensed in W. D.Ellis (Ed.), *A Source Book of Gestalt Psychology*. New York: Harcourt Brace.

30. Ellis, J., Kedem, G., Lyerly, T., Thielman, D., Marisa, R., Menon, J., & Voelcker, H. (1991). The ray-casting engine and ray representations: a technical summary. *International Journal of Computational Geometry and Applications.*, **1(4)**, 347-80.

31. Elliot, S., Miller, P., Abouaf, J., Espinosa-Aguilar, D., Alexander, S., Barnard, D., Kakert, P., Kalwick, D., Kelm, K., Koch, M., Williamson, M. (1998). *Inside 3D Studio Max 2, Volume I.*. Indianapolis, IN: New Riders Publishing.

32. Falcidieno, B., & Giannini, F. (1987). In: Marechaled, editor. *Extraction and organization of form features into a structured boundary model.* EUROGRAPHICS'87. Amsterdam, Elsevier. p249-259.

33. Falcidieno, B., & Giannini, F. (1989). Automatic recognition and representation of shape based feature in a geometric modeling system. *Computer Vision, Graphics and Image Processing*, **48**, 93-123.

34. Farin, G. (1993). *Curves and Surfaces in Computer Aided Design: A Practical Guide*. San Diego, CA: Academic Press.

35. Feldman, J. (1997). Curvilinearity, covariance, and regularity in perceptual groups. *Vision Research*, **37(20)**, 2835-2848.

36. Foote, R., Mirchandani, G., Rockmore, D., Healy, D. & Olson, T. (2000). A wreath product group approach to signal processing. *IEEE Transactions in Signal Processing*, **48(1)**, 102-132.

37. Gallistel, R. C. (1990). *The Organization of Learning*. Cambridge, Mass: MIT Press.

38. Gallistel, R. C. (1999). Coordinate transformations in the genesis of directed action. In B. O. M.Bly & D. E.Rummelhart, *Cognitive Science*, New York: Academic pp1-42.

39. Ge, Q. J.& Ravani, B. (1995). Geometric construction of Bézier motions. *ASME Journal of Mech. Design*, **116**, 749-755.

40. Gentner, D. (1983). Structure-mapping: A theoretical framework for analogy. *Cognitive Science*, **7**, 155-170.
41. Giblin, P. J., & Kimia, B. B. (1999). On the local form and transitions of symmetry sets, and medial axes, and shocks in 2D. *Proc. of ICCV*, 385-391, Greece.
42. Giblin, P. J., & Kimia, B. B. (2000). On the local form and transitions of symmetry sets, and medial axes, and shocks in 3D. *Proc. of CVPR*, IEEE Computer Society.
43. Giblin, P. J., & Sapiro, G. (1998). Affine-invariant distances, envelopes and symmetry sets. *Geometriae Dedicata,* **71**, 237-261.
44. Gips, J. & Stiny, G. (1980). Production systems and grammars: a uniform characterization. *Environment and Planning B,* **7**, 399-408.
45. Goldmeier, E. (1936/1972). Similarity in visually perceived forms. English translation. *Psychological Issues,* 1972, **8 no. 1**, Monograph 29, New York: International Press. Cambridge, Mass: MIT Press.
46. Golubitsky, M., Stewart, I., Buono P-L. & Collins, J. J. (1998). A modular network for legged locomotion. *Physica D*, **115**, 56-72.
47. Greeno, J. G.& Simon, H. A. (1974). Processes for sequence production. *Psychological Review*, **74**, 187-198.
48. Han, J., Regli, W. C., & Brooks, S. (1998). Hint-based reasoning for feature recognition: status report. *Computer-Aided Design*, **30(13)**, 1003-1007.
49. Han, J., & Requicha, A. A. G. (1995). Integration of feature based design and feature recognition. In: ASME International Computers in Engineering Conference, Boston.
50. Hartman, P. (1957). The highway spiral for combining curves of different radii. *Transactions of the American Society of Civil Engineers*, **122**, 389-409.
51. Hartquist, E. E. (1998). Processing ray-representations in parallel: options, algorithms and results. Technical Report CPA98-4, Sibley School of Mechanical Engineering, Cornell University.
52. Hartquist, E. E., Menon, J. P., Suresh, K., Voelcker, H. B., & Zagajac, J. (1999). A computing strategy for applications involving offsets, sweeps, and Minkowski operations. *Computer-Aided Design*, **31(3)**, 175-183.
53. Hassenstein, B. & Reichardt, W. (1956). Systemtheoretische analyse der zeitreihenfolgenund vorzeichenauswertung bei der bewgungsperzeption der russelkafers. *Chlorophanus. Z. Naturf.,* , **IIb**, 513-524.
54. Hayes, P. J. (1985). Naive physics I: Ontology for liquids. In Hobbs, J. R.& Moore, R. C. (Editors). *Formal Theories of a Commonsense World*. Norwood, NJ: Ablex.
55. Hayes, P. J., & Leyton, M. (1989) Processes at discontinuities. *International Joint Conference on Artificial Intelligence*. Detroit. p1267-1272.
56. Henderson, M. R., & Anderson, D. C. (1984). Computer recognition and extraction of form features: a CAD/CAM link. *Computers in Industry*, **5**, 329-339.
57. Heo, H-S., Kim, M-S, & Elber, G. (1999). The intersection of two ruled surfaces. *Computer-Aided Design*, **31(1)**, 33-50.
58. Hoffmann, C. M. (1989). *Geometric and Solid Modeling*. San Mateo, California: Morgan Kaufmann.
59. Hoffmann, C. M., & Juan, R. (1993). Erep - an editable, high-level representation for geometric design and analysis. In: Wilson, P. R., Wozny, M. J.& Pratt, M. J., editors. Elsevier, Amsterdam. p129-164.
60. Horn, B. (1986). *Robot Vision*. Cambridge, Mass: MIT Press.
61. Hoschek, J. & Lasser, D. (1993). *Fundamentals of Computer Aided Geometric Design*. (Translated by L. L.Schumaker.) Wellesley, Mass: A. K.Peters.

62. Ishida, Y. (2000) Symbol system for symmetry of physical systems. *Computers and Mathematics with Applications,* **39**, 221-240.
63. Ishida, Y. & Kotovsky, K. (1995) Symmetry analysis on symmetry cognition on multilevel figures. *Computers and Mathematics with Applications,* **30**, 93-102.
64. Jablan, S. V., (1995). *Theory of Symmetry and Ornament.* Mathematics Institute, Belgrade. Can be downloaded from the web at http://www.emis.de/monographs/jablan/index.html.
65. Johansson, G. (1950). *Configurations in event perception.* Stockholm, Sweden: Almqvist and Wiksell.
66. Jones, M. R. (1974). Cognitive representations of serial patterns. In B. Kantowitz (Editor). *Human information processing: Tutorials in performance cognition.* Potamac, MD: Erlbaum.
67. Jones, M. R. (1981). A tutorial on some issues and methods in serial pattern search. *Perception & Psychophysics,* **30(5)**, 492-504.
68. Joshi, S. & Chang, T. C. (1988). Graph-based heuristics for recognition of machined features from a 3D solid model. *Computer-Aided Design,* **20(2)**, 58-66.
69. Jüttler, B. (1994). Visualization of moving objects using dual quaternion curves. *Computer Graphics,* **18(3)**, 315-326.
70. Jüttler, B., & Wagner, M. G. (1996). Computer aided design with spatial rational B-spline motions. *ASME Journal of Mech. Design,* **118**, 193-201.
71. Jüttler, B., & Wagner, M. G. (1999). Rational motion-based surface generation. *Computer-Aided Design,* **31(3)**, 203-213.
72. Kanade, T. (1981). Recovery of the three-dimensional shape of an object from a single view. In J. M.Brady (Ed.), *Computer Vision.* New York: North Holland.
73. Karger, A. & Novak, J. (1985). *Space Kinematics and Lie Groups.* London: Gordon and Breach.
74. Kim, M. J., Kim, M. S.& Shin, S. Y. (1995). A C2-continuous B-spline quaternion curve interpolating a given sequence of solid orientations. *Proceedings of Computer Animation '95,* Geneva, Switzerland 72-81.
75. Kim, Y. S. (1992). Recognition of form features using convex decomposition. *Computer-Aided Design,* **24(9)**, 461-76.
76. Kimia, B. B., Tannenbaum, A. R., & Zucker, S. W. (1995). Shapes, shocks, and deformations, I. *International Journal of Computer Vision,* **15**, 189-224.
77. Kirupaharan, N. & Dayawansa, W. P. (2001). Theory of reference frames and biological control. *Mathematical and Computer Modeling,* **33(1-3)**, Jan-Feb, 193-198.
78. Kotovsky, K. & Simon, H. A. (1973). Empirical tests of a theory of human acquisition of concepts for sequential events. *Cognitive Psychology,* 4, 399-424.
79. Krohn, K. & Rhodes, J. (1965). Algebraic theory of machines, I. Prime decomposition theorem for finite semigroups and machines. *Transactions of the American Mathematical Society,* **116**, 450-464.
80. Kumar, V., Burns, D., Dutta, D., & Hoffmann, C. (1999). A framework for object-modeling. *Computer-Aided Design,* **31(9)**, 541-556.
81. Kyprianu, L. K. (1980). *Shape classification in computer-aided design.* PhD thesis, Univesity of Cambridge, UK.
82. Lee, J. Y., & Kim, K. (1998). A feature-based approach to extracting machining features. *Computer-Aided Design,* **30(13)**, 1019-1035.
83. Lenz, R. (1990). *Group Theoretical Methods in Image Processing.* Berlin: Springer Verlag.

84. Leymarie, F., & Kimia, B. B. (2000). Discrete 3D wave propagation for morphology and skeletons from unorganized samples. In J. Goutsias, L.Vincent, and D. Bloomberg (Editors). *Math. Morphology and its Applications to Image and Signal Processing*, Vol 18 of *Computational Imaging and Vision*, p351-360, Palo Alto, CA: Kluwer.

85. Leymarie, F., & Levine, M. D. (1992). Simulating the grassfire transform using an active contour model. *IEEE Trans. on PAMI*, **14(1)**, 56-75.

86. Leyton, M., (1974). *Mathematical-logical postulates at the foundations of art.* Tech Report, Mathematics Department, University of Warwick.

87. Leyton, M. (1984). Perceptual organization as nested control. *Biological Cybernetics,* **51**, 141-153.

88. Leyton, M. (1986a). Principles of information structure common to six levels of the human cognitive system. *Information Sciences,* **38**, 1-120. Entire journal issue.

89. Leyton, M. (1986b). A theory of information structure I: General principles. *Journal of Mathematical Psychology,* **30**, 103-160.

90. Leyton, M. (1986c). A theory of information structure II: A theory of perceptual organization *Journal of Mathematical Psychology,* **30**, 257-305.

91. Leyton, M. (1987a). Nested structures of control: An intuitive view. *Computer Vision, Graphics, and Image Processing,* **37**, 20-53.

92. Leyton, M. (1987b). Symmetry-curvature duality. *Computer Vision, Graphics, and Image Processing,* **38**, 327-341.

93. Leyton, M. (1987d). A Limitation Theorem for the Differential Prototypification of Shape. *Journal of Mathematical Psychology,* **31**, 307-320.

94. Leyton, M. (1988). A Process-Grammar for Shape. *Artificial Intelligence,* **34**, 213-247.

95. Leyton, M. (1989). Inferring Causal-History from Shape. *Cognitive Science,* **13**, 357-387.

96. Leyton, M. (1992). *Symmetry, Causality, Mind.* Cambridge, Mass: MIT Press.

97. Leyton, M. (1999). New foundations for perception. In Lepore, E. (Editor). *Invitation to Cognitive Science.* Blackwell, Oxford. p121 - 171.

98. Leyton, M. (2001). *Group Extensions.* Book submitted.

99. Liu, Y. (1990). *Symmetry Groups in Robotic Assembly Planning.* PhD thesis, Univesity of Massachusetts, Amherst, MA.

100. Liu, Y. & Collins, R. (2000). A computational model for repeated pattern perception using frieze and wallpaper groups. *Computer Vision and Pattern Recognition Conference,* CVPR'2000.

101. Liu, Y. & Popplestone, R. (1994). A group theoretical formalization of surface contact. *International Journal of Robotics Research,* **13(2)**, 148-161.

102. Mach, E. (1897). *The Analysis of Sensations.* English Translation, 1959. New York: Dover.

103. Marsella, S. & Schmidt, C. F. (1992). On the application of problem reduction search to automated composition. In M. Balaban, K. Ebcioglu, & O. Laske (Eds.). *Understanding Music with AI: Perspectives on Music Cognition.* Cambridge, MA: MIT Press/AAAI Press. p238-256.

104. Mazzola, G. (1990). *Geometrie der Töne - Elemente der Mathematischen Musiktheorie.* Basel: Birkhäuser.

105. Mazzola, G. (1993-1996). *Geometry and Logic of Musical Performance I-III.* Reports for Schweiz. Nationalfonds, Univ. Zürich.

106. Mazzola, G. (2002). *The Topos of Music.* Basel: Birkhäuser.

107. Menon, J. P., Marisa, R., & Zagajac, J. (1994). More powerful solid modeling through ray representations. *IEEE Computer Graphics and Applications,* **14(3)**, 22-35.

108. Meyer, B. (1997). *Object-Oriented Software Construction*. New Jersey: Prentice Hall.

109. Miura, J. & Ikeuchi, K. (1998). Task-Oriented Generation of Visual Sensing Strategies in Assembly Tasks. *IEEE Transactions on Pattern Analysis and Machine Intelligence*, **20(2)**, 126-37.

110. Mortenson, M., (1997). *Geometric Modeling*. New York: Wiley.

111. Murakami, T. & Nakajima, N. (2000). DO-IT: deformable object as input tool for 3-D geometric operation. *Computer-Aided Design*, **32(1)**, 5-16.

112. Murray, R. M., Li, Z. & Shastry, S. S. (1993). *A Mathematical Introduction to Robotic Manipulation*. Boca Raton, FL: CRC Press.

113. Noll, T. (1995). Fractal Depth Structure of Tonal Harmony. In R. Bidlack (Editor). *ICMC-Proceedings*. Banff: The Banff Centre for the Arts.

114. Noll, T. (1997). *Morphologische Grundlagen der abendländischen Harmonik*. Musikometrika 7, Bochum: Brockmeyer.

115. Park, F. C., Bobrow, J. E.& Ploen, S. R. (1995). A Lie group formulation of robot dynamics. *International Journal of Robotics Research*, **14(6)**, 609-618.

116. Park, F. & Ravani, B. (1995). Bézier curves on Riemannian manifolds and Lie groups with kinematic applications. *Proceedings of the 1994 ASME Mechanics Conference*, **70(1)**, 15-20.

117. Paul, R. P. (1981). *Robot manipulators*. Cambridge, Mass: MIT Press.

118. Peternell, M., Pottmann, H., & Ravani, B. (1999). On the computational geometry of ruled surfaces. *Computer-Aided Design*, **31(1)**, 17-32.

119. Pletinckx, D. (1989). Quaternion calculus as a basic tool in computer graphics. *Visual Comput.*, **5**, 2-13.

120. Popplestone, R., Liu, Y. & Weiss, R. (1990). A group theoretic approach to assembly planning. *Artificial Intelligence Magazine*, **11(1)**, 82-97.

121. Pottmann, H., Peternell, M., & Ravani, B. (1999). An introduction to line geometry with applications. *Computer-Aided Design*, **31(1)**, 3-16.

122. Regli, W. C. (1995). *Geometric algorithms for recognition of features from solid models*. PhD thesis, Univesity of Maryland.

123. Reichardt, W. (1961). Autocorrelation, a principle for the evaluation of sensory information by the central nervous system. In W. A.Rosenblith (Ed.). *Sensory Communication*. Cambridge, Mass: MIT Press.

124. Requicha, A. A.G., & Rossignac, J. R. (1992). Solid modeling and beyond. *IEEE Computer Graphics and Applications*, **12(5)**, 31-44.

125. Restle, F. (1970). Theory of serial pattern learning: Structural trees. *Psychological Review*, **77**, 481-495.

126. Restle, F. & Brown, E. R. (1996). Organization of serial pattern learning. In G. H.Bower (Editor) (eds.) *The psychology of learning and motivation: Advances in research and theory (Vol 4)*.. New York: Academic Press.

127. Richards, W. Jepson, A. & Feldman, J. (1996). Priors, preferences, and categorical percepts. In D. Knill & W. Richards (Editors.) *Perception as Bayesian Inference*. New York: Cambridge University Press

128. Rock, I. (1975). *An Introduction to Perception*. New York: Macmillan.

129. Röschel, O. (1998). Rational motion design - a survey. *Computer-Aided Design*, **30(3)**, 169-178.

130. Rosch, E. (1975). Cognitive reference points. *Cognitive Psychology*, **7**, 532-547.

131. Rosch, E. (1977). Human categorization. In N. Warren (Ed), *Studies in Cross-Cultural Psychology, Vol 1*. London: Academic Press.

132. Rosch, E. (1978). Principles of categorization. In E. Rosch & B. B.Lloyd (Eds), *Cognition and Categorization*. Hillsdale, NJ: Lawrence Erlbaum.

133. Rossignac, J. R. (1990). Issues on feature based editing and interrogation of solid models. *Computer and Graphics*, **14(2)**, 149-172.

134. Schmidt, C. F. *Computation and Cognition*. Book on web. http://www-rci.rutgers.edu/ ⌣cfs/472_html/home472.html

135. Schmitt, P. R. (1996). *Reactive path shaping: local path planning for autonomous robots in aisles*. MSc thesis, Georgia Institute of Technology, GA.

136. Schoemake, K. (1985). Animating rotation with quaternion curves. *ACM Siggraph*, **19(3)**, 245-254.

137. Schönberg, A. (1967) *Fundamentals of Musical Composition*. London and Boston: Faber & Faber.

138. Schwarzenberger, R. (1980). *N-Dimensional Crystallography*. London: Pitman.

139. Sederberg, T., & Parry, S. (1986). Free-form deformation of solid geometric models. *Computer Graphics (SIGGRAPH '86 Proceedings)*, **20(4)**, 151-160.

140. Segal, D. (1983). *Polycyclic groups*. Cambridge: Cambridge University Press.

141. Shah, J. J. (1991). Conceptual development of form features and feature modelers. *Research in Engineering Design, 2*, 93-108.

142. Shepard, R. N. (2001). Perceptual-cognitive universals as reflections of the world *Behavioral and Brain Sciences*, **24 (3)**. Reprinted from *Psychonomic Bulletin & Review*, 1994, **1**, 2-28

143. Shubnikov, A. & Koptsik, V. A. (1974). *Symmetry in Art and Science*. New York: Plenum Press.

144. Siddiqi, K., Shokoufandeh, A., Dickinson, S., & Zucker, S. W. (1999). Shock graphs and shape matching. *International Journal of Computer Vision*, **35(1)**, 13-32.

145. Simon, H. A.& Kotovsky, K. (1963). Human acquisition of concepts for sequential patterns. *Psychological Review*, **70**, 534-546.

146. Sims, C. C. (1994). *Computation with Finitely Presented Groups*. Cambridge: Cambridge University Press.

147. Stiny, G. & Gips, J. (1978). *Algorithmic Aesthetics*. Berkeley, CA: University of California Press.

148. Toogood, R., (1999). *Pro/Engineer Tutorial*. IN: SDC Publications.

149. Vandenbrande, J. H. (1990). *Automatic recognition of machinable features in solid models*. PhD thesis, Univesity of Rochester, Rochester, NY.

150. Vandenbrande, J. H.& Requicha, A. A.G., (1993). Spatial reasoning for automatic recognition of machinable feature in solid models. *IEEE Transactions on Pattern Analysis and Machine Intelligence*, **15(12)**, 1-17.

151. Vishwanath, D. (2001). *The epistemological status of computer vision*. Tech report. Center for Cognitive Science. Rutgers University.

152. Voelcker, H. B.& Requicha, A. A.G., (1993). Research in solid modeling at the University of Rochester: 1972-1987, Fundamental Developments of Computer-Aided Geometric Modeling. In: Piegl, L., (editor). London: Academic Press Ltd., pp. 203-54.

153. Walton, D. J.& Meek, D. S. (1998). Planar G^2 curve design with spiral segments. *Computer-Aided Design*, **30(7)**, 529-538.

154. Walton, D. J.& Meek, D. S. (1999). Planar G^2 transition between two circles with a fair cubic Bézier curve. *Computer-Aided Design*, **31(14)**, 857-866.

155. Washburn, D. & Crowe, D. (1988). *Symmetries of Culture*. Seattle: Univ of Washington Press.

156. Witkin, A. P.& Tenenbaum, J. M. (1983). On the role of structure in vision. In A. Rosenfeld & J. Beck (Eds), *Human and Machine Vision, Vol 1*. Hillsdale, NJ: Erlbaum.

157. Zagajac, J. (1997). *Engineering analysis over subsets.* PhD thesis, Cornell University, NY.
158. Zefran, M. & Kumar, V. (1998). Interpolation schemes for rigid body motions. *Computer-Aided Design,* **30(3)**, 179-189.

Index

Lecture Notes in Computer Science

For information about Vols. 1–2118
please contact your bookseller or Springer-Verlag

Vol. 1962: Y. Frankel (Ed.), Financial Cryptography. Proceedings, 2000. XI, 379 pages. 2001.

Vol. 2119: V. Varadharajan, Y. Mu (Eds.), Information Security and Privacy. Proceedings, 2001. XI, 522 pages. 2001.

Vol. 2120: H.S. Delugach, G. Stumme (Eds.), Conceptual Structures: Broadening the Base. Proceedings, 2001. X, 377 pages. 2001. (Subseries LNAI).

Vol. 2121: C.S. Jensen, M. Schneider, B. Seeger, V.J. Tsotras (Eds.), Advances in Spatial and Temporal Databases. Proceedings, 2001. XI, 543 pages. 2001.

Vol. 2123: P. Perner (Ed.), Machine Learning and Data Mining in Pattern Recognition. Proceedings, 2001. XI, 363 pages. 2001. (Subseries LNAI).

Vol. 2124: W. Skarbek (Ed.), Computer Analysis of Images and Patterns. Proceedings, 2001. XV, 743 pages. 2001.

Vol. 2125: F. Dehne, J.-R. Sack, R. Tamassia (Eds.), Algorithms and Data Structures. Proceedings, 2001. XII, 484 pages. 2001.

Vol. 2126: P. Cousot (Ed.), Static Analysis. Proceedings, 2001. XI, 439 pages. 2001.

Vol. 2127: V. Malyshkin (Ed.), Parallel Computing Technologies. Proceedings, 2001. XII, 516 pages. 2001.

Vol. 2129: M. Goemans, K. Jansen, J.D.P. Rolim, L. Trevisan (Eds.), Approximation, Randomization, and Combinatorial Optimization. Proceedings, 2001. IX, 297 pages. 2001.

Vol. 2130: G. Dorffner, H. Bischof, K. Hornik (Eds.), Artificial Neural Networks – ICANN 2001. Proceedings, 2001. XXII, 1259 pages. 2001.

Vol. 2131: Y. Cotronis, J. Dongarra (Eds.), Recent Advances in Parallel Virtual Machine and Message Passing Interface. Proceedings, 2001. XV, 438 pages. 2001.

Vol. 2132: S.-T. Yuan, M. Yokoo (Eds.), Intelligent Agents. Specification. Modeling, and Application. Proceedings, 2001. X, 237 pages. 2001. (Subseries LNAI).

Vol. 2133: B. Christianson, B. Crispo, J.A. Malcolm, M. Roe (Eds.), Security Protocols. Proceedings, 2001. VIII, 257 pages. 2001.

Vol. 2134: M. Figueiredo, J. Zerubia, A.K. Jain (Eds.), Energy Minimization Methods in Computer Vision and Pattern Recognition. Proceedings, 2001. X, 652 pages. 2001.

Vol. 2136: J. Sgall, A. Pultr, P. Kolman (Eds.), Mathematical Foundations of Computer Science 2001. Proceedings, 2001. XII, 716 pages. 2001.

Vol. 2138: R. Freivalds (Ed.), Fundamentals of Computation Theory. Proceedings, 2001. XIII, 542 pages. 2001.

Vol. 2139: J. Kilian (Ed.), Advances in Cryptology – CRYPTO 2001. Proceedings, 2001. XI, 599 pages. 2001.

Vol. 2140: I. Attali, T. Jensen (Eds.), Java on Smart Cards: Programming and Security. Proceedings, 2001. VIII, 255 pages. 2001.

Vol. 2141: G.S. Brodal, D. Frigioni, A. Marchetti-Spaccamela (Eds.), Algorithm Engineering. Proceedings, 2001. X, 199 pages. 2001.

Vol. 2142: L. Fribourg (Ed.), Computer Science Logic. Proceedings, 2001. XII, 615 pages. 2001.

Vol. 2143: S. Benferhat, P. Besnard (Eds.), Symbolic and Quantitative Approaches to Reasoning with Uncertainty. Proceedings, 2001. XIV, 818 pages. 2001. (Subseries LNAI).

Vol. 2144: T. Margaria, T. Melham (Eds.), Correct Hardware Design and Verification Methods. Proceedings, 2001. XII, 482 pages. 2001.

Vol. 2145: M. Leyton, A Generative Theory of Shape. XVI, 554 pages. 2001.

Vol. 2146: J.H. Silverman (Eds.), Cryptography and Lattices. Proceedings, 2001. VII, 219 pages. 2001.

Vol. 2147: G. Brebner, R. Woods (Eds.), Field-Programmable Logic and Applications. Proceedings, 2001. XV, 665 pages. 2001.

Vol. 2149: O. Gascuel, B.M.E. Moret (Eds.), Algorithms in Bioinformatics. Proceedings, 2001. X, 307 pages. 2001.

Vol. 2150: R. Sakellariou, J. Keane, J. Gurd, L. Freeman (Eds.), Euro-Par 2001 Parallel Processing. Proceedings, 2001. XXX, 943 pages. 2001.

Vol. 2151: A. Caplinskas, J. Eder (Eds.), Advances in Databases and Information Systems. Proceedings, 2001. XIII, 381 pages. 2001.

Vol. 2152: R.J. Boulton, P.B. Jackson (Eds.), Theorem Proving in Higher Order Logics. Proceedings, 2001. X, 395 pages. 2001.

Vol. 2153: A.L. Buchsbaum, J. Snoeyink (Eds.), Algorithm Engineering and Experimentation. Proceedings, 2001. VIII, 231 pages. 2001.

Vol. 2154: K.G. Larsen, M. Nielsen (Eds.), CONCUR 2001 – Concurrency Theory. Proceedings, 2001. XI, 583 pages. 2001.

Vol. 2156: M.I. Smirnov, J. Crowcroft, J. Roberts, F. Boavida (Eds.), Quality of Future Internet Services. Proceedings, 2001. XI, 333 pages. 2001.

Vol. 2157: C. Rouveirol, M. Sebag (Eds.), Inductive Logic Programming. Proceedings, 2001. X, 261 pages. 2001. (Subseries LNAI).

Vol. 2158: D. Shepherd, J. Finney, L. Mathy, N. Race (Eds.), Interactive Distributed Multimedia Systems. Proceedings, 2001. XIII, 258 pages. 2001.

Vol. 2159: J. Kelemen, P. Sosík (Eds.), Advances in Artificial Life. Proceedings, 2001. XIX, 724 pages. 2001. (Subseries LNAI).

Vol. 2161: F. Meyer auf der Heide (Ed.), Algorithms – ESA 2001. Proceedings, 2001. XII, 538 pages. 2001.

Vol. 2162: Ç. K. Koç, D. Naccache, C. Paar (Eds.), Cryptographic Hardware and Embedded Systems – CHES 2001. Proceedings, 2001. XIV, 411 pages. 2001.

Vol. 2163: P. Constantopoulos, I.T. Sølvberg (Eds.), Research and Advanced Technology for Digital Libraries. Proceedings, 2001. XII, 462 pages. 2001.

Vol. 2164: S. Pierre, R. Glitho (Eds.), Mobile Agents for Telecommunication Applications. Proceedings, 2001. XI, 292 pages. 2001.

Vol. 2165: L. de Alfaro, S. Gilmore (Eds.), Process Algebra and Probabilistic Methods. Proceedings, 2001. XII, 217 pages. 2001.

Vol. 2166: V. Matoušek, P. Mautner, R. Mouček, K. Taušer (Eds.), Text, Speech and Dialogue. Proceedings, 2001. XIII, 452 pages. 2001. (Subseries LNAI).

Vol. 2167: L. De Raedt, P. Flach (Eds.), Machine Learning: ECML 2001. Proceedings, 2001. XVII, 618 pages. 2001. (Subseries LNAI).

Vol. 2168: L. De Raedt, A. Siebes (Eds.), Principles of Data Mining and Knowledge Discovery. Proceedings, 2001. XVII, 510 pages. 2001. (Subseries LNAI).

Vol. 2170: S. Palazzo (Ed.), Evolutionary Trends of the Internet. Proceedings, 2001. XIII, 722 pages. 2001.

Vol. 2172: C. Batini, F. Giunchiglia, P. Giorgini, M. Mecella (Eds.), Cooperative Information Systems. Proceedings, 2001. XI, 450 pages. 2001.

Vol. 2173: T. Eiter, W. Faber, M. Truszczynski (Eds.), Logic Programming and Nonmonotonic Reasoning. Proceedings, 2001. XI, 444 pages. 2001. (Subseries LNAI).

Vol. 2174: F. Baader, G. Brewka, T. Eiter (Eds.), KI 2001: Advances in Artificial Intelligence. Proceedings, 2001. XIII, 471 pages. 2001. (Subseries LNAI).

Vol. 2175: F. Esposito (Ed.), AI*IA 2001: Advances in Artificial Intelligence. Proceedings, 2001. XII, 396 pages. 2001. (Subseries LNAI).

Vol. 2176: K.-D. Althoff, R.L. Feldmann, W. Müller (Eds.), Advances in Learning Software Organizations. Proceedings, 2001. XI, 241 pages. 2001.

Vol. 2177: G. Butler, S. Jarzabek (Eds.), Generative and Component-Based Software Engineering. Proceedings, 2001. X, 203 pages. 2001.

Vol. 2180: J. Welch (Ed.), Distributed Computing. Proceedings, 2001. X, 343 pages. 2001.

Vol. 2181: C. Y. Westort (Ed.), Digital Earth Moving. Proceedings, 2001. XII, 117 pages. 2001.

Vol. 2182: M. Klusch, F. Zambonelli (Eds.), Cooperative Information Agents V. Proceedings, 2001. XII, 288 pages. 2001. (Subseries LNAI).

Vol. 2184: M. Tucci (Ed.), Multimedia Databases and Image Communication. Proceedings, 2001. X, 225 pages. 2001.

Vol. 2185: M. Gogolla, C. Kobryn (Eds.), «UML» 2001 – The Unified Modeling Language. Proceedings, 2001. XIV, 510 pages. 2001.

Vol. 2186: J. Bosch (Ed.), Generative and Component-Based Software Engineering. Proceedings, 2001. VIII, 177 pages. 2001.

Vol. 2187: U. Voges (Ed.), Computer Safety, Reliability and Security. Proceedings, 2001. XVI, 261 pages. 2001.

Vol. 2188: F. Bomarius, S. Komi-Sirviö (Eds.), Product Focused Software Process Improvement. Proceedings, 2001. XI, 382 pages. 2001.

Vol. 2189: F. Hoffmann, D.J. Hand, N. Adams, D. Fisher, G. Guimaraes (Eds.), Advances in Intelligent Data Analysis. Proceedings, 2001. XII, 384 pages. 2001.

Vol. 2190: A. de Antonio, R. Aylett, D. Ballin (Eds.), Intelligent Virtual Agents. Proceedings, 2001. VIII, 245 pages. 2001. (Subseries LNAI).

Vol. 2191: B. Radig, S. Florczyk (Eds.), Pattern Recognition. Proceedings, 2001. XVI, 452 pages. 2001.

Vol. 2192: A. Yonezawa, S. Matsuoka (Eds.), Metalevel Architectures and Separation of Crosscutting Concerns. Proceedings, 2001. XI, 283 pages. 2001.

Vol. 2193: F. Casati, D. Georgakopoulos, M.-C. Shan (Eds.), Technologies for E-Services. Proceedings, 2001. X, 213 pages. 2001.

Vol. 2194: A.K. Datta, T. Herman (Eds.), Self-Stabilizing Systems. Proceedings, 2001. VII, 229 pages. 2001.

Vol. 2195: H.-Y. Shum, M. Liao, S.-F. Chang (Eds.), Advances in Multimedia Information Processing – PCM 2001. Proceedings, 2001. XIX, 1149 pages. 2001.

Vol. 2196: W. Taha (Ed.), Semantics, Applications, and Implementation of Program Generation. Proceedings, 2001. X, 219 pages. 2001.

Vol. 2197: O. Balet, G. Subsol, P. Torguet (Eds.), Virtual Storytelling. Proceedings, 2001. XI, 213 pages. 2001.

Vol. 2200: G.I. Davida, Y. Frankel (Eds.), Information Security. Proceedings, 2001. XIII, 554 pages. 2001.

Vol. 2201: G.D. Abowd, B. Brumitt, S. Shafer (Eds.), Ubicomp 2001: Ubiquitous Computing. Proceedings, 2001. XIII, 372 pages. 2001.

Vol. 2202: A. Restivo, S. Ronchi Della Rocca, L. Roversi (Eds.), Theoretical Computer Science. Proceedings, 2001. XI, 440 pages. 2001.

Vol. 2205: D.R. Montello (Ed.), Spatial Information Theory. Proceedings, 2001. XIV, 503 pages. 2001.

Vol. 2207: I.W. Marshall, S. Nettles, N. Wakamiya (Eds.), Active Networks. Proceedings, 2001. IX, 165 pages. 2001.

Vol. 2208: W.J. Niessen, M.A. Viergever (Eds.), Medical Image Computing and Computer Assisted Intervention – MICCAI 2001. Proceedings, 2001. XXXV, 1446 pages. 2001.

Vol. 2209: W. Jonker (Ed.), Databases in Telecommunications II. Proceedings, 2001. VII, 179 pages. 2001.

Vol. 2210: Y. Liu, K. Tanaka, M. Iwata, T. Higuchi, M. Yasunaga (Eds.), Evolvable Systems: From Biology to Hardware. Proceedings, 2001. XI, 341 pages. 2001.

Vol. 2211: T.A. Henzinger, C.M. Kirsch (Eds.), Embedded Software. Proceedings, 2001. IX, 504 pages. 2001.

Vol. 2212: W. Lee, L. Mé, A. Wespi (Eds.), Recent Advances in Intrusion Detection. Proceedings, 2001. X, 205 pages. 2001.

Vol. 2213: M.J. van Sinderen, L.J.M. Nieuwenhuis (Eds.), Protocols for Mulitmedia Systems. Proceedings, 2001. XII, 239 pages. 2001.